The Tall Buildings Reference Book

As the ever-changing skylines of cities all over the world show, tall buildings are an increasingly important solution to accommodating growth more sustainably in today's urban areas. Whether it is residential, a workplace or mixed use, the tower is both a statement of intent and the defining image for the new global city.

The Tall Buildings Reference Book addresses all the issues of building tall, from the procurement stage through the design and construction process to new technologies and the building's contribution to the urban habitat. A case study section highlights the latest, the most innovative, the greenest and the most inspirational tall buildings being constructed today.

A team of over fifty experts in all aspects of building tall have contributed to the making of the *Tall Buildings Reference Book*, creating an unparalleled source of information and inspiration for architects, engineers and developers.

Dave Parker was Technical Editor of *New Civil Engineer* magazine for 12 years before leaving in May 2006 to become a freelance author and journalist. His interests include microgeneration and structural safety, and he successfully campaigned for the setting-up of the CROSS confidential reporting scheme in the UK.

Antony Wood is Executive Director of the Council on Tall Buildings and Urban Habitat and an Associate Professor at the Illinois Institute of Technology College of Architecture. His field of specialism is the sustainable design of tall buildings. His PhD explored the multi-disciplinary aspects of skybridge connections between tall buildings.

The Tall Buildings Reference Book

Edited by
Dave Parker
and Antony Wood

Routledge
Taylor & Francis Group
LONDON AND NEW YORK

Council on Tall Buildings and Urban Habitat

First published 2013 by Routledge

2 Park Square, Milton Park, Abingdon, Oxfordshire OX14 4RN
52 Vanderbilt Avenue, New York, NY 10017

Routledge is an imprint of the Taylor & Francis Group, an informa business

First issued in hardback 2019

This book has been made possible in part with kind assistance from Furthermore: a program of the J.M. Kaplan Fund.

British Library Cataloguing in Publication Data
A catalogue record for this book is available from the British Library

Library of Congress Cataloging-in-Publication Data
The tall buildings reference book / edited by Dave Parker and Antony Wood.
pages cm
Includes bibliographical references and index.
1. Tall buildings.
NA6230.T34 2012
720'.483—dc23
2012002547

ISBN: 978-0-415-78041-4 (hbk)
ISBN: 978-0-203-10646-4 (ebk)

Typeset in Optima by Alex Lazarou

Contents

PART SIX: CASE STUDIES

Contributors

Tom Akerstream has more than 30 years of experience in energy management and was the Energy Advisor responsible for ensuring that Manitoba Hydro Place is the most energy-efficient and sustainable office tower in North America. Tom was recently awarded the prestigious Award of Merit by the CSA for his work in energy efficiency and sustainability.

Warren Alexander-Pye is a Senior Project Manager and a core member of the Mace Construction Management team. He spent three years working on the Shard at London Bridge Tower and also has experience on a number of project management, construction management and pre-construction commissions, including two years of construction delivery at Canary Wharf, London.

Jason D. Averill is the Leader of the Engineered Fire Safety Group in the Engineering Laboratory at the National Institute of Standards and Technology, Gaithersburg, Maryland. Since joining NIST in 1998, Mr Averill has focused his research on quantitative assessment of fire hazards to occupants and responders by blending numerical modelling with fire experimentation, ranging from bench to real scale.

Clyde N. Baker, Jr. is a Senior Principal Engineer with AECOM. He graduated with a BS degree in Physics from William and Mary College in Virginia and with both BS and MS degrees in Civil Engineering from MIT. Over the past 57 years he has served as the geotechnical engineer or consultant on most of the high-rise buildings in Chicago and six of the eleven tallest buildings in the world, including the Petronas Towers, Taipei 101 and Burj Khalifa.

William F. Baker is the Structural Engineering Partner for Skidmore, Owings & Merrill LLP. Throughout his career, Bill has dedicated himself to the structural innovation of tall buildings, most notably developing the "buttressed core" structural system for the Burj Khalifa, currently the world's tallest man-made structure. He was the 2008 recipient of the CTBUH Fazlur R. Khan Lifetime Achievement Medal.

Sara Beardsley is a Senior Architect at Adrian Smith + Gordon Gill Architecture (AS+GG), a firm specializing in high-performance supertall buildings. She has contributed to high-rise projects around the world, including the proposed modernization of the 110-story Willis Tower and FKI Headquarters in Seoul, Korea, and has lectured widely on improving the energy performance of existing tall buildings.

Jeffrey Boyer is founder of eFlippr LLC and former head of AS+GG's Clean Technology group prior to his involvement with HOK and the US Green Building Council on Project Haiti. Jeff's laterally integrated whole building design experience defies traditional categorization, ranging from advanced system energy modeling and environmental physics analysis to macroeconomic studies of renewable energy and on-site power generation internationally. Jeff's latest pursuit, eFlippr, is a "next level" manufacturing hub that brings designers, manufacturers and consumers together in a reverse auction for dramatic cost and energy savings.

Bob B. Brennan is the past President and Chief Executive Officer of Manitoba Hydro. A Chartered Accountant, he has an extensive and diversified background in the electric utility industry. The Institute of Chartered Accountants of Manitoba elected Bob as a Fellow of the Chartered Accountants (FCA).

Bill Browning, Partner at Terrapin Bright Green, is one of the green building and real estate industry's foremost thinkers and strategists. His expertise has been sought out by organizations as diverse as Fortune 500 companies, leading universities, non-profit organizations, the US military and foreign governments. His client projects include Disney Hong Kong, Luscasfilm's Letterman Digital Center, the Greening of the White House, the Sydney 2000 Olympic Athletes' Village, and a 118-story mixed-use tower in Kuala Lumpur, Malaysia. Bill was a founding member of the US Green Building Council's Board of Directors.

Joseph Burns is both an engineer and an architect, and has more than 30 years of experience designing structures as well as conducting investigations and renovations. He specializes in the design of complex structural systems, including earthquake engineering and dynamic analysis.

Rick Cook is a founding Partner at both COOKFOX Architects—a firm devoted to creating environmentally responsible high-performance buildings, including the LEED Platinum Bank of America Tower (CTBUH 2010 Best Tall Building Americas award winner)—and Terrapin Bright Green. Over the past 25 years as a New York City architect, he has built a reputation for innovative, award-winning architectural design. In 2006, Rick and Bob Fox joined with Bill Browning and Chris Garvin to form Terrapin Bright Green LLC, an environmental consulting firm committed to improving the human environment through public and corporate policy, environmental performance strategies, and green development.

Roger Courtenay is a Principal and Vice President in AECOM's Washington DC offices. His practice focuses on sustainable public design, encompassing historic and cultural landscapes and facilities, museums and memorials, and public and private civic realms, including smart streets and parkways, embassies and consulates, parks and open space, and large scale commercial urban design.

Steve Edgett is President of Edgett Williams Consulting Group, Inc., a US consulting firm specializing in vertical transportation. In his 30-year career, Mr Edgett has worked on major projects in over 20 nations throughout the world. In 2010, his firm completed the construction documents for the 600-metre Shanghai Center and it is currently working on a number of 400-metre buildings in China.

Ian Eggers's career development saw him join Mace, having previously worked for Bovis on various prestigious projects, including Broadgate. During his time at Mace, Ian rose to the position of Managing Director of Construction Management and the subsidiary business Mace Living. Some of the highlights of Ian's career at Mace include the successful negotiation of the contract to build the Shard at London Bridge Tower and also the accolade of having been the youngest ever winner of the coveted Construction Manager of the Year Award. Having left Mace, Ian is currently enjoying success running his own construction business, Rise.

Stephen Engblom is Senior Vice President of AECOM and is the Americas Practice Leader for AECOM's Design + Planning and Economics practices. An accomplished urban designer and architect, he has over 20 years of professional experience on urban regeneration projects in the Americas, Asia, Europe, the Middle East and Africa.

Gail Fenske is Professor of Architecture in the School of Architecture, Art and Historic Preservation at Roger Williams University. She is the author of *The Skyscraper and the City: The Woolworth Building and the Making of Modern New York* (University of Chicago Press, 2008). She is a registered architect and received her PhD in the history, theory and criticism of architecture from MIT.

Roger Frechette is a registered professional engineer with 22 years of experience. His work includes many significant national and international projects, from energy master plans, laboratories, hospitals, academic buildings and corporate offices to a wide range of government buildings and museums. Roger's recent projects include the 71-story Pearl River Tower in Guangzhou, China and the Burj Khalifa in Dubai.

Chris Garvin, Partner at Terrapin Bright Green, is an accomplished practitioner and active voice in the sustainable design community. Since moving to New York in 1998, he has specialized in environmental architecture and materials research, while serving on numerous advisory boards and in organizations that advocate for sustainability in design. He currently serves on the Board of Directors for the Urban Green Council, the New York chapter of the US Green Building Council, and on the Advisory Board for Mayor Michael Bloomberg's Office of Long-Term Planning and Sustainability.

Russell Gilchrist, a native of Scotland, registered architect, currently design director for Gensler in their Beijing office, having previously worked largely in the USA and UK as director at AS+GG and SOM (Chicago), Rogers Stirk Harbour Partnership, and Foster + Partners (London). His award-winning project work includes Pearl River Tower in Guangzhou, Protos Winery, Penafiel in Spain, 88 Wood Street in London, Reichstag Parliament Building in Berlin, Maison du Japon in Paris and the opera house in Glyndebourne, Sussex.

Alistair Guthrie is a Director and Leader of global buildings sustainability at Arup. He is an Arup Fellow and Honorary Professor of Environmental Design in the School of Built Environment Nottingham University. He joined Arup in 1979 and has been involved in leading the design of arts and culture, education, commercial and high-rise buildings. He has worked extensively in the USA, Europe and Asia. His particular expertise is in the application of building physics and micro climate design in finding sustainable solutions in the built environment.

Malcolm Hannon is Managing Director of MLogic. He has successfully managed numerous construction logistics commissions in the open market for a variety of clients. Malcolm has brought his experience and expertise to the management of collaborative construction logistics on a host of complex projects including the Shard at London Bridge Tower and Heathrow Terminal 5A.

Steve Hanson is a Principal based in AECOM's San Francisco office and has practised landscape architecture since 1989. His portfolio of work encompasses some of the firm's most prestigious projects, including public realm design for the World Trade Center site in New York, the acclaimed Tokyo Midtown project, and more than a dozen large new mixed-use high-rise developments in China.

Robert Henderson is a Senior Engineer with Arup's Buildings San Francisco division and has been involved in the design and construction of several tall buildings around the world, including the Shard at London Bridge Tower, Parc 1, Seoul, and Yongsan landmark Tower, Seoul. He has been involved in the design of a range of projects, including arts and culture, education, commercial and transport buildings.

Nicholas Holt is a Director at Skidmore, Owings & Merrill LLP (SOM). He joined SOM in 1995 and oversees the technical efforts for all projects completed by the New York office. He is a senior leader in SOM's High Performance Design Initiative, an internal think tank that is developing the standards and goals by which SOM will meet the net zero carbon, energy, waste and water goals of the coming years. He is also a Principal at the Center for Architecture Science and Ecology (CASE), a research collaboration co-founded by SOM and Rensselaer Polytechnic Institute.

Peter Irwin, a Principal of Rowan, Williams, Davies and Irwin, Inc., joined the company in 1980 and served as president from 1999–2008. His experience in wind engineering dates back to 1974 and includes extensive research and consulting in wind loading, aeroelastic response, wind tunnel methods and instrumentation, as well as supervising many hundreds of wind engineering studies of major structures. He has worked on the Petronas Towers in Kuala Lumpur, the Taipei 101 building, Two International Finance Centre in Hong Kong and the Burj Khalifa tower in Dubai.

Richard Keating, FAIA was educated at the University of California, Berkeley. Joining SOM in Chicago, he became one of the firm's youngest partners and, after eight years in Chicago, he created the Houston office and later took over the Los Angeles office. Subsequently he formed his own firm, recently acquired by Jacobs, where he now serves as the Managing Director for Architecture and Design. He is well known for his high-rise office towers in Texas, California and Korea, as well as numerous corporate projects throughout the US and Asia.

Tony A. Kiefer received his BS and MS degrees in Engineering at the University of Illinois. He has over 27 years experience in foundation engineering. As a Principal Engineer at AECOM, he has been responsible for more than 50 high-rise buildings in Chicago and consultant for some of the world's tallest buildings currently under construction, such as the Busan Lotte Town Tower, the Doha Convention Center and Tower, and the Kingdom Tower.

Ron Klemencic is President of Magnusson Klemencic Associates, an award-winning structural and civil engineering firm headquartered in Seattle. He has worked in 20 countries and 22 US states; some of his recent projects include Doha Convention Center and Tower in Qatar, Rosewood Abu Dhabi in UAE, One Rincon Hill in San Francisco, the 300 East Randolph (Blue Cross Blue Shield) Vertical Expansion in Chicago, and Museum Plaza in Louisville, Kentucky.

Graham Knapp is a specialist wind engineer at the French national building technology centre, CSTB, where he uses wind tunnel tests and computational models to predict the interaction between strong winds and the built environment. He is the newsletter editor and former executive committee member for the UK Wind Engineering Society.

Simon Lay is former chair of the CTBUH Working Group on Fire & Safety in High-Rise Buildings and a member of the CTBUH steering committee, as well as author of many papers and articles on fire engineering and high-rise design. He has developed holistic design strategies for over 3000 storeys of high-rise residential, hotel and office design, including projects that exceed 750 metres in height and numerous mixed-use schemes that exceed 250 metres in height.

Richard Marshall is a Partner of Buro Happold and leads their Tall Buildings Expert Community. He graduated from the University of Canterbury in 1992 and joined Buro Happold in 1997. He has worked on dozens of tall building projects in Asia, the Middle East, the US and Europe while based out of a number of Buro Happold offices.

Dave Parker was Technical Editor of *New Civil Engineer* magazine for 12 years before leaving in May 2006 to become a freelance author and journalist. His interests include microgeneration and structural safety, and he successfully campaigned for the setting up of the CROSS confidential reporting scheme in the UK.

Mark Pauls is the Building Energy Management Engineer for Manitoba Hydro Place. He was involved in mechanical and electrical commissioning of the building, and currently monitors and optimizes energy consumption for the office tower.

Jason Pomeroy is an award-winning architect and academic, and founding Principal of Pomeroy Studio, a design studio of international architects, urbanists, designers and theorists at the forefront of the green built environment agenda. His research at Cambridge considered the socio-spatial functions of sky courts and sky gardens within the vertical city model of the twenty-first century. In addition to leading the Studio he lectures and publishes widely, and is the author of *Idea House: Future Tropical Living Today*. He is an Adjunct Professor at Mapúa Insitute of Technology and an Honorary Professor at the University of Nottingham, and also sits on the editorial board of the Council on Tall Buildings and Urban Habitat.

Dennis Poon, Vice Chairman of Thornton Tomasetti, has more than 30 years of experience in the structural engineering of a wide variety of building types, from long-span structures to supertall buildings, including Taipei 101 and the two tallest buildings in China, the 632-metre Shanghai Tower and the 660-metre Ping An IFC Tower. He specializes in seismic design and optimization of structural systems.

Fahim Sadek is the Leader of the Structures Group in the Engineering Laboratory at the National Institute of Standards and Technology, Gaithersburg, Maryland. Since joining NIST in 1996, Dr Sadek has focused his research on modelling of the nonlinear static and dynamic response and collapse of structures subjected to extreme loads. His research specialisms include mitigation of progressive structural collapse, structural dynamics and earthquake engineering, and wind engineering and structural reliability.

Swinal Samant is an Associate Professor at the Department of Architecture and Built Environment at the University of Nottingham. She obtained her Diploma in Architecture (1992), M.Arch (1998) and PhD (2011) and has extensive practice, research and academic experience in India and the UK. Swinal's research on "Environmental Sustainability" spans across the fields of "Building Sciences" and "Urban Design" and has made a considerable impact within these fields.

Ken Shuttleworth founded Make in 2004, having worked on some of the world's most iconic and groundbreaking architectural landmarks while at Foster + Partners. Make is a creative and imaginative international practice with innovative and environmentally-responsible design as its hallmark. The practice currently has a broad range of projects underway, ranging from residential towers, sporting venues and office developments to hospitals and urban masterplans. Recently completed projects include Handball Arena for the London 2012 Olympics, Weihai Pavilion in China, The Cube in Birmingham, an office refurbishment for HSBC Private Bank in Geneva, and the Gateway Building for the University of Nottingham.

S. Shyam Sunder is the Director of the Engineering Laboratory at the National Institute of Standards and Technology, Gaithersburg, Maryland. Among his many leadership roles at NIST, he led the US federal building and fire safety investigation of the World Trade Center disaster in the aftermath of the terrorist attacks of September 11, 2001. He received a doctorate in structural engineering and served on MIT's faculty before joining NIST in 1994.

Werner Sobek studied Civil Engineering and Architecture at the University of Stuttgart. Professor at the University of Stuttgart since 1995, he has also held the Mies van der Rohe Chair at IIT in Chicago since 2008. His engineering and design consultancy, Werner Sobek, was founded in 1992 and now has branches in Stuttgart, Cairo, Dubai, Frankfurt, Istanbul, London, Moscow, New York and São Paulo. He was the 2005 recipient of the CTBUH Fazlur R. Khan Lifetime Achievement Medal.

Phil Solomon is the Operations Director within Mace leading the Construction Management and Major Projects planning teams. He has a wide background of planning and preconstruction engineering experience on projects with a high public profile and where quality is important. He has contributed to many complex projects, including several of the skyline towers in the City of London.

David Taylor is a writer on architecture and built environment matters. He edits *New London Quarterly*, an award-winning magazine which tackles key issues across all sectors as they affect the English capital. He has also written and contributed to a number of books including *Plus*, a book about architects Jestico + Whiles; *Volume*, a book about architects Stanton Williams; *Continued*, a book about architects Buckley Gray Yeoman; *1001 Buildings You Must See Before You Die*; *Liverpool: Regeneration of a City Centre*; and *Architecture & Commerce: New Office Design in London*.

Scott Thomson is the President and Chief Executive Officer of Manitoba Hydro. He is a Director on the Boards of the Canadian Electricity Association, the Canadian Gas Association, CIGRE Canada and a number of volunteer boards. He is also a volunteer pilot for Angel Flight of BC, an organization that transports cancer patients from remote locations for treatment.

Richard Tomasetti is a Founding Principal of Thornton Tomasetti, a member of the National Academy of Engineering, and Adjunct Professor at Columbia University. He was the 2012 recipient of the CTBUH Fazlur R. Khan Lifetime Achievement Medal, and has contributed to the engineering of some of the world's tallest structures, including Plaza 66 in Shanghai and Taipei 101. *ENR* magazine cited him for developing the stressed skin tube structure for high-rise buildings.

Steve Watts is a Director of Davis Langdon, an AECOM company, based in London, UK. A cost consultant by profession, he leads the firm's global Tall Buildings Group, which enjoys extensive experience of over 500 tall building projects. Steve has personally worked on many towers, including the HSBC headquarters at Canary Wharf, the Leadenhall Building in the City of London and the already iconic Shard at London Bridge Tower. He has been UK Country Leader for the CTBUH for a number of years while chairing the Council's Finance and Economics working group, and in September 2011 he was elected to its Board of Trustees.

Claus Peter Weller studied Architecture at the University of Stuttgart and was a member of the facade team at Werner Sobek Stuttgart until 2010. During his PhD he specialized in innovative facade frames.

Chris Wilkinson is a principal and founder of Wilkinson Eyre Architects, one of the UK's leading architecture practices and twice winner of the RIBA Stirling Prize, in addition to receiving more than 150 international design awards. Among Chris's recent projects is the Guangzhou International Finance Center, winner of the 2011 CTBUH award for Best Tall Building Asia & Australasia and the Lubetkin Prize in 2012.

Thomas Winterstetter studied Civil Engineering at the Universities of Dortmund and Essen. He is CEO of Werner Sobek Stuttgart, where he also directs the company's facade team. Other responsibilities include the general management of WSGreenTechnologies and lectureships at the University of Stuttgart.

Antony Wood is Executive Director of the Council on Tall Buildings and Urban Habitat and an Associate Professor at the Illinois Institute of Technology College of Architecture. His field of specialism is the sustainable design of tall buildings. His PhD explored the multidisciplinary aspects of skybridge connections between tall buildings.

Acknowledgments

Dave Parker and Antony Wood would like to express their gratitude to all those who contributed to this book, especially the 47 authors who gave their time and expertise without financial reward. Thanks are also due to the architectural and structural engineering practices that supplied the data for the case studies, without which this book would be incomplete.

The editors would also like to give special thanks to Taylor & Francis project manager Caroline Mallinder. Her dedication and organizational skills were vital to the success of what turned out to be a particularly complex undertaking.

About the CTBUH

The Council on Tall Buildings and Urban Habitat, based at the Illinois Institute of Technology in Chicago, is an international not-for-profit organization supported by architecture, engineering, planning, development, and construction professionals, designed to facilitate exchanges among those involved in all aspects of the planning, design, construction, and operation of tall buildings.

Founded in 1969, the Council's mission is to disseminate multi-disciplinary information on tall buildings and sustainable urban environments, to maximize the international interaction of professionals involved in creating the built environment, and to make the latest knowledge available to professionals in a useful form.

The CTBUH disseminates its findings, and facilitates business exchange through: the publication of books, monographs, proceedings, and reports; the organization of world congresses, international, regional, and specialty conferences and workshops; the maintaining of an extensive website and tall building databases of built, under construction, and proposed buildings; the distribution of a monthly international tall building e-newsletter; the maintaining of an international resource center; the bestowing of annual awards for design and construction excellence and individual lifetime achievement; the management of special task forces/working groups; the hosting of technical forums; and the publication of the CTBUH Journal, a professional journal containing refereed papers written by researchers, scholars, and practicing professionals. The Council actively undertakes research into relevant fields in conjunction with its members and industrial partners, and has in place an international "Country Representative" network, with regional CTBUH representatives promoting the mission of the Council across the globe.

The Council is the arbiter of the criteria upon which tall building height is measured, and thus the title of "The World's Tallest Building" determined. CTBUH is the world's leading body dedicated to the field of tall buildings and urban habitat and the recognized international source for information in these fields.

For Guy Parker

1972–2009

Introduction

Tall Trends and Drivers: An Overview

Antony Wood

That there is a need for this book is indisputable. The tall building typology has witnessed more rapid growth in the past two decades than in the preceding hundred years, and yet the vast majority of current books on the subject fall into the glossy, coffee-table book category—high on image, low on content, especially technical content. All aspects of tall buildings—form and architectural design, relationship to urban habitat, structural systems, elevatoring techniques, construction methods, environmental strategies, attitudes towards facades—have seen major development in the first decade of the twenty-first century, and thus this book is more than just timely in its intended capture of the "state of the art"; it is absolutely necessary.

The boom in tall building construction over the past decade or so is unprecedented in the history of mankind. Although there have been intense periods of tall building construction in specific geographic areas throughout history—late nineteenth-century Chicago, art deco New York, or post–Second World War Europe, for example—the boom of the 1990s and 2000s has reached across virtually the entire globe—from Brisbane to Beijing, Rio to Riyadh, Toronto to Tokyo. It seems that almost all cities globally have been developing their urban habitat skyward.

At the same time, many characteristics of high-rise construction have changed fundamentally from what they were for most of the twentieth century. A number of emergent trends are now evident.

Trends

Trend 1: An Increase in Number

Research by the Council on Tall Buildings and Urban Habitat illustrates the recent explosive growth in the number of tall buildings. Plotting the number of skyscrapers 200 meters or higher[1] completed each year since 1960 (Figure 0.1) shows a steady build-up during the 1990s, followed by exponential growth from the mid-2000s onwards. Although there has been a very definite drop in skyscraper activity in most western countries recently due to the global economic crisis of 2008–09 (which is reflected in the dip in global numbers completed during 2012 compared to the year previously), in the longer term, this is generally being offset by activity in Asia—and China specifically: we now expect the global number of tall buildings completed

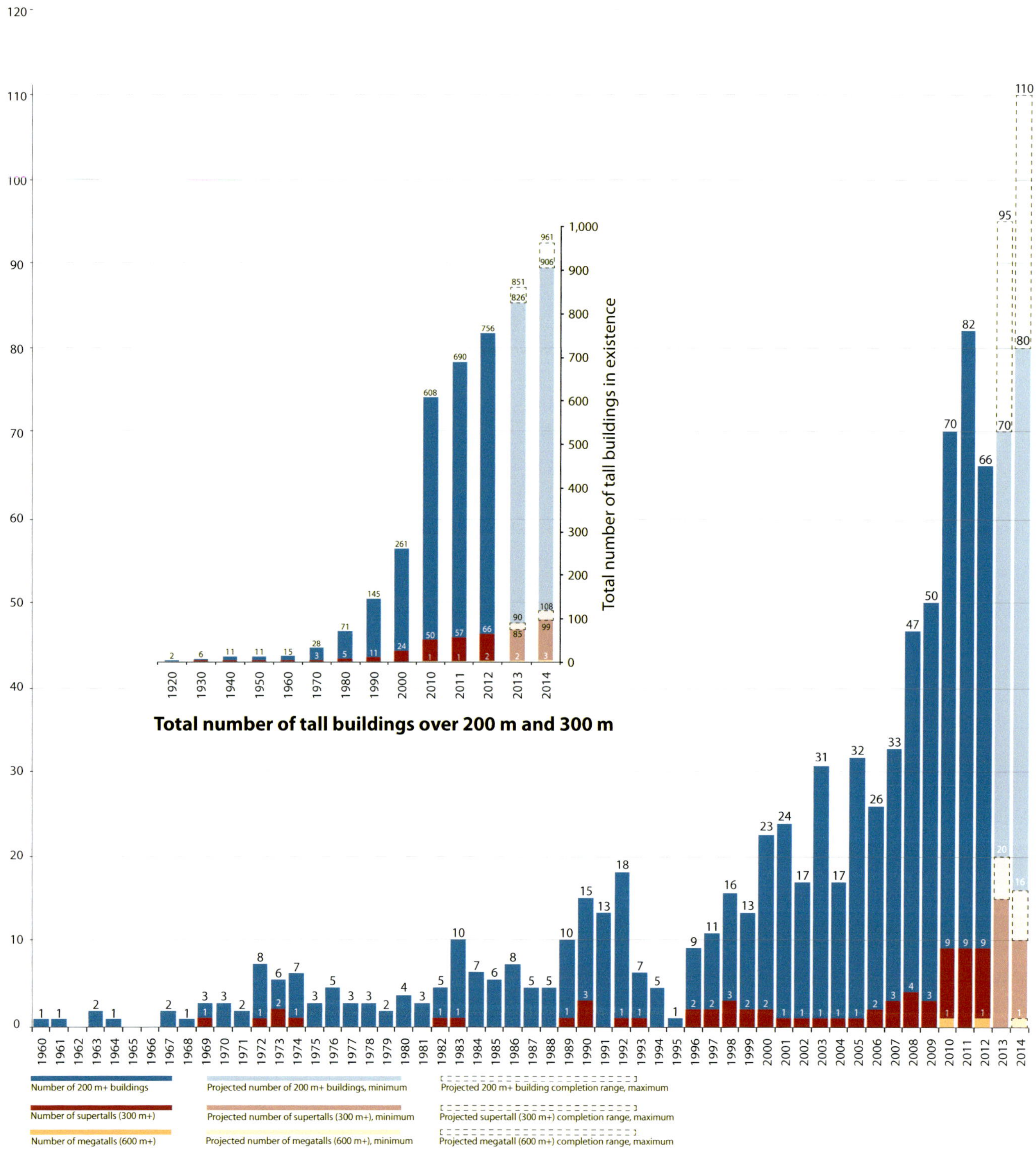

0.1 Tall buildings completed each year over 200 meters, 300 meters, and 600 meters since 1960. The inset shows total numbers of completed buildings by decade, revealing exponential growth rates (data as of January 2013). 2013–14 building completions are predicted from projects in advanced construction; totals after 2001 take into account the destruction of the World Trade Center Towers 1 and 2

Graphic: © CTBUH

each year to keep climbing for the foreseeable future. It is also worth noting that although there was a dip in the number of building completions in 2012 relative to the year previous, 2012 still saw the third highest number of tall building completions over 200 meters in history.

The detailed statistics are quite staggering. At the time of writing (January 2013), some 56 of the current 100 tallest buildings in the world have been completed since the end of 2005. In addition, a further 24 buildings are expected to enter the list by the end of 2013, which will translate into a 68 percent change in the "World's 100 Tallest" in just eight years (this change takes into account those buildings completed since 2005 but subsequently pushed out of the list by newer, taller buildings).

To translate statistics into actual buildings, Figure 0.2 (overleaf) shows the current 20 tallest buildings in the world. Of these, eleven have been completed since the end of 2005—55 percent.

Trend 2: An Increase in Height

Tall buildings are indisputably getting taller, in terms both of the tallest and of the global average height. As the graph on the average height of the world's 100 tallest buildings over the past eight decades (Figure 0.3 overleaf) shows, the average height has more than doubled in that time, and increased by 13 percent in the period 2000–2010 alone.

At the "World's Tallest" end of the scale, the year 2010 witnessed an incredible feat, with the completion of the 828-meter tall Burj Khalifa in Dubai. At no previous time in the history of the "World's Tallest" (see Figure 0.4 overleaf) has any building surpassed its predecessor by more than 68 meters (221 feet), but Burj Khalifa achieved a leap of an unprecedented 320 meters (1,050 feet) over the previous world's tallest, Taipei 101. The total height of the Burj Khalifa is just five meters shy of the equivalent height of putting the Empire State Building on top of Petronas Towers (both formerly "World's Tallest" buildings).

Continuing this trajectory, the next "World's Tallest"—expected to be the Kingdom Tower in Jeddah, Saudi Arabia—will be over 1,000 meters in height. It is worth putting this into perspective, however: although "supertall"[2] buildings are now achieving incredible heights, at the time of writing (January 2013) there are still only 69 buildings of 300 meters or more completed globally. Thus, though the recent "World's Tallest" achievements tend to skew the impression of tall building height, the reality is that the completion of a supertall building over 300 meters is still a significant urban and technological achievement.

Trend 3: A Change in Location

The predominant location of the tallest buildings in the world has been changing rapidly. Whereas as recently as 1990, 80 of the "World's 100 Tallest" were located in North America, now that figure is only 23, with the shift predominantly to Asia (45, with 31 in China alone), and the Middle East (27, with 20 in Dubai alone) (see Figure 0.5 overleaf).

Trend 4: A Change in Function

There has also been a major move away from the office function which dominated the "100 Tallest" list for many decades. We are now seeing residential and mixed-use functions, up from 12 to 45 buildings in just the last decade (see Figure 0.6 overleaf).

The rapid urbanization of developing countries (see the section on "Drivers") partially explains why many of these buildings are now residential in nature rather than commercial—to accommodate growing city populaces. There are other reasons for the shift away from the office function however, especially, in the case of mixed-use, the commercial incentive to "hedge bets" on fluctuating demand for office–residential–hotel functions by including them all in the building program. The trend also makes sense given that, if great height is the main objective of the project, it is easier to achieve this with a residential rather than an office function. Residential floor plates tend to be much smaller than office ones—an advantage when subjecting materials to wind and other pressures almost a kilometer in the sky—and also require less floor area–consuming elevators and other vertical services to support the function. In other words, if the creation of the "World's Tallest Building" is the primary motivator, then it is easier to do it with a function that will put fewer people in continual occupation at the top of the tower and thus reduce the size of the floors needed to house them and the services needed to support them.

Trend 5: A Change in Structural Material

The change in structural material has also been very significant over the past few decades. The proportion of all-steel structure buildings in the "100 Tallest" list has fallen from 90 percent as recently as 1970 to just 17 percent, shifting in favor of concrete or composite structures (see Figure 0.7 overleaf).

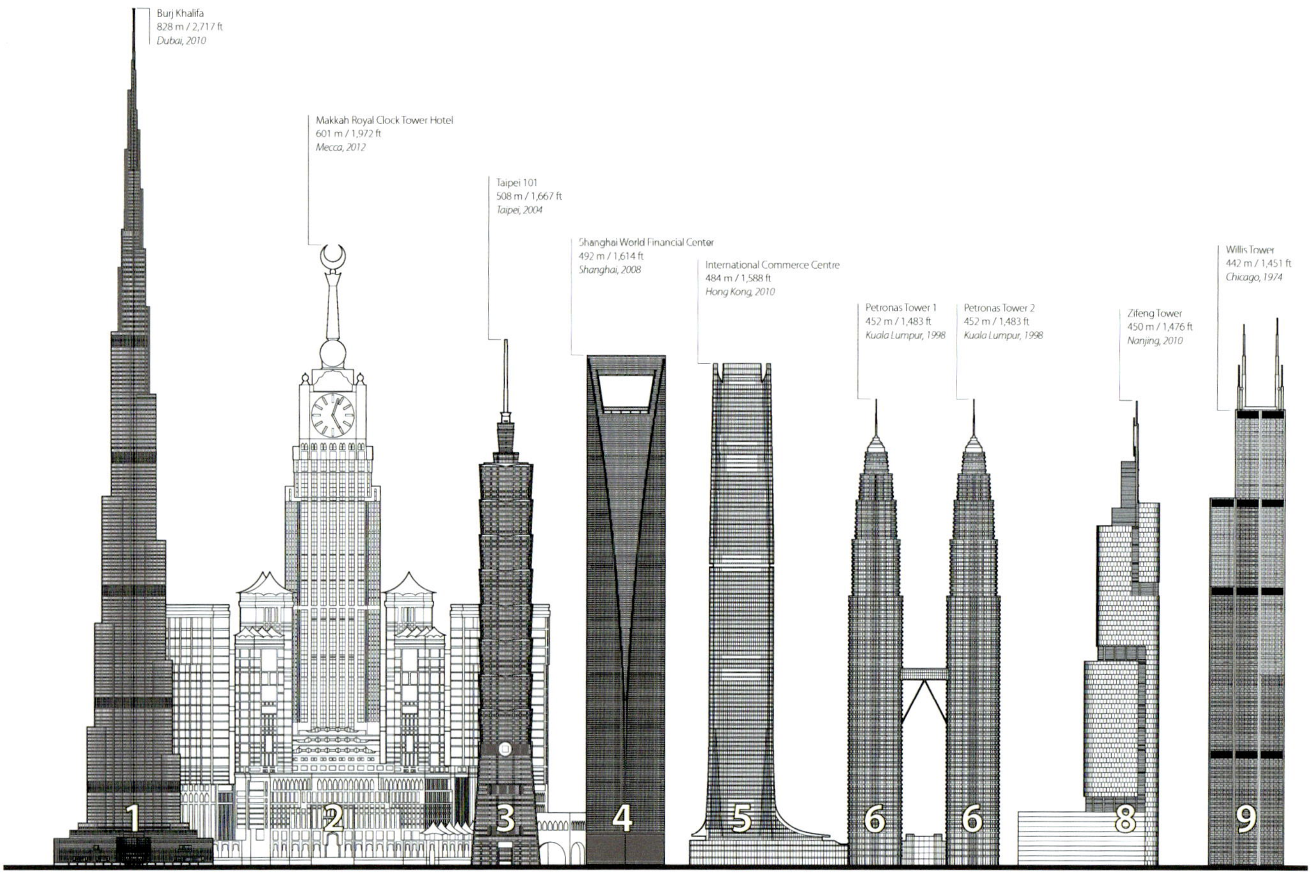

0.2 Diagram of the world's tallest 20 buildings according to the CTBUH height criterion of "height to architectural top" (data as of January 2013)
Graphic: © CTBUH

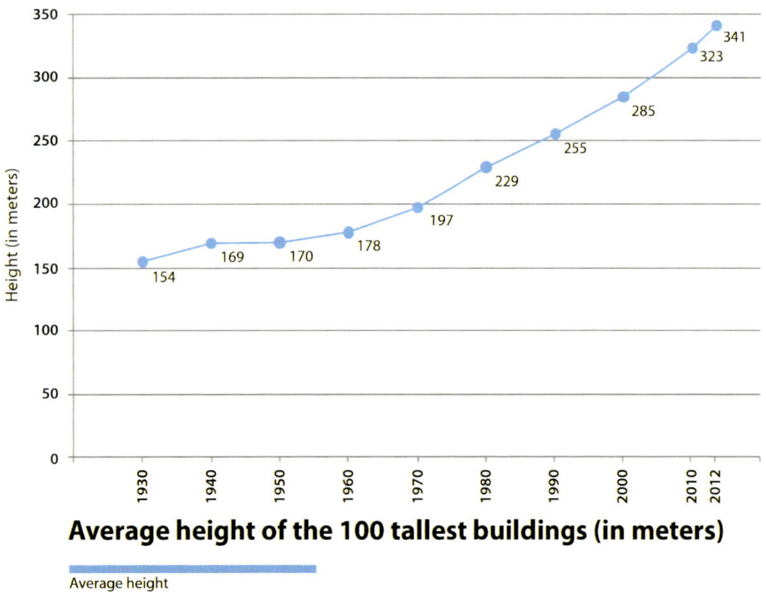

0.3 Average height of the 100 tallest buildings each decade, 1930–2012 (data as of January 2013)
Graphic: © CTBUH

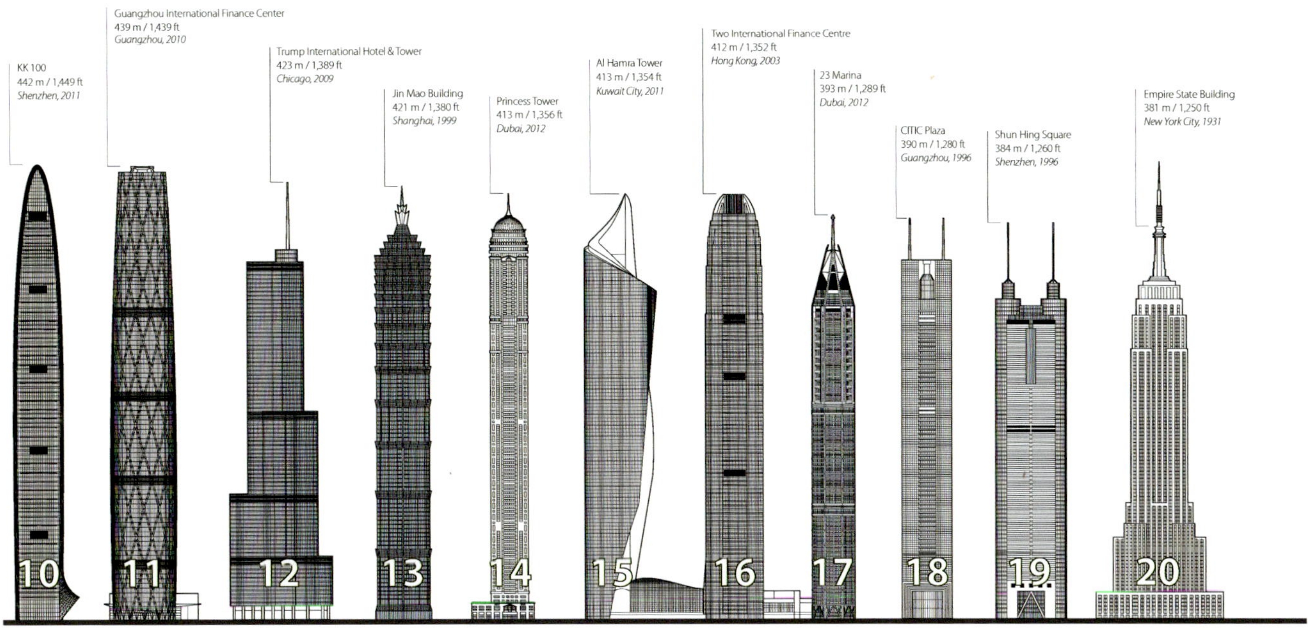

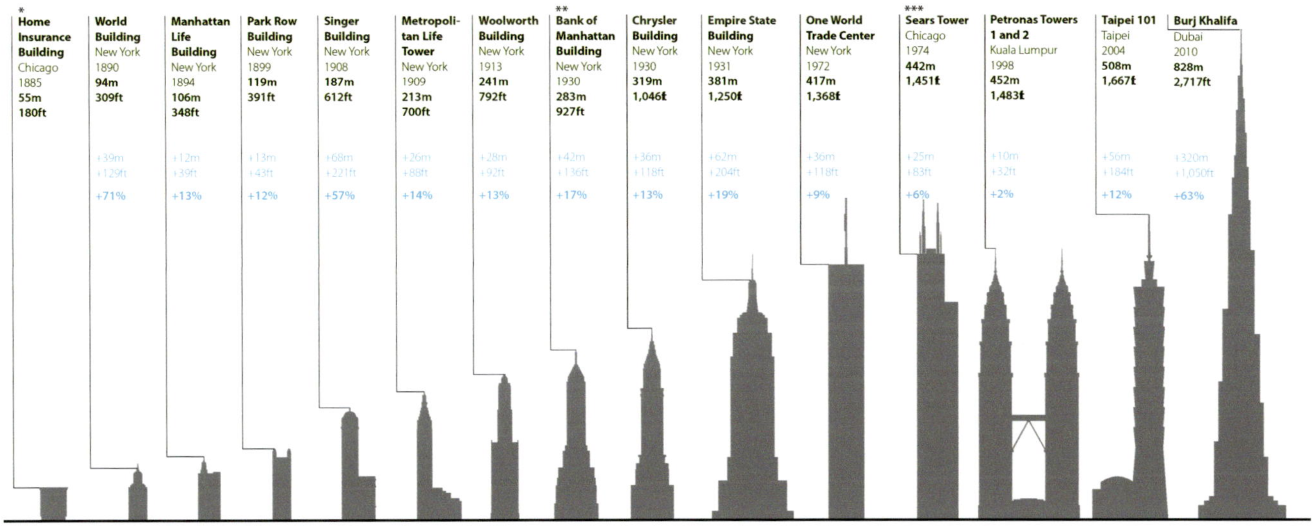

0.4 History of the "world's tallest building" (data as of January 2013)
Graphic: © CTBUH

* While the Home Insurance Building was never the tallest building in the world, it is considered the first skyscraper constructed (having a framed/non-loadbearing facade construction) and thus the first "tall building" as defined by CTBUH.

** Now known as the Trump Building, "Bank of Manhattan building" was the building's title when it was the "world's tallest."

*** Now known as Willis Tower, "Sears Tower" was the building's title when it was the "world's tallest."

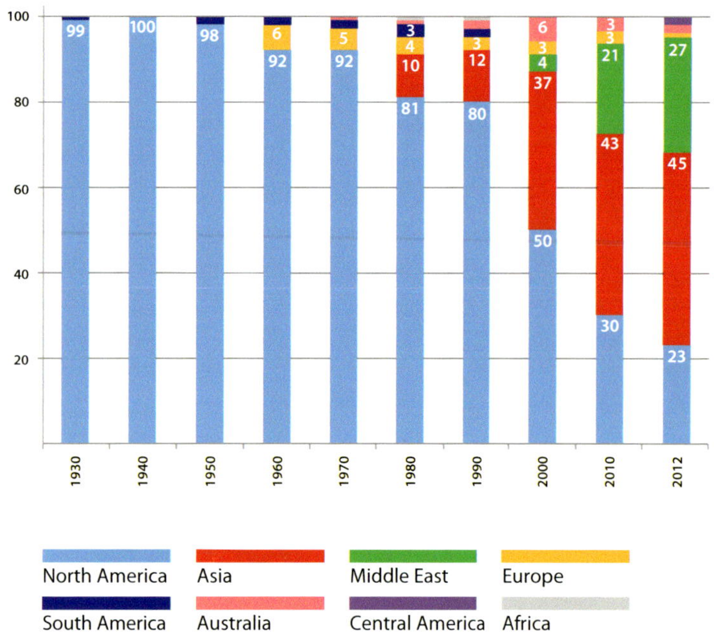

0.5 100 tallest buildings by location, 1930–2012 (data as of January 2013)
Graphic: © CTBUH

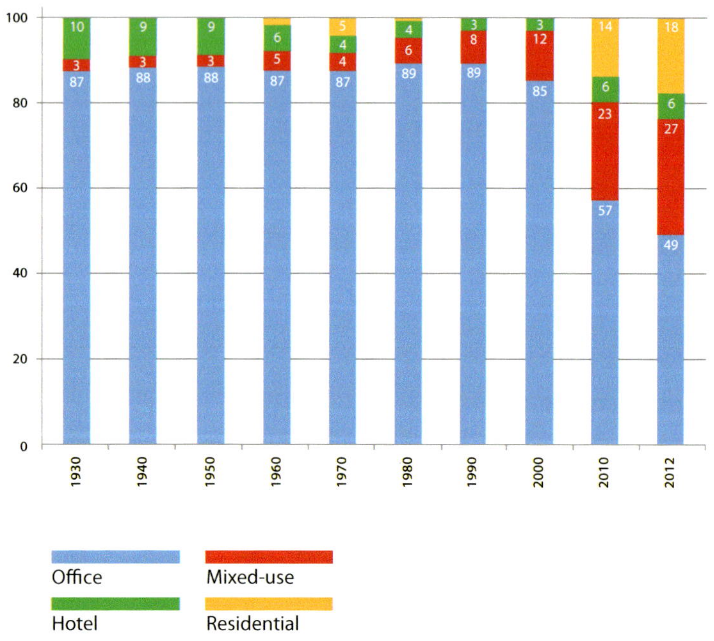

0.6 100 tallest buildings by function, 1930–2012 (data as of January 2013)
Graphic: © CTBUH

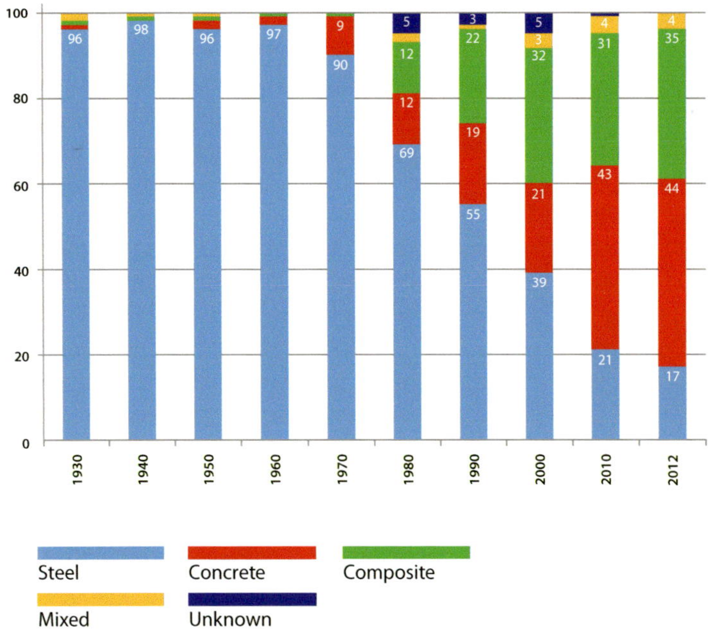

0.7 100 tallest buildings by structural material, 1930–2012 (data as of January 2013)
Graphic: © CTBUH

The reasons for the trend towards concrete or composite structures in the world's tallest buildings are multi-layered. Partly it is a product of the characteristics of the developing countries where these projects are located, which are much more likely to have technological expertise in concrete than they are in steel. Cost is also a significant factor, with concrete believed to be cheaper. The aforementioned change towards residential and mixed-use functions is also influential, since the fire, acoustic and cellular requirements of "living" lend themselves better to concrete construction than to open-plan-enabling steel. There are also many who believe that the increased performance required of the structure at great height—the damping of movement (especially in residential towers) as well as the transfer of vertical loads—can be more adequately handled by steel and concrete acting together, rather than by one material alone.

Drivers

There are a number of interrelated reasons for all these changes. Drivers 1–3 described here can perhaps be considered as "traditional" drivers, which have influenced tall building development through much of its 120-year-plus history. Drivers 4–6 are perhaps more relevant to the massive growth experienced over the past decade or so.

Drivers 1 and 2: Land Prices and Return on Investment

Historically the higher cost of land typical of city centers—where the vast majority of tall buildings are located—has always been a driver for tall buildings. The higher land cost drives both the developer's need to offset that cost by creating more floor area for sale or rent, and the opportunity for a greater return on investment by developing tall. It should be noted that the greater return on investment through providing more floor area is obviously offset by the higher construction costs of the required high-performing materials and systems at height, such that eventually a "height threshold" of financial return versus cost will be reached. This height threshold will be specific to project and location, with issues such as material costs and the premium or return on "iconic-ness" playing a part.

Drivers 3 and 4: Building as Corporate Branding and Skyline as Global Branding

Whereas tall buildings have been used throughout their history as marketing tools to portray the vitality of a corporation, now they are increasingly being used to portray the vitality of a city or country on a competitive world stage. This is reflected in the titles of the buildings themselves—previously endowed with names such as Woolworth, Sears or Petronas, they are now more likely to be named Shanghai Tower, Taipei 101, or Chicago Spire. Buildings are being used to brand a city, since many cities, especially in developing countries, believe it necessary to have a signature skyline to be considered successful and thriving.

Drivers 5 and 6: Rapid Urbanization and Climate Change

There are other, perhaps more compelling, reasons for the increase in tall buildings than just corporate or urban branding. It is believed that there are now almost 200,000 people urbanizing on this planet every day (United Nations 2007, 2008; see also http://esa.un.org/unup), requiring a new city of about one million inhabitants every week. But with this migration from rural to urban happening predominantly in developing countries with large populations, such as China, India, Brazil and Indonesia, these people are not flocking to new cities but rather to existing ones, putting significant strain on existing urban space and infrastructure.

A Note on the Future ...

Whether in new or existing cities, the question of how these new urban inhabitants are accommodated is a challenging one. It is increasingly being recognized that the American model of a dense downtown working core with massive, ever expanding low-rise suburbs is an unsustainable one, due to the increased infrastructure (roads, power, lighting, waste handling, etc.) needed, as well as the energy expenditure and carbon emission implications of the home–work commute. Thus it is increasingly being recognized that cities need to become denser to create more sustainable patterns of life—to reduce the horizontal spread of infrastructure networks and to make land use more efficient, partly in order to retain "natural" land for agricultural purposes.

Though tall buildings are not the only solution to achieving high density in all cities, they can be part of the solution for some cities. This urban density driver, coupled with the city symbolism/iconic driver, has certainly been influential in the escalation of the number of tall buildings being built and planned in developing countries.

The trends towards greater numbers and height of tall buildings and the changing characteristics of the typology are only one part of the equation, however. Questions of the appropriateness of developing vertically, in social, urban, cultural, and environmental terms, are increasingly being asked, and often the answers are only partial. The afterword to this book will address these issues specifically, but first the following chapters capture the current "state of the art" in global tall ...

References

United Nations (2007) *World Population Prospects: The 2006 Revision*, Department of Economic and Social Affairs, Population Division. New York: United Nations.

United Nations (2008) *World Urbanization Prospects: The 2007 Revision*, Department of Economic and Social Affairs, Population Division. New York: United Nations.

Notes

1 Please note that the focus of this research on buildings over 200 meters in height is driven by the need to ensure accuracy of data, rather than the suggestion that 200 meters is the threshold for determination of a "tall building." Even projects over 200 meters in height can be difficult to stay abreast of, especially in rapidly developing markets such as China, thus accuracy of data would diminish at lower height thresholds, though the trends would largely be the same. For more on CTBUH definitions of a tall building and its criteria for measuring tall building height, see http://criteria.ctbuh.org

2 A supertall building can be defined as any building over 300 meters (about 984 feet) in height. Incidentally, in January 2012 the CTBUH introduced the term "megatall" to describe buildings that are over 600 meters (about 1,969 feet) in height. As of January 2013, only two buildings can claim this title, Burj Khalifa in the UAE and Makkah Royal Clock Tower in Saudi Arabia.

Part One

Why Build Tall?

Introduction

Antony Wood

As described in the introduction to this book, the typology of tall buildings has witnessed unprecedented growth and change over the past two decades. There are numerous drivers for the act of building tall, some of which have already been mentioned: land price and the desire for a greater return on financial investment; the desire for an icon to "brand" or promote a corporation or a city; and, increasingly, the need for denser cities as a response to climate change and for more sustainable patterns of life. Part One of this book explores these themes, focusing in more detail on the question "Why Build Tall?"

Chapter 1, "A Brief History of the Twentieth-Century Skyscraper," examines the development of the tall building from its beginnings in late nineteenth-century Chicago through to the start of its major export around the world a hundred or so years later. In doing so, it highlights many of the differing motivations behind the seminal buildings of the period.

Chapter 2, "Aesthetics, Symbolism and Status in the Twenty-First Century," continues the historical perspective where Chapter 1 leaves off, looking specifically at that period in the latter part of the last century and the first decade of this in which the tall building began its major shift from being mainly a North American product to proliferate in almost every major city around the world.

Chapter 3, "A Client's Perspective," delivers a detailed insight into the motivations for building tall for just one client, in this case the Manitoba Hydro Electric company, with their award-winning headquarters building in Winnipeg, Canada. Here the sub-theme which underpins much of this book is introduced—how building tall needs to embrace the best advances in sustainable design and construction and how, in doing so, it can contribute both to society and to tackling the challenges of climate change. The story of this building's commissioning shows how consideration of sustainability in its widest and broadest sense influenced many of the client decisions made.

Such is the importance of the financial aspects of tall buildings that Chapter 4, "Tall Building Economics," is dedicated to the myriad complex issues surrounding this subject, taking the reader through the basics of tall building elemental costs and the drivers that influence them, as well as wider issues such as development funding and value creation.

In the final chapter of Part One, Chapter 5, "Is Refurbishment a Better Option?" a detailed review is made of the options available for existing buildings. Often the refurbishment chapter in a book like this is pushed to the end, once all the "new-build" topics have been exhausted. Not so in this book. The investment in every tall building in terms of embodied energy and carbon as well as in terms of time, professional expertise and financial cost (not to mention the difficulty of demolishing them) means that refurbishment is a scenario that virtually all tall buildings will face at some point in their life cycle—most more than once. Thus, in Chapter 5, we cover the basics of refurbishment, especially from an environmental standpoint, and examine some salient case studies.

Chapter 1

A Brief History of the Twentieth-Century Skyscraper

Gail Fenske

From the time of its appearance in the late nineteenth century, the skyscraper represented the quintessentially intractable building type. It violated with arbitrary, uncontrolled heights historical notions of urban decorum, effected congestion, polluted the atmosphere with coal burning, cast shadows, and, as an instrument of advertising, disrupted the profile views of cities. The image of the city, previously dominated by communal symbols such as a spire or a dome, fell subject to the will of an individual or enterprise.

Still, even in the nineteenth century, architects and builders showed evidence of concern for the skyscraper's relationship to the urban environment, particularly as demonstrated by their regard for urban civility in design or for improving the public domain, whether as defined within the skyscraper itself or through its contribution to a city's surroundings. When the twentieth century drew to a close, the renewed interest in such a relationship distinguished designs for prominent skyscrapers around the world. The emphasis of architects and builders had shifted from the quest for height to a new set of competitive criteria, many of which were aimed at enhancing the aesthetic, spatial, and environmental experience of the skyscraper and the city.

Early Skyscrapers in Chicago and New York

The skyscraper emerged within the context of two distinctive urban environments, those of Chicago and New York. Chicago, from its founding in 1803, distinguished itself as a frontier city, the primacy of which stemmed from its geographic location astride the great waterways of the American mid-continent. The city's first plan of 1834, organized on a grid suggested less a community than a real estate lottery. In 1848 the Illinois and Michigan Canal opened, linking the city with the port of New Orleans, in addition to the port of New York via the Erie Canal. Shortly thereafter, the railroads—ten trunk lines converged on the city by 1856—further accelerated the pace of the city's development. By the end of the Civil War, the city's population had grown almost tenfold and it achieved renown as the livestock, lumber, and grain center of the world.

Founded as a Dutch trading colony in 1623, New York had risen to become by 1820 the nation's center of banking and finance and, during the 1850s, achieved distinction as the nation's pre-eminent import–export center and key port of entry for European luxury commodities. In New York, commercial activity clustered around Broadway and the adjacent Ladies' Mile, both

1.1 John Kellum & Son, A.T. Stewart's second store, New York, 1859–62, 1868

1.2 Gilman & Kendall, Architects and George B. Post, Equitable Building, business hall, New York, 1889
Image: Courtesy AXA Equitable Life Insurance Company

of which served as the settings for the ornate cast-iron framed and cast-iron embellished buildings that pre-saged the skyscraper. Similarly, Chicago's State Street—or "that great street," as it was known after the merchant Potter Palmer improved and transformed the street with his Palmer House Hotel—served as the main axis of commercial development. Given New York's status as an import–export center, the city's retailers sought distinction on Ladies' Mile by placing primacy on a store's permeability, both visually, as goods in show windows enticed passersby, and spatially, through accessibility to sidewalk crowds. In the interiors, they emphasized spectacular and luminous multi-story spaces. In service of such objectives, the new technology of iron seemed to offer limitless potentials. The most notable of New York's "commercial palaces," A.T. Stewart's second store (1859–62), illustrated the capacity of the new iron construction to alter the character of an entire urban district and to promote a new standard of urban civility (Figure 1.1).

Chicago's Great Fire of 1871, a disaster of epic proportions, destroyed most of the buildings in the city's downtown, but by the early 1880s the city had begun to rise phoenix-like from the ashes to become *the* metropolis of the Midwest. Clearly circumscribed by the Chicago River, cable car lines, and then the elevated lines of the 1890s, Chicago's downtown provided little in the way of space for horizontal expansion. By the early 1880s, the profit motive of the land speculator had driven the heights of buildings skyward, creating a new "ceiling" of height in office buildings, now called "elevator buildings", because builders used the elevator to provide access to upper floors.

New York's elevator buildings, which dated from the spectacular French Second Empire–styled Equitable Life Assurance Society Building (begun 1868, enlarged 1875–76, 1886–89) designed by Arthur Gilman and George Post, stood out in their urban surroundings with showy exteriors. Elisha Graves Otis had demonstrated the first elevator in 1853, but only in the 1860s did the new technology, whether steam powered or hydraulic, reach an advanced stage of development. The Equitable featured the earliest use in an office building of two steam-powered elevators, but, more important, its post-1889 interior, as the headquarters of the world's wealthiest life insurance company, competed on spatial terms with Stewart's second store: forty offices on two levels surrounded a monumental hall (Figure 1.2). The Equitable had widened the building to incorporate that interior along with an arcade of shops covered by a glass skylight, creating what contemporaries called a "micro city." Consequently, from the outset, fostering the quality of spatial graciousness within the urban domain—surely directed towards clients for the Equitable's product as well as visitors—took pre-eminence in certain proposals for office buildings.

The Western Union and Tribune companies constructed elevator buildings in the city during 1872–75 that stood out as the city's tallest. Both had recently risen to prestige and power in the two leading communications industries of the day—the telegraph (Western Union built the nation's first telegraph line) and the

1.3 Burnham and Root, Rookery, Chicago, 1885–86, interior
Image: The Rookery Building, covered court, Chicago, IL, 1885–1888. Burnham and Root, architects. From *Inland Architect*, Ryerson and Burnham Library Collection, The Art Institute of Chicago. © The Art Institute of Chicago

1.4 Holabird & Roche, Tacoma Building, Chicago 1886–89
Image: The New York Public Library, Astor, Lenox and Tilden Foundations

newspaper—and both utilized speculative finance to ensure their prominence on the urban scene. But the Tribune's "mean accommodations" suggested little in the way of the Equitable's enlightened approach to interior planning. Stylistically ornate and showy exteriors, moreover, achieved their effect as advertising through their very contradiction of existing conventions of urban decorum.

George Post's New York Produce Exchange (1881–84) demonstrated that a building designed for commercial use might indeed contain an element of urban civility. Post, both an architect and an engineer, organized four stories of offices around a magnificent skylight exchange hall, utilizing "cage construction" for the inner court walls with the aim of opening up the entire interior to natural light. Post's design may have inspired Burnham and Root's Rookery in Chicago (1885–86), one of a series of office buildings financed by Peter and Shepherd Brooks of Boston (Figure 1.3). Noted for the grace and elegance of its iron-framed light court, the Rookery introduced a new level of graciousness and urbanity into the congested and rapidly modernizing downtown. Angled, perforated wrought iron beams, ornamentation in lace-like filigree, open balconies seeming to hang in midair, and a theatrical Piranesian stair announced a new cosmopolitanism in Chicago that alluded to the newest Paris department stores, among them the Bon Marché.

Post had experimented with cage construction in the light court of New York's Produce Exchange, but William Le Baron Jenney used the construction as a nearly complete skeleton throughout the upper stories of the Home Insurance Building in Chicago (1883–85), albeit still as a cobbled-together arrangement of cast iron columns, wrought iron box columns, and wrought iron and steel beams with connections bolted throughout. This showed little in the way of a systematic approach to wind bracing, and with walls partially self-supporting, it was hardly a "pure" demonstration of "skeleton construction."

Jenney's students, William Holabird and Martin Roche, by contrast, took a systematic approach to the steel frame and its requisite terra-cotta cladding, making it an object of scrutiny in their Tacoma Building (1886–89; Figure 1.4). Rather than embed the iron skeleton within a clothing of masonry, they explored a rational and complementary relationship between frame and cladding. Given the lightness of the construction, they utilized two strategically placed masonry walls as reinforcement against the lateral forces of the wind.

During the late 1880s and early 1890s, Chicago and New York's architects and engineers vigorously explored the problem of wind bracing, proposing an array of inventive solutions. Bradford Gilbert made significant strides with his tall and narrow Tower Building in New York (1889–91), employing metal diagonals inspired by Gustav Eiffel's wrought iron armature for the Statue of Liberty, thereby demonstrating iron's capacity to withstand the forces of tension, shear, and compression. The engineer Corydon T. Purdy developed an alternate method of wind bracing for his Old Colony Building in Chicago (1893–94)—the heavier and "planar" portal arch system. Both types of bracing and variations upon them would be used to support the construction of the tallest skyscrapers in New York.

The Early 1890s: Chicago's Building Boom

During the 1890s, the term "skyscraper" came into popular use, as Chicago's fireproofed, steel-framed construction irrevocably altered the skylines of Chicago and New York. Chicago experienced a building boom during 1889–93, and by the early 1890s the city had become an object of fascination both in America and abroad, presenting a bold, new, and dramatic example of the effect of the tall building on the city. Many critics nonetheless viewed the city as crass and reprehensible; Rudyard Kipling found it "inhabited by savages" (Mayer and Wade 1969: 192). Burnham and Root's Masonic Temple (1891–92) towered over Randolf Street, with its status as the city's tallest building only serving to reinforce the perception of the city as chaotic and overbuilt. As something of an antidote to the over-congested city, John Wellborn Root designed the Masonic Temple's internal light court as an exhilarating vertical space. It served as the central focus for the building's many uses, among them office space, shops, a hotel, and a rooftop observatory, which comprised in their own right a small "city." Root's design challenged the concept of the skyscraper as just an ordinary investment commodity.

Chicago's heightened level of prosperity during the late 1880s and early 1890s encouraged the construction of notable tall buildings such as Burnham and Root's Monadnock (1889–90) and D.H. Burnham and Company's Reliance Building (1894–95), but it also marked the flowering of civic and cultural spirit which manifested itself prominently in the World's Columbian Exposition of 1893. In their Marquette Building (1893–94), Holabird & Roche combined new and widely shared sentiments about urban civility and the proper decorum required of a dignified place of white-collar work, employing simplicity, standardization, and efficiency in design to develop a "classic solution" to the problem of the skyscraper. Shortly after the Exposition opened, Chicago's City Council passed a law restricting building heights to 130 feet (40 meters), or about 12 stories. The law would change prior to 1920 to accommodate heights up to 264 feet (80.5 meters), but nevertheless it served the aim of bringing a semblance of order and decorum to the city's skyline.

Louis Sullivan of Adler & Sullivan, already noted Chicago architects on account of their Chicago Auditorium project of 1886–87, might well have taken Holabird & Roche's Tacoma Building (see Figure 1.4) as the starting point for the Wainwright Building in St. Louis (1890–91; Figure 1.5). Sullivan chose instead, however

1.5 Adler & Sullivan, Wainwright Building, St. Louis, 1890–91

1.6 Louis Sullivan's "High Building Question" of 1891 showed a proposal for ameliorating Chicago's overbuilt condition—a setback skyscraper city that predicted New York's Zoning Resolution of 1916

Image: Illustration from *The Graphic*, 1891

to imbue the skyscraper with a poetic, even spiritual dimension, inspired by his belief in nature's capacity to counteract the overtly industrial character of the typical American downtown. As he stated in his "Tall Office Building Artistically Considered" of 1896, he accepted the "facts of the case," but still believed in architecture's potential to change the city (Sullivan 1979). His setback skyscraper city of 1891, an effort to ameliorate Chicago's overbuilt condition, predicted New York's Zoning Resolution of 1916 (Figure 1.6).

In designing the Wainwright Building, Sullivan focused on the skyscraper's exterior—virtually dismissing the interior offices as "the facts of the case"—with the aim of bringing nature into the city in a profound and visible way. He read the works of Ralph Waldo Emerson and Walt Whitman, and came to view the artist as a creature of instinct rather than reason. He would fulfil the tenets of Emersonian individualism—as a poet, man of vision, and Whitman's "seer." Sullivan emphasized the skyscraper's height with verticals, introducing his concept of functionalism while making it a "proud and soaring thing" (Sullivan 1979: 206). In the process he naturalized the industrially-based type with intricate foliate patterns—in his view, making the skyscraper a fresh, vital, and spontaneous work of art. In doing so, he conceived the most elaborate flora-inspired system of ornament yet developed in the history of architecture and imbued the skyscraper with the life and vitality that many identified with the natural world.

New York around 1900

Builders of skyscrapers in New York around 1900 pursued a wholly contrasting set of aims. The 1890s efforts to impose a height restriction paralleled those in Chicago, but prior to the implementation of the Zoning Resolution of 1916 these failed repeatedly, despite concerted discussion and debate. A litany of complaints put together by the city's architects and planners derided the frenzy of tall, unregulated construction and claimed the skyscrapers' effects on the urban environment as no less than menacing, causing dark and windy streets, abetting congestion, encouraging conflagration, and incubating disease. A booming construction economy fueled the frenzy—and after a brief hiatus during the economic downturn of 1893, the upward curve of real estate investment continued and the drive for height intensified.

Bruce Price's American Surety Building (1894–96), conceived as a "campanile," marked the acceptance of steel skeleton construction in New York. By the late 1890s, George Post's St. Paul and R.H. Robertson's Park Row buildings rose to heights of 315 and 391 feet (96 and 119 meters), respectively. Post placed box columns of steel well inside the inner faces of the St. Paul's piers to facilitate inspection and replacement in case of corrosion, and employed Purdy's portal arch system of wind bracing, probably for the first time in New York. Robertson's Park Row set yet another world record for height. Both projects firmly established the city as a center of technological experimentation. In 1898, the noted journalist and "muckraker" Lincoln Steffens condemned the "every man for himself" attitude that had failed to put the urban commonwealth first (Landau and Condit 1996: 276–7). But by 1900, New York had surpassed Chicago in exploring the potentials of record-breaking heights and in the process had created the world's first "signature skyline."

From the 1870s, the construction of elevator buildings in New York, notable among them the aforementioned Western Union and Tribune Buildings, attested to the ambition of the city's builders to advertise. But the tendency exhibited itself especially strongly in George Post's cage-framed World Building (1899–90). It set the stage for the new type of skyscraper by which New York distinguished itself after 1900: showy, publicity-seeking, and extravagantly competitive in height. The World's interiors showed little in the way of distinction, with the exception of its founder Joseph Pulitzer's office, located just beneath the skyscraper's gilded dome and noted for its opulence, spaciousness, and panoramic views. But its colorful, ornate exterior and conspicuous crown set the pattern for the city's post-1900 and still more iconic skyscrapers: the Singer, Metropolitan Life, and Woolworth towers. In seeking height and urban profile, these deployed the newest (though historically based) stylistic flourishes—modern French, Venetian Renaissance, Gothic—to forge memorable advertising trademarks that stood out in skyline views.

Ernest Flagg designed the Singer Tower of 1906–08 (Figure 1.7) as a slender "campanile" of 612 feet (186.5 meters) that soared nearly twice as high as the Park Row Building. Singer had recently expanded abroad, with sales offices in London and Glasgow, and had achieved notable success in Russia, so was eager to proclaim its new role in the world commercial marketplace. The Singer Tower's engineer, Otto Francis Semsch, ingeniously exploited the newest techniques for resisting lateral forces, among them cross bracing throughout the tower and eyebars anchoring the tower's columns to the foundation piers to combat the forces of uplift. For Flagg, however, the Singer Tower's importance resided in its serving as the basis for a specific proposal to regulate building heights and congestion.

1.7 Ernest Flagg, Singer Tower, New York, 1906–08
Image: MNYC/Irving Underhill, 1913

1.8 Illustration by Richard Rummell in *King's Views of New York* (1911) showing "future New York"

New York's skyscrapers, Flagg argued, should be constructed with an even cornice line, its height determined by the width of the street, and with towers limited to twenty-five percent of the site, but potentially rising to an indefinite height. The resulting "city of towers" Flagg conceived as the antithesis of contemporary illustrations in popular literature showing a future city ruled by the frenzied chaos of uncontrolled land speculation, among them Richard Rummell's 1911 view of lower Broadway for *King's Views of New York* (Figure 1.8).

The Metropolitan Life Tower (1908–09), designed by Pierre Le Brun of Napoleon Le Brun & Sons, and the Woolworth Building (1910–13), by Cass Gilbert, might have complemented Flagg's proposed "city of towers"—but Francis Kimball's City Investing Building (1906–08), with walls that rose straight upwards from the sidewalk, making it the city's most egregiously bulky building to date, and D.H. Burnham & Company's Equitable Building (1912–15), illustrated the persistence of the tendency illustrated in *King's Views*. The Equitable's extraordinary size and bulk cast a vast shadow across surrounding streets and buildings, which only served to aggravate the mounting pressure for an ordinance that controlled height and bulk. An in-depth study of the problem, *The Report of the Heights of Buildings Commission* of 1913, formed the basis for the Zoning Resolution of 1916.

Cass Gilbert's composition for the Woolworth Building, an office block from which extended a graceful Gothic tower, alluded to the Flemish free-trading cities of the Middle Ages, noted among them Bruges (Figure 1.9). Gilbert aimed to identify New York, the "corporate headquarters for the nation," romantically and nostalgically with the Middle Ages—while suggesting his client's, the retailer Frank Woolworth's, sense of civic obligation (Fenske 2008: 48). Gilbert relied on the engineering skills of Gunvald Aus for the design's 792-foot (241.5-meter) tower, the world's highest. Aus proposed combatting the lateral forces of the wind with the most extensive system of portal arch bracing ever designed or constructed.

More importantly, Gilbert followed the example set by the earlier Singer and Metropolitan Life towers by incorporating a gracious public lobby and, in addition, an array of public and semi-public interior spaces

1.9 Cass Gilbert, Woolworth Building, New York, 1910–13
Image: Library of Congress

1.10 Cass Gilbert, Woolworth Building, New York, 1910–13
Image: The New York Historical Society

ranging in use from a shopping arcade to a health club and a proposed downtown club to a spectacular pinnacle observatory (Figure 1.10). The spaces varied in character from a "Pompeian bath" to a "medieval German rathskeller" and "Elizabethan English banking hall," and as such simulated the experience of cosmopolitan travel while offering tenants the comforts of a "city within a city." Visitors to the Woolworth might crown the experience—theatrical in its own right—with a trip to the pinnacle observatory for a panoramic view, making the skyscraper an urban destination and sightseeing attraction that rivaled the city's most famous landmarks. In addition to providing the graciousness and urbanity of such an array of public and semi-public attractions, the Woolworth predicted the character and scale of New York's setback skyscrapers of the 1920s, setting a new standard for the skyscraper's relationship with the city.

New York as a Skyscraper Metropolis: the 1920s

During the 1920s, New York produced a spate of setback skyscrapers rising to unprecedented heights, resulting in a new type of city—a quintessential modern metropolis—the visual identity of which caught the imagination of the architects, artists, photographers, and filmmakers who made it known throughout the world. The city's center of gravity had shifted to Midtown, where a substratum of bedrock lay conveniently close to the surface, and now the setback buildings, many with slender towers, seemed to literally rise out of that bedrock and to become lighter and more ethereal as they reached for the sky. This new urban character of the 1920s metropolis can be largely ascribed to the Zoning Resolution of 1916.

Ralph Walker's Barclay-Vesey Building (1923–26), the vertical piers and nature-inspired ornament of which owed a debt to Sullivan, illustrated with great clarity the formal implications of the new law. But the Finnish architect Eliel Saarinen's entry for the Chicago Tribune Tower competition of 1922, well enough received within the American architectural community to be considered the "unofficial winner," just as strongly influenced the future city. It served as inspiration for one of the earliest examples of the setback type, Raymond Hood's American Radiator Building (1923–24) and throughout the 1920s exerted a powerful influence over the imaginations of architects.

That the Zoning Resolution of 1916 had implications for city planning extending beyond the straightforward—even prosaic—regulation of height and bulk can be credited in large part to the drawings of Hugh Ferriss. In his *Metropolis of Tomorrow* (1922–29), Ferriss projected with drawings of great visual force the artistic potentials of a well-regulated metropolis of the future (Figure 1.11). Clusters of setback skyscrapers resembling great mountain masses covered several city blocks. They stood in a powerfully ordered relationship to each other, rose above the machinery of the transportation infrastructure connecting them below, and on their setbacks provided terraces for gardens open to sunlight and fresh air, suggesting an array of new types of public spaces elevated as glamorous social settings and spectacular outlooks high above the city. Ferriss's accompanying text called for humanistic values in city planning. He had envisioned such a metropolis of the future in contradistinction to what he called "crowding towers," or the very nightmare of an overbuilt city, virtually devoid of human scale, and so submerged in noise, darkness, and chaos as to be barely habitable (Ferriss 1929: 62).

1.11 "The Business Center," from Hugh Ferriss's *Metropolis of Tomorrow*, 1929
Image: Avery Library

The 1920s setback skyscrapers of New York, or "vernaculars of capitalism," that altered entire districts (the Garment District among them) functioned as kind of a stage set for the new series of dazzling landmarks built to record-breaking heights (Willis 1995: 16). For a short time, the 649-foot (198-meter) Chanin Building (1926–29) at 42nd Street held the record in Midtown, its design more a slab than tower, due to its site at the intersection of three height districts defined by the Zoning Resolution. The skyscraper's lower stories featured the new ornamental fashion later called "Art Deco." But even before the Chanin Building's completion, William Reynolds proposed the tower that would become the Chrysler Building (Figure 1.12).

Reynolds, a builder of Coney Island's "Dreamland," aimed to outdo Woolworth in the competition for height, so he commissioned the architect William Van Alen to design a still taller tower at Lexington Avenue and 42nd Street. Unable to realize the project, Reynolds sold the site and Van Alen's design to the automobile tycoon, Walter Chrysler, and Van Alen reconfigured the skyscraper's iconic crown as a fantastic heaped-up terraced dome, spoked like a wheel, destined to signify both the corporation and the automobile while resonating with the theatrical energy of New York's latest Broadway spectacles. At its base, the Chrysler Building linked directly to the subway and Grand Central Station and at its pinnacle featured the "Cloud Club," along with an "observation lounge" that sported faceted, inward-sloping walls indebted to German Expressionist films, and light fixtures designed as miniature Saturns.

For the critic Douglas Haskell, the Chrysler Building's theatricality comprised mere "stunts and effects,"

1.12 William Van Alen, Chrysler Building, New York, 1928–30
Image: Library of Congress

but Van Alen and Chrysler had intended to play to popular audiences. This became evident as the city's competition for height reached fever pitch and Van Alen excited news publicity over his personal rivalry with H. Craig Severence, his former partner in practice. Van Alen and Severance had split on less than amicable terms and with his latest design, for the still-taller Bank of Manhattan Company on Wall Street, Severance sought revenge. He topped off his tower at 71 stories and 927 feet (282.5 meters). Van Alen struck back: from a shaft hidden within the Chrysler's crown he raised a 185-foot (56.5-meter) steel spire to 1,048 feet (319.5 meters), proclaiming it "the highest piece of stationary steel in the world". For the next Beaux-Arts ball, Van Alen dressed up as his design, earning the title, "the Ziegfeld of his profession" (Stern, Gilmartin, and Mellins 1987: 605–606). Severance's Bank of Manhattan Company nonetheless created a striking identity for Wall Street and the entire financial district, when joined on the skyline with three other new, equally slender and elegant towers—the Cities Service Building, the City Bank Farmer's Trust Building, and One Wall Street—an arrangement still viewed by many today as the "classic" skyline.

By 1930, Chicago, too, claimed a distinctive skyline. More than 30 spires had punctured the former 260-feet (79-meter) height limit, with the highest being the 612-foot (186.5-meter) Board of Trade Building. Chicago's zoning law of 1923, by contrast to New York's, did not allow towers of unlimited height; it specified instead that a tower built over twenty-five percent of the lot should not exceed one-sixth of the maximum cubic area of the building. While this had the effect of limiting towers to 17 to 20 stories, Chicago had nonetheless developed a new urban identity as a "city of towers."

The new skyline recalled the towers of Manhattan, despite Frank Lloyd Wright's effort to reform the Chicago skyscraper with his 1924 design for the National Life Insurance Company, a project unprecedented in its attention to qualities of natural illumination—via "wall screens" of sheet copper and glass—and human scale. In New York, the race for height continued, culminating in the Empire State Building, the mooring mast of which soared to 1,250 feet (381 meters), over 200 feet (61 meters) higher than the Chrysler's spire. The Empire State stood isolated in its command of the skyline, given its location just south of the energetic scene of construction in Midtown, suggesting in its independence and singularity the capacity of such an aesthetically graceful and distinctive landmark to shape the very identity of a city.

In light of the downturn in construction activity resulting from the economic chaos of the 1920s and culminating in the Wall Street stock market crash of 1929, the urban phenomenon represented by Rockefeller Center stands out as an anomaly (Figure 1.13). Begun in 1926 as the most grandly scaled project of its day, Rockefeller Center occupied several blocks along Fifth Avenue. From the outset, its builders took an "enlightened" approach in proposing how a group of skyscrapers might be arranged with attention to the public domain, in order to collectively support the life of the city. Initially conceived as a home for the Metropolitan Opera, Rockefeller Center eventually took shape as a group of office buildings incorporating a carefully orchestrated sequence of public spaces at street level, and above, elevated promenades linked by bridges crossing over city streets. John D. Rockefeller, who had joined the project by 1928, is credited with forming the "Metropolitan Square Corporation" and with appointing the engineering firm, Todd, Robertson, Todd to carry out financial studies, as well as the project's "Associated Architects," including Raymond Hood and Harvey Wiley Corbett, to advance the project's design.

The final version of the project, based on "scheme G3," featured the RCA Building as a salient vertical

1.13 Rockefeller Center, New York, 1929–40—roof garden schemes, 1932
Image: Rockefeller Center, Inc.

marker replacing the Metropolitan Opera House. Rooftop attractions suggesting an imagery of Babylonian hanging gardens fulfilled Ferriss' imagery of a "metropolis aloft." At the foot of the RCA Building, the "agora" materialized as a primary center of the city's public life: with the sculpture *Prometheus* it drew city crowds towards a café in the summer or skating rink in the winter. It combined plazas at sidewalk level with lower level shopping arcades, widely separated towers, and elevated terraces to proclaim a new spatial luxury, the very hallmark of urbanity. That luxury challenged the darkened mood of the Depression with a ray of light, this in turn emblematized by the finest space of the complex's interior, Radio City Music Hall. For modernist critics such as Sigfried Giedion, Rockefeller Center heralded a new ideal in urban planning. A widely recognized *tour de force*, it continues to serve as a reminder that the structure of the city can be transformed not just for the sake of an individual but rather for the sake of the community as a whole.

The 1950s: The Postwar Skyscraper and the "Return to Earth"

By contrast to the builders of the 1920s metropolis, who had embraced the day's speculative fervor and competition for height, those who shaped the 1950s city exemplified, in the historian Winston Weisman's phrase, a "return to earth" (Weisman 1950: 197). The characteristic corporate work environment of the 1950s, which aimed to support typewriters and the new electronic calculators and to facilitate a rationalized paper flow, required large, unencumbered interior spaces, along with the newly available building systems that guaranteed a higher standard of human comfort, principally air conditioning and fluorescent lighting. Such open work environments, now possible with structural designs that supported column-free spaces allowing an abundance of natural light, took precedence over the quest for height.

Just as important, corporations such as the Seagram Company, who employed Ludwig Mies van der Rohe as their architect, and the Lever Company, who chose Mies's emulators, Skidmore, Owings & Merrill, aimed to cultivate within the city a corporate image of *noblesse oblige* (Figures 1.14 and 1.15). Consequently, rather than aiming to build the tallest or the showiest of skyscrapers, each deployed an austere and elegant modernism with the intention of creating instead a dignified and reserved yet powerful presence in the city. This new corporate modernism, strongly identified with the Museum of Modern Art's "international style" exhibition of 1932, traced its lineage back to the 1920s experiments of the European *avant-garde*, among them Walter Gropius's entry into the Chicago Tribune Tower competition and Le Corbusier's steel, concrete, and glass skyscrapers for the "Ville Contemporaine," both of 1922. Perhaps equally important, the earliest such skyscraper in the United States, Howe & Lescaze's PSFS Building (1929–32) in Philadelphia, with its asymmetrical composition, accentuation of structure, and image of crispness and efficiency, signaled the acceptance of such an *avant-garde*, European modernism within the American business community.

Mies van der Rohe's design for the Seagram Building (1953–54; see Figure 1.14) had a nobility, subtlety, and reticence that struck just the right note with the Seagram Company's president, Samuel Bronfman, who aspired to improve the company's public image as a corporation. The company's less than noble past had included bootlegging during Prohibition. Having recently completed his 860–880 Lakeshore Drive Apartments—two skyscrapers in which he developed the prototypical metal and glass system he would continue to employ in future skyscrapers—Mies now perfected his elemental

1.14 Mies van der Rohe, Seagram Building, New York, 1953–54
Image: © Bettmann/CORBIS

1.15 SOM, Lever House, New York, 1951–52
Image: SOM/Hedrick Blessing

modern vocabulary for the skyscraper, using his I-beams in bronze as prominently placed and powerfully iconic verticals. Equally important, he situated his modern monument, or corporate "temple," within a piercingly empty plaza set back 90 feet (27.5 meters) from the lot line, further fulfilling Bronfman's aim of *noblesse oblige*: Seagram had given up otherwise buildable space for the purpose of enhancing the public life of the city.

Recognizing the positive urban implications of such a plaza, New York City planning authorities, engaged in 1961 to revise the Zoning Resolution of 1916, viewed the Seagram project as a model for the new ordinance, which ultimately resulted in the city's FAR (floor area ratio)-based code. Mies van der Rohe's influential design and the code, both of which emphasized the significance of public space, would significantly shape the future of the city.

Skidmore, Owings & Merrill's Lever House (1951–52; see Figure 1.15), which preceded the Seagram Building in breaking the standard pattern of development along New York's Park Avenue, showed its designers utilizing a comparable strategy to that of Mies van der Rohe. By contrast, however, SOM's design relied for its stunning visual effect on the delicate balance of two volumes, a horizontal slab and a vertical slab, each spare in its detailing, and sheathed in an environmentally sensitive heat-absorbing blue-green glass curtain wall—the earliest use of such a complete curtain wall—to project an unforgettable image of lightness, grace, and ethereality.

Lever Brothers produced household products for cleaning and hoped that its customers would identify the "cleanliness" of the building's exterior with those products. Like the Seagram Building, Lever House suggested that its builders had little concern for besting the competition with the tallest or the showiest skyscraper. Instead, they had given up to the city space on which it would have been possible to "build to the limit." This surely served as a sound strategy for cultivating a positive corporate image—one associated with urban civility.

The Supertall Skyscrapers of the 1960s and 1970s

The renewed quest for height by the builders of skyscrapers in the 1960s represented a bold departure from the 1950s' "return to earth." Frank Lloyd Wright predicted the trend with his "Mile High Skyscraper" in 1956, an imaginative but wholly impractical project. Still, within a decade, the engineer Fazlur Khan, working in association with the architect Bruce Graham of Skidmore, Owings & Merrill (SOM), conceived a new structural paradigm that realistically supported spectacular height, as demonstrated by the John Hancock Center (1965–70) and Sears Tower (1968–74) in Chicago (Figures 1.16 and 1.17).

Khan's breakthrough entailed conceptualizing tall buildings in a wholly new way, as "tubes" rather than as framed structures. That SOM, so strongly indebted to the formal innovations of Mies van der Rohe, should provide the setting for Khan's dramatic engineering innovations may seem uncharacteristic, but since its origins the firm had gained a reputation for providing a collaborative framework for both architects and engineers. John Merrill, an engineer who had joined the firm in 1930, and Myron Goldsmith, an architect and engineer engaged in research at the Illinois Institute of Technology, had strongly influenced SOM's direction towards engineering innovation.

1.16 SOM, John Hancock Center, Chicago, 1965–70
Image: Marshall Gerometta

SOM's Inland Steel Building in Chicago (1956–58), completed just six years after the Lever House, showed the firm continuing to refine the curtain wall, albeit for a modestly scaled building of nineteen stories. In addition, Bruce Graham and Walter Netsch, the architects responsible for the project, placed the design's fourteen structural columns on the exterior, to visibly express their function as supports, but also to provide for completely column-free office floors. Inland Steel had prospered during America's burgeoning postwar consumer economy and the building's satin-finished stainless steel cladding set off against green-tinted dual glazing served as a tribute to the company's product.

SOM's Chase Manhattan Bank in New York (1957–61), designed with the engineers Weiskopf & Pickworth, represented, through its exposed exterior columns set off against a curtain wall, a still further development of the paradigm that the firm had conceived for Inland Steel. Inland Steel, not unlike the Seagram Building, had fronted a plaza, but Chase Manhattan advanced the tower-and-plaza concept insofar as it became the first skyscraper to adopt the name of a plaza for its address. David Rockefeller, chairman of the building's development committee, determined to reverse the 1950s view that lower Manhattan had devolved into an economic "backwater," combined two blocks of ground, closed the street between, and turned one block into public space while erecting the skyscraper on the other. Chase Manhattan Bank benefitted from the aura of urban distinction identified with its predecessors, but despite its spectacular height of 60 stories and 800 feet (244 meters), its structural system, essentially a frame system—albeit heavily braced against wind-induced tension, shearing, and compression with 3,000 diagonal members and thousands of rivets—showed little in the way of progress over that of the Empire State Building, designed almost thirty years earlier.

The Chase Manhattan experiment inspired SOM's Myron Goldsmith to ask: how can the engineer build in new ways with less structural weight in support of great height? In his research at the Illinois Institute of Technology, Goldsmith evaluated the effects of great height on what he called the "super-frame" with regard to stiffness

1.17 SOM, Sears Tower, Chicago, 1968–74
Image: SOM/Hedrich Blessing

and deformation. Goldsmith's assistant, the young Fazlur Khan, utilized new methods of computer calculation to analyze three-dimensionally the effects of wind-induced stresses on such a super-frame. Khan tested his own engineering breakthrough, the "tube system," for the first time in his reinforced concrete De Witt Chestnut Apartments in Chicago (1962–65).

In their design for the 100-story, 1,127-foot (343.5-meter) John Hancock Center of 1965, Khan and Graham put Khan's "tube system" to a spectacular test (see Figure 1.16). Khan used computer analysis to determine exactly how a tube supported by columns in conjunction with giant cross-braces would respond three-dimensionally and dynamically to the forces of the wind. The skyscraper's exterior, noted for its "oil derrick" aesthetic, presented viewers with a clear visual diagram of the tower's structural behavior. Evenly spaced stiff floors, functioning as rigid diaphragms to distribute lateral loads to the structural components of the tube, contributed to the design's highly efficient use of structural steel. As the world's first supertall skyscraper and arguably also the first true "multi-use" skyscraper, the Hancock Center incorporated spaces housing a range of activities—among them offices, apartments, commercial space, an observatory, a restaurant, and parking—within a single structure.

The Hancock Center quickly garnered the prestige of "address," but critics asked: is it good urbanism? The plaza at its base served neither as an impressive entry nor as a lively gathering space. Those who occupied the tower discovered little in the way of any true relationship to the city. On top of that, the repetitive floor plates that ensured the tower's efficient behavior as a trussed tube also determined the spatial relationships of the tower's interior, allowing little in the way of spatial variation, much less the creation of significantly scaled public spaces in which tenants might gather and so foster a sense of community. Jane Jacobs's *Death and Life of Great American Cities* (1961) defined the characterizing features of a viable urban neighborhood—among them sidewalks, parks, diversity, and small blocks—none of which described the urban environment either within or around the Hancock Center.

Khan and Graham designed the Sears Tower (now the Willis Tower) in 1968–74, setting a new height record at 110 stories and 1,451 feet (442.5 meters) (limited only by Federal Aviation Authority guidelines) and proposing a still bolder relationship between the skyscraper and the city (see Figure 1.17). In structural concept the Sears Tower was a "bundle of tubes", connected integrally with belt trusses to behave structurally as a unit. Each of the nine tubes represented a square in plan and provided at different floors a variety of options for office shapes and areas. But Graham above all else sought a distinctive form for the skyline—"one more mountain ... for this city on the plains"—that functioned no less rationally as a tube-generated structure than the Hancock Center (Ali 2001: 231). For Sears Roebuck and Co., which had been based in Chicago since 1893, the tower served as a compelling symbol. While it showed little in the way of a positive relationship with the urban surroundings, its extravagant height in its own right offered some recompense: offices on the top floors have always been occupied and the observatory, which receives 1.4 million visitors per year, affords legendary views.

In New York, the World Trade Center's twin towers (1962–67, 1968–73) represented the day's most grandiose deployment of the "tube concept" for the purpose of attaining great height, with the towers rising to 1,362 and 1,368 feet (417 meters). The project's engineer, Leslie Robertson, wrote in 1962 that he had utilized the "tube concept" for the design, perhaps for the first time, even though Khan had dated the origin of the idea to 1961: "Faz and I independently thought of this idea" (Ali 2001: 43). The Sears Tower rose nearly 100 feet (30.5 meters) higher after its completion in Chicago one year later, but the World Trade Center contained nearly twice as much gross rentable floor area and its infrastructure of transit and shopping arcades covered twelve city blocks. When conceiving the project, the Port Authority of New York's director, Guy Tozzoli, instructed the project's architect, Minorou Yamasaki: "President Kennedy is going to put a man on the moon: you're going to figure out a way to build me the tallest buildings in the world" (Glanz and Lipton 2003: 108). Tozzoli justified the towers' great height by the publicity value they would have for their tenants, as well as the Port Authority, which aimed to centralize the city's many aspects of shipping and merchandising of both imports and exports, and so to bolster the fortunes of its declining port.

To create the World Trade Center's site, builders tore down 164 structures—to local protests—and with them hauled away the old local infrastructure of water, sewer, and gas lines. The old two-block Hudson Terminal with access to the IRT, BRT, and IND subway lines and the Port Authority's trans-Hudson rail system served as the project's transportation node. Initially, critics hailed the towers, but, as the 1960s ended, the culture changed, and Wolf van Eckardt judged them "a fearful instrument of urbacide," Ada Louise Huxtable called them "graceless," while others derided their "Buck Rogers" quality, noting that such new technologies often bring with them new vulnerabilities (Gillespie 1999: 175). The way in which the towers met the sidewalk—rather abruptly in an open, windswept plaza—represented for many the very antithesis of Jacobs's urban "neighborhood."

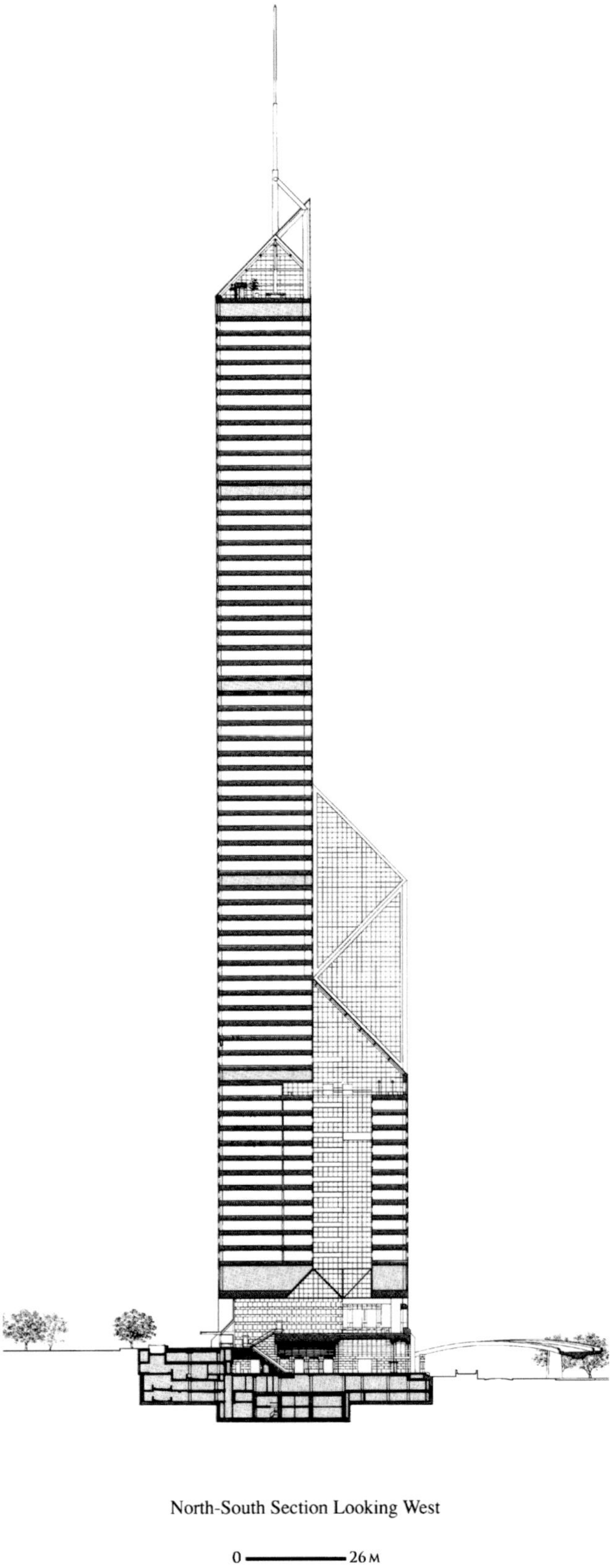

1.18 I.M. Pei, Bank of China, Hong Kong, 1984–90, section drawing
Image: Pei Cobb Freed & Partners

In skyscrapers designed during the 1970s and 1980s around the world, the "tube concept" inspired structural permutations and possibilities that even Khan or Robertson may not have imagined. Furthermore, many showed greater delicacy than did their predecessors in their approach to the human scale and to spaces designed to support human activity within a public or semi-public domain. They included the Citicorp Center, New York, 1970–77, by Hugh Stubbins and the engineer William Le Messurier, and the Bank of China, Hong Kong (1984–90), by I.M. Pei and Leslie Robertson. I.M. Pei conceived the Bank of China as a "bundle of tubes" comprising four vertical shafts, each triangular in plan and stopping at a particular story as the building rose, resulting in a profile that appeared "carved away," so taking the shape of a fantastic rock crystal rising 72 stories into the sky in a geometric play of progressive dislocation (Figure 1.18).

Although Robertson admitted that the details at the structure's corners were problematic, a kind of "hocus pocus" in which nothing intersects—in part the consequence of the client's desire to avoid large "X" shapes, a symbol of failure for Feng Shui geomancers, across groups of floors—the Bank of China nonetheless suggests an integration of architecture and engineering not unlike that of the Hancock Center or Sears Tower. The faceted, crystalline result provided Hong Kong with an icon for the city that it deployed in a range of visual media, but more importantly it produced an array of delightful interior spaces, among them the atria of the banking hall and upper stories, which are filled with sunlight and greenery, afford spectacular views, and offer tenants places to gather as a respite from the routine of office work.

The "Green Skyscraper" and the Potentials of Urbanity

In the 1990s designers had begun exploring a number of new directions for skyscraper design: a growing environmental consciousness had given rise to the "green" skyscraper; the heightened attention to urban neighborhoods enriched earlier notions of civility; and the aim of reinvigorating the public domain encouraged them to seek a sense of community within the skyscraper and the spaces of its immediate surroundings.

In 1989–92, Ken Yeang proposed a striking reconceptualization of the skyscraper typology with his IBM Tower, or Menara Mesiniaga, in Selangor, Malaysia (Figure 1.19), which he called a "bioclimatic skyscraper." Exterior plantings, "skycourts," east- and west-facing windows to minimize sun gain, and a rooftop

1.19 T.R. Hamzah & Yeang, IBM Tower, Subang Jaya, Selangor, Malaylasia, 1989–92
Image: © 2012 T.R. Hamzah & Yeang Sdn. Bhd.

sun terrace shaded by fins and louvers were among the features that Yeang combined in his innovative, environmentally sensitive design.

Yeang's skycourts began their ascent from a three-story high planted mound and spiraled up the face of the building to intersect along the way with spatially gracious atria featuring greenery. These he envisioned establishing a connectedness with nature while channeling the flow of air, providing shade, and supporting an oxygen-rich atmosphere. Yeang premised his bioclimatic skyscraper on a rigorous program of research and development, comprising "sunpath projects," "windrose projects," "cladding and skin projects," and "lifestyle patterns based on ecological principles." At the heart of Yeang's program lay the aims of "energy reduction" and devising buildings as "open systems."

Like Yeang, Foster + Partners aimed to reinvent the skyscraper. The firm's Commerzbank, Frankfurt (1993–97), met resistance as the tallest skyscraper in Europe at the time, rising 60 stories and 850 feet (259 meters) (Figure 1.20). Norman Foster, consequently, thought it prudent to introduce compensatory amenities and energy-conserving measures within the design. Working

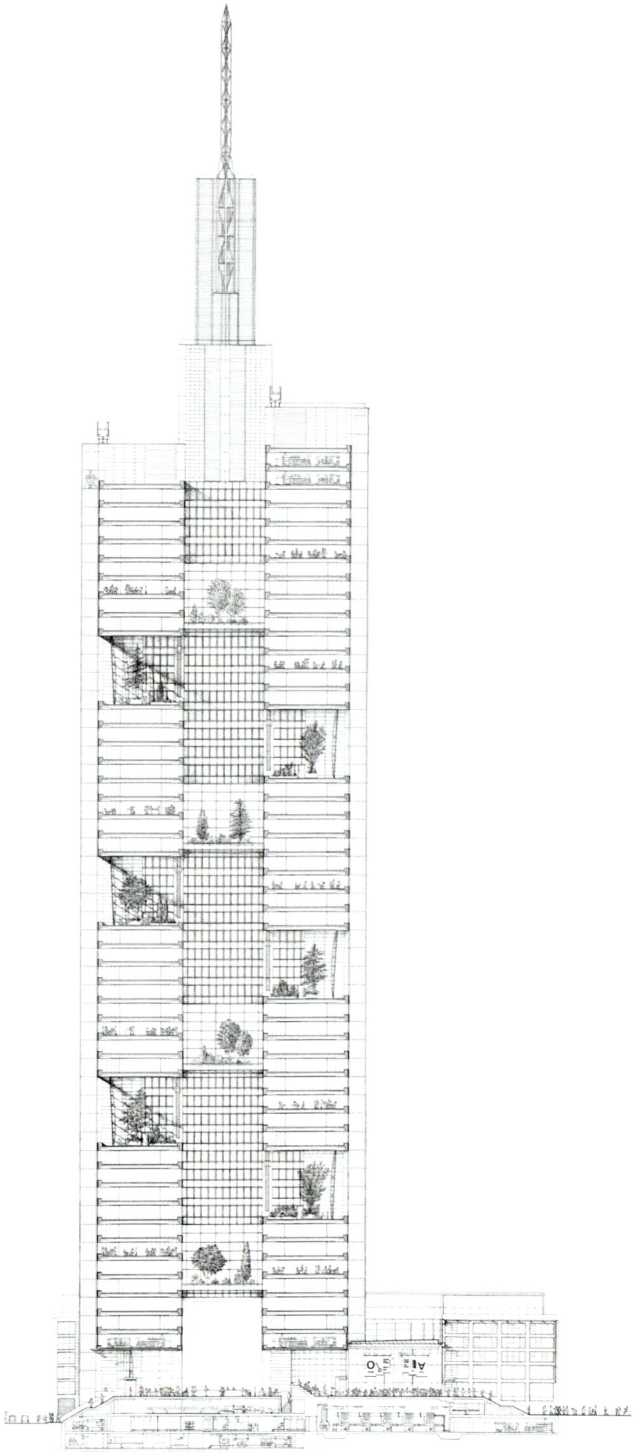

1.20 Foster + Partners, Commerzbank, Frankfurt, 1993–97, section drawing
Image: Foster + Partners

1.21 Cesar Pelli, Petronas Towers, Kuala Lumpur, 1993–98
Image: Thomas Wanhoff/Flickr, under the Creative Commons licence

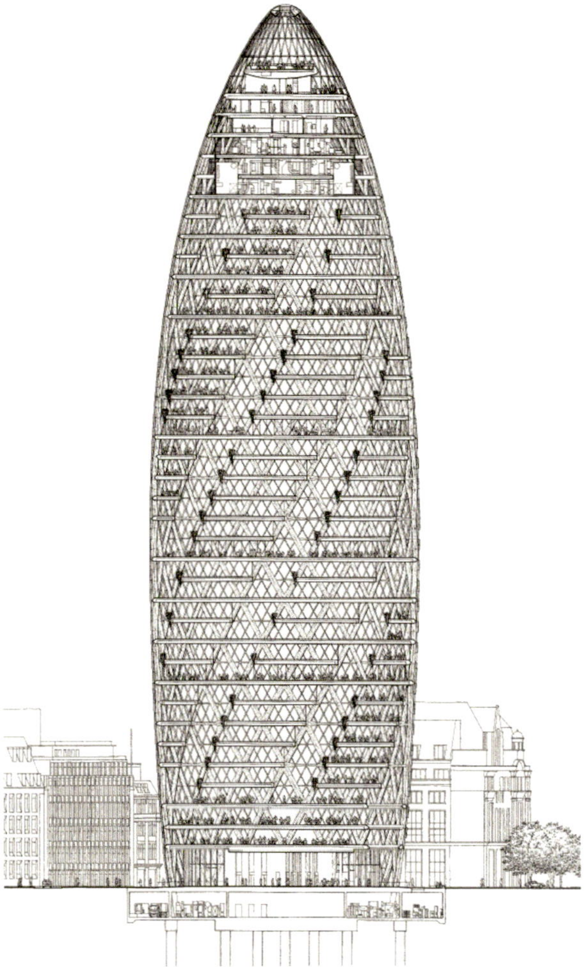

1.22 Foster + Partners, 30 St Mary Axe, London, 1997–2004
Image: Foster + Partners

with Arup, he restructured the typical skyscraper's interior, turning it "inside out" to open it up to the ambient environment. He utilized a triangular plan with structural service cores (elevators, utilities, structural supports) at each corner, framed a vertical shaft of space as a central ventilating atrium, and incorporated nine thematically planned gardens, each four stories high and spanning the full distance between the structural service cores. The design provided offices with an optimal amount of sunlight while offering occupants significant semi-public spaces in which to gather beyond the confines of their workspaces. It incorporated double-glazed windows with motorized controls to circulate fresh outside air. Commerzbank instantly achieved fame as a "green skyscraper" and today remains a classic example of the type.

The needle-like supertall skyscraper paralleled the emergence of the "green skyscraper." Cesar Pelli's Miglin-Beitler Tower of 1989 for Chicago, higher than the World Trade Center but with ten times less floor area, and designed to accommodate a host of small entrepreneurial business ventures as opposed to large corporations, typified the new idea of the skyscraper as a spire. Pelli would take his own idea a step further in the Petronas Towers in Kuala Lumpur, Malaysia (Figure 1.21). The Petronas Towers commanded attention as the world's tallest, at a height of 1,483 feet (452 meters), at the time of their completion in 1997. Working with Charles Thornton of Thornton Tomasetti Engineers, Pelli devised a structural system comprising core walls and perimeter columns made of high-strength concrete. Pelli wrote persuasively about the power of the skyscraper to symbolically mark a particular geographic place within a global civilization—his towers echoed the slender towers of Malaysian mosques and a city gate, while symbolizing Kuala Lumpur's arrival on the global economic scene.

As the twentieth Century began to draw to a close, some builders abandoned "height for height's sake." Norman Foster + Partners, in consultation with Arup as engineer, designed 30 St Mary Axe (1997–2004; Figure 1.22) in the middle of renewed controversy in London over building tall. At 590 feet (180 meters) and 40 stories, with a circular core of steel and a "diagrid" of interlocking steel members that spiral upwards around the outer edge of the tower to meet at its apex, 30 St Mary Axe is noted especially for the volumetric relationships that characterize its interior: six triangular light wells pierce each circular floor plate, each of which is in turn rotated slightly so that the wells spiral upward around the surfaces of the tower, conducting air upwards vertically while reducing energy consumption. The tower's skin dissipates heat gain in the air space between two layers of glass. Equally important, the light wells serve a community function: as gathering spaces, they offer multiple overlooks fostering social interaction. The tower's shape, an aerodynamic form that reduces the downward wash of wind gusts, has the benefit of complementing the skyline with a modern yet elegant landmark—frequently shown juxtaposed with the dome of St. Paul's Cathedral—affectionately called "the Gherkin."

In the twenty-first century, the quest for height continues, as illustrated by the completion of the Burj Khalifa in Dubai, at 2,717 feet (828 meters), and the many plans to build still taller in cities around the world, especially in China, South Korea, and the Middle East. Yet, as shown in the works of other designers, among them Renzo Piano's 2012 London Bridge Tower—which features a spacious winter garden and upper level atria—the competition has shifted to more socially aware criteria: ecological responsibility, the creation of community, urban civility, and the preservation of the public domain.

Bibliography

Books

Ali, Mir M. *Art of the Skyscraper: The Genius of Fazlur Kahn.* New York: Rizzoli, 2001.

Billington, David. *The Tower and the Bridge: The New Art of Structural Engineering.* New York: Basic Books, Inc., 1983

Bluestone, Daniel. *Constructing Chicago.* New Haven: Yale University Press, 1991.

Boyer, M. Christine. *Manhattan Manners: Architecture and Style, 1850–1900.* New York: Rizzoli, 1985.

Bruegmann, Robert. *The Architects and the City: Holabird & Roche of Chicago, 1880–1918.* Chicago: University of Chicago Press, 1997.

Condit, Carl. *The Chicago School of Architecture.* Chicago: University of Chicago Press, 1964.

Domosh, Mona. *Invented Cities: The Creation of Landscape in Nineteenth-Century New York and Boston.* New Haven: Yale University Press, 1996.

Fenske, Gail. *The Skyscraper and the City: The Woolworth Building and the Making of Modern New York.* Chicago: University of Chicago Press, 2008.

Ferriss, Hugh. *Metropolis of Tomorrow.* New York: Princeton Architectural Press, 1986 (first published 1929).

Flowers, Benjamin Sitton. *Skyscraper: The Politics and Power of Building New York City.* Philadelphia: University of Pennsylvania Press, 2009.

Gissen, David. *Big and Green: Toward Sustainable Architecture in the 21st Century.* New York and Washington, DC: Princeton Architectural Press and National Building Museum, 2002.

Gillespie, Angus Kress. *Twin Towers: The Life of New York City's World Trade Center.* New Brunswick, NJ: Rutgers University Press, 1999.

Glanz, Jim and Eric Lipton. *City in the Sky: The Rise and Fall of the World Trade Center.* New York: Henry Holt and Company, 2003.

Hoffmann, Donald. *Frank Lloyd Wright, Louis Sullivan and the Skyscraper.* Mineola, New York: Dover Publications, 1998.

Jacobs, Jane. *The Death and Life of Great American Cities.* New York: Random House, 1961.

Jordy, William H. *American Buildings and Their Architects: The Impact of European Modernism in the Mid-Twentieth Century*, vol. 5. New York: Oxford University Press, 1972.

King, Moses. *King's Views of New York 1896–1915.* New York: Arno Press, 1977.

Krinsky, Carol Herselle. *Rockefeller Center.* New York: Oxford University Press, 1978.

Landau, Sarah Bradford and Carl Condit. *Rise of the New York Skyscraper, 1865–1913.* New Haven: Yale University Press, 1996.

Mayer, Harold M. and Richard C. Wade. *Chicago: Growth of a Metropolis.* Chicago: University of Chicago Press, 1969.

Merwood-Salisbury, Joanna. *Chicago 1890: The Skyscraper and the Modern City.* Chicago: University of Chicago Press, 2009.

Moudry, Roberta, ed. *The American Skyscraper: Cultural Histories.* New York: Cambridge University Press, 2005.

Mujica, Francisco. *History of the Skyscraper.* Paris: Archaeology and Architecture Press, 1929.

Pelli, Cesar and Michael Crosbie. *Petronas Towers: The Architecture of High Construction.* London: John Wiley & Sons, 2001.

Report of the Heights of Buildings Commission. New York: Board of Estimate and Apportionment, 1916.

Riley, Terence and Guy Nordenson. *Tall Buildings.* New York: The Museum of Modern Art, 2003.

Shultz, Earle and Walter Simmons. *Offices in the Sky.* Indianapolis: Bobbs-Merrill, 1959.

Solomonson, Katherine. *The Chicago Tribune Tower Competition: Skyscraper Design and Cultural Change in the 1920s.* New York: Cambridge University Press, 2001.

Sorkin, Michael and Sharon Zukin. *After the World Trade Center.* New York: Routledge, 2002.

Stern, Robert A.M., Gregory Gilmartin, and Thomas Mellins. *New York 1930: Architecture and Urbanism between the Two World Wars.* New York: Rizzoli International Publications, 1987.

Stravitz, David. *The Chrysler Building: Creating a New York Icon, Day by Day.* New York: Princeton Architectural Press, 2002.

Tauranac, John. *The Empire State Building: The Making of a Landmark.* New York: Scribner, 1995.

Willis, Carol. ed. *Building the Empire State.* New York: W.W. Norton & Company and The Skyscraper Museum, 1998.

Willis, Carol. *Form Follows Finance: Skyscrapers and Skylines in New York and Chicago.* New York: Princeton Architectural Press, 1995.

Yeang, Ken. *Bioclimatic Skyscrapers.* London: Artemis, 1994.

Articles and Chapters

Bletter, Rosemarie, Haag. "The Invention of the Skyscraper: Notes on Its Diverse Histories," *Assemblage* 2 (February 1987).

Bruegmann, Robert. "The Marquette Building and the Myth of the Chicago School," *Threshold* (Fall 1991).

Clausen, Meredith L. "Paris of the 1880s and the Rookery," in John Zukowsky, ed., *Chicago Architecture, 1872–1922: Birth of a Metropolis.* Chicago and Munich: The Art Institute of Chicago and Prestel, 1987.

Fenske, Gail and Deryck Holdsworth. "Corporate Identity and the New York Office Building, 1895–1915," in David Ward and Olivier Zunz, eds., *The Landscape of Modernity: Essays on New York's Built Environment.* Baltimore: Johns Hopkins University Press, 1997.

Jordy, William H. "The Tall Buildings," in Wim de Wit, ed., *Louis Sullivan: The Function of Ornament.* New York: W.W. Norton & Company, 1986.

King, Anthony. "Worlds in the City: Manhattan Transfer and the Ascendance of Spectacular Space," *Planning Perspectives* 11 (1996): 97–114.

Robertson, Leslie and Saw-Teen See. "Structural Systems," in A. Eugene Kohn and Paul Katz, eds., *Building Type Basics for Office Buildings.* New York: John Wiley & Sons, 2002.

Siry, Joseph. "Adler and Sullivan's Guaranty Building in Buffalo," *Society of Architectural Historians Journal* 55 (March 1996): 6–37.

Sullivan, Louis H. "The Tall Office Building Artistically Considered" (first published 1896), in *Kindergarten Chats and Other Writings.* New York: Dover Publications, 1979.

Van Zanten, David. "Twenties Gothic," *New Mexico Studies in the Fine Arts* 8 (1983): 19–23.

Weisman, Winston. "Skyscrapers: The Return to Earth," *Architectural Review* (March 1950): 197–202.

Chapter 2

Aesthetics, Symbolism and Status in the Twenty-First Century

Chris Wilkinson

When, in his 1896 article 'The Tall Office Building Artistically Considered', Louis Sullivan coined the phrase 'form ever follows function', he could not possibly have imagined some of the recent developments in form-finding for supertowers. Twists, curves, folds, tapers and cantilevers are just some of the geometric contrivances being explored in the quest for innovative new forms.

Of course, he was keen to divest buildings of the need for unnecessary architectural ornament, but his aesthetic wisdom held true for almost a century of 'Modernism', during which the form of most buildings was dictated by fairly simple structural principles. All this has been challenged more recently by new advances in sophisticated three-dimensional architectural form-finding programmes and speedy stress analysis modelling programmes for structural engineers.

It is clear that architecture is at a new turning point, at which almost all imaginable forms are possible, but decisions still have to be made on what is appropriate.

This situation is particularly relevant to the design of towers and supertowers, because with the huge financial investment required for these buildings comes recognition and status, but this can only be fully realized with a bespoke, individual architectural identity. As a result, exciting new architectural solutions are being explored at a time when they can be technically realized. This is entirely consistent with the evolution of the skyscraper, which has always been linked to new developments in construction technology and the expression of status and power.

Making a Statement

The symbolism of a tower relates very much to wealth, but this is relative and can always be judged against other towers in the vicinity, the city or the rest of the world, which is why there is so much competition in relation to height and identity. In the exclusive world of high-rise and super-high-rise it is not enough simply to have more than one hundred storeys – it is the overall visual impact which is vitally important. This esoteric problem raises the polemic of the architecture of towers and how it should be judged. Whilst aesthetics and beauty are fundamental, other criteria such as innovation, individuality, status, symbolism and context are important. Normal judgements related to scale and proportion can become challenging once buildings and their components substantially exceed human dimensions. The visual appearance is more likely to be assessed in terms of form, structure and materials used for the enclosing skin. In these areas, the scope has greatly increased in the past decade and this has led to experimentation and innovation.

For instance, the extruded vertical form, be it rectangular or curved in plan, is being replaced by more complex geometries offering more interesting silhouettes on the skyline. Structural gymnastics facilitate options for folded crystalline forms or curvilinear organic shapes. Developments in curtain walling offer more freedom to explore these new forms, while also providing more energy efficiency and more sustainable solutions. Height is now more a factor of economics and aspirational intent than a technical restriction.

Breaking the Bonds

In earlier times, the height of buildings was severely limited by insufficient engineering skills, the use of natural materials, problems with fire safety, water pressure and the number of steps to climb due to access without lifts. As outlined in Chapter 1, the first real skyscrapers emerged in the USA only in the late nineteenth century.

For most of the twentieth century, New York was the capital of the skyscraper world, epitomising capitalism and the age of technology. The bright, shiny towers competing with each other for daylight and a share of the skyline were less important individually than as a composition.

The breakthrough in terms of architecture came in the early 1950s, with two projects that established an 'international style' for high-rise buildings, becoming a visual template for others to follow. These were SOM's Lever House, designed in 1951, and the Seagram Building, designed in 1957 by Mies Van der Rohe in collaboration with Philip Johnson. Both were rectangular, simple forms clad entirely in repetitive curtain walling and they sat nearly opposite each other on Park Avenue, New York, both within their own urban piazza. In a way, they brought many of the ideals of the Bauhaus to New York, but on a larger scale and with more commercial appeal.

Developers have seen the design of these two buildings, with their rectangular floor plates and central structural core, as representing the perfect office floor arrangement, due to flexibility and efficient net-to-gross floor area ratio. The preferred size of typical floor plans for high-rise towers has increased over the years, to suit modern office working practices, and the external appearance has taken on different iterations, but this basic form remains a popular option today for medium-high-rise towers in cities around the world.

With new technology, however, there has been a move towards more complex shapes in super-high-rise towers. This is mainly due to the inescapable fact that traditional rectangular frames with their central core need some form of additional external bracing to cope with the extra wind loads associated with super-high-rise towers This can be achieved with a close grid of rigid columns on the perimeter, as pioneered on the New York World Trade Center Towers; the adoption of super columns with outriggers, as in the Hong Kong International Financial Centre; some kind of external diagrid frame structure, as used on the John Hancock Center in Chicago; or the 'tube in a tube' approach, as pioneered by Fazlur Khan of SOM in the Sears Tower, Chicago.

Shaped by the Wind

A derivation of the structural concept has been developed in the recently completed Burj Khalifa in Dubai, also designed by SOM, which is now the undisputed tallest building in the world, with an impressive height of 828 metres. It uses stepped, tubular multiples arranged in a 'Y' shape which buttress the hexagonal central structural core, in what has been termed a 'buttressed core' system. The chief designer, Adrian Smith, has claimed that the flower *Hymenocallis* is the inspiration for the shape, but it is also influenced by the practicalities of maximising view and light to the apartments and hotel rooms. The building's surface area of over 1.8 million square feet (167,225 square meters) is clad with a repetitive glass and stainless steel curtain walling system that follows the curvature of the tubular forms and is broken every 30 storeys with a double-floor band of special horizontal cladding enclosing the plant rooms (Figure 2.1).

The architecture is hard to assess because the design is so dominated by its height and shape, but there is a clarity to the concept which has an integrity based on sound engineering principles. Stepping the tubes creates a tapering spiral form that seems to emphasize the building's height in a 'constructivist' manner, and there are similarities to Frank Lloyd Wright's visionary project for the Mile High Tower, which was sadly never constructed.

2.1 Cladding on the Burj Khalifa, Dubai
Image: SOM/Nick Merrick

2.2 Taipei 101
Image: Taipei Finance Corp.

Standing Alone

Spirals occur frequently in nature and in the Burj Khalifa the geometry provides an elegant balance. The bulk of the accommodation is at the lower levels, which makes the base seem extremely broad. This has little or no effect on the building's context, however, because it sits very much in its own domain, relatively unrelated to the rest of the city. In this respect, it is not really an urban building, more a self-contained metropolis.

Construction of the Burj Khalifa is undoubtedly an incredible achievement, which has focused world attention on this comparatively small but ambitious United Arab Emirates principality. It is unfortunate, therefore, that the time it took to plan and construct such a major project affected its fortunes severely, due to the world's financial recession. There is nothing new here, as the same thing happened when the Empire State Building was launched during the great recession of the 1930s. It lay mostly empty for many years, earning the nickname the 'Empty State Building'.

Perhaps it is too early to judge the full significance of the Burj Khalifa, but with stories still emerging about poor conditions for the construction workers and concerns over its enormously high carbon footprint, the project has been described as a 'monument for an era of credit-fuelled over-consumption, both irresponsible and unsustainable'. Located on the edge of the desert, with a potential occupancy of 35,000 people, the tallest building in the world makes a symbolic statement about wealth and ambition – but its feasibility must remain questionable.

Symbolism and Tradition

Taipei 101 (Figure 2.2), which held the coveted 'world's tallest' spot from 2004 to 2010, has a very different aesthetic, being steeped in symbolism related to local traditions. Its architect, C.Y. Lee, was keen to establish the building as an icon of modern Taiwan by incorporating meaningful Asian references and symbols. Its very shape is intended to evoke the rhythms of a Chinese pagoda, which in spiritual terms brings the sky and earth together. Its 101 floors relate to the new millennium, since it was the first major skyscraper completed after the year 2000. In Chinese culture the number eight is associated with prosperity and good fortune, so the building is constructed of eight upturned boxes of eight floors, each box shaped like a Chinese moneybox, representing abundance. The digital unit of the byte, made up of eight bits, is also referenced. If this were not enough symbolism, there are four large discs mounted on the faces of the tower at podium level and gold coin emblems over the entrances, representing wealth, and 'ruyi' motifs appear throughout the structure, offering fulfilment and protection.

There is a philosophical question over the relevance of these historic and cultural references in a high-tech supertower that is on such an entirely different scale to the indigenous Taiwanese buildings. It may be seen as an important way of providing some local significance to what is otherwise an American concept, born out of New York and Chicago.

Similar questions were raised about the Petronas Towers in Kuala Lumpur, designed by Cesar Pelli, when they were completed in 1998. In describing the concept, the architect stated 'I tried to express what I thought were the essences of Malaysia, its richness in culture and its extraordinary vision for the future. The building is rooted in tradition and about Malaysia's aspiration and ambition.'

Instead of a smooth, clear form, the facade, with its indentations and articulations, resembles an Islamic monument more than a modern skyscraper. The geometry of the plan form is based on two interlocking squares, creating an eight-pointed star with eight semicircles inserted into the inner angles, which soften the external appearance. Upper storeys of the towers are tapered, with five setbacks, and the uppermost walls slope into a peak that finishes with a spire.

It is said that 'Architecturally, these forms reflect the paramount Islamic principles of unity within unity, harmony, stability and rationality', but however the symbolism is interpreted, the buildings have a strongly distinctive look far removed from that of the stereotypical American skyscraper.

When completed in 1998, the Petronas Towers took over from the Willis Tower as the world's tallest and helped to establish Kuala Lumpur as a major international city. Much of their visual prominence rests on the power of identical twin towers joined by a slender pedestrian skybridge, on the forty-first and forty-second floors (Figure 2.3). This elegant link, so dramatically

2.3 The skybridge on the forty-first and forty-second floors of the Petronas Towers, Kuala Lumpur
Image: Antony Wood

portrayed in the film *Entrapment* starring Sean Connery and Catherine Zeta-Jones, is unique and makes a strong statement about the building's identity.

This point has been reinforced by the architect of the Towers, Cesar Pelli, who has stressed the importance of the void between the buildings as 'delineating an invisible axis of symmetry', which has also been related back to the writings of Lao Tse, the ancient Taoist philosopher.

Another project that takes inspiration from traditional Asian aesthetic is the Jin Mao Tower in Shanghai, by SOM. While it has a thoroughly modern programme, with a thirty-eight-storey hotel on top of fifty storeys of office, its visual appearance draws on traditional Chinese architecture, with a stepped pagoda-like facade that tapers as it rises up to a central spire in a similar way to the Petronas Towers.

There are also similarities with Taipei 101 in the Jin Mao Tower's symbolic references to the number eight, and in fact the name 'Jin Mao' translates literally as 'Golden Prosperity'. In order to maximise the relationship with the number eight, the building is designed with eighty-eight floors in sixteen segments around an octagonal core.

This is a distinctive and extremely successful building that was one of the first to signify China's prominence in super-high-rise towers. In terms of pure function, however, the intricacies of the highly articulated cladding of this building may prove to be a problem during its lifetime, for any increase in surface area and in the number of junctions between materials leads to higher maintenance costs and inefficiencies. The same caveat applies to the Petronas Towers and Taipei 101.

Alternative Approaches to Aesthetics

In many ways, there is a case for treating the design of super-high-rise towers more like the design of an aeroplane, with a smooth aerodynamic skin giving efficiencies in surface area, weathering and maintenance. One particular building that follows this aesthetic is the Guangzhou International Finance Center, by Wilkinson Eyre (Figure 2.4; see also Case Study 7). With its powerful industrial base, Guangzhou is China's third most important city, as well as one of its most ancient and historic. A symbol of Guangzhou's emerging international strength, the new building rises to a height of 437.5 metres without a mast. Its elegantly simple form is only achieved with the use of complex geometry. The plan is a triangle made up of three trochoids (a trochoid is the curve described by a fixed point on a circle as the circle rolls along a straight line) set out asymmetrically,

2.4 Guangzhou International Finance Center
Image: Wilkinson Eyre

each with a radius of 5.1 kilometres, with the widest point at a third of the height, tapering to the narrowest point at the top.

In a similar arrangement to the Jin Mao Tower, the lower sixty-nine floors accommodate office uses and the upper thirty-three floors contain a hotel, with sightseeing on the one hundred and third floor and a helicopter landing pad above that. The hotel lobby on the seventieth floor opens up to a vast atrium, tall enough to accommodate St Paul's Cathedral plus its dome.

Lateral stability is achieved with an exoskeletal frame of composite construction, set out on a grand order, with the diamond forms created by the circular tubes extending 48 metres high over twelve floors and the nodes over two floors. A total floor area of 285,000 square metres is enclosed by a simple skin of frameless glass units spanning floor to floor with flush joints. The flat glass units are repetitive, with the only variations occurring on the curved corners, where the panels are still flat but half the normal width, and on the plant/refuge double floors, where narrow panels are set out in a hit-and-miss arrangement to allow for ventilation.

Care was taken over the choice of glass, in order to maintain transparency while still achieving the necessary environmental performance, meaning the structural diagrid can be clearly seen from outside. In this way the main architectural components of skin and structure that emphasise the form can be clearly read, without distractions.

Another interesting tower, soon to be completed in Guangzhou, is the Pearl River Tower by SOM. The building was originally intended to bring forward new standards for the sustainability of towers by generating more energy than it uses, however that ambition is sadly not now going to be realised, for various reasons. The tower's distinctive form, which consists of three pillow shapes stacked on top of each other, is designed to draw horizontal wind forces through two central gaps housing wind turbines that can power the tower's heating, cooling and ventilation systems. In addition, the tower has a high-performance facade with integrated louvres and natural ventilation to cool under-floor voids, as well as a geothermal (ground source energy) system in the caissons, to improve the efficiency of the chillers. With these and other devices, the designers are hopeful of providing a zero-carbon tower that will be the first of its kind and provide a template for others to follow.

The Rise of China

Many of the recent and proposed supertall towers in China explore new geometric forms to achieve a distinctive and innovative identity. For instance, the Shanghai World Financial Center in Shanghai's Pudong area, designed by KPF, succeeds in achieving the desired 'wow factor' with an impressively razor-cut shape which tapers as it reaches up to the sky and finishes with a rectangular opening at the top. Much significance has been attributed to this powerful void, which started out as a circular shape but was changed because of its possible symbolic reference to the traditional Japanese rising sun emblem.

With a height of 492 metres, and close to the Jin Mao Tower and the other buildings which make up the financial district, this tower makes an impressive statement on the Shanghai skyline. It is destined, however, to be dwarfed by Gensler's 632-metre high Shanghai Tower, scheduled to be completed in 2014.

This new tower is designed with a 'twist', enclosed within an outer skin of glass curtain walling – it seems that twists are currently extremely popular in tower designs, having first been tried and tested by Santiago Calatrava in his Turning Torso apartment building in Malmö, and with the Infinity Tower in Dubai by SOM, featuring a 90-degree spiral to a height of 330 metres, also under construction. While the architecture of the Shanghai Tower is likely to be a little more brash than its neighbours, it will be one of the tallest buildings in the world and there is no doubt that it will make a spectacular addition to what will be an unrivalled complex of three supertowers (Figure 2.5). With this in mind, it is hard to imagine a stronger statement of Shanghai's powerful economic position.

2.5 Shanghai's financial district, showing the cluster of three supertowers. The Shanghai Tower is on the left, the Shanghai World Financial Center on the right and the Jin Mao Tower behind
Image: Gensler

2.6 CCTV Headquarters, Beijing
Image: OMA

In China, as in the rest of the world, supertowers have great symbolic status and there is considerable rivalry between cities. Ultimately the heights are controlled by central government, but it seems to be a strategic aim to encourage financial investment in high-rise inner city development. Aspirations still appear to be rising, as evidenced by the ever-increasing height of buildings. It is interesting that recently completed towers, such as the Guangzhou International Finance Center and the Shanghai World Finance Center, are in the 400 metres-plus range, but the planned next generation will be in the 600 metres-plus range. For instance, the Shanghai Tower's planned height of 632 metres will in turn be eclipsed by the proposed Pingan International Finance Center in Shenzhen, designed by KPF, with a height of 648 metres. This is, at the time of writing (in 2011), the tallest building under construction; when completed it will be the tallest in China and the second tallest in the world.

Of course, height is not the only measure and status can be achieved in other ways, such as distinctiveness. The Beijing CCTV Headquarters by OMA (Figure 2.6), for instance, takes a completely new approach to architectural form-making by abandoning conventional structural principles. Two 'L'-shaped towers are set out on a leaning angle, joining at top and bottom. This unusual composition, which resembles a rectangular Mobius strip, makes for an exciting urban intervention on the edge of Beijing's financial district.

The complexities of the building's structure were resolved with inventive engineering by Arup, working closely with the East China Architectural Design & Research Institute (ECADI). Although the shape may at first seem visually challenging, it offers something fresh and distinctive and soon becomes familiar and hugely enjoyable. The richness of the architecture is enhanced by the extraordinary cladding arrangement, in which the glazing mullions broadly follow the pattern of forces within the structure on bracing lines.

Inevitably, there are many more extraordinary schemes under consideration around the world. The quest for innovative designs never stops, with the architecture of towers constantly evolving to challenge technology and excite man's desire to build high. The process continues.

References

Sullivan, Louis H (1979, first published 1896) 'The Tall Office Building Artistically Considered', in *Kindergarten Chats and Other Writings*. New York: Dover Publications.

Chapter 3
A Client's Perspective

Bob B. Brennan, Scott Thomson, Tom Akerstream and Mark Pauls

In 2009, the 115-meter tall, 22-storey Manitoba Hydro Office Tower was completed in Winnipeg, Canada and was immediately heralded as one of the most sustainable buildings not only in North America but indeed the world. The unique story of how the building came to be is presented in this chapter by the owner/developers of the building. It gives a valuable insight into a client's perspective on developing a tall building according to a particular agenda.

Client and Brief

Manitoba Hydro is a Crown Corporation and the major electrical and natural gas utility for the province of Manitoba, Canada. Headquartered in Winnipeg, the company has over $12 billion in capital assets and more than six thousand employees. Its transmission and distribution network provides electricity for half a million customers in the province, and it exports to over thirty other utilities in North America. Until it embarked on the Manitoba Hydro Place project (see Case Study 4), the company had never commissioned a new office tower. Its main construction focus over the previous forty years has been hydroelectric facilities and transmission lines. However, the experience gained from these projects was crucial in the procurement of the new headquarters building, which had very ambitious targets to meet.

In 2002, Manitoba Hydro purchased Winnipeg Hydro to become the sole electrical utility in the province. As part of the purchase agreement, Manitoba Hydro agreed to build a new head office in downtown Winnipeg. Before the deal, Manitoba Hydro did not provide electricity to Winnipeg's downtown area, and operated from a low-rise head office as well as over twenty smaller offices scattered around the suburbs of Winnipeg. The amalgamation of the two utilities became the genesis of a project to bring Manitoba Hydro's staff together in a single, larger office, and to locate this building in a downtown that was in the midst of a slow but steady process of regeneration.

Moving two thousand employees from suburban offices to a downtown location presented significant challenges and required buy-in from the staff. Winnipeg, once referred to as the "Chicago of the North," had suffered from urban sprawl and its associated issues, and was in the intermediate stages of renewal. Manitoba Hydro employees were accustomed to short commutes to work and large, inexpensive parking lots. A move downtown would undoubtedly encounter concerns about longer commutes and expensive parking. As such, it was felt that the new head office had to be a world-class building that supported the productivity and health of its occupants, while reflecting the core values of the company. A unique building, it was believed, required a relatively unique approach to the building process.

Four decades of construction experience on hydro-power projects did provide a number of useful guidelines that were central to the procurement process. For example, the importance of involving all stakeholders early in the project was fully appreciated. In addition, the company understands the value of the collaborative approach and that it is absolutely essential that building should never be a linear process in which a design is handed sequentially from one expert to another.

Internal Organization

The process began with the assignment of three Manitoba Hydro employees to the project—one to act as project manager, a second as project coordinator, and a third to function as a full-time energy and sustainability advisor. This third position played a key role, championing the two corporate goals of energy efficiency and sustainability throughout the design, construction and commissioning of the building. These three individuals initiated the design process by establishing six general objectives, as follows:

1. Create a high quality of space for employees that is second to none.
2. Be the most energy efficient office building in North America.
3. Be as sustainable as possible.
4. Contribute to the urban fabric of downtown Winnipeg.
5. Showcase signature architecture.
6. Be cost effective.

To bring all employees on board with the move downtown, the project team created an Internal Advisory Team to function as a conduit between employees and the project team, to keep all employees informed throughout the entire process and ensure that the project team adequately addressed employee concerns. A second group, the External Advisory Team, was created shortly after. As a Crown Corporation, Manitoba Hydro is ultimately responsible to the public of Manitoba. Throughout the process, it had to balance the company's goals for the project with its corporate and fiscal responsibility to ratepayers. The External Advisory Team consisted of a mix of downtown business owners and the general public and provided public feedback.

Site Selection

With the major stakeholders in place, the next step was to select a site that supported the goals of the project. A number of sites were available in the downtown area and all were thoroughly evaluated against the six key goals. At this time, the structure of the building, whether low-rise or a tower, was not yet determined. An advocate architect was added to the team to guide the company through the design process. The site selection process took over six months, with the evaluation of numerous potential sites against hundreds of criteria, including public transit connectivity, sunlight exposure and solar potential, wind effects, parking concerns, geothermal (ground source energy) potential, site profile, potential problems associated with nearby development, and safety issues. When a shortlist of potential sites had been developed, they were presented to the public in a downtown exhibition. The public was invited to provide comments to the project team, and these comments were taken into consideration prior to the final site selection.

Integrated Design Team Selection

During the finalization of the site selection it was determined that the only way to meet the design objectives was to utilize an integrated design process. An integrated design process meant that all disciplines involved in the design would be brought in to contribute from the onset. Manitoba Hydro was counting on the synergistic effect of having the key players involved early, both to create new approaches to design and to save money through having the different disciplines working harmoniously. With this in mind the selection of the design team began.

The design architect was the first position filled, as it was felt to be important to allow the design architect input in the selection of the rest of the integrated team. A key priority in the selection process was the collaborative abilities of the consultants. The intention was to put the best possible team of consultants together, so it was essential for Manitoba Hydro, as the client, to lead the selection process. In a typical design process, the design architect, once selected, assembles a team of familiar consultants. In this case the idea was to attract consultants who were the "best of the best," with experience of working on "state-of-the-art" design in the past.

A Request for Qualifications was issued internationally to attract potential design architects. This specifically required experience in the goals set out by the project team charter, such as high-quality office environments, energy efficiency, and sustainable design. Interested architects were also asked to include as part of their proposals a single-page description of their design philosophy. This proved to be a valuable and insightful portion of the submissions. From this single page, it was possible to discern how well the potential architects matched Manitoba Hydro's corporate philosophy and its vision for the project, and how well they would lead an integrated design process.

The request attracted the major architectural firms of North America and Europe, in large part due to the building's stated, aggressive energy and sustainability goals. There were no doubts about the technical abilities of any of the applicants, and the final decision ultimately came down to the architect who was felt to best understand the local context and who would thrive in a collaborative setting. Each of the eight shortlisted firms underwent a full-day interview.

Selection of the Rest of the Design Team

The design architect assisted with the selection of the rest of the design team. While the design architect was present and provided comment throughout the selection process, the decisions were ultimately left to the three Manitoba Hydro team members and the advocate architect. In the end, the assembled integrated design team consisted of a complete range of building professionals, including design architect, architect of record, climate engineer, mechanical, electrical and structural engineers, quantity surveyors, construction manager, commissioning agents, and numerous others. The team included members from across Canada and a climate engineer from Germany. The geographic locations of the design team created challenges that were addressed in large part by a series of design charrettes and the use of a shared electronic file storage system which contained all of the project files and drawings.

Design Charette

Once the overall team was selected, it assembled at a resort near Winnipeg for the first design charrette, which lasted three days. The idea was for the team to get to know each other in a casual atmosphere and to have everyone participate in all aspects of the design with the entire team working to the same criteria. A key outcome of this first charrette was defining a Design Charter, which focused the early project goals into six specific targets:

1. Provide a healthy and effective work environment for employees.
2. Reduce energy consumption by 60 percent relative to the Model National Energy Code for Buildings of Canada.
3. Achieve a gold rating in the LEED sustainability system.
4. Be a positive contribution to downtown Winnipeg.
5. Showcase signature architecture.
6. Be cost effective and on budget.

Once the Design Charter was finalized, the entire project team and all the members of Manitoba Hydro's executive signed it. This document became a key resource throughout the design process and the framework through which difficult decisions were evaluated.

The first charrette presented unique challenges, which were resolved over subsequent meetings. For example, with over thirty members of the design team it was difficult to solicit input from every member of the team in an efficient manner. The importance of effective facilitation of the meetings became immediately apparent. It was found to be extremely important to have an authoritative facilitator who could ensure that the relevant parties were heard, and that discussion remained focused.

A municipal meeting–type arrangement was adopted, whereby an inner circle of individuals key to a specific area of discussion were allowed to contribute to the exchange of ideas. Remaining members of the design team participated in the outer circle, were kept fully aware of the discussion, and could provide input to their representative in the inner circle. This arrangement balanced the need for efficient discussion with broad communication throughout the team. It facilitated a tremendous synergy among the team members and functioned very well for the context.

Design Development of a Tall Building

It quickly became apparent that Winnipeg's climate would prove to be a challenge to the energy efficiency goals. Winnipeg is the coldest major city in the world,

3.1 Manitoba Hydro Place, aerial view

with temperatures varying over 70 degrees Celsius throughout the year (–35 degrees Celsius in the winter and +35 degrees Celsius in the summer). At the same time, however, the climate offered tremendous possibilities to the design team. Locally abundant sunshine, especially during the coldest months, and predominant southerly winds were elements of the climate that could be reflected in the building.

The design took shape over a dozen design charrettes. Between the group meetings, consultants went away to study ideas in further detail and prepare additional information for the next charrette. The form of the building was a decision clearly stemming from the collaboration of the design team. Hundreds of forms were discussed and debated, with wide-ranging criteria, from aesthetics to acoustical impacts. Energy modeling was used as a design tool to evaluate the energy impacts of the various forms. These energy targets were very influential on the final form of the structure.

Two key decisions emerged very early. The first was that to attain the unprecedented energy target there would have to be a paradigm shift, from traditional energy-efficient practices to climatically responsive design. And the second was that the only way to meet the Design Charter was to build a high-rise tower. Advantage had to be taken of natural processes, such as wind effects, passive solar heating and lighting, and the stack effect. A tall vertical tower enabled two thousand employees to be housed in a structure with narrow floor plates, yielding excellent outside views and natural daylight. The height allowed for optimal use of natural ventilation and a potential solar chimney reliant on the stack effect.

Over countless iterations, the form was eventually narrowed down to three options—the Daylight Tower, the Comfort Tower, and the Solar Tower—each meeting the design goals in different ways. The three options were once again presented to the public in numerous public consultations. The Manitoba Hydro executive and employees also had input in selecting their favorite form. In all cases the third option, the Solar Tower, came out ahead. In plan view the Solar Tower has the form of a capital 'A'. It takes the most economical, rectangular shape of a high-rise office tower, splays it open on the south and collapses it on the north (see Figure 3.1). The form is ideal for maintaining the key services in the core while maximizing the thermal solar energy potential at the south end and minimizing exposure and subsequent heat losses at the north end.

The significance of the role played by Manitoba Hydro's energy and sustainability advisor during this process cannot be emphasized enough. The advisor was an advocate for efficiency and sustainability throughout the project and constantly considered design decisions for their impact on these goals. Manitoba Hydro wanted to demonstrate that energy-efficient buildings do not have to sacrifice a high-quality office space for employees. The energy and sustainability advisor also ensured that the efficiency measures were inherent in the architecture of the building itself. Too often energy efficient measures are added to a design, and are all too easily removed subsequently, during the value engineering phase. At Manitoba Hydro Place, however, the integrated design process allowed energy efficiency to be embedded in the architecture.

3.2 Winter garden, Manitoba Hydro Place

For example, there are three six-storey winter gardens at the south end of the tower floors (Figure 3.2). These winter gardens are architecturally interesting spaces, with six storeys of south-facing glazing, wood-trimmed staircases and 22-meter high water features stretching from floor to ceiling. It certainly is not obvious that these spaces are essentially mechanical rooms, acting as natural air conditioners for ventilation air. These "lungs of the building" use southern exposure to heat up the ventilation air, and the water feature provides the humidity control. The height of the space allows air to rise naturally, due to the stack effect, and be drawn into six raised floors with minimal energy use. These winter gardens are a result of the design goal of embedding energy efficiency directly into the building, and could only be achieved in a high-rise building.

Mockup

As the client and eventual occupant of the end product of this integrated design process, Manitoba Hydro was somewhat cautious about the unique and cutting-edge design elements. While the company wanted to lead by example, it was not willing to accept the "bleeding edge" risk associated with new technologies. To allay such concerns, connections were made with a local community college, where it was agreed that a full-scale mockup of a complete tower floor bay with several key systems would be built in conjunction with the college's Architectural Technology department. This allowed the design team, to monitor the performance of, for example, the proposed double-facade curtain wall, a unique design which placed a double wall on the outside with a single wall on the inside as a response to Winnipeg's cold climate. This is believed to be the only system of its kind in the world. In addition to verifying the basis of the design, the mockup served to strengthen relationships with the college and provided valuable experience to students who participated.

The relationship also yielded other benefits. For example, the association with the college led to the installation of a unique moisture-sensing system that enables any future moisture damage in the new building's green roof to be pinpointed. These unique collaborations were mutually beneficial and ensured that Manitoba Hydro Place reflected the state of the art in building technology.

Continued Consultation

Throughout the design and construction process, the Internal and External Advisory Teams continued to provide valuable insight from the public and employees. Input from downtown businesses proved critical in determining how best to enhance the downtown and ensure that no harm was done. Even the form of the office tower was designed with urbanism in mind. The lower element of the building is a three-storey podium which fills a city block and fits into the scale and the context of the immediate neighborhood. The lower level is not office space, but consists of tenant spaces designed to add vitality at the street level. The tower height reflected key concerns from neighbors downtown. There is a vibrant park across the corner of Manitoba Hydro Place, where in the summer months many office workers enjoy the sunshine and vendors sell their wares. When a shading study determined that the proposed tower would put this popular park in the shade during noon hour in the summer, the location of the tower was adjusted to accommodate this existing positive aspect of the downtown.

Results

Manitoba Hydro Place (Figures 3.3 and 3.4) was officially opened in fall 2009. While construction was not without its challenges, the end product exceeded the company's expectations. The quality of space for the employees is remarkable, with excellent air quality, plenty of natural light and access to views for everyone, resulting in improved productivity, most notably through reduced absenteeism. Manitoba Hydro have embraced the move downtown, adding a positive impact that has not gone unnoticed. Increased sales and development in the surrounding area have been observed by the downtown business group. The architecture has become a much-photographed, iconic symbol of Winnipeg, and

3.3 Manitoba Hydro Place

3.4 Manitoba Hydro Place

has won awards in North America and Europe. From a sustainability perspective, the project team is very pleased with the outcomes. Existing buildings on the site were deconstructed, with over 94 percent of the materials being re-used or recycled in the new building or off-site. The deconstruction cost less than half as much as a typical demolition, which would have seen all of the materials end up in landfill.

Employees have embraced alternative transportation options as well. At the old head office, over 95 percent of employees parked in large parking lots, while downtown at Manitoba Hydro Place, more than 70 percent take public transit, and underground parking for bicycles is full during the summer months. The site was specifically selected to ensure the maximum number of public transport options. The sustainability goal in the Design Charter has been surpassed. Manitoba Hydro is the first LEED Platinum certified office tower in Canada. Actual energy data has shown that the energy efficiency is an almost 70 percent reduction in energy use relative to the Model National Energy Code. Interest in the building is so great that more than ten thousand individuals were given tours in the first year and a half of operation alone. Manitoba Hydro is really proud of this building, and of the fact that the design team achieved all of the goals of the Design Charter in an elegant building that marries architecture and engineering in a high-rise tower.

Going Forward

The commitment to the project goals did not end at occupancy. For example, an Energy Advisory Group and a full-time energy management engineer have been put in place to monitor energy consumption and engage in ongoing commissioning and optimization to ensure that the aggressive energy targets and comfort levels for the building are met. Connections continue to be forged with local universities and further study of the building encouraged, taking advantage of more than 25,000 monitoring and control points available in the building management system. It is hoped that Manitoba Hydro Place will continue to reflect the company's commitment to innovation, and to function as a living laboratory where various building technologies and control strategies are tested and evaluated.

In the end, while Manitoba Hydro as a company will continue to benefit from the building as a corporate headquarters, the real legacy of the project may be the connections made from the integrated design process responsible for this result. The Manitoba Hydro Place project has brought together a very talented group of consultants, and these individuals, as well as the building community at large, will use this paradigm-shifting building as a stepping stone to the next generation of climatically responsive, high-quality, high-rise office buildings.

Chapter 4
Tall Building Economics

Steve Watts

Introduction

What constitutes 'tall' varies according to perspective, be that an architectural, town planning, technical, fire and life safety or other perspective. From a financial viewpoint it is no more straightforward, with judgements being more relative than absolute. There are a number of reasons for this, but the predominant factors are location and building form, which to an extent are interdependent.

It is also a view that can change over time. Whereas tall buildings have been the dominant feature of many international skylines for many years, this has not been the case in some established cities, for instance London, which provides a good example of the impact of location, not only in comparison to other global cities (Figure 4.1) but also within the Greater London boundary. The 'fat, functional' towers that characterise Canary Wharf in the docklands area of East London provide a counterpoint to the 'articulated, landmark' towers of the City

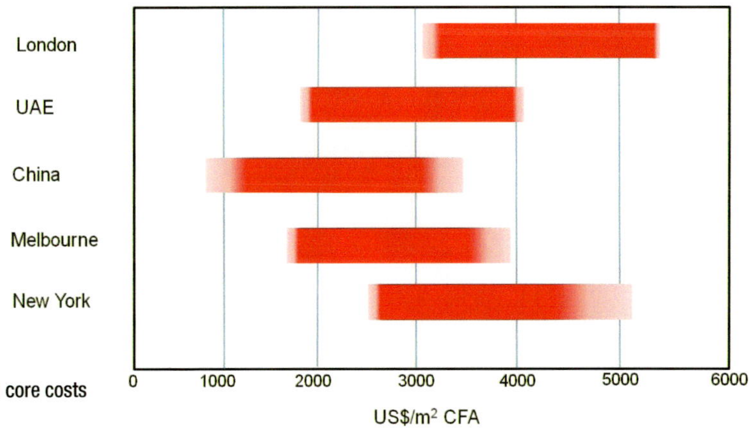

4.1 Global comparison of shell and core costs
Graphic: Davis Langdon

4.2 (top) The future City of London versus (bottom) Canary Wharf
Images: (top) Hayes Davidson and Nick Wood; (bottom) Stephen Earnshaw

(Figure 4.2), with different cost profiles to match their very different forms.

New approaches to tall building typologies – in new locations and with alternative ways of introducing a mix of uses – are bringing into focus their respective cost and value drivers and how form as much as height determines these. This is not to underestimate the influence of height, particularly with the advent of the supertall tower. Clearly, the taller and more complex the building, the greater the challenges – including the financial challenges – and it is interesting to note the results of research undertaken by Miller (2000), covering forty case studies of infrastructure projects ranging from oil rigs to sea ports, which suggested that the common characteristics of such a project are that it:

- is a product of negotiated compromise
- is an integrated part of the infrastructure network
- is subject to contested externalities
- is crafted over many years
- is exposed to political risk
- complies across multiple regulatory frameworks
- involves large irreversible commitments.

There is a resonance with tall buildings here, the strength of which will depend on height, form and location. The more like a large engineering project a tall building is, the greater the challenges across the key measures of cost, time and efficiency; the more susceptible it will be to global trends; and the greater will be the relationship between the building and its immediate environment.

Whilst rapid urbanisation and changing demographics are making the case for appropriate high-rise development, the twin current challenges of global recession and climate change are placing pressure on the fundamentals that underpin such developments and adding to the scrutiny of proposals. There are those who suggest that this should prompt the removal of some of the design excesses of the past (with certain commentators calling for an 'architecture of austerity'). Nevertheless, there is still a call for iconic architecture.

What is clear is that both the value and the cost of tall architecture, iconic or otherwise, need to be understood, and, though the detailed economies of sustainability are beyond the remit of this chapter, the triple bottom line of sustainability – economic, social and environmental (or profit, people and planet) – needs also to be fully understood.

The Challenges

Tall buildings present a number of fundamental challenges, from what can be a difficult navigation through the town planning process to creating a building that not only satisfies various and many interested parties, but also stacks up financially, whilst securing finance on appropriate terms.

Regulatory taxes

Chicago and New York witnessed the birth of the modern skyscraper in the early part of the twentieth century. Developers in those cities moulded their creations within statutory constraints: the former city being characterised by a plateau of high roof lines as plots were completely filled both horizontally and vertically, working to height restrictions; the latter seeing a 'stepped' effect as tall building forms accommodated setbacks to reflect different floor area restrictions at certain heights (Figure 4.3 overleaf). Carol Willis, in her seminal book *Form Follows Finance* (1996) describes this with clarity and insight.

In an attempt to move away from the plethora of similar, 'boxy' buildings and improve the skyline, authorities in Chicago allowed the tops of buildings to be articulated by permitting height to be added beyond the normal limit, over 10 per cent of the building footprint. Unsurprisingly, given that office buildings are built equations as much as they are architecture, developers then sought economic benefit by building more rentable area as efficiently as possible within the additional permitted volume. The 'wedding cake' style of New York was also a response to statutory constraints based on similar principles.

Eighty or so years later and in sensitive locations such as the historic parts of central London, all building projects face a 'regulatory tax'. Such strict regulations impose certain restrictions on tall building proposals that inform not only the quality of design in terms of specification, materials and detailing, but also the fundamental shape of the building. British Land's Leadenhall Building in the City of London was developed into its wedge shape by the Richard Rogers Partnership (now Rogers Stirk Harbour + Partners) to allow for the protected view of St Paul's Cathedral from Fleet Street (Figure 4.4).

Notwithstanding their architectural and town planning merits, all these examples illustrate the search for profit banging its head against regulation. The increased scrutiny of a tower's architectural, environmental and economic impact means that time, effort and money have to be expended to ensure a successful passage

4.3 Skylines of (top) Chicago in 1913 and (bottom) New York in the 1930s
Images: The Skyscraper Museum, New York

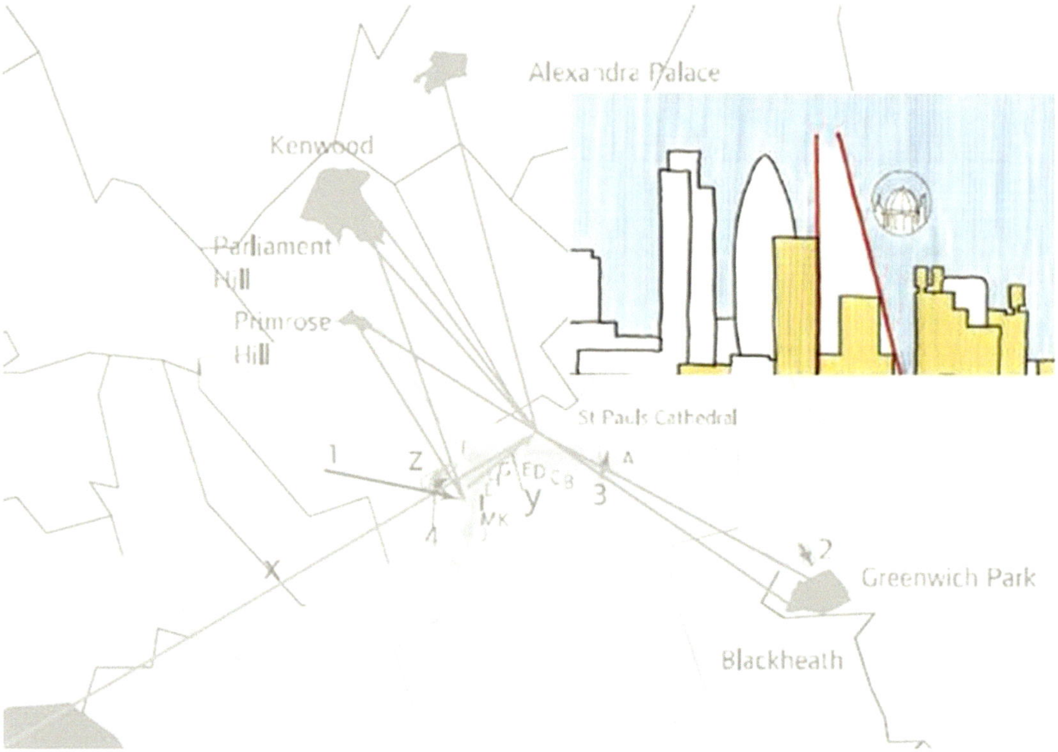

4.4 Form follows regulation: the sloping south facade of The Leadenhall Building avoids the viewing corridor for St Paul's Cathedral
Graphic: Rogers Stirk Harbour + Partners

through the town planning process. In sensitive locations like London this investment can be huge, with a significant number of stakeholders to be satisfied and convinced of the wider benefits of the scheme.

Cost, time and area efficiencies

Towers inherently cost more per unit floor area, take longer to develop and build, and are less efficient in net:gross[1] floor area terms than low-rise developments. Figure 4.5 represents the key financial metrics for a commercial tall building development. Their interrelationship is intense and complex, but critical to the success of the project. The challenge is to squeeze every possible efficiency out of each and all – every dollar saved, every day taken out of the schedule and every square metre of lettable area created within the same envelope will ease the pressure on the bottom line.

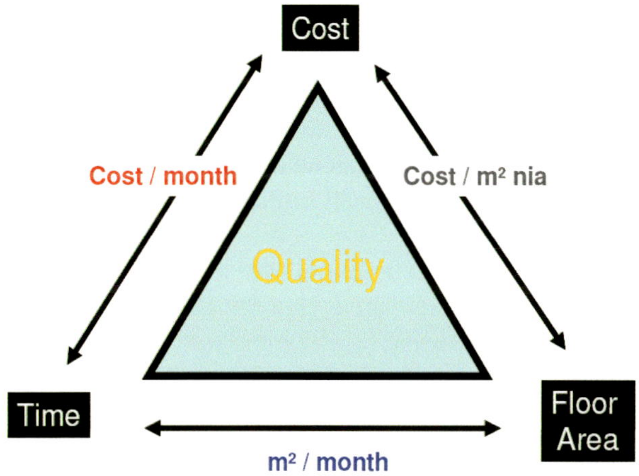

4.5 The fundamental financial drivers of high-rise buildings
Graphic: Davis Langdon

Cost

Generally, the increased expense of a tower is due to, in no particular order:

- its increased wind loadings and heavier frames
- the greater numbers, capacities and speeds of lifts
- the larger capacities and the space needed to accommodate MEP (mechanical, electrical, plumbing) plant, services distribution over long vertical

Element	Central London shell and core costs (£/ft² gross internal area) Low-rise	High-rise
Substructure	18	20
Superstructure	36	45 (40–70)
Facades	30	52 (50–65)
Internal walls and finishes	18	23
MEP services	35	42
Vertical transportation	8	18
On-costs	35	50
Total	180	250

Table 4.1 Notional comparison of Central London low-rise and high-rise shell and core costs, showing typical elemental build-ups
Source: Davis Langdon

distances (with higher hydraulic pressures), and the systems and controls associated with a sophisticated environmental strategy

- the logistics of constructing a tall building, including extensive cranage or hoisting and the increased time on site
- the risks associated with scale and uniqueness
- the higher quality of design and materials that reflect a landmark building, including the pressures to incorporate an attractive interaction with the streetscape and an architectural statement at the pinnacle
- the reduced cost efficiencies of smaller floor plates, including a higher wall-to-floor ratio.

The level of the cost premium over lower rise buildings will depend on the relative importance of the above factors, and the extent to which the scheme responds to the cost drivers outlined later.

The variety of architectural and engineering solutions inherent in the typical 'landmark' tall building results in a large range of costs, something that can fundamentally drive costs higher but which also presents opportunities to limit the cost penalties of high-rise building. Table 4.1 shows not only the potential significant premiums across superstructure and facades, but also the large cost ranges of these two elements.

Time

Getting over the planning hurdles combined with ensuring that the economics stack up (and, increasingly importantly, that funding is secured) means that the gestation period of tall buildings can be extensive. And once on site, it takes longer to build vertically than horizontally. This additional time costs money, both in construction and finance. It also involves uncertainty, not only in predicting future construction costs but also in assessing demand and value.

This puts the spotlight on the construction schedule, with pressure to bring the completion date forward as much as possible. Pace needs to be considered at the outset through strategic planning and buildability reviews.

As with cost, complexity and articulation drive the tall building schedule as much as, if not more than, height. Consideration should be given to process and methodology as much as design: coordinating delivery and cranage slots, maximising labour efficiency through generous hoisting and welfare facilities, etc. In other words, every aspect of how the building is put together and how the works are managed needs to be planned and optimised.

The significant effects of a longer development and leasing period require forensic input, and often money spent on accelerating the schedule is more than repaid in saved finance charges and/or advanced income streams (and, just as importantly, reduced risk).

Floor areas

Tall buildings are less efficient than low-rise buildings because:

- their structure is larger
- more core area is taken up by risers due to larger services distribution systems
- there are greater numbers and sizes of elevators with more space given to associated lobbies and circulation.

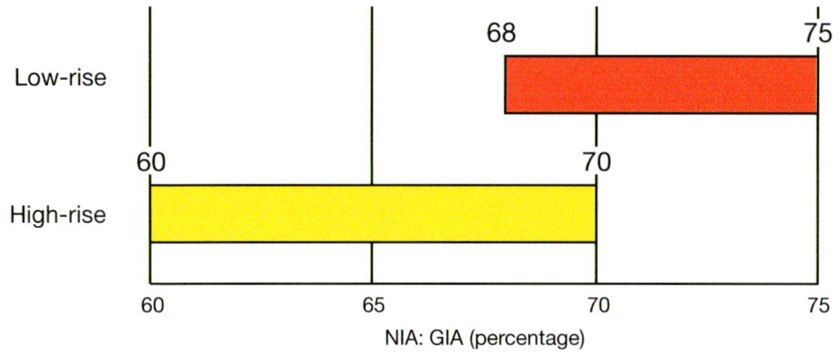

4.6 Overall net:gross floor area efficiencies, low-rise versus high-rise (rationalised ranges)
Graphic: Davis Langdon

There are three basic measures of floor area efficiency – overall net:gross, above ground net:gross and typical floor net:gross – and towers are adversely affected across all three, though the impact is just as much related to size and regularity of floor plate as it is to the number of floors.

Overall net:gross floor area efficiency is the most common benchmark as it is the factor most aligned with viability – the rental income from the building is based on the net area (or a variant of it) and the value of the development is sensitive to relatively small changes in this equation.

A typical total efficiency of, say, 65 per cent versus 70 per cent will mean the loss of 50,000 square feet (4,645 square metres) out of a building with a total gross floor area of 1 million square feet (92,903 square metres), which in Central London could mean a difference in capital value of £50 million, all else being equal. This is a simplification, of course, but it does frame the challenge of high-rise from an area efficiency perspective.

Funding

Despite a continuing 'tale of two worlds' as the centre of global economic gravity shifts to the emerging nations, the world is an ever more connected place, made more so by the emergence of some 'mega-trends' and 'meta-forces' that will have an impact at the local scale: step changes across economics, population and resources. The taller the building, the more susceptible it will be to these trends. Already the global economic crisis that came to the fore in 2008, with a continuing and severe debt overhang across the developed world, has put a significant constraint on development finance. For tall buildings, the level of funding is greater, the payback period is longer and the risk profile is more difficult.

Basel III, the latest revision of the Basel Accords, will force banks to maintain higher liquidity and lower leverage ratios, and limit exposure to real estate. Banks are therefore focussing on the risks they hold in property, which means that risk-weighted premiums will find their way into finance charges. In other words, not only is finance being squeezed, the finance that is available is becoming more expensive.

This difficulty in accessing capital is beginning to shape the nature of high-rise developers in some locations and therefore influence the attitude to tall building typologies. It is also prompting the increasing prevalence of joint ventures as developers seek partners to share the financial burden and risk of high-rise development. Funders (sovereign-wealth backed or otherwise) are investing more time and keeping a keener eye on the proper due diligence of such schemes. They are also seeking substantial pre-lets to kick-start development. This does, however, help to ensure that the tall building business case is as robust as possible, and that backers have the best change of obtaining the long-term trophy asset that they desire.

Economic cycles

It is difficult to legislate for the vagaries of economic and development cycles. In the eighteenth century the economist Cantillon (1755) described the effect of low interest rates in stimulating money supply and therefore creating a magnet for investment; this applies in particular to capital-intensive projects like tall buildings.

This cheap finance is causal in Andrew Lawrence's (1999) analysis of economic cycles and supertall tower construction – the basis of the now infamous *Skyscraper Index*, increasingly cited by the media following completion of Burj Khalifa. Lawrence suggested a relationship between completion of the world's tallest buildings and widespread financial collapse. For example, the Chrysler and Empire State Buildings were finished one year into the Great Depression of the 1930s; New York's World Trade Center and Chicago's Sears Tower were completed in the midst of American stagflation; Kuala Lumpur's Petronas Towers were nearing completion as the Asian financial crisis struck; and, of course, the Burj

Khalifa was officially opened to fanfare and fireworks in January 2010 as the credit crunch firmly gripped the world.

In many ways, the skyscraper in the twentieth century replaced the factory and the railroad, just as the information and service sectors replaced heavy industry and manufacturing as the dominant sectors of the economy in developed countries. It could be said that the modern commercial tower is the nexus of global capitalism and commerce, so it should not be surprising that it is an important manifestation of the business cycle, just as large engineering projects were in previous times.

And it is perhaps this business cycle – or to be precise, its non-alignment with supertall towers' development cycles – which holds the key to the Skyscraper Index, and therefore represents one of the greatest challenges of speculatively-built towers. The uncertainties of the property cycle make most commercial developments a tough call. With towers, the decision is tougher still due to the time taken to get them off the ground and constructed. The taller, more complex tall buildings can span two or even three property cycles, making it difficult if not impossible to get the timing right.

The drivers for tall building development

Given the significant challenges, the drivers for high-rise building need to be compelling. They range from the macro to the micro scale, with an increasing connection between the two as height, size and complexity increase.

Innovation

The modern tall building was made possible by innovative technology – the development of the structural steel frame and the invention of the elevator in particular – which is why it is the first factor to be mentioned here. Advances in curtain walling then enabled the vision of Ludwig Mies van der Rohe to be realised, albeit many decades after he first sketched a tower enveloped in a glass skin.

Since then, there has been little in the way of fundamental technological advance – except perhaps the power of the computer, which has enabled the design of more complex structures and forms. Indeed, design proposals have become so extraordinary as to be accused of irrational exuberance, passing the point at which the value of an architectural statement exceeds its cost or risk – though 'value' has a number of interpretations.

Whilst innovation can be a driver for the supertall and for the 'super-iconic', the Burj Khalifa is testament to the fact that this can be just as much about brain power (of the architect and engineer) as anything else. Expertise can create something truly groundbreaking, yet cost effective; as Kurt Lewin (1952: 346) said, 'there is nothing so practical as a good theory'.

At some point in the future, a fundamental advance will create a completely new tall building typology, underpinned by a different cost profile – a step change that could emanate from deep and pervasive computing, robotics, nanotechnology or some other mega-trend. The development equation in its simplest form – value less cost creates profit – will remain a constant, though, as will the importance of cash flows (hence the importance of time).

Economic growth and the development equation

Properly planned urbanisation tends to increase affluence, which in turn requires the development of Grade-A commercial space, together with housing for a burgeoning population. Demand for land increases, pushing up prices and increasing the amount of occupiable space that must be developed to provide an acceptable return.

For the developer, by far the most important (if not the only) point of creating a tower, as with any commercial building, is to make money, and a tall building essentially does this by creating more floor space on a plot, increasing the value of the development.

To use an over-simplified version of the development equation to make the point (for example, the possible effects of tenant inducements are ignored), let us assume a total gross internal floor area of 300,000 square feet (27,871 square metres) (see Table 4.2). At overall net:gross efficiency of, say, 72 per cent, provides a net internal floor area of 216,000 square feet (20,067 square metres) (assuming, reasonably, that net internal area is the same as lettable area). At £45 per square foot, that would provide an annual rental income of £9,720,000. Applying a typical investment yield of, say, 6.00 per cent – meaning the developer or investors want their money back in a little over 16½ years (a theoretical requirement given the likelihood of a sale before that time) – turns that rent roll into a capital value of £162 million.

On the cost side, the building may average a cost of £225 per square foot over its total gross internal area, encompassing: demolitions and enabling works; the base building plus a developer's fit-out; and incoming services and external works. Add to that professional and other fees and all other developer's costs, including finance changes, and the cost of development totals £91.4 million.

Element		Low-rise	High-rise
1	Gross internal floor area (ft²)	300,000	1,200,000
2	Net:gross efficiency (%)	72	67
3	Net internal floor area (ft²)	216,000	804,000
4	Rental value (£/ft²)	45	55
5	Annual rent roll (£ p.a.)	9,720,000	44,220,000
6	Yield (%)	6.00	6.25
7	Years purchase (100/yield)	16.67	16.00
8	Gross development value (5 × 7) (£)	162,032,000	707,520,000
9	Construction costs (£)		
	300,000 ft² GIA at £225/ft²	67,500,000	—
	1,200,000 ft² GIA at £300/ft²	—	360,000,000
	Professional fees at 13%	8,775,000	46,800,000
	Insurances at 1%	762,000	4,000,000
	Legal, letting and other fees, marketing at 5%	3,375,000	18,000,000
	Surveys, miscellaneous costs	1,000,000	3,000,000
	Planning gain, infrastructure costs, air rights	2,500,000	7,500,000
	30 months at average 7% p.a.	7,500,000	—
	50 months at average 7% p.a.	—	60,000,000
	Total	91,412,000	499,300,000
10	Land/acquisition costs (£)	30,000,000	30,000,000
11	Profit (8 – 9 – 10) (£)	40,620,000	178,220,000

Table 4.2 Summary development appraisal – residual profit calculation
Source: Davis Langdon

If the developer paid, say, £30 million for the land, then the profit for his efforts would be £40.6 million.

Theoretically, however, the greater the area that can be developed on that plot of land, the greater the profit, despite an increase in development costs associated with height and a penalty in floor area efficiencies. What Table 4.2 shows is that if the developer is the first to create a tower on the site then he has the potential to earn a windfall profit. In other words, if he is able to persuade the planners of the outstanding merits of the tower proposal and therefore obtain permission to build four times as much on the same site, once purchased, then his profit could multiply almost five-fold (with land cost remaining unaffected).

Re-working the low-rise appraisal to reflect a high-rise development of 50-plus storeys shows the annual rent roll increasing from £9.7 million to £44.2 million, assuming 804,000 square feet (74,694 square metres) of net internal area (given a reduced net:gross efficiency of 67 per cent but an increase in rental value for a more 'attractive' building). The investment yield might be tweaked upwards to reflect risk but the value of the development rises spectacularly, to over £707 million.

In this second scenario, construction costs per square foot have now increased by over 30 per cent to £300 per square foot GIA, with associated premiums for all developer's costs, particularly finance changes. Despite total construction costs therefore rising to £499 million, a profit margin that was around £40 million is increased between four and five times, to £178 million.

This would happen just once, of course, with such windfall profits going into the land value, and the next buyer having to develop at least as much to justify the purchase price.

City competition and signposting

In 1991, Saskia Sassen introduced the concept of the 'global city', meaning cities that function in four new ways: as highly concentrated command points for the world economy; as key locations for specialised service firms; as sites of production and innovation; and as markets for products and innovations. In other words, these cities need to provide and support a multitude of facilities and functions, and they are in a form of competition with other cities in the same league.

It is in the context of these complex issues and the competitive world environment that the pros and cons of tall buildings are considered for historic environments and emerging economies alike. Indeed, as cities and megacities increase in number, size and power, and globalisation brings markets closer together, then competition between global cities can only become more intense, not least because they also represent the economic engines that drive the progress of their respective nations.

As one global city becomes more successful than another it will increase in size, reducing availability of prime sites for buildings and creating a demand-supply imbalance. To accommodate larger space requirements and avoid excessive sprawl, a successful global city will ultimately need tall buildings.

When tall buildings emerged in the American urban landscape in the early part of the twentieth century, they were clustered to concentrate large groups of people within walking distance of each other and of major public transportation systems. And, despite the argument put forward in recent years that developments in communication technology and working practices should have reduced the need for physical concentration of economic activities, there would seem to be a continuing demand for commercial hubs and economic clusters (or 'good addresses') in which businesspeople congregate, communicate and concentrate – even when in competition. Economists such as Edward Glaeser (in his 2011 book *The Triumph of the City*) eloquently put forward the argument that cities and density enable progress.

There is also a good sustainability argument for placing accommodation in locations that make best use of the infrastructure and transport network. Of course, the city's infrastructure will have an ultimate capacity, and would have to be extended and enhanced at that point. In this sense, the tower and the city support each other.

Like businesses, cities need to market themselves, and one way to signpost growing prosperity is through the skyline: the development of quality landmark towers can signify an emergence on the world stage. Even single buildings can go a long way to putting their host cities 'on the map' – as the iconic status of 30 St Mary Axe, affectionately known as the 'The Gherkin' by taxi drivers, the public and commentators, does in London. Similarly, the *Financial Times* still uses an image of the Commerzbank Tower as its symbol for Frankfurt; and the Petronas Towers, Sears (now Willis) Tower and Empire State Building are strongly associated with their cities, respectively Kuala Lumpur, Chicago and New York.

Incentivisation

Governments can address all the above factors by providing specific incentives for cluster development in less vibrant areas, such as Canary Wharf in London and the Pudong area of Shanghai, both of which are prime examples of enterprise zones with tax advantages and significant investment in infrastructure. They also created 'magnet' towers – One Canada Square and Jin Mao respectively – that acted as a catalyst for further development (Figure 4.7).

Encouragement can also be given for single buildings, an often-cited example being Commerzbank Tower in Frankfurt, for which the local green party, in power at the time, granted certain subsidies to reward the sustainability features of the scheme.

Value creation

There are various ways in which buildings generate value, and the taller the proposal the sharper the focus on some of these:

- *Capital value* or exchange value of the completed development
- *Life-cycle value* associated with future running costs
- *Operational value* generated by effectiveness in business processes
- *Brand value* that benefits corporate image (note in this regard the purchase of the naming rights to the famous Chicago monument, Sears Tower, by the Willis Group in 2010)
- *Contingent value*, or the multiplier effect of a development on its surroundings, as described above
- *Civic value* to the wider community.

4.7 Before and after: the magnetism of Jin Mao Tower, Shanghai
Images: SOM/© China Jin Mao Group Co., Ltd

While development appraisals still focus on short-term profit, driven by the scheme's capital value, occupiers are increasingly assessing buildings for their whole-life costs, not the least of which are salary and related costs – so tall buildings should be 'sold' for their ability to attract and retain employees and increase productivity. The 'people' and 'planet' elements of the triple bottom line may be more difficult to accurately assess than the 'profit' element, but their impacts are potentially far greater (this is the subject of the 'Future Office' study mentioned in the section on 'The Economic Future of Tall Buildings'). Companies have long understood the potential for strengthening their brand through the occupation and ownership of landmark towers, which can go as far as the naming of the building itself.

While the financials of taller, more complex towers are more difficult to stack up, the buildings can present a 'destination' that enhances the value of the wider estate. Burj Khalifa has an obvious effect on its surrounding buildings, including the many that form part of the larger development by the Emaar Group. Similarly, the Shard at London Bridge has had a significant and positive effect on commercial and, in particular, residential values in the Bermondsey district of South East London, acting as a catalyst for the continued regeneration of the borough and providing a springboard for profitable development in the local area (including land purchased and owned by the same developer).

Demand

No matter the driver for tall building development, there needs to be a demand for the product. Different products satisfy different demands. For example, in London, the British Council for Offices undertook research in 2002 that concluded that there was a demand for tall buildings that fell into two distinct categories:

- Large occupiers, who would look to take advantage of consolidation, with a strengthened corporate identity and culture.
- Small occupiers, who would enjoy the benefits of a quality environment, together with the prestige and visibility of occupying a landmark building.

These distinct types of consumer lend themselves to very different types of tall buildings in London, the former, such as global financial institutes, requiring the large buildings and large floor plates that peripheral business districts (like Canary Wharf) offer, while the latter are suited to taking space in the more slender, articulated landmark towers of the City and central districts.

Economists would argue that capitalist arrangements provide the most efficient market economies, and so cities around the globe are inhabited by towers that have been developed to meet specific market demands, for example the 'dormitowers' of Leeds, in northern England, which satisfy the need for good quality student accommodation in a city boasting five universities, or the residential towers containing expensive apartments that adorn the Hirandani township in the northern suburbs of Mumbai.

Ego

Although almost without fail the business case must stack up, there is no doubt that ego can play a part in the creation of a tall building, and certainly in determining the height and form of a tower. Ego may take the form of personal vanity ('I want my building to be the tallest in the region') or corporate prestige – why else would the plan of the NatWest Tower in London, now Tower 42, be in the same configuration as the company's logo? Such 'whims' can influence other features, such as the number of storeys: '100 would have been the perfect number' said Harace Lin, the developer of Taipei 101 in Taiwan, 'but I want it to be more than perfect' (quoted in Binder 2008).

Perhaps stretching the definition of ego, there are examples of beautiful and successful buildings whose designs were not completely driven by economic factors, but by issues such as 'contextual architecture'. In a world increasingly determined by the global project – in which highly international teams come together to create and deliver world-class buildings – icons such as the Jin Mao Tower, with its floor plates and facades reflecting the ancient Chinese pagoda, stand out as being designed with local culture very much in mind. Indeed, it was the designer of that building, Adrian Smith, who said, 'Let the cities come first and the ego of the individual support, but not overpower' (2007: 9).

Yet ego is not necessarily a bad thing. It is sometimes said by those who have been close to a difficult and challenging tower project that it wouldn't have been possible without the ego – if you wish to call it that – of a very determined entrepreneur behind it.

Tall Building Cost Drivers

The fundamental financial drivers

No matter the building use, the fundamental measures that underpin tall building viability are cost, time and floor area, as shown in Figure 4.5.

There is an intense connectivity between these metrics, but also a complex one, for example the relationship between cost and height is not a straightforward one. This is partly because certain factors come into play at different heights – for example, lifts and structure will be affected by height at a lower threshold than, say, electrical services – but, critically, shape and form (in both the vertical and horizontal planes) are at least as important as height. And the array of very different architectural and engineering solutions means that not only do towers generally suffer adversely in all three areas, but the range of their costs, programmes and net:gross efficiencies is also greater.

Thus, the tall building's financial challenges are also its opportunities: to address every one of the three financial drivers to push them towards the better end of their respective ranges. And economies of scale mean that attention to detail can provide leveraged rewards – or leveraged losses. Cost savings or cost premiums are multiplied many times over where they are applied to components or details that occur throughout the building.

Key cost drivers

Shape

Shape – particularly the size and configuration of the floor plate – determines the efficiency of the building and speed of construction, and is often more important than height as a cost driver. This, together with the fact that it has a profound effect upon the cost of the superstructure and facade (which together can contribute 50 per cent or more of total net elemental costs), is why it appears at the top of the list in Figure 4.8.

1 Shape
2 Structural solution
3 Facade specification
4 Environmental strategy
5 Vertical transportation
6 Market Conditions

4.8 Key tall building cost drivers
Graphic: Davis Langdon

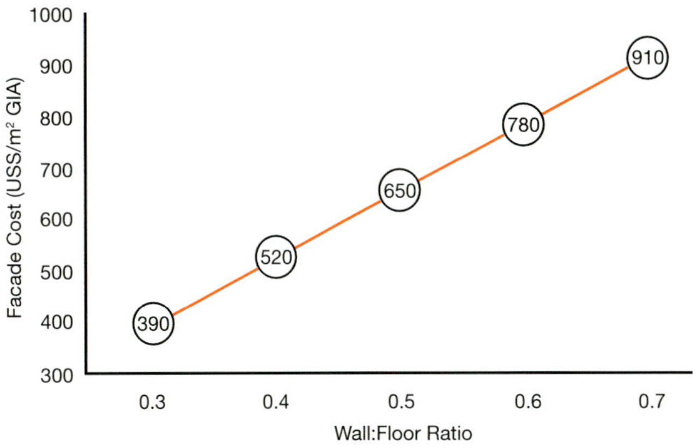

4.9 The financial implications of wall:floor ratio
Graphic: Davis Langdon

The shape of a tall building goes a long way to determining the two key ratios that influence the cost and value sides of the equation:

- Wall:floor ratio
- Net:gross floor area ratio.

Wall:floor ratio is one of the principal implications of shape. It represents the amount of wall area that has to be constructed for every unit of floor area, so from a cost perspective the lower it is the better. For example, a wall:floor ratio of 0.40 (typical for a low-rise office building but also quite possible for a high-rise building) means that the building possesses 40 square metres of facade for every 100 square metres of gross floor area. A ratio of 0.60 (not unusual for iconic high-rise buildings) raises the relative extent of the facades to 60 square metres for every 100 square metres of floor space. Figure 4.9 shows the effects of this, assuming a common average facade cost of $1200 per square metre. All else being equal, the additional 50 per cent of envelope for an increase in the wall:floor ratio from 0.40 to 0.60 increases the cost of the external walls, when expressed in terms of the total gross internal area of the building, from $520 per square metre to $780 per square metre. For a building totalling 100,000 square metres, this equates to an additional $26 million. Given that facade can represent over 20 per cent of the total shell and core cost of a tower, this could result in a 10 per cent addition to total construction costs. This is one of the reasons why it is difficult to provide rules of thumb for cost-versus-height questions.

Structural solution

The form and structure of the tall building are interrelated in their influences and responses. The size and

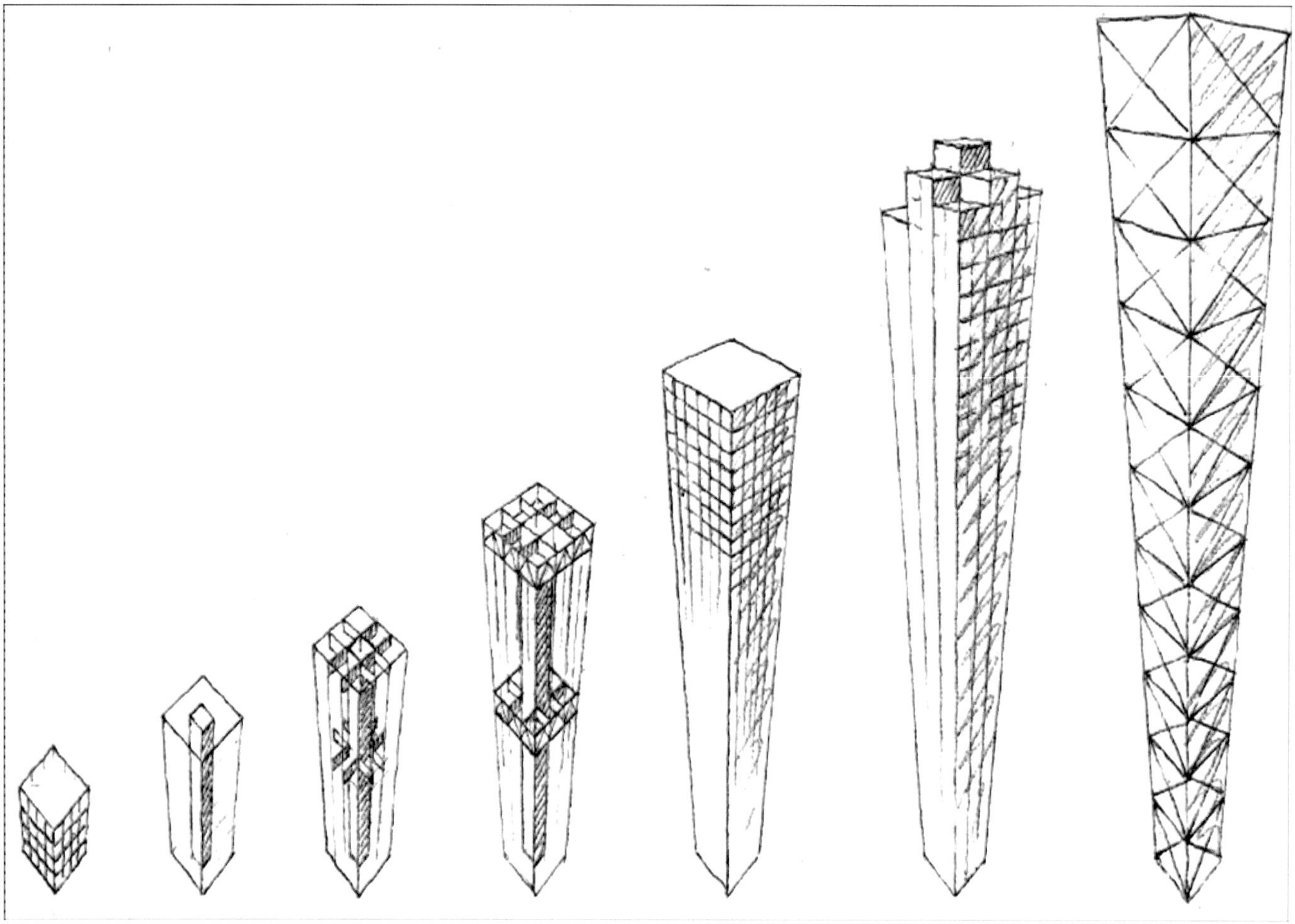

4.10 Theoretical structural systems according to height thresholds
Graphic: Ron Slade, WSP

shape of the floor plate will fundamentally impact the location and configuration of the core, the design of which plays a key role in maximising net:gross floor area efficiencies.

Height generally reduces these efficiencies, as core and structural zones expand relative to the overall floor plate to meet the requirements of vertical circulation and to resist wind loads. The size and regularity of the floor plate will largely determine the extent to which these effects are integrated across both net:gross floor area and wall:floor ratios.

Some of the tallest buildings in the Asia-Pacific region demonstrate that height and architectural intent can be achieved alongside commercial targets, through large and regular floor plates (with central concrete cores that provide lateral stability) and extruded forms that provide the added benefit of avoiding additional learning curves during construction.

Superstructure cost is essentially the result of weight and complexity, the former determining the quantity of material and the latter influencing the price per tonne (for steelwork) or per cubic metre (for concrete), and both can vary substantially. For example, the weight of steel in an all-steel tower can range from less than 150 kilograms per square metre to in excess of 250 kilograms per square metre, and the average all-in price per tonne can differ by as much as 100 per cent.

Whilst it is clearly valid to focus on making the frame as materially efficient as possible, it is important to note that the material can constitute only around 25 per cent or less of the total price per tonne of steelwork. The complexity of the structural frame will directly affect the fabrication and erection costs, which typically represent around 70 per cent of the cost. Thus, adequate focus should be placed on erection methodology (and temporary works), including piece count, standardisation of members, etc. And these factors should be addressed relatively early in the process because they can influence the fundamental design of the superstructure and therefore the architectural concept.

Having said that, the continuing price volatility of commodities is affecting such cost make-ups, and with some suggesting that the world may be at a major inflection point in terms of commodities and resources, the comparison of the supply, demand and cost of fundamental structural options may take a different course.

The shape of the proposal fundamentally drives both frame weight and simplicity/complexity, and the variety of tall building forms therefore explains the large range of superstructure costs. The overriding importance of shape means that certain forms can create a structurally efficient solution that contradicts the theoretical change in superstructure systems at different thresholds (Figure 4.10).

Facade specification

There is an intense – and sometimes political – relationship between high-rise aesthetics and performance, which is nowhere more evident than in their facades. The form and envelope of a tower create its identity, and its external walls play a critical role in its passage through the town planning process. But they also have to satisfy a number of performance criteria, all of which exist in a certain tension. There are five key parameters in respect of which the facades of a tower should be designed and assessed.

1. Architectural intent Given the prominence and profile of tall buildings, allied to their potential longevity, it is understandable that an architect's focus will veer towards how it looks: not only the form of the envelope, but its articulation, its materials and its detailing (inside and out).

Architectural intent, in response to the client's brief, represents the catalyst for any project. Exploring new products and technologies is to be encouraged, but scale prompts a risk-averse approach, so creativity has to be tempered by pragmatism. Few wish to be an expensive, high-profile 'guinea pig'.

2. Performance Providing natural light to the floor space is achieved by glass, which means that solutions gravitate towards curtain walling technology as the facade system, assisted by the fact that such system solutions are capable of accommodating the diverse range of other project-specific criteria.

But maximising natural light attracts the inevitable risk of solar overheating and heat loss, or heat gain in hot climates, thus facade solutions must be flexible in order to address project-specific performance nuances, all of which are possible with an aluminium unitised curtain walling 'base chassis' solution.

High-performance coated glass units, often combined with solid spandrel panels, represent a cost-effective solution in parts of Western Europe and the US, whereas in cold climates the introduction of triple-glazed sealed units comes to the fore. Mirrored glass addresses intense solar gain risk in very hot climates, and there are a variety of solutions combining vision glass and solid panels in a range of facing materials that likewise contribute to achieving the necessary balance between maximising light transmittance ('LT' values) and minimising solar gain ('G' values) while taking cognisance of reducing heat loss ('U' values), down drafts and heat gain for hot climates.

These solutions help to keep the facade depth to a commercially advantageous 250–300 millimetres. It is possible to maintain this depth using double-skin facade technology, where the glazing can be clear (single glazed externally and double or possibly triple glazed internally, depending upon climate). These systems address the solar gain risk by way of motorised blinds located within the cavity, which react to the sun path via sensors and close and open automatically, thereby maximising light transmittance and views out, as appropriate.

Innovations in double-skin technology have now seen the advent of closed-cavity-type facades (compared to ventilated cavity solutions that require cleaning access to the cavity via opening frames within every module). Double-skin facades negate any need to introduce external shading (brise soleil) or solidity (for example through spandrel panels) and thus manage the impact that external shading would have upon cost and risk for a tall building (through impeded cleaning access and wind noise).

Double-skin facades are not appropriate for all climates, so ultimately facade solutions will be heavily influenced by local climactic conditions, on a project-by-project basis, springboarding from the initial design intent.

Other influencing factors for choosing the right facade solution include accommodating project-specific criteria such as wind pressures and particular facade solutions.

3. Buildability and maintainability Building at height in the vast majority of cases necessitates prefabricated facade solutions; that is, units that can be lifted into position on a floor-by-floor basis, directly from the floor plate, avoiding the need for any external access and taking the pressure off the tower cranes, thus dovetailing with the critical path aspect of the construction schedule.

An added benefit of this approach is maximisation of quality and management of risk, with off-site fabrication carried out in controlled workshop conditions allied with tried and trusted logistical planning methodology to 'feed' the building with facade 'elements' on an as-required basis.

The importance of ease and cost of maintenance is magnified by the expanse of facade on a tall building. It is therefore critical to assess at the design stage how, and how frequently, the facade will be cleaned and panels replaced, and the service and design date of materials and components.

4. Procurement Ensuring that building envelope design solutions can be built by a wide range of appropriately skilled specialist contractors should be a prerequisite for any project. This reduces risk and maximises the opportunity for completion, irrespective of market conditions. Design teams should identify likely facade contractors for their project at the earliest opportunity and engage with them to seek specialist input and maximise design efficiency without detriment to architectural intent, programme or quality.

The balance between early engagement of capable specialists and maintaining competitive leverage is a difficult but not impossible task, and the means of its assurance will differ by project and by geography.

5. Cost The importance of wall:floor ratio was noted earlier. Figure 4.11 compares a selection of the tallest buildings in the Asia-Pacific region with some of the proposals for London and demonstrates the potential variance in this ratio. Facade cost per square metre of a building's gross floor area is the product of two things: wall:floor ratio (as noted earlier); and the elemental cost of the external walls, which is driven by specification, articulation, repetition and detailing.

Put simply, an expensive facade allied to a shapely tower can create a facade that is almost three times the cost of a simpler facade on a more regular tall building. Hence the point that a team cannot spend too much time on attempting to optimise building form and facade performance and cost in the early stages of design.

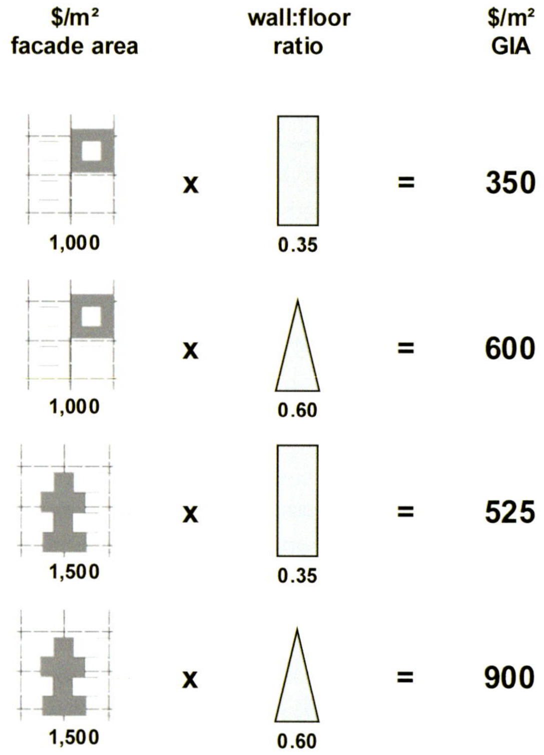

4.11 The dual importance of facade specification and building shape
Graphic: Davis Langdon

Environmental strategy

Though MEP services, or environmental strategy, are not placed in the top three cost drivers (see Figure 4.8), partly due to the relatively narrower range of potential costs, they are an absolutely key element of design that must be thought through at the outset, not least because the design of the services installations and the tower's facades have to be considered and developed together to ensure they provide a coordinated response to the scheme's environmental strategy.

The mechanical engineer will, among other things, contribute data on insulation properties ('U' values) and solar gain performance ('G' values) to help the architect, facade consultant or cost consultant to assess the most effective and efficient ways of achieving blended success across the five facade criteria mentioned earlier. Whether solar gain is addressed through mechanical means (ventilated facades) or physical means (simpler facades with solidity and/or shading components) is a key option to be worked through in the early stages of design.

Another principal focus of the MEP strategy with a potentially considerable impact on cost and area efficiencies is the space required for plant and its location. For example, a decentralised air handling plant will cost more (principally because of the increased number of air-handling units) and could impact the facades, but may avoid additional cost in the basement or release valuable space above ground that otherwise would have been taken up by intermediate plant floors.

Space allocated for tenants' plant is also an important factor. The problems in installing items of plant such as generators or chillers post-completion means that allocating space (which is already at a premium) for future occupiers is difficult if not impossible, particularly if the intention is for the building to be let to multiple tenants. Future upgrades can be catered for by designing in capacity, which increases the base build cost of the

MEP systems but avoids difficult subsequent conversations with tenants and their agents.

Height means having to maintain hydraulic pressures for water and coolant, requiring pressure breaks and multiple plant. The number and rating of valves requires careful thought because of their expense, and needs to be considered alongside the heating, ventilation and air-conditioning strategy for the tenants' spaces, and any flexibility needed in this respect. The key is to arrive at the most cost-effective combination of heat exchange/risers and pressure ratings, with resilience an additional consideration. System pressures also drive options for other systems such as sprinklers and wet risers.

Vertical transportation (VT)

Compared to the elements of the tower considered so far, its elevators represent a relatively small proportion of the total construction cost. However, they are an integral part of the core design and their strategy will help determine the efficiency of the building as well as its operational effectiveness.

There are a number of fundamental options, their relative appropriateness depending principally on building height, floor plate size and configuration, and use or occupational densities. These options are essentially:

- straightforward zoning of floors, with transfer levels
- sky lobbies served by express elevators
- double-deck elevators
- a combination of these.

Sophisticated controls like destination hall call are becoming the norm for taller buildings, at relatively little cost. The VT industry, however, is one characterised by continual product improvement, with the latest developments (such as double-deck elevators) having to be balanced against their untested performance, also restricted commercial leverage whilst they remain single-source options.

Achieving optimum travel times and waiting intervals for the building's projected population will inherently achieve best fit between the number, size and speeds of elevators with both cost and area implications having to be assessed. Sometimes, a solution may be chosen because of its reduced area take, with, for example, double-decker elevators being chosen for buildings that are not particularly tall for this reason. Architecture can also come into play, for example with the proposed use of transfer floors for amenities or with the need to carefully consider the entrance lobby requirements in a double-deck solution.

On the other hand, ease of use and operational issues are equally important, including how the VT strategy influences the way people flows are dealt with at the entrances to the building.

The criticality of an effective and efficient VT solution, and the number of variables involved, means that early expert analysis (and, just as importantly, interpretation) is essential, working alongside the design team and cost consultant to ensure a coordinated approach.

Market conditions

Height brings risk, and therefore cost. More pertinently, articulation, irregularity and, above all, lack of precedent create uncertainty and risk. The more unusual the shape and the more it changes with height, the greater the adverse effect. Combine this with a slender form, and the cost:height relationship becomes a logarithmic one. The project will also be more difficult and complicated to manage and construct, causing a stretching of the schedule.

Put all that into an overheating market – one in which a limited supply of contractors and specialists will often avoid risk or price it 'fully' – and one has the 'perfect storm' of an inefficient, risky (albeit trophy) building with a high price and supply-side capacity constraints. The need to mitigate schedule effects, in particular, has tended, dependent on location and local industry customs, to encourage procurement strategies to become more sophisticated over the years, in an attempt to overlap design and construction, overcome capacity constraints and offload various risks to the supply chain.

All of this has a significant impact on the level, and robustness, of a project's 'on-costs' – cost consultant-speak for main contractor organisational costs (tower cranes, tools and equipment, hoardings, management staff, accommodation, security, etc.), overheads, profit, risk allowances and contingencies. With benchmarks showing a range of 20–40 per cent for these items, risk – how the design drives it, the market's perception and appetite for it, and the procurement strategy's response to it – is evidently of paramount importance in the economics of tower development.

Keys to Success

'Success' means different things to different people. From the commercial perspective of a property development, the success of a tall building will be measured in units of currency, as profit (though not necessarily of the tower in isolation, considering its possible 'contingent' value). The extent of this will be determined by the scheme's performance in respect of the 'golden triangle' outlined earlier (see Figure 4.5), but this is just the starting point.

The keys to success for the most part reflect common sense, but perhaps unsurprisingly, given the fragmented nature of property markets and construction industries, common sense is not always allowed to prevail. It is critical that these keys to success are put in place at the start of a tall building project. Experience shows that they can be collated under the headings that follow.

Produce the right concept

Ensure the team understands and appreciates the project's potential cost (and value) at the outset, enabling the height and form of the tower to be optimised, and the core developed as soon as possible. Clarify the brief at the same time.

This is the most critical stage, particularly for a tall building project, during which the fundamental economic drivers will be addressed, and the foundations laid for a commercially successful tower. Even more than for a lower rise scheme, it will need considerable time, energy and focus to instil the metrics of the golden triangle in the concept design, creating the right height and shape.

Persistently apply the concept principles

Continue to optimise every aspect of the design, checking the 'big ticket' items but also focusing at the micro-scale, so that economies of scale can be maximised through the honing of details.

Check progress and precedent through relevant benchmarking – of design, construction methodology and costs – and peer reviews (including external 'audits' where appropriate).

Create the conditions for success

Gather the right set of consultants, with appropriate experience and skills, yet engender an atmosphere that encourages new ideas. Above all, ensure teamwork at every level so that all the implications of alternatives are considered and every decision is a collaborative one. Where possible, co-locate the team, particularly at the delivery stage.

Put the right project governance in place, with clear roles and responsibilities, fast decision making and true agreement on key milestones such as information release.

Design it properly

Complete and coordinate the design before tendering, or at least be absolutely clear on what needs to be completed and bought at a later stage. Clarify the respective design responsibilities of the design team and the contractor(s).

Engage trade specialists in a focused and coordinated way in the earlier stages of design, at which they are valuable in 'sense-checking' design approach, and considering methodology and buildability issues and ways of making the constructors' lives easier, to avoid unnecessary costs.

Get the procurement strategy right

Align the strategy and the building contract with the local market, playing to the respective strengths of the client, design team and contractors. Be absolutely clear and honest about the extent and nature of risks that rest with the contractors. Above all, sell the project to the market. Having put in the hard work to create an efficient, cost effective and buildable tower, take the time to demonstrate this to potential contractors, to have the best chance of competitive and realistic prices.

The Economic Future of Tall Buildings

Investors, developers, owners and occupiers around the world are facing an unprecedented economic environment, one that is having a profound effect upon the tall building market – and one which may even change certain dynamics for good.

The conditions and concepts of economics are not fixed – and with any property development there are many factors in play – but tall building developments are increasingly tied to global and regional economics and politics, which continue to exhibit sustained uncertainty.

There are a number of widespread trends that appear likely to extend beyond the short term and have a lasting effect upon the drivers for building tall. These include:

- *The 'tale of two worlds'* While developed nations contend with economic tightening, many emerging markets have undergone remarkable growth, driven by investment and domestic demand.
- *Debt reconciliations* The global financial crisis has left an enormous burden of debt, with governments facing huge challenges in reconciling public finances.

- *New growth frontiers* Beyond the BRIC countries (Brazil, Russia, India and China), there are new 'championship economies' with great construction growth potential – countries with large populations, rapid urbanisation and industrialisation, and changing industries.
- *Inflation* There are strong inflationary pressures caused by high domestic demand in emerging economies, monetary policies (near-zero interest rates and quantitative easing) and rising commodity prices.
- *The global cost of capital* While the credit crunch made funding more expensive and harder to obtain, interest rates generally remain low for a number of reasons, including economic weakness in mature economies and central banking policies aimed at stimulating growth. However, there are growing expectations that the balance of global savings and investments will reverse, making capital more expensive.

The combination of these factors is creating a unique global situation. Overriding themes of economic, social and environmental sustainability are not only bringing into sharper focus the fundamental cost and value drivers of tall buildings, but should encourage longer, wider and deeper views of high-rise economics.

In the light of these contextual issues, Davis Langdon – alongside a team that includes Chelsfield Partners, Aedas, Hilson Moran Partnership and WSP Cantor Seinuk – has undertaken a 'Future Office' study (see Mason 2012) with the aim of creating a cost-effective and sustainable commercial tower, an exercise that has prompted such questions as:

- Is investment directed towards the right outcomes, for example could money spent on expensive facades be redirected towards tall buildings that provide a flexible, adaptable, comfortable and effective space – all of which is of real value to the business occupier?
- Should more emphasis be placed on longevity, depreciation, replacement costs or adaptability?
- Could a more cost-effective tower make its city more competitive?
- Is sustainability better achieved through architecture than through expensive technical fixes (some of which anyway possess arguable 'green' credentials when factors such as embodied carbon are properly considered)?

This work reveals the varying results of cost-benefit analyses according to perspective: profit, people or planet. Figure 4.12 (overleaf) shows the Sustainability Evaluation Matrix that summarises in a traffic light system the relative performance of various options against the three perspectives. Critically, one of the results of the study was that a different answer was always achieved depending on the value perspective applied. The importance of wider value perspectives is joined by the need for collaborative assessments and a framework for long-term improvement (which will rely on much more extensive analysis and collection of operational data) to form the key themes that emerge from the study. This provides a compelling agenda for the future, competitive business case for tall buildings.

The challenges and drivers of high-rise development remain valid, but their relative importance will change, not only with time but also according to location. Cities like London, which possess iconic but relatively expensive towers, are testing alternative skyscraper forms – indeed they are re-evaluating the appropriateness of some tall building proposals conceived before the financial crisis. Elsewhere, megacities like Mumbai – which India will be relying on to help propel it to its predicted position as one of the world's top three economic powers – are grappling with more fundamental challenges, including how an array of tall buildings can solve density issues, overcome infrastructure shortcomings, avoid sprawl, provide quality accommodation for businesses and people, and advertise growing prosperity and development to the wider world.

Tall building typology will of course continue to develop, to respond to particular demands, to reflect changing working and living patterns, and as a result of progressive thinking on how different users can work in a building together and how a building and its city can support each other.

For economic and practical reasons, there has been a general shift in emphasis around the globe from vertical to horizontal, with civil engineering projects such as transport systems and water engineering projects taking precedence. But strong economic growth in the BRIC countries and championship economies, along with a rise in population and migration to cities, has put increasing pressure on and exposed the importance of urban policy. This underlines the connection between the tall building and its environment, and the need to consider both.

There will be a step change in technology at some point in the future that will have a dramatic effect on how towers are designed, built and operated – but that is beyond the scope of this chapter. Such dramatic advancements will impact the economics of a tall building, but some things will not change: the grounding principles of the development equation and the

Brand: | **People**

Cladding	Structure	Services	Occupant Comfort	Adaptability	Average Score
Curtain Wall	Long Span Steel	Fan coil - Localised AHU	0.4	1.2	0.9
		Fan coil - Centralised AHU	0.9	1.3	1.1
		Chilled Beam - Localised AHU	0.6	1.3	1.0
		Chilled Beam - Centralised AHU	1.0	1.4	1.2
	Long Span Concrete	Fan coil - Localised AHU	0.7	1.0	0.9
		Fan coil - Centralised AHU	1.1	1.1	1.1
		Chilled Beam - Localised AHU	0.8	1.1	1.0
		Chilled Beam - Centralised AHU	1.3	1.2	1.2
	Short Span Steel	Fan coil - Localised AHU	0.5	1.0	0.8
		Fan coil - Centralised AHU	1.0	1.1	1.0
		Chilled Beam - Localised AHU	0.6	1.1	0.9
		Chilled Beam - Centralised AHU	1.1	1.2	1.1
	Short Span Concrete	Fan coil - Localised AHU	0.8	0.9	0.9
		Fan coil - Centralised AHU	1.2	1.0	1.1
		Chilled Beam - Localised AHU	0.9	1.0	1.0
		Chilled Beam - Centralised AHU	1.4	1.1	1.2
Precast	Long Span Steel	Fan coil - Localised AHU	0.6	1.0	0.8
		Fan coil - Centralised AHU	1.1	1.1	1.1
		Chilled Beam - Localised AHU	0.7	1.1	0.9
		Chilled Beam - Centralised AHU	1.2	1.2	1.2
	Long Span Concrete	Fan coil - Localised AHU	0.9	0.8	0.8
		Fan coil - Centralised AHU	1.3	0.9	1.1
		Chilled Beam - Localised AHU	1.0	0.9	0.9
		Chilled Beam - Centralised AHU	1.5	1.0	1.2
	Short Span Steel	Fan coil - Localised AHU	0.7	0.8	0.8
		Fan coil - Centralised AHU	1.1	0.9	1.0
		Chilled Beam - Localised AHU	0.8	0.9	0.9
		Chilled Beam - Centralised AHU	1.3	1.0	1.1
	Short Span Concrete	Fan coil - Localised AHU	1.0	0.7	0.8
		Fan coil - Centralised AHU	1.4	0.8	1.1
		Chilled Beam - Localised AHU	1.1	0.8	0.9
		Chilled Beam - Centralised AHU	1.6	0.9	1.2

Maximum Deviation

4.12 Sustainability Evaluation Matrix
Graphic: Aedas

Profit

Net Area (m2)	Capital Cost £/m2	Whole life cost £/m2/60 years
95,451	1,572	4,168
93,831	1,615	4,060
95,451	1,572	3,976
93,831	1,615	3,952
94,917	1,604	4,156
93,297	1,647	4,049
94,917	1,604	3,964
93,297	1,647	3,941
95,455	1,615	4,169
93,835	1,658	4,061
95,455	1,615	3,977
93,835	1,658	3,953
94,952	1,485	4,151
93,332	1,518	4,043
94,952	1,485	3,959
93,332	1,518	3,935
94,281	1,572	4,201
92,661	1,615	4,092
94,281	1,572	3,972
92,661	1,615	3,947
93,747	1,593	4,189
92,127	1,636	4,080
93,747	1,593	3,960
92,127	1,636	3,936
94,285	1,615	4,202
92,665	1,647	4,093
94,285	1,615	3,973
92,665	1,647	3,948
93,782	1,464	4,185
92,162	1,507	4,075
93,782	1,464	3,956
92,162	1,507	3,931
3%	12%	6%

Planet

Embodied Whole Life kg CO2e/m2	Operational Carbon kg CO2/m2/year
740	57.8
767	57.8
722	56.7
749	56.7
730	57.8
763	57.8
712	56.7
745	56.7
873	57.8
909	57.8
855	56.7
891	56.7
631	57.8
662	57.8
613	56.7
644	56.7
731	59.0
761	59.0
713	56.6
743	56.6
715	59.0
751	59.0
697	56.6
733	56.6
862	59.0
898	59.0
844	56.6
879	56.6
622	59.0
653	59.0
604	56.6
634	56.6
34%	4%

need to instil the conditions for success in the project team to ensure that the business case is maintained to the end. And height will remain just one factor in those economics.

References

Binder, G. 2008. 'The International Skyscraper: Observations', *CTBUH Journal*, 1.

British Council for Offices. 2002. *Tall Office Buildings in London: Giving Occupiers a Voice*. London: BCO.

Cantillon, R. 1755. *Essai sur la Nature du Commerce en Général [Essay on the Nature of Trade in General]*.

Glaeser, E.L. 2011. *The Triumph of the City: How Our Greatest Invention Makes us Richer, Smarter, Greener, Healthier, and Happier*. New York: Macmillan.

Lawrence, A. 1999. *The Skyscraper Index: Faulty Towers*. Property Report. Dresdner Kleinwort Waserstein Research, January 15, 1999.

Lewin, K. 1952. *Field Theory in Social Science: Selected Theoretical Papers*. London: Tavistock.

Mason, J. 2012. 'The Office Towers of the Future', *New London Quarterly*, 6. Available at: http://aedasresearch.com/files/publications/future_office1.pdf (accessed 1 September 2012).

Miller R.L.D. 2000. *The Strategic Management of Large Engineering Projects: Shaping Institutions, Risks, and Governance*. Cambridge, MA: Massachusetts Institute of Technology.

Sassen, S. 1991. *The Global City: New York, London, Tokyo*. Princeton, NJ: Princeton University Press.

Smith, A. 2007. *The Architecture of Adrian Smith: SOM: Towards a Sustainable Future*. Melbourne, Australia: Images Publishing.

Willis, C. 1996. *Form Follows Finance: Skyscrapers and Skylines in New York and Chicago*. New York: Princeton Architectural Press.

Notes

Please note that all figures in the chapter are correct as of mid-2012. They are presented to demonstrate relative performance rather than absolute answers that represent every situation.

1 Net internal area:gross internal area, or NIA:GIA. Terms and definitions vary across the world, but the fundamental principles of net and gross area remain valid.

Chapter 5
Is Refurbishment a Better Option?

Sara Beardsley and Jeffrey Boyer

Demolition and reconstruction are not always the only route to a better tall building. Many tall buildings, even the oldest ones, have soundly built structures and exterior construction—making refurbishment the preferred alternative to tearing down. Buildings are refurbished for many reasons, from unexpected damage, maintenance issues or desired changes in style to energy issues and the need to modernise outdated systems. Whole-building refurbishment can be more complex than the initial process of constructing the building. However, refurbishment is able to take advantage of the original structural shell of the building while modernizing everything from MEP (mechanical, electrical, plumbing) systems to curtain walling and interiors, usually at significantly less cost than it would take to rebuild completely. Analysis of a recent landmark tower project showed that refurbishment would cost no more than $300 million, as opposed to an estimated demolition and rebuild cost in excess of $1 billion. Refurbishment can also be more environmentally friendly than new construction in terms of overall embedded energy.

The history of the skyscraper dates back to the late 1800s, but most supertall buildings worldwide range in age from the present to about eighty years old. In contrast to the monuments of the more distant past, existing tall buildings remain in use as part of the urban and economic fabric of cities today. The owners of these buildings have many challenges and opportunities, not only in caring for these buildings and ensuring occupant safety, but also in keeping them viable as financial assets.

Only a small percentage of the overall existing building stock is made up of buildings constructed during the current era of energy awareness. Although intelligent design of new buildings is a positive step, it is even more critical to refurbish *existing* buildings to reduce energy usage and carbon emissions in cities. Tall buildings have proven to be an exemplary form of development in terms of better resource use and reducing environmental impact through greater density in cities. Intelligent refurbishment of existing tall buildings in particular can make a meaningful impact on carbon emissions.

This chapter addresses the opportunities for current building stock through environmental planning, retrofitting, upgrades and green facilities management as a pre-eminent approach to sustainable development, and presents a four-phase systematic approach for implementation. Also discussed are important issues associated with refurbishing existing tall buildings, including historic preservation concerns, changes in building codes, construction logistics and occupant safety.

5.1 Existing high-rise buildings range in age, material type and height, forming an urban fabric that gives character to downtown areas
Image: Sara Beardsley, AS+GG

Energy Efficiency: A Phased Approach

Older buildings are a very important part of cities around the world. Many are historically significant, and form the varied fabric that gives cities their unique identities. However, older buildings were also constructed at a time when energy codes either did not exist or were less stringent than they are today. Many existing buildings therefore consume several times as much power per square meter as comparable newer buildings. If their performance is not improved, older buildings will continue to tax the power grid disproportionately, which will make it very difficult for cities to meet carbon reduction benchmarks.

Construction began on the 104-storey Willis (formerly Sears) Tower in August 1970, when energy prices were at their lowest level, in real value terms, practically since the beginning of the modern petroleum era. By the tower's completion in 1974, the United States was in crisis as a result of the Arab oil embargo, with energy prices at record highs. The paradox of building, especially in relation to tall buildings, is that design is a past event in response to a historical context. How, then, do we create flexible, adaptable buildings? More importantly, how do we make the existing building stock, which has a far greater ecological impact, sustainable?

Abandoning existing buildings is not the answer. Existing buildings represent millions of kilowatt hours of embodied energy in terms of material production and transportation, component manufacture, construction and erection. If an older building is simply torn down, the loss of embodied energy is massive. Instead, it is better to save elements—such as the building's structure, shell and even, where feasible, its equipment—as part of

an energy-saving retrofit. Quite often the oldest existing buildings, constructed in the era before electric lighting, heating and air-conditioning, have architectural features that remain viable tools for saving energy today.

The payoff can be dramatic. High-performance buildings use an average of 40 percent less water and energy than conventional buildings. These efficiencies translate into immediate and ongoing savings in operating costs. Greening of existing buildings—which produces a cleaner, healthier working environment—can also be a powerful marketing instrument.

It is important to note that no single strategy is a cure-all. Each building is unique, based on its site, architecture, systems, use and the era in which it was built. The challenge is to look at each building independently and identify the measures that make the most sense, based on overall goals and existing conditions. Greening existing buildings, especially large-scale high-rise buildings, can be a daunting task. It is a long-term process, and best done using a phased approach.

Phase 1: Plan for the Future

Establishing an environmental management plan must begin with establishing and maintaining a procedure to identify the environmental impact of the activities within building. By quantifying energy consumption, water,

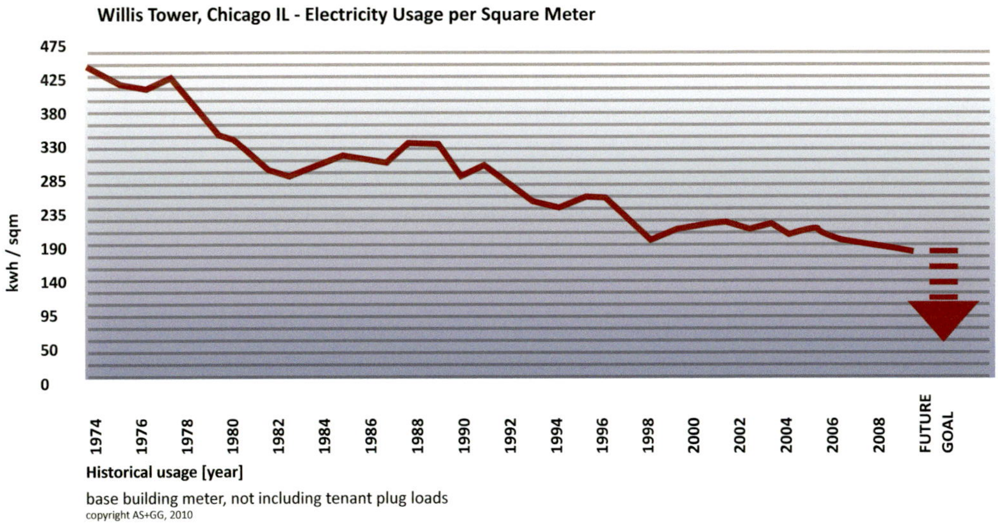

5.2 Willis Tower has reduced its energy use by over 35 percent since construction using low-cost, minimal impact retrofits. A recent plumbing retrofit has saved 10 million gallons of water annually. The owners of the tower have recently adopted a long-term plan to make another 40 percent reduction through a complete modernization project
Image and graphic: Sara Beardsley, AS+GG

waste and transportation early in the process, a strategic framework can be developed and championed by the design and management teams.

This process requires co-operation between building owners, managers, building engineers, design professionals, and even tenants. Data must often be collected from many sources for a building's performance to be understood. In addition to energy performance, issues such as occupant comfort, air quality, maintenance, waste disposal and operational practices must all be taken into account to plan for a comprehensive program to improve performance. Often the most valuable information can come from the team members who have worked on or managed a building for the longest period of time, because they can bring to bear an understanding of the building's past and how it has evolved over the years.

As a strategy team begins to plan for the future of an existing building, it is important to establish goals, priorities and budgets related to sustainable improvements. Minor improvements can often be implemented as part of a current operational budget or scheduled maintenance program. For major upgrades, it may be necessary to pursue additional funding sources over longer periods of time. Overall priorities and realistic phasing should always be considered.

There are many rating systems available for existing buildings, the most common of which are BREEAM (Building Research Establishment Environmental Assessment Method), LEED (Leadership in Energy and Environmental Design) and Energy Star®. These systems are often used when retrofitting in order to benchmark performance. Before undertaking the process to achieve a rating with one of these systems, it is important to understand the prerequisites and how the rating system credits are tailored to existing buildings.

One of the first steps in assessing a building's energy performance is to determine how much electricity, natural gas and water it uses. This can be done through the auditing of energy bills. However, data collection can be a complicated process if there are multiple tenants, who may be on separate meters. If a building is not metered properly, it may be necessary to update or modify metering practices to better track usage in different areas. Once energy use is determined, the energy profile of the building should be studied. Buildings will have different energy profiles depending on their age, layout and occupancy use. Often the best indicator of where energy savings potential exists is identifying where most of the energy is used. Design professionals can assist with forming energy models that can target areas for energy savings.

Strategies in this planning phase include:

- Assign an energy manager and establish a strategy team
- Audit current environmental impact (energy, water, waste, transport)
- Evaluate current use patterns of the building (equipment, size of staff, hours of operation)
- Develop SMART (Specific, Measurable, Attainable, Relevant, Time-bound) sustainability goals and objectives
- Formalize an awareness program with buy-in from key stakeholders
- Assess communication/educational opportunities
- Develop a budget and assess available aid and financial assistance
- Research historic and projected energy/operations cost data
- Examine opportunities to partner with companies and institutions to help realize sustainability objectives
- Determine whether a rating product such as BREEAM is appropriate
- Solicit input from employees through interviews and focus groups.

Phase 2: Initiate an Environmental Plan (Minor Upgrades)

When considering the scale of tall buildings, some of the simplest improvements can yield an impressive impact. Lighting, on average one of the largest consumers of energy in the built environment, is just one of many simple environmental management strategies that can be implemented with a low capital investment in the short term that will yield significant long-term benefits.

The energy, water and waste performance of buildings can also be significantly improved through simple operational and behavioral adjustments. Allowing for an increased temperature range for air-conditioning, shutting off lights and power-strips at night, introducing a green housekeeping policy, implementing recycling programs and encouraging sustainable transportation through bicycle storage facilities can dramatically change the operational characteristics of buildings. It is also important for owners to create building standards, so that when a space is renovated, policies such as day-lighting, use of recycled materials and low-VOC (volatile organic compound) paints can be implemented. In commercial buildings, tenant participation in sustainable practices should be encouraged. Through tenant education programs, building occupants can be helpful when implementing measures such as lighting operation, use of blinds, and recycling. Occupants are able to take ownership in daily greening practices and, at the same time, building managers and owners can benefit from positive marketing.

In some cases, large percentages of energy have been saved simply by optimization of existing equipment. Commissioning, which is the regular monitoring and optimization of building systems to ensure maximum performance, can have benefits that far outweigh the costs. Practices such as cleaning equipment, balancing, changing filters on a regular schedule and modifying the sequences of startup and shutdown are low-impact measures that can save energy.

Lighting fixture upgrades and reduction in density can cut the lighting energy used per square meter by three to four times in buildings originally designed to mid-twentieth-century standards. If density reductions are too costly, programs that decommission some of the lights can save energy. Changes to fixture types (replacing T8 fluorescent bulbs with T5s, for example) can also have a large impact. LED technology is also becoming more cost effective for use in buildings, depending on the application.

Mechanical retrofits to variable frequency equipment can allow systems that once operated all the time to only use energy when required. Often, the simplest place to start with minor mechanical upgrades is the mechanical plant rooms, where building owners have unfettered access and do not risk disrupting tenants.

Exterior wall maintenance is important for energy performance as well as safety in high-rise buildings. Over time, the breakdown of seals and gaskets can increase air and water infiltration, which can negatively affect thermal comfort as well as the overall longevity of the system. A proactive maintenance program can not only save energy but also increase the lifespan of the wall system.

One of the largest loads, especially in hospitals and commercial buildings, is internal equipment. This includes computers, copy machines, appliances—anything and everything that is plugged into an electrical outlet. Although equipment is not often part of the architecture of a building, building owners should create policies and incentives for building occupants to use more efficient equipment. It is also important for owners to make sure that equipment is shut down when not being used, and is sized appropriately for users' needs. Computer systems—especially data centers and equipment for trading floors—consume a tremendous amount of power and give off heat. To the extent possible, this equipment should be concentrated so that the excess heat it gives off is confined to specific, small spaces.

In the future, 'thin client' solutions may allow large personal computers and data storage space to be separate from the office environment, reducing the cooling loads in office space. A thin client, also called a 'lean' or 'slim client', relies on a central server to provide the majority of the computing power to fulfill the traditional role of a desktop machine. There are a number of variations on thin client; some include distributed storage while others are similar to a mainframe and terminal system without storage. These systems save energy when data centers are located in more hospitable climates where cooling energy needs are lower and when they are co-located with renewable energy to offset their direct demands. A thin-client system also dramatically lowers the internal heat gain of the office building and thus reduces the cooling necessary to maintain a comfortable working environment.

5.3 Jin Mao Tower, completed in 1998, was able to make a 20 percent energy saving by recommissioning its mechanical systems
Image: Travis Howe, AS+GG

Strategies in Phase 2 include:

- Minor envelope improvements (seals, gaskets)
- Window energy control film
- Re-evaluate lighting operation schedule and efficient distribution
- Upgrade to high efficiency/efficacy lamps
- Reduce lighting density
- Re-evaluate temperature set-points and HVAC control strategy
- Upgrade to Energy Star equipment
- Upgrade to variable speed drives/ECM smart motors
- Night purge operation/air side economizer
- Vestibules/partitions stack effect management
- Rebalancing and damper upgrades
- Bike storage facilities
- Low-emitting adhesives and sealants
- Low-emitting paints and coatings
- Pollutant source control
- Enhanced refrigerant management
- District chilled water service
- Improved waste management (waste-to-profit network)
- Integrated pest management
- Green housekeeping policy implementation
- Renewable energy credits
- Carbon offset markets
- Tenant educational programs
- Public communication/education program
- Life-cycle cost assessment
- Carpooling network.

Phase 3: Expand the Plan (Major Refurbishment Projects)

While great strides can be made through the implementation of a basic environmental plan, many buildings require more extensive, impactful renovation. Such modernization projects can have significantly increased rewards, in terms of energy savings and revenue in payback. They are also more complicated and require higher levels of investment, planning and phasing, and evaluation of multiple approaches by design professionals. Major refurbishment, however, can make it possible for older buildings to perform at a level equivalent to, or even better than, a comparable current building. Renovations can allow existing buildings to keep up with the current market, in which tenants and occupants demand comfortable, energy-efficient spaces.

When a significant renovation is made of any system, there is great potential for energy savings. It is important, then, to design the renovation in the most sustainable way possible, to position the building positively for its long-term future.

Building envelope upgrades are the most architecturally challenging renovations. These can consist of repairs to the facade joints and seals, re-glazing, increasing insulation value, or re-cladding a building completely. Envelope upgrades are often the last modifications to take place because they can be the most costly. However, the building envelope is the most important element of all in terms of energy, because the overall heating and cooling loads of building are primarily determined by the performance of the envelope. It can be argued that building envelope upgrades should come *first*, rather than last, because mechanical renovations can then take into account building envelope improvements. There also can be significant positive impacts to occupant comfort, such as increased daylight, glare reduction and increased thermal comfort, which are difficult to quantify but beneficial to the long-term success of the building.

The technology of building envelopes has evolved significantly over the years. Many mid- to late-twentieth-century high-rise buildings lack a thermal break (a barrier inside the connection detail which prevents heat from transferring from the outside to the inside of a building or vice versa). Solid materials such as stone or

5.4 Mid-century modern facades such as this one have large, single-pane, dark-tinted glass
Image: Kevin Nance

5.5 Construction photograph of the 36-storey 1 Indiana Square
Image: Thornton Tomassetti Engineers

5.6 Madou Plaza, constructed in 1965
Image: Marc Detiffc

metal may have low insulation values. Glazing installed before the energy crisis of the 1970s was often single-pane instead of insulated, even in cold or temperate climates. Glass coating technology, such as Low-E, has improved significantly over the last ten to twenty years, and is now able to let in a large amount of natural daylight while also protecting the interior from excessive solar gain. Many buildings with coating technology from the 1980s and 1990s lack natural daylight due to the dark tints and excessive amounts of reflectivity used to meet energy standards before higher-performance coatings were available. These types of windows increase the need for artificially lit interiors.

In a major facade renovation, important factors to consider that may affect cost and feasibility include structural anchorage and capacity, wind loading, insulation value, thermal breaks, solar heat gain and orientation, blast and noise requirements, phasing, maintenance, and historical and aesthetic impacts. It is often necessary to re-clad, re-glaze or repair a building's facade whilst the building is occupied, which requires architects and contractors to design a system that can be removed and installed quickly, with minimal impact on occupants. Facade projects may need to be phased over a number of years due to cost concerns. In these cases, it can be critical to match existing glass or finish colours so that the differences can go unnoticed.

5.7 Madou Plaza after refurbishment
Image: ASSAR Architects

5.8 Existing tall buildings are often heated and cooled by large perimeter heating systems located near the glass, to offset temperature gains and losses through the facade
Image: Sara Beardsley, AS+GG

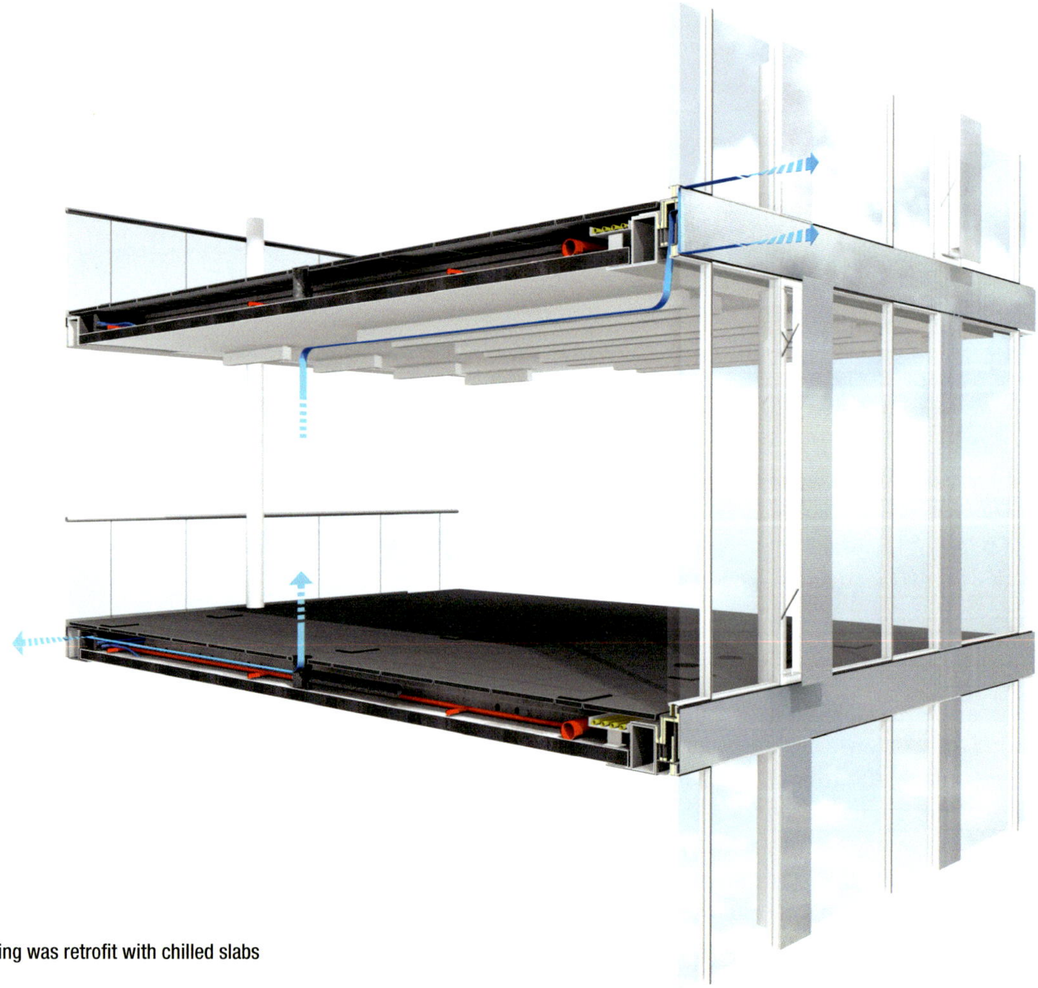

5.9 In this example the building was retrofit with chilled slabs
Image: CePeZed

Renovations to mechanical, electrical, elevator and plumbing equipment can result in significant savings in water, electricity and gas usage. For example, older toilets can consume up to five times more water than today's equivalents. Modern elevator systems use up to 40 percent less power than early to mid-twentieth-century systems. Older mechanical systems often run at a constant volume rate, consuming a larger amount of energy than do today's variable volume systems, which change their power output according to demand.

Modernized equipment can often be smaller in size than older systems, allowing usable space to be recaptured in buildings. Upgraded mechanical equipment can save significant amounts of floor space and shaft space, especially if a system can be downsized due to improved envelope performance. Older elevator equipment and some types of mechanical and electrical systems have components that can be completely removed during modernization and replaced with smaller digital equivalents.

During equipment modernizations, it is always important to consider which systems are truly outdated and which can be successfully retrofitted in order to extend their useful life and increase efficiency without incurring excessive costs. There are times when wholesale replacement is the best alternative. This can happen where a completely different system type is deemed suitable for the building, when retrofit is infeasible due to equipment age or when the old equipment is in such poor condition that it cannot run efficiently or reliably for the long-term future.

Lighting systems in high-rise buildings are typically updated as interior spaces are renovated. Interior renovations are required to meet current energy standards, and so there may be a wide variety of lighting designs within a single building. During a significant renovation, it is important to consider the overall lighting loads in a building. Building-wide standards such as new fixtures, daylighting, occupancy and dimming controls can save a significant amount of energy within an entire building. If a building owner can begin to control levels of lighting and offer incentives for the renovation of individual floors, a building as a whole can be brought up to current standards for lighting energy and save large percentages of the overall electrical load. Metering of electricity on individual floors is also important, so that occupants can see the environmental impact and cost of over-lit spaces.

In any significant modernization, it is important not only to consider the building itself but also to keep the surrounding site in mind. Sustainable site design can reduce the urban heat island effect and stormwater run-off through implementation of landscaped and pervious areas, especially in parking lots or large stone plazas. The site may also have available area for new systems such as renewable energy or rainwater recapture.

Green roofs can be challenging in existing buildings due to structural capacity. In most cases, the superstructure can support the additional load, but the secondary structure may need reinforcement before a green roof can be installed. There are many types of green roof systems for use in various applications. 'Extensive' green roof systems, as opposed to the heavier 'intensive' systems, are relatively lightweight and better suited to existing buildings. Tray systems are easy to install, but can be a challenge in high-wind areas. Plant types may be limited in the lighter weight systems due to soil depth, but these options do allow existing buildings to realize beneficial stormwater reduction. In retrofits where green roofs are cost prohibitive, light-colored roofing materials can be used to lower the surface temperature and reduce the urban heat island effect.

Strategies in Phase 3 include:

- High-performance glazing replacement
- Integral solar shading
- High-performance chiller/boiler replacement
- Occupancy sensors
- Carbon dioxide monitoring
- Indoor VOC sensors
- Regenerative braking elevators
- Daylight-responsive controls
- HVAC fit-out replacement or upgrade
- Heat recovery
- Desiccant dehumidification
- Piping layout improvement
- Site improvements for stormwater management
- Rainwater collection system
- Low-flow fixtures
- No-flow urinals
- Smart-grid enabled appliances.

Phase 4: Toward Neutral Operations

In the past decade the subject of carbon-neutral operations (building operation using power from a carbon-neutral source) has come to the forefront of discussions of retrofits. Utility grids in most countries are primarily supplied by power generation that contributes to carbon emissions and pollution, such as coal or natural gas. As efficiently retrofitted buildings move toward 'carbon neutrality', the stress on the utility grid can be reduced. The potential to remove buildings from the grid completely, and even have buildings contribute stored energy back to the grid during peak periods, is an emerging idea that could help power utilities use cleaner energy for more of the time.

Neutral operations are new and challenging even on clean sites that lend themselves well to renewable or storage systems. In existing buildings, the challenges can be even more significant than in new construction. Existing buildings are limited by an existing footprint and land area, and structural as well as space constraints. The energy intensity of existing buildings is typically greater than new buildings, and so there is more load to overcome in using renewable energy systems to remove an existing building from the power grid. For this reason, energy reduction measures are critical before neutral operation programs are attempted.

Existing buildings do have advantages in that their load profiles are known and can be documented. The performance of existing systems is known, so opportunities for energy storage or off-peak use can be assessed in a very meaningful way. As buildings are renovated, design professionals can find creative ways to reallocate space for energy systems and share energy between new and existing buildings to minimize the overall impact to the power grid.

Technologies to be used toward neutral operations in existing buildings are similar to those employed in new construction: solar, wind, geothermal, energy storage and cogeneration (see Chapter 13). Solar energy in general is best used in buildings with a large proportion of unobstructed, south-facing roof area, and is therefore challenging for high-rise buildings. Facade retrofits are the most viable alternative when roof area is not available. Photovoltaics can be integrated into facades either as thin film on the glass or as solid spandrel panels or sun shades. In a retrofit, wind turbines can be either roof- or facade-mounted, but structural issues such as vibration, load capacity and movement need to be considered. Existing buildings can be studied using CFD (computational fluid dynamics) modeling or anemometers to identify the best locations for turbine installations.

Ground- and water-source energy systems, which use heat pumps to extract energy from the ground below or adjacent to the building or bodies of open water nearby, can be difficult to retrofit to existing buildings because they often require large networks of underground ducts or pipes. The same applies to aquifer cooling systems and the like. Existing buildings may not have access to below-grade areas, or it may be excessively costly to excavate. However, an existing building on a large site or near a body of water may be able to capitalize upon ground- or water-source energy.

Cogeneration plants are not carbon neutral but can reduce the carbon footprint of an existing building. Battery or ice storage can be used to store excess energy generated by renewable elements such as photovoltaic panels and cogeneration, or to store energy during off-peak times so that it can be used during peak conditions.

Gray-water systems can be particularly challenging in existing buildings due to the lack of piping or lack of space for additional piping. However, these challenges can be creatively overcome during modernizations. The recycling of rainwater or condenser water is often the most feasible way of limiting potable water use.

Many owners of existing buildings have been able to reduce their impact on carbon emissions by purchasing renewable energy credits from a power provider. This energy is typically generated at an off-site wind or solar installation. Although there are losses in power transmission from the power source to the building, renewable energy can be generated very efficiently when it is built on a large scale in an ideal location. For some existing buildings, one can make a reasonable argument that supporting off-site renewables is a more efficient and cost effective way of achieving a neutral operation than attempting to harvest energy on-site.

Refurbishment Challenges

Refurbishing buildings is an important specialty in the design and construction profession. Existing buildings have unique challenges based on their age, condition or use, as well as the scope of the renovation. Renovations of existing tall buildings, especially occupied ones, must involve extra planning and care on the part of the contractors, owners and design team.

Calculating Payback Periods

Major renovations are usually not short-term investments, and access to financing is one of the main hindrances to getting them completed. At first glance, the most obvious way to calculate return on investment (ROI) for sustainable renovation projects is simple payback (i.e., investment cost divided by annual energy savings equals payback period, in years), which often results in payback periods of several years or even decades. However, there comes a time in the life of each building where major renovations and upgrades are absolutely necessary rather than optional. Sometimes the cost of not renovating a building can actually be greater in the end, because older buildings become outdated, expensive to maintain and are unable to compete with the newer, more efficient buildings on the market.

5.10 Westraven Tower in Utrecht. Tall buildings can pose special challenges in terms of logistical and regulatory concerns
Image: Jannes Linders

Questions to ask when contemplating a major renovation project include:

- What is the remaining useful life of the system being renovated, and what is the projected lifespan of the new or retrofitted system?
- How much can be saved in annual maintenance labour costs?
- How much is spent yearly on custom replacement parts for an outdated system, and how much longer will these parts be available?
- What is the cost *premium* for a more efficient renovation, versus basic like-for-like replacement?
- Is it possible to retrofit an old system instead of replacing it completely?
- Will the renovation increase rental revenues, occupancy rates or the overall resale value of the building?
- Will the renovation improve factors like thermal comfort, noise and air quality?
- Can usable space be regained by replacement of larger equipment with smaller, more efficient equipment?
- Does renovation of one system allow for savings on the capital or operational cost of another system?
- Are there synergies between systems that can be beneficial?
- Will cost be saved by carrying out the refurbishment of several systems at the same time?
- What is the projected change in the cost of utilities over the payback period?
- Are incentives available, such as tax credits, carbon credits, accelerated depreciation, zoning bonuses, grants for rehabilitating landmark buildings, or low-interest loans?
- What is the overall energy saving goal of the renovation, and is the available financing of the project in line with the goal?

Regulatory and Safety Concerns

Because building codes and safety regulations are ever-changing, existing buildings can have unique problems with local code regulations. One challenge many older building owners face during renovation (depending on how significant the refurbishment project) is the requirement to bring the building into compliance with current regulations. Common examples include compliance with life safety codes requiring sprinklers and fire alarm coverage in buildings, requirements related to accessibility, and mechanical system requirements having to do with outside air. Refurbishment projects offer great opportunities to perform these upgrades, but complying with current codes can also add cost and additional scope to refurbishment projects. In cases where compliance is not technically or structurally feasible, design professionals are required to work with code officials to develop solutions. Upgrades can also be challenging in historical buildings where the desire to preserve a certain feature of a building can come in to conflict with current regulations. Common challenges include monumental stairs, lobbies and older toilet rooms that do not meet accessibility standards.

During adaptive re-use projects, it is not unusual for an existing building to change its occupancy classification. For example, many existing older buildings are located in urban areas that are becoming more residential, and due to their layouts may not lend themselves to modern office space. Therefore a refurbishment of an older office building may include changing its use to residential or hotel. In these cases, special care must be taken to ensure full compliance with codes and regulations.

Energy codes were created in the late twentieth century, which means that many older buildings were not at the time of construction required to comply with energy benchmarks. This is the reason that many older buildings are currently allowed to consume more energy per square meter than newer buildings. However, as building codes and regulations evolve, many authorities are starting to require that existing buildings' energy use be monitored and their systems upgraded to reach energy benchmarks set by more modern regulations. Governmental incentives for energy retrofits are important in assisting buildings to comply with current standards.

Hazardous materials abatement is a significant cost and safety factor in the refurbishment of buildings. The main concern for historical buildings from the early twentieth century is typically the use of lead in paint and certain types of sealants. In mid-century buildings, asbestos-containing materials (ACMs) were widely used in fireproofing and in the tile and mastic in resilient floors. These materials are not necessarily problematic if left in place, but can create a significant hazard during refurbishment because disturbing them can contaminate the air. The harmful effects of asbestos-containing materials came to light during the construction of the north tower of the World Trade Center. All levels below the 40th floor had asbestos-containing fireproofing, while its use was discontinued in floors above. Many existing buildings constructed earlier than 1970 currently have asbestos-containing fireproofing in place.

Testing for hazardous materials is therefore necessary before beginning the refurbishment of older buildings. Specialty consultants are available to do this work. Abatement of these materials can be very costly and time consuming due to the necessity of workers creating containments and wearing protective clothing. Because of the amount of protection required for construction workers and occupants in areas with ACMs, it is common to fully abate all hazardous materials before an area is renovated. The presence of hazardous materials often discourages owners from doing even simple retrofit projects in occupied buildings. Rather than retrofitting small areas, sometimes the best practice is to renovate and abate entire floors of buildings in a phased approach.

Logistical Concerns

When a new high-rise building is constructed, it is typically possible for contractors to use the entire site as a staging area for materials. As the building is constructed, materials are hoisted using dedicated construction elevators and tower cranes. Much of the major mechanical equipment is brought in through large floor openings or areas where the exterior wall has not yet been constructed.

In an existing tall building, the movement of materials is a significant challenge. There are often limitations on the size of materials that can be brought in due to the size of the loading dock, elevators, stairs and door openings. There have been cases where it has been necessary to remove portions of the building facade to bring in large components. In other instances, the solution has been to design the system in smaller elements that can be brought into the building easily.

Another challenge can be access to the facade. In older tall buildings, the window washing and building maintenance system may not have the necessary capacity for the hoisting of materials during an exterior renovation. In addition, the building maintenance system may not have the capacity needed for the number of workers involved in an exterior refurbishment. It is not uncommon for new hoisting systems to be attached to existing buildings during a large facade maintenance project so that it can be completed in a timely manner.

Moving materials within an existing building, especially an occupied one, must take into account service lift size, speed and availability, as well as availability of loading docks. The need for construction workers to have continuous access to materials can conflict with the needs of an operating building. For this reason, many refurbishments are done either at night or on weekends, which can incur additional labor costs. However, these costs can sometimes be balanced by the greater efficiency of the workers. This should be considered during project phasing and budgeting.

Owners of occupied buildings being refurbished must plan carefully to minimize tenant disruption as much as possible. Occupant disruption such as limited access, system shut-downs, security concerns and noise complaints can have negative financial consequences for a large building. However, tenants can be relatively accommodating if proper planning and communication are in place. Minor upgrades such as sprinkler systems, fire alarm systems and even mechanical distribution or window replacement can be done in an occupied space in coordination with tenants.

Some buildings, such as hospitals, schools, courthouses and prisons, have particularly sensitive tenants. In these cases, it may be necessary to vacate the areas being refurbished for the duration of the project. In commercial buildings, tenant "churn" presents an opportunity to upgrade floors. For example, when Sears Roebuck vacated the lower floors of the Willis Tower, the building owner upgraded the lighting and air systems before new tenants moved into these floors.

The Future

As the costs of energy rise, there will be a drive to discover new ways to reduce energy use. Design professionals and government officials are beginning to recognize the great potential existing buildings have for reducing energy use through necessary retrofits. In a city where 70 percent of carbon emissions come from buildings, a saving of 50 percent of electricity in all buildings would cut carbon emissions by 35 percent. This could result in a reduction of the number of power plants required and therefore eliminate the need to construct more, and allow renewable energy sources to be used to provide a greater proportion of the overall power supply as the city expands. Widespread decarbonization may be a long way in the future. The treatment of existing buildings is one of the great challenges we now face—and the owners and operators of the largest buildings have an opportunity to be leaders in the movement to retrofit responsibly.

Part Two

Human, Social and Urban Issues

Introduction

Antony Wood

While a book of this nature will naturally be seen as pro-tall buildings, we need to acknowledge that by no means everyone around the world is convinced of this. Many cities reject their incorporation (or, at least, only allow them in a "controlled" zone at the periphery of the city) in order to preserve the "character" of the often historic urban center. Even in cities where the tall building has been enthusiastically embraced, there is often strong and vocal dissent related to specific proposals at the neighborhood level. Many of the stated arguments have great merit, while others are limited to the murmurings of NIMBY (Not In My Back Yard)-ism. Part Two's opening chapter, Chapter 6, "Tall Buildings in the Public Domain," tackles this issue head on, and examines the often fractious relationship between tall buildings, the public and the domain they inhabit.

This issue is taken further in Chapter 7, "Masterplanning Urban Habitats," and Chapter 9, "Islands in the Sky, Cities on the Ground: The Public Realm and Its Impact on Tall Buildings," in which specific case studies are used to present good and bad examples of the high-rise urban domain, specifically in relation to the impact on the ground plane. Between these two chapters, Chapter 8, "Building Tall: The Future of Cities," encapsulates much of what this part of the book is about: where skyscrapers are heading in the future, and how they relate to cities and the people who inhabit them.

Much of the discontent with tall buildings around the world can be brought down to two factors: their often negative impact on the urban domain (visually as well as physically) and the shortcomings of their internal environments. It is this latter subject which occupies the final two chapters of Part Two, Chapter 10, "Internal Environment and Planning," and Chapter 11, "Atria: A Vital Ingredient of Sustainable Tall Buildings." These chapters demonstrate that we need to move away from the quest to maximize every square inch of saleable or rentable floor space and start introducing support spaces and functions in tall buildings to give a higher quality environment. Such spaces need to provide the opportunity for a sense of "community" to develop as it would more naturally at the ground plane. Thus the urban spaces and functions that occur in the city—the parks, gardens, squares, sidewalks, shops, restaurants, schools, doctor's surgeries, etc.—need to be brought up into the sky.

Beyond their social dimension, these spaces can also be used for the creation of a more sustainable internal environment in a broader sense. Atria can be used to assist natural ventilation and bring natural daylighting deep into the building plan, helping reduce energy consumption and reliance on mechanical systems. Perhaps even more important, however, are the less tangible benefits that we are only now beginning to appreciate. Reducing a building's energy consumption by 30 percent might achieve a saving of less than 1 percent in a tenant's overall operation costs, where staff salaries make up the bulk of the costs. Reducing staff costs by 30 percent would obviously be a much greater incentive to the typical business occupier—but as studies are now showing, this is effectively achievable through the provision of healthier, more sustainable environments, in which happier employees equate to increased productivity, often on a significant scale.

Chapter 6

Tall Buildings in the Public Domain

Richard Keating

The desire of man to extend his influence and to overcome the forces of nature has existed since the first boulder was upended in a Neolithic world. From funereal markers and solar calendars for the annual planting and harvesting cycles to supertall buildings, throughout history these "stelae" have had a social purpose.

To build against the natural force of gravity is the one thing that man can do to symbolize his ability to think and affect nature. As such this symbolism has been translated from one epoch to the next: as reverence to the heavens in the Gothic period, in symbols of individual and then corporate power in the twentieth century, and as pure technological achievement reflecting the status of emerging nations in our modern times.

In today's world, the tall building is a source of pride and object of awe even to those that never make it past the front door. At the same time, in other countries and situations they represent all that is evil about architecture

6.1 Stonehenge
Image: Richard Keating

6.2 Burj Khalifa, Dubai, architect SOM
Image: SOM/Nick Merrick

6.3 Sky garden, Meritz Insurance Building, Seoul, architect Richard Keating Architecture
Image: Richard Keating

and planning. Interestingly enough, the latter attitude is found in what otherwise would be considered one of the most progressive cities, Los Angeles, and the former in regions that may be somewhat socially repressive, for example some countries in the Middle East.

Recent images of the Burj Khalifa exhibit the dramatic spirit of that building, designed by Skidmore, Owings & Merrill (SOM). It is nothing less than a stunning expression of mankind's technical achievement. While the formalism created by revolving setbacks on ever-diminishing floors denies a fully rational use of the interior, and the curtain wall is almost, if not exactly, identical to other recent buildings by SOM, the height is arresting. Emerging from the relentless flatness of the desert, it emphasises the conquering of gravity in a way that would be denied to a similar building set within a larger city. It is interesting to speculate how such a feat of engineering by Bill Baker and Adrian Smith, partners of SOM, would be regarded by Nat Owings, founding partner of the firm, who at the end of his life expressed clearly his belief that tall buildings went against the nature of human existence.

This dichotomy is worth understanding. Why is it that this perception exists both in the minds of NIMBYs in Los Angeles and among major figures of architectural history like Nat Owings? Is it the object itself that insults, or what it represents? Does it matter what the internal function is? What is it that varies from one society to another to produce differences in prevailing opinion? Is it to do with democracy and autocracy?

6.4 Sky garden, 3161 Michelson, Irvine, CA, architect Richard Keating Architecture
Image: Richard Keating

Tall Buildings in America

The force of NIMBY-ism parallels the lack of political leadership in most cities in the U.S. Motivated solely by the need to be re-elected, most politicians cannot think beyond their constituency, nor their own time in office, in order to lead urban development properly. With that in mind, and the ever-increasing force of demography and crowding, it is undeniable that urbanity in the U.S. is headed for a troubled future. Issues of sprawl, density, transit and open space are inextricably interlinked and communal. Unfortunately the one glaring deficiency of a democracy—not allowing thinking beyond the individual needs of individual voters—will affect us for the foreseeable future. We are in a situation where we must find the means to think and act communally to

6.5 Triple towers over transit with landscaped structural links
Image: Richard Keating

resolve the growing problems within our urban fabric. In general, the issue of traffic is the primary negative voiced by the individual and heard by the politician in connection with tall buildings. The issues of the scale of large buildings, including bulk and shadows, come a distant second in terms of levels of concern.

Current efforts to promote transit-oriented development (TOD) are a very positive step and this is likely the approach that must ultimately prevail over concerns about excessive traffic. One can make a strong case that, for the functions of office work, if a tower can be accessed solely by transit and literally be the aboveground manifestation of a transit station, concerns about traffic can be nullified. In fact, in today's green revolution there is a lot that could and should be said about the densification of functions along transit systems which makes tall buildings more pertinent than ever as solutions. Swedish and other European town planning concepts have led the way for the past fifty years, and a great deal can be learned from these very livable cities that can be applied to the American model.

In the U.S., there are two types of cities: those that were mature prior to the evolution and effects of the automobile and those that were not. While this categorization parallels the division of Sun Belt versus Rust Belt (Los Angeles or Houston versus New York or Chicago), it is also an issue on the suburban fringes of pre-car cities. As densification and TOD take hold in the U.S., it is largely a matter of using the genetic code of the earlier, pre-car cities as a stem cell implant for the Sun Belt cities. Los Angeles shares its post-car lineage with Atlanta, Houston, and Dallas. With less than extensive transit systems and the excesses of sprawl, such cities will have a difficult time making the adjustments needed to develop into the new organisms that they must become.

During times of decreasing mobility and increasing congestion, this will become a critical issue for such cities, in particular. Their economic viability and even energy consumption will be affected. The single largest means by which we can save energy is to limit sprawl. Sustainable cities are synonymous with densification, and densification is a resultant pattern of movement systems. Our American cities will not have the open opportunities of Abu Dhabi's Masdar City, but can certainly adapt using lessons learned from such experiments. The differences will be derived from existing population needs, which, in turn, will result in high-rise structures, and in accommodating new urban forms through building in the third dimension.

6.6 Twisting towers, Dubai
Image: tvsdesign

Speaking at the 2010 Council on Tall Buildings and Urban Habitat conference in Mumbai,[1] local architect Charles Correa made a convincing argument that major housing uses of very tall buildings would deny land-use for open space, schools, services, etc. to all but the very few, so the potential of tall buildings for housing use is limited, compared to office use. Housing in tall buildings could be made more appropriate, however, if we can move to a truly legitimate use of skybridges as garden space.

Communal-use spaces that afford an outdoor environment provide obvious benefits to all users. Towers with major outdoor spaces well above grade may be attractive to look at, but they are often very difficult to use due to wind, and there are all sorts of landscaping difficulties in such an environment. In the future, variations on this theme should be explored so that increased utilization becomes possible. At present these spaces have remained relatively private, but with a higher density and a mix of uses in towers, garden spaces could be made more useful and meaningful to the occupants.

Mixed use represents more of a viable opportunity for building vertically, and one wonders why, if the reality is that tall buildings by necessity must be associated with transit, uses such as healthcare, higher education, and religious and cultural functions have not had more play.

Despite the concerns, it seems inevitable that densification of American cities will ultimately prevail, if only to keep up with demographic change. Beyond planning issues, it is perhaps more to the point to be concerned about scale, bulk and wind. Numerous tall buildings around the world are located at significant transportation hubs, but almost always result in scale-less, windy plazas and unwelcoming spaces at their base. Ideally, the transit stop should be arranged so that the user arrives in the sunlight, in a park. Access to the vertical elevators of the surrounding towers can associate with this open space yet still provide weather-protected linkage.

An alternative version of this concept involves surrounding the transit park with multiple towers, structurally linked above with additional sky gardens and parks. Towers linked with landscaped skybridges provide a structural system for lateral loads, at the same time as focusing on the links between as spaces for users, rather than each tower as a single object. It is perhaps worth remembering that Owings called his biography *The Spaces in Between*. Given the need to reserve the ground plane for livability and recreation, one can almost imagine Frank Lloyd Wright's Broadacre City, with its sentinels of tall buildings at transit intervals, as a legitimate model once more.

Ultimately, however, the central issue in creating a tall building is power, meaning the ability to finance and marshal the resources needed to build such an object, which sets up a dividing line between those who can and those who can't. In a democracy this hits nerves when the object is unusually distinct from surroundings that can be seen as part of the community. In other words, a single tower standing alone would be perceived as an object apart from the norm of a city's patterns.

Transco (Williams) Tower in Houston is seen as a monument to the builder, in contrast to the similar-scaled towers downtown, Wells Fargo and Chase, which are seen as part of the business community. Even the old World Trade Center in New York stood apart from the rest of the city enough to be identified as a unique object worthy of attack. Sears (Willis) Tower in Chicago is also set apart from the rest of the city, but somehow its bundled tubes derive from familiar forms nearby, and the building seems to be more about a natural extension of the urban mass of Chicago and less about itself as an object (as far as this is possible for a 110-story tower).

Sears is also one of the very few asymmetrical towers, which by definition is less self-centered. The object most identified as a statement of power and a symbol of wealth and privacy becomes the focus of dislike or even outright attack. This can come simply from grumpy NIMBYs or from all-out terrorists. As America redefines its need for density based on communal issues, and the citizenry begins to understand that the tall building has broader values that are positive for society, NIMBY opposition can be reduced. Unfortunately, terrorists will be drawn to symbols of power and of the "other" in any form, whether the Murrah Federal Building or the tallest building in the financial center of America. For that matter, one can also imagine a terrorist attack on other symbols of America, such as Disneyland or the Golden Gate Bridge, having an equally potent effect.

Presently in Korea, China, Saudi Arabia and India there are buildings on the boards or under construction based on the old models. All of these are impressive for the fact that they can be built, and deserve engineering achievement awards each time a new amount of mass is raised in the air beyond the immediate past record. They are also very attractive and remarkable as objects. However, it is also clear that forms have become somewhat analogous to retail perfume bottles, each reaching for a gestalt based on imagery unto itself. Not part of their cities, only markers. When only a commodity and only an image or seductive form, the tower is by definition remote from all but its occupants, who share the exclusivity of unique, distant views.

Tall building designers should explore a vision of tall buildings that embraces the larger issues of the city and

6.7 Towers imposed on FLW Burnham plan Architecture
Image: Richard Keating

its citizenry in multiples throughout the urban fabric. While maintaining the drama of man's ability to defy gravity, these buildings embrace opportunities for distant views and, at the same time, wholly new energy-use solutions. Acting as a grid of transit stations, solar collectors, local power sources and waste collection and water towers, these buildings would take on significantly different designs and imagery. By purposely incorporating mixed usage of an even wider variety than housing, hotel and office use, and making them publically accessible beyond the usual sky lobby viewing platform, they can become more part of a society's perception of itself. It is this vision of tall buildings that architects and engineers must evolve to create a positive symbol that will allow people to live thoughtfully on the land. In the future, the tall building needs to become part of a system, as opposed to an individual statement.

It is impossible to separate building from planning and urbanity. As we go through the wrenching adaption caused by the post-automobile society, we will also redefine the character of the tall building beyond its current private statement of power so that it is more acceptable to our own communities, and simultaneously the world.

Note

1 See the CTBUH Conference Video at: http://www.ctbuh.org/TallBuildings/VideoLibrary/ConferenceVideos/2010_Mumbai_P1_Correa/tabid/2129/language/en-GB/Default.aspx.

Chapter 7
Masterplanning Urban Habitats

Stephen Engblom

For the ancient Greeks, the metropolis was a balance of "staying" and "going," personified by the complementary relationship between the gods Hermes, the restless and clever son of Zeus and the Titan Maia, and Hestia, whose seat on Mount Olympus was in the center, at the hearth. Like the action of a compass, Hermes' role as traveler and delimiter of space (at doors, crossroads, and the entrances to cities) cannot be understood without a center point: Hestia and her hearth. A "perfect city," in this paradigm, balances "staying" and "going," creating an urban habitat in which individuals can explore and exchange ideas amid society, then return home for restoration at the hearth.

For centuries, cities were created on this Hellenic model: the grid of city streets represented movement, while blocks of buildings provided places for living and working. In the second half of the twentieth century, perhaps as a reaction to the cities of the industrial era, Modernist architects and engineers imagined a different city paradigm, one in which "gardens" were provided for living and working and roads were hidden. For the first time in history, staying and going were separated. At the 1939 World's Fair, the GM Pavilion gave a glimpse into a world of cloverleaf highways and suburban houses situated in leafy gardens.

For seventy years, economics and technology allowed man to realize this suburban stratification at a time of unprecedented urbanization and population growth. Despite the good intentions behind it, the sprawling city first envisioned by Modernists has had unintended environmental, economic and social consequences: an unprecedented rate of loss of virgin land resources; global water and air quality challenges; eroded social fabric and exacerbated energy challenges; and devaluation of much of the built urban fabric. All of these issues demand that the balance between staying and going expected of cities be reimagined.

The subject of this chapter is thus the challenge posed to the architects and engineers of tall buildings and the urban habitats that surround them: re-examining how "moving" and "staying" systems may be put back together. Without reverting to nostalgia, how can transportation systems link up with modern structures to avoid (and resolve) the problems inherent in the Modernist approach? Examples of cites which have successfully balanced transportation and urban fabric, like Hong Kong, Vancouver, and New York will be considered, as well as places like Shanghai and Dubai, which boast amazing new buildings and transport systems, yet have largely missed the opportunity to link the two.

Also surveyed will be a third type—exemplified by São Paulo, Brazil; Los Angeles and San Diego, California; and Houston, Texas—which are now actively pursuing rebalancing.

The Development of Cities

Cities continue to attract new residents, new investment, and cultural and technological innovation. Tall buildings and dense urban habitats have the potential to advance social, economic, and environmental sustainability when concurrently developed with infrastructure and public realm improvements.

By most significant metrics, density results in environmental savings. In one year, residents of a typical American suburb consume 8.9 tons of oil per person and 2.5 kilowatts of electricity per household, in comparison to dense Manhattan's 3.0 and 1.5, respectively (Choa 2010). Similarly, contrast cities with and without effective transit infrastructure and the same pattern is apparent: London's 3.0 tons of oil per person, per year versus Abu Dhabi's 6.8. If one takes into account the many other benefits of city living—proximity to work, open space, shopping, dining, and entertainments—and the lasting appeal of these amenities for employees in the tech and creative sectors, density appears to be inevitable, with the only question being how to accomplish it in the smartest and most integrated manner. Here, tall buildings can play a substantial role, especially as new mechanisms for "joined-up" urban planning or public/private development emerge and are capitalized upon.

How tall buildings can help to restore both environmental and social balance in the twenty-first century has been the topic of broad discussion in recent years, in part because of the impact growing taller has had on many of our cities, and in part because, to accommodate a growing urban population globally, increasingly dense development seems inevitable. Urban planners face a range of challenges, falling between two extremes: the disconnected, poorly-served tall-building development that now demands infrastructure to support it, and the vast opportunity presented by new developments where public/private partnerships or other arrangements encourage full integration.

This chapter examines a range of cities and their efforts to find a balance between moving and staying. Although each city's situation is unique in terms of its original plan, its culture, the extent to which its government and development community coordinate decisions, and various other factors, it is instructive to look at examples from both extremes to inform planning choices made along the continuum between them.

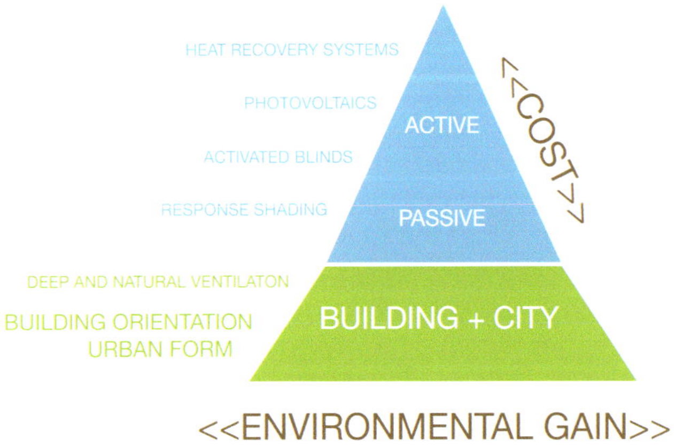

7.1 Characteristics of urban density have demonstrated environmental sustainability benefits
Image: © 2011 AECOM

U.S. Cities: Manhattan, Los Angeles, and Houston

The purest form of this balance is best exemplified by Manhattan, where a grid of streets and underground transportation connect blocks of buildings where universally accessible homes, offices, institutions and parks provide an urban realm in which people live, work, and play. Widely recognized as one of the most convenient and sustainable places on earth (not to mention as one of those with the highest real estate values), Manhattan is not a success because of the soaring beauty of its architecture, but because in New York's early urbanization, the development strategy laid out a universal grid that organized the island in a way that would allow for movement as well as development. In contrast to the discredited Modernist ideal of "garden towers" detached from the street as they reach for the sky New York's buildings reach into the city below and beyond them—they are underpinned by one of the oldest and most-used transportation systems in the world. Because of its flexible underlying grid and the fundamental integration of transit and development that this permits, New York has continued to grow and thrive. In addition, its success is characterized by the mix of uses to which its tall buildings are put.

In the twentieth century, Modernist city planners wanted to separate where people stayed and how they moved into separate systems. Pulling Hermes and Hestia apart, post-World War II cities began to be organized around a new infrastructure model—the car and highway. The city center (where people worked) and the garden suburbs (where people lived) was the dominant model. Soon, however, real estate developers looked to challenge that model and began constructing tall buildings at the far reaches of the cities. The result of this decentralization was a web of congested automobile highways connecting new multi-nucleus developments. With growing populations and burgeoning economies, congestion ensued, and "solutions" included widening highways and building more parking lots to enable residents from far-flung suburbs to access urban amenities. These approaches neglected to address the underlying problem: single-use density, estranged from people's places of employment and "play," is inherently unsustainable.

After decades of trying to widen highways to keep up with decentralized development, cities such as Los Angeles, California and Houston, Texas are currently retrofitting light rail and bus systems to connect and serve speculative tower developments sited at their far edges. These places face a double economic challenge: historically, they constructed tall buildings as part of speculative models aimed at decentralizing the city, and now they are having to build expensive transit systems to support these multi-nucleus development nodes. There is a compelling correlation between transit and economics extant in these cities: in Los Angeles, neighborhoods supported by the first transportation nodes are experiencing the greatest booms in development, while many of the most far-flung residential and office parks have been hit hard by the recent economic crisis. In Houston, the neighborhoods where re-investment in transit has occurred over the past ten years have also been the areas with the greatest investment in high-value working and living environments. The downtown core is now connected to the Medical Center and Astrodome.

Hong Kong

It is instructive to contrast these examples of cities which are playing "catch up" to a place that has historically strong parallels between transit and real estate development. Hong Kong—often cited as an exemplary transit-oriented city—was, in August 2010, home to thirty-one of the world's one hundred tallest residential buildings.[1] It is also among the world's most "vertical" cities, with a vast population living or working above the fourteenth floor (Skyscraper Museum 2008–09).[2] The island's biggest real estate developer is the rail company MTR Corporation Limited (MTRC). Two of the world's tallest building complexes—the International Commerce Centre (ICC) atop the Kowloon Tong Center and the International Finance Center (IFC) atop the Hong Kong station—are both directly above the MTRC's Airport Express train line. After arriving at the Hong Kong airport, the express train delivers travellers to these tall buildings in fifteen minutes and twenty-three minutes, respectively.

In Hong Kong, the colocation of these tall building clusters and stations is no accident. The rail companies are given the right to develop multi-use towers on top of the station locations in exchange for developing the rail lines. The towers, most of which combine office, commercial, and residential uses, are constructed afterwards, along with shopping malls. The highly effective and efficient method of combining transit with multi-use high-density results in virtually immediate tenancy and patronage, alongside increased mobility and ridership. This strategy has made Hong Kong one of the densest places on earth as well as one of the most commuter efficient.

Mainland China

Mainland China has not adopted Hong Kong's transit-oriented development strategy as a precedent for its urban development in the past twenty years. In fact, China still disallows the development of any private real estate atop public infrastructure. Until recently, the country has looked to U.S. postwar urbanism as a model. To a first-time visitor, Beijing is reminiscent of Houston, with a vast network of ring roads and far-flung pockets of speculative development demanding long commute times of its residents. Despite enormous investment in transit during the past decade, joining together nodes of development in Beijing continues to be a challenge. In Shanghai, the superfast magnetic levitation ("maglev") train, with a commercial operational speed of 431 kilometers per hour, purportedly cost 9.93 billion yuan (approximately $1.5 billion, U.S.). Speeding visitors to the middle of a low-density neighborhood on the edge of Pudong, the Shanghai Maglev then requires riders to transfer to a cab that will fight its way through traffic towards the tall building clusters in Pudong or Puxi.

Dubai

Hong Kong also stands in stark contrast to Dubai, where the government has made a high priority of constructing new roads and extending public transport, in part

7.2 Hong Kong's ICC (left) and IFC (above) are on opposite sides of the City's harbor, and are the real estate value developed by the MTR Corporation (a rail company) based on air rights granted by the government in exchange for building the territory's world-renowned Airport Express line
Images: © 2011 AECOM; photography by Vorrarit Anantsorrarak

because of severe traffic congestion and an expected increase in population from 1.1 million to roughly 3 million by 2017 (rudi.net, undated). An elevated rail system connects various neighborhoods to Dubai International Airport, but the islands of development where the tallest buildings have been constructed can only be accessed by car, and only then by a circuitous route (Kamin 2010).

With many ambitious mass transit projects planned, Dubai is clearly aware of the need to retrofit its density with more appropriate means of "people-moving," and is moving these projects forward at an aggressive pace. But the vast expense and necessarily reactive nature of such an undertaking provides a clear example of the benefits of planning high density, whether new or infill, either adjacent to existing transit or in tandem with new transit developments.

The economic savings that can be accomplished by coupling dense real estate development with transit development, while significant, are perhaps dwarfed by the environmental benefits. Not only are the carbon savings of reduced car use alone a significant motivator, higher density buildings also generally consume less energy per square foot due to their condensed surface area, larger boilers, and centralized gas systems. The opportunity to concentrate daytime (office) and nighttime (residential) energy consumption in a single location (a multi-use tall building) is another benefit, allowing for district-level energy planning that capitalizes on synergies between developments. For example,

7.3 (top) The evident horizontal separation of transportation and buildings in Dubai. (bottom) Although this image shows adjacency of tall buildings and transportation infrastructure, failure to plan their better integration will challenge their future success
Images: © 2011 AECOM; photography by David Lloyd

heat gained and rejected by certain buildings during daylight hours can be fed back to buildings that require it in the evenings. Another innovation that is gaining ground is the use of waste material for biofuel and biogas, which could result in tremendous savings through eliminating the need to pump the waste for miles outside of the community that has generated it, and instead turning it into a locally used energy resource.

San Francisco

In San Francisco, California, the Transbay Development Plan, which includes thirteen towers and three "super-tall" buildings, is an excellent example of "joined-up" thinking with regards to both energy and transportation. Located in a new district south of Market Street that is meant to complement the city's existing financial district north of Market (home to the TransAmerica pyramid and

7.4 These two images show the impact on the San Francisco skyline of the new towers associated with the Transbay Terminal development, which will deliver real estate value and population density, and also enhance regional sustainability via transportation and energy infrastructure improvements

Images: © 2011 Skidmore, Owings & Merrill; courtesy San Francisco Planning Department

7.5 San Francisco's new, transit-rich downtown core is integrated with human-scale streetscapes
Image: © 2011 AECOM; rendering by Robin Chiang

the Bank of America Tower, to date two of the tallest buildings in San Francisco), the new-generation towers are largely driven by two downtown San Francisco needs: to improve transportation and to reverse the migration of jobs to suburban Bay Area office parks.

Through a public/private partnership initiated by the Transbay Joint Powers Authority and the City of San Francisco, the development of the new towers (including one that is 50 percent taller than the existing height limit) are the core real estate value that underpins the rebuilding of the Transbay bus terminal as a multi-modal transit hub, which will connect San Francisco to regional high speed rail (CalTrain). The terminal is already located on the Bay Area Rapid Transit (BART) line and at the terminus of the Market Street spine for the city's MUNI buses, and the city's Director of Citywide Planning, David Alumbaugh, anticipates an increase in ridership to work from an already impressive 75 percent to 85 percent.

Replacement of the Transbay Terminal transportation complex is necessary as the current structures are seismically challenged. The new hub will also pay for a new tunnel that will bring CalTrain services to the new downtown district from their current, ineffective location on the far edge of downtown. This replacement will not only serve new tenants and residents of the Transbay towers, but also improve connectivity for the wider San Francisco Bay Area. In addition, the high-density infill commercial development is paving the way for a whole new generation of sustainable energy planning for the city's South of Market neighborhood, because of its complementary use and adjacency to a generation of recently completed and in-the-works residential towers. Because of the density of the residential district and the new commercial district, new energy efficiencies can be realized. During the day, the office towers utilize the energy, and at night the residential population can take advantage of the network.

Vancouver

Of course, many older tall buildings, while serving as anchor points in the urban fabric or points of architectural or historic importance, must be effectively "planned around"; this approach may proceed best by looking at the underlying system of transportation and service provision. Again, looking to a success story can help to reshape traditional notions of constraints. As with Manhattan, Vancouver, British Columbia's original underlying grid has allowed for density, and its planning history is replete with examples of development that leverage the value of the transportation infrastructure. Using greenhouse gas (GHG) emissions to measure climate impact, Vancouver is already a North American leader, with an average of only 1.5 GHG tons per person per year, just behind global leaders like Oslo, Norway (Busby Perkins + Will 2010).

The most recent example of Vancouver's commitment to transit planning is the city's investment in the Canada Line, a new rail line from the airport in the wake of the city being awarded the 2010 Winter Olympics. Building on ideas forwarded by former Mayor Sam Sullivan, who trademarked the term "EcoDensity" (the concept that transit supports density, density supports higher-level amenities, and that ultimately those amenities create more livable and sustainable communities), Vancouver's city council approved plans to develop mixed-use density within a radius of 500 meters of the Canada Line's major eight-station nodes. Each of these nodes draws on livability precepts (such as the European "high street") to maximize walkability and quality of life. (In addition, outside of the downtown core, intersections bisected by rapid bus routes and therefore also ripe for development were examined.)

Driverless, automated, and with headways of about a minute per train, the Canada Line was completed just in time for the (February) 2010 Winter Olympics, with each node recommended for higher density development. New up-zoning regulations accompanied investment in the new line, and the maturity of the transit system makes it far easier to develop than that of many other cities.

For the past twenty years or more, Vancouver has been lauded as a great transit-oriented city, and it is now exporting some of its "lessons learned" abroad. The city's former planning director, Larry Beasley, has been hired in Abu Dhabi as part of its push to position itself as the "more sustainable emirate." It is not unreasonable to expect that the Abu Dhabi Plan 2030 will incorporate the latest thinking about "smart density," along the lines of the Vancouver model.

Mexico City

In cities with long-extant planning challenges, tall buildings may offer solutions—especially in the form of public/private partnerships for which development can spur needed infrastructural changes. Where Vancouver wisely or fortuitously avoided being split by a highway, Mexico City has struggled for decades with development along its central axis of Avenida Chapultepec, a boulevard with a route dating back to a pre-Columbian Aztec alignment of the city's aqueduct. Over the centuries, the corridor has continued to play an important role in the life of the city. In the 1970s, the city's first subway line was installed under Avenida Chapultepec, and topped by a highway without an attendant real estate strategy to induce private real estate development along this corridor of huge public investment. Currently, the highway is not only a partition between communities, but also an area plagued by crime. Ironically, this degraded area of the city is located only blocks from some of its most expensive real estate. Currently, work is underway to redress the lack of incentives to investment along Chapultepec, and to create a better physical environment. Moving the highway underground one level and replacing it with a linear public park will no doubt create a better physical environment, but without tall buildings lining this avenue, the neighborhood lacks the economic catalyst to be transformed into a vibrant urban heart.

In an echo of the Transbay model, the financial strategy for the Avenida Chapultepec gives the private sector the opportunity to bond, earning air rights (essential for tall building development) along the corridor. In this arrangement, the public sector will grant the air rights and the private sector will undertake the sublimation of the highway and the development of an underground bus station. Above ground, the linear park—the People's Street—will provide open space for residents and workers from the new towers and the neighborhoods beyond, while it is anticipated that the mixed uses of the towers themselves will draw in residents and visitors and make new jobs available.

São Paulo

Tall buildings isolated on urban fringes result in the environmental and social challenges discussed earlier: neither transportation networks nor cities' cores function at their best without the density offered by tall buildings and urban amenities. One city that has long embraced tall buildings is São Paulo, Brazil, the world's seventh-largest metropolitan area, with over fourteen million residents. For decades, easy development opportunities at the urban edge and beyond have created a dispersed populace that, regardless of economic level, now faces traffic jams and social segregation, despite the city having a historic core that is well served by parks and transit networks. After fifty years of vacating the center for peripheral development, a shift to re-examine and re-populate the center is underway.

The city has commissioned a study to regenerate its historic core and to cultivate a true "city lifestyle" in the Nova Luz district. By capitalizing on the fact that this neighborhood is well served by two (soon to be three) transit lines and regional rail, is rich in historic city fabric, and surrounded by museums and parks, São Paulo is betting that the development community will realize the extreme value of this land and partner with the city to invest in new high-value environments for working and living that will attract the best, brightest, and most creative. In short, São Paulo wishes to offer an environment that re-balances staying and going in a complete neighborhood, rather than requiring long commutes between home, work, and leisure. This pivotal project, aimed at reversing the cycle of sprawl, potentially represents a turning point in Brazil's continuing urbanization.

With the economic backdrop of the first ten years of the twenty-first century, the growing global awareness of sustainability issues, and steady population growth, the future of development may lie in tall buildings served by multiple uses and effective public transit. While there is clearly no one-size-fits-all approach, the disparity in the economic profiles, energy consumption, and quality-of-life metrics between cities that have paired tower development with infrastructure planning and those that have not is compelling.

Mega-cities are the dominant form of urbanization on the planet today, and tall buildings, properly integrated, can clearly contribute to sustainable densification. Many cities are already trending in this direction. In San Diego, a southern California city where a single-family home might be an hour or more from work, there are no more greenfield opportunities: open spaces are protected as critical ecological habitat corridors. This presents a huge challenge to developers and city planners alike, because the city's many neighborhoods (spread across 963.7 square kilometers, not including its greater metropolitan area or conurbation with Tijuana, Mexico) lack an integrated transit system. Like many of its counterparts in the U.S. and worldwide, the city is now facing the economic and environmental consequences of sprawl, and the only way to ward off a host of related issues and finance necessary infrastructure is to create hubs of density in the urban fabric— which is to say, well-integrated and vibrant pockets of livable density.

A New Hellenistic Ideal

Neither a married couple nor a brother and sister, Hermes and Hestia are both gods of the earthly rather than the celestial, presiding over basic human needs, for warmth and companionship as well as for commerce and adventure. For too long, city planning practices have been testing the bond between these interrelated imperatives by stretching it to the breaking point. In both Hellenistic thought and recent urban history, the lesson is the same: cities function better economically, environmentally, and socially when density ("staying") and transportation ("going") are in balance. After the twentieth century's technological explosion allowed superhighways and supertowers to pull cities apart, the challenge for the next century is for professionals involved in the planning, design, engineering, and development of cities to combine their disciplines in a collective endeavor. Although the avatars of chariot and hearth are now mass transit and tall buildings, we would do well to remember the benefit of maintaining balance and harnessing synergies between the two.

References

Busby Perkins + Will (2010). "Busby Perkins + Will on EcoDensity." Online. Available at: http://www.railvolution.org/rv2010_pdfs/20101910_10am_GrowBetter_Nielsen.pdf (accessed October 13, 2011).

Choa, Christopher (2010). "Dencity" (ecard). Online. Available at: ecards.aecom.com/dencity (accessed September 15, 2010).

Kamin, Blair (2010). "In Dubai, you can't get there from here; architectural feats undercut by shoddy urban planning." *Chigaco Tribune*, January 8, 2010. Online. Available at: http://featuresblogs.chicagotribune.com/theskyline/2010/01/in-dubai-you-cant-get-there-from-here-architectural-feats-undercut-by-shoddy-urban-planning.html (accessed January 8, 2010).

rudi.net (undated). "Trains, planes and automobiles: transport planning in Dubai." Online. Available at: www.rudi.net/node/17350 (accessed September 14, 2010).

Skyscraper Museum (2008–09) *Vertical Cities: Hong Kong/New York*, exhibition.

Notes

1 See the CTBUH's global tall building database, *The Skyscraper Center*, at http://skyscrapercenter.com/create.php?search=yes&page=0&type_building=on&status_COM=on&list_continent=&list_country=&list_city=&list_height=&list_company=&completionsthrough=on&list_year=2010
2 See http://www.skyscraper.org/EXHIBITIONS/VERTICAL_CITIES

Chapter 8

Building Tall: The Future of Cities

Ken Shuttleworth and David Taylor

How will tall buildings shape the cities of the future? This is, of course, a complex question, the answer to which will inevitably be a heady mix of projections, conjecture, opinions, and fact. But one thing is for certain: tall buildings will have a big say in the future of urban environments, as environmental issues come ever more to the fore and the world's population keeps on growing.

Much can be learnt from a reappraisal of the major tall structures of the past. The 53-storey Commerzbank project in Frankfurt, built in 1997 to designs by Foster + Partners, was the world's first ecological office tower and the tallest building in Europe, and it is still remarkably current in terms of its technologies and its ethos. The kind of natural ventilation it allowed is still in its infancy when it comes to towers. This will become much more common in the next ten years.

Why mention an old building in a forward-looking section of this book? Because it is important in the wider context of building tall that Commerzbank set some lasting environmental standards, the project explored the nature of the office environment, developing new ideas for its ecology and working patterns. Most notably, the design included 'sky gardens' spiralling around the building, bringing daylight and fresh air into the central atrium and acting as the visual and social focus for village-like clusters of offices. This was a broad theme that would be picked up in 30 St Mary Axe (commonly referred to as 'The Gherkin') in the City of London.

Tall buildings may represent one of the best ways to help deal with population growth across the world. And yet tall building solutions are not always appropriate; sometimes eight-storey buildings coupled with squares and streets can be better contributors to the urban landscape. High-density living like this contrasts strangely with some areas of Hong Kong, where huge towers of housing create a sense of claustrophobia and light is hard to come by, especially at low levels. This dissonance is partially a cultural thing – people in the UK tend to favour the house over the apartment market. One thing many nations strive to do when it comes to building tall, however, is to concentrate on designing structures that sit over public transport interchanges, thereby enabling them to contribute even more fully to their sustainability credentials by cutting the need to travel.

The same focus on tall building as sustainable structure is also applied, of course, in ensuring that a tall building has a wide mix of uses. Schemes such as the Renzo Piano-designed Shard at London Bridge Tower exemplify this, with a mix of uses including hotel,

8.1 Morello London, East Croydon Station
Image: Millar Hare

residential, and office. China in particular is developing many examples of mixed-use towers, with residential on top of office, on top of hotel. In the UK today, residential and office mixes in towers are relatively uncommon, largely because of the differences in lease lengths.

Mixed use has another important ecological pay-off, of course, since it means heat can be moved around a building – perhaps using heat from the offices to warm up water for the residential component, for instance. Such a principle opens up all sorts of sharing possibilities across uses.

Another important issue in the tall building scenario is context and difference. In a world where homogeneity reigns – where a Starbucks in London is identical to a Starbucks in Madrid, or Dubai, or Kuala Lumpur – there should be a place for design to reflect the unique nature and richness of where it is; a sense of place. Buildings must be culturally aware and reflective of their particular local environments in order for them to make a difference.

By contrast, an identikit, one-size-fits-all approach is one of the architect's worst sins. Canary Wharf in London Docklands and La Défense in Paris come to mind here: soulless places with tall buildings that place utility above visual appeal. So, when it comes to somewhere like China, for example, the better designers should engage in a subtle quest for references.

But context or a new identity cannot be easily created, despite the success and worldwide acclaim that came about for the City of Bilbao through Frank Gehry's Guggenheim Museum, for example. That is a low-rise form, but it served to put the city on the map, and other cities in other countries have tried to elevate their world standing by means of commissioning architects to design and build taller and taller structures as emblems of their international standing or aspirations. The idea of constructing a tall building or tower as a status symbol has been a major factor in cities in some parts of the world for many decades.

But this idea is flawed. Putting a tower up just for its iconic status, to create a marker, is highly questionable. Dublin, for instance, does not need a tower to be iconic or a better place. Perhaps this 'mine's bigger than yours' trend is on the wane across the globe. Buildings will most likely veer away from the 'funny shapes' encouraged over the last decade by advances in computer-aided design and a desire to be different. They will be replaced by regular, straightforward forms, square boxes which must nevertheless add something to the skyline.

These buildings are likely to be beautifully detailed, but much simpler. If the market is allowed to do what it wants to do with little in the way of planning controls, the inevitable result will be the boxes which typify

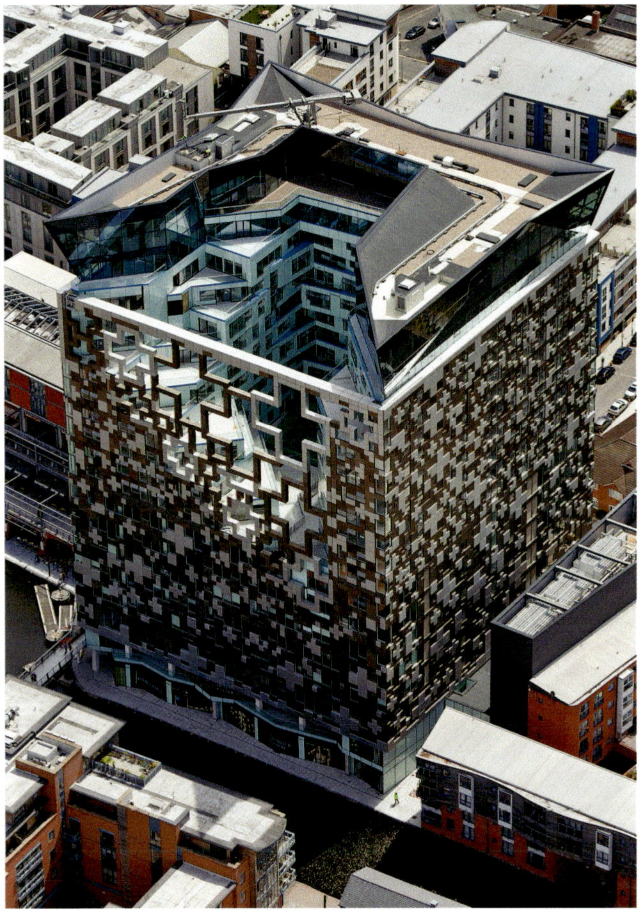

8.2 The Cube, Birmingham, UK
Image: Chris Wade, Crew Photography

development at Canary Wharf in London and, to a degree, New York. But in the new era we are back to doing more straightforward, simpler, more rectilinear buildings.

Perhaps the most notable of all of history's changes to the tall building was the transition from solid masonry buildings in Chicago to glass curtain wall buildings, when air-conditioning came to the fore. As soon as air-conditioning arrived, all buildings were made skin-tight with curtain walls, single glazed. But that is about to change again: the environmental agenda is now stronger than it was even a year ago. The evidence for this is all around – the various governmental missions to achieve zero carbon by 2016 for residential, with further-off targets for other sectors, and the LEED agenda. But it would be wrong to consider the green movement to be a peculiarly Western principle, with China, too, pushing harder for more environmentally responsible buildings.

The main driver in this trend seems to be that people want to work in such green buildings. And clients want

8.3 Aura, Beijing, China
Images: Make

to attract the best talent to work for them. Therefore, making an attractive building for them to spend a large part of their daily lives in represents a powerful tool, as well as an outward display of a social and environmental conscience. The kinds of office buildings companies commission cannot, therefore, just be a 'bunker with slit windows'. Firms want to be seen to be green in order to attract a younger workforce that is perhaps more imbued with environmental thinking than its older, more senior employers. The logical extension of this is a move towards achieving the first zero-carbon office building within the next five years.

How does the tall building fare in the attractiveness-to-prospective-employee stakes? Does it score any points over a campus or a low-rise 'place' as a lure for the best talent? There are certain companies, particularly in the financial or legal sectors, which want to have a symbol of their wealth, to put down a marker. 30 St Mary Axe was one of those and the owner now markets its 'brand' based on that building. But in the future the tall building will not be wearing the new green consciousness on its sleeve. Skylines are unlikely to feature a raft of towers with wind turbines on them, for instance.

In the future, architects will look more at the carbon usage and footprints of buildings in use, and, as they are demolished and rebuilt, at the carbon costs of using aluminium, for example, rather than other materials. This approach will apply to towers too, with closer scrutiny of whether there is a need to demolish and rebuild in the first place. It may be a better and more environmentally viable goal to seek out more re-skinning and refurbishment solutions: to improve cladding, change air-conditioning systems and upgrade finishes rather than tear the building down and start again. Going down the refurbishment route may not save money, but it does save time by leaving the frame up.

Looking to the future, there is almost no limit to how tall one could go if one chose: a kilometre, two kilometres, ten kilometres – although these heights are unlikely to be achieved for many decades. But as pressure on land becomes universally more intense, so the need for more farmland to feed a growing population grows ever more pressing. As a result of climate change, increased flooding due to more extreme weather and rising sea levels will reduce available land, driving people into tighter environments. Should the population double or triple, the only way to go is up. Perhaps in a thousand years' time everyone will be living in towers on tops of hills in the middle of lakes – who knows?

Tall buildings are the future because as populations grow an adequate solution must be found. This is the lesson of the last ten years or so –buildings have to be responsive and much more reflective of their multiple roles. They have got to do more than just one job.

As buildings get taller and taller, there is more scope for linking tall buildings – much like the 'streets in the sky' principle but in a more successful format than was experimented with in the 1960s, with gardens, bridges and other amenities many storeys up. And as technology moves on, the relationship between where people live and where they work will blur even more, potentially resulting in more mixed-use towers which achieve a balance between humans as social, gregarious animals and virtual reality techniques allowing meetings to be held over the internet, reducing the need to travel.

Ultimately, tall buildings are an emotive subject that divides public opinion. While some feel they can be an alien presence in the community, and even that they are an ominous, looming, shadow-creating presence on the landscape at an inhuman scale, others love their symbolism, their majesty, their views. Designed with ingenuity, aesthetic ability and innovation, tall buildings have the power to lift the heart and spirit and bring joy into people's lives. And architects, engineers and other consultants cherish them for the technical challenge they represent. Part of the goal is to make the next tall building project as economic, slender and beautiful as possible. But the main challenge for the future of the tower is to contribute fully to the city through being enhanced and enriched by environmentally responsible design.

Chapter 9

Islands in the Sky, Cities on the Ground: The Public Realm and Its Impact on Tall Buildings

Steve Hanson and Roger Courtenay

Because tall buildings often make money while landscapes rarely, if ever, do, the success of a city's public spaces is often directly linked to successful tall buildings. However, from the perspective of the quality of urban life, tall buildings are only as "good" as the public realm spaces that they enable. Whether these spaces are traditional plazas or green parks, or elements that contribute to human interaction and health in newer ways (such as the "sky lobbies" of Asia's super-tall buildings, or the resurrection of the World Trade Center site), their relationship with buildings of any scale cannot be ignored. While an architect may focus narrowly on the iconic or aesthetic quality of a tall building as its own object, the urban designer or landscape architect views them as a source of users (social fabric) and of funding (financial health) for the urban environment.

The precise relationship between open space and development density is still working itself out, and the evolution of that relationship has deep roots. Many of the earliest known tall structures had sacred importance, creating vertical links between humans and their sensed connections with supernatural elements both in the sky and underground. These were presumably sited in areas with geographical features, or natural or cultural histories, that indicated transcendence or potential interaction with the sacred (La Boda, Ring and Salkin 1996). The civilizations that allowed for the agricultural and social structures that sustain social hierarchies were the same civilizations that enabled the engineering of durable tall construction. In other words, the reasoning behind building tall was already interconnected with microclimate, geology, geomorphology, sociology, commerce, and culture: all things that contribute to a sense of place.

The pyramids of the Egyptians are the most obvious early examples of tall buildings constructed for reasons centered on religion and power, and the Potala Palace, a seventeenth-century, thirteen-story grand monastery in Llasa, Tibet, was perhaps the tallest occupied building in the world throughout the nineteenth century (Buckley and Strauss 1992). But it was in the West that tall buildings took root and first flourished. A church or cathedral steeple could act as a wayfinding landmark as well as a symbol of community pride. But the typical four- to five-story merchant's house of a trade center such as Amsterdam (the Dutch having cornered world financial markets for much of the seventeenth century) is a good illustration of the reasons why land surrounding dense development became highly coveted very early. Warehousing, business, family and servants' residences, and a manufacturing or repackaging plant could all be squeezed into a small footprint close to the core, close to the canal, and close to other businesses. Many of the components necessary to support even taller buildings were already in place: access to employment, trade, and resources, as well as the public spaces demanded by the human psyche which strictly utilitarian streets and alleys would not include – the urban square and the agora.

Landscape-integrated development (with evolving models of use, function, and scale) continued to focus tall buildings in mercantile and financial centers, where they arguably served well up to the beginning of the twentieth century. However, the 1902 construction of Manhattan's Flatiron building – which was not only architecturally distinct from its neighbors in terms of shape, but vastly different in height (Alexiou 2010) – necessitated a new relationship with landscape in order to make the challenges and opportunities of a taller building type and increased density acceptable within the urban fabric.

The Rockefeller Center complex, located between 48th and 51st Streets in Midtown Manhattan, is a notably successful example of the tall building/human scale equation. A massive undertaking of the 1930s, it was essentially complete in nine years and in its earliest form comprised fourteen buildings over six blocks. The scale and stepped setbacks of the building's articulation, and the scale of the spaces created, provide a rich diversity of space and place within the six-block precinct. Safe, wide pedestrian corridors predominate, but are frequently balanced by vegetation, whether trees or robust shrub plantings in planters, all conceived as natural extensions of the architectural design's elegant geometry. Rather than being overwhelmed by tight spaces and dimensions, pedestrians are offered observation opportunities at multiple grade-separated vantage points. The attention paid to articulating the pedestrian environment energizes the person-to-building relationship.

Today, programming of public space is a fine art at the Center – and its spaces, places and venues were designed with this in mind, as part of the total envelope of experience. An extensive indoor-outdoor art program is integrated into the open space. The precinct retains a vibrant and continuous pattern of vital public spaces, successful retail, and numerous other civic-minded facilities, including gardens, theaters, and a skating rink and observation deck. In filling spatial opportunities, leaving nothing to opportunity or chance, Rockefeller Center establishes a socially responsible ethic of design and development.

The World Trade Center site in New York, infamously leveled by terrorist attacks in 2001, represents a case in which integration with the urban context is proving successful, even under the watchful eye of the public and the project's many stakeholders. A melding of public realm and building development by the Port Authority of New York & New Jersey and private developer Silverstein Properties, Inc., what was once "Ground Zero" will soon become seven buildings, on 16 acres, thoroughly engaged with the public realm. The 9/11 Memorial is already the development's center, with an extensive park and related above- and below-ground buildings dedicated to the tragedy and to commemorating the lives of those who died as victims and as rescuers. Surrounding the memorial park is a suite of pedestrian-scaled places that are more "secular" in purpose: streets that are open to the public, pedestrian lanes, and a number of new public open places, including the plaza around the Port Authority Trans Hudson (PATH) station and a small neighborhood park in front of 7 World Trade Center, the first office building to open on the site in May 2006.

Much of the core of this area is served by access to subway and bus, with a largely contiguous retail area linking most of the buildings. The integration of above- and below-grade public and public/private areas has been a key element of the plan and design. The (very tall) buildings meet the ground at a succession of levels at-grade and below; are directly linked to all modes of transportation; and are seamlessly tied into the movement and life of the city. As a singly conceived precinct, paving, site furnishings, lighting, and service delivery will be united in an attractive package that honors the citizen and the city, and in doing so creates an expressive setting for tall buildings. The buildings' lobbies are also large in scale and transparent, inviting the outside in and offering a grand canvas for public-space art. The realms of the buildings, of the sacred space they embrace, and of the city will mesh and invigorate one another.

While the sort of precinct that the World Trade Center site is patterned on is not new, the way in which this tall buildings project is linked with city infrastructure and urban fabric will be closely monitored by the press and the surrounding community because of its historic and emotional significance. The project is unique – few tall building developments spring from the need to memorialize while simultaneously jump-starting economic growth – and yet exemplary. Occupancy rates at the World Trade Center will match those of comparable Lower Manhattan sites, but the public realm will be disproportionately large and carefully linked to the grid of the city. There will be no yawning, empty plazas here; only useful, attractive, energized city streets accompanied by spaces for contemplation or "just for art."

9.1 The masterplan for the World Trade Center site
Image: AECOM Design + Planning

9.2 World Trade Center development – view of district streetscape at the Wedge of Light
Image: AECOM Design + Planning; courtesy Port Authority of New York & New Jersey

9.3 World Trade Center development – view of streetscape at the HUB Plaza
Image: AECOM Design + Planning; courtesy Port Authority of New York & New Jersey

Despite the example of the World Trade Center and some of its Asian counterparts, some developers and governments are betting that the public realm is irrelevant, and continuing to build taller and taller without reference to the lessons of history. Students and practitioners of urban design, however, argue that integration is still relevant, because without redeeming social and cultural value, such edifices are largely supported by artificially-driven real estate markets.

No doubt the question will be settled eventually, but perhaps the best way to illustrate the urban design perspective is to review a place where some of the world's tallest and most beautiful buildings are currently being built, at a fevered pitch, with little or no regard for the public realm at street level. Pity, for example, the tourist who is dumped out on foot on the point of land created by a sharp bend in the Huangpu River in Pudong, central Shanghai. Just across the river is the Bund waterfront, with its revitalized public promenade and ten-story buildings: an example of urban success even if only measured by the vast numbers of people who visit it. On the tourist's side of the river, however, there is a different reality: six-, eight-, and ten-lane roadways carve up the peninsula, making pedestrian movement between mega-blocks virtually impossible. High quality of life in this part of Shanghai is only to be had high up in the air, and at high cost. There are some wonderful spaces inside these iconic, fog-piercing monoliths, and a few of them are remarkably beautiful "objects" in their own right. But a guest at the Park Hyatt high atop the World Financial Center who ventures outside for a look around is likely soon to retreat back upstairs to the lobby bar. Around the building and all its neighbors, the ground is completely sterile, with little or no usable green space, or retail or pedestrian connectivity. (As mentioned earlier, the "sky lobby" – or elevator interchange in a super-tall building – is replacing civic space on the ground as a space for human interaction, at least among the elite denizens of these towers. See Chapter 25 for more on this.) There is much to admire about such buildings, but the only reason to leave them at all on foot is to re-enter and access another elevator to go to an observation deck at the very top.

9.4 Aerial rendering of the World Trade Center development
Image: Silverstein Properties, Inc./dbox

9.5 The bustling Bund waterfront contrasts sharply with Pudong's imposing skyline
Images: Steve Hanson

From the point of view of urban design, a district of tall buildings is a failure. Some planners and owners, however, have recognized the issues already discussed. Currently, a project is underway to create an "elevated landscape" – a pedestrian realm two stories above the street that will presumably connect all of the major Pudong tall buildings at an upper floor. This may be an arcane and expensive solution to a sterile and uninviting ground plane, but it at least recognizes the need for a public realm and attempts to create a surrogate. It is to be hoped that this elevated landscape will be a green or at least a pleasant pedestrian experience, similar to the Highline in Manhattan, although of course much larger and built to fill a specific need for connectivity rather to create a destination out of a derelict industrial remnant.

This failure is also a reminder, when considering the public realm, that most tall buildings around the world are connected to the larger city below-grade already, through subway systems. It is possible, for example, to travel from home in one tall building in Hong Kong to one's office and back again without ever emerging into daylight and open air. This phenomenon enables the development of superblocks as "islands" completely disconnected from the street. Roppongi Hills in Tokyo is another example of this island approach. Cars, including taxis, are sent into an underground maze for an unceremonious dropoff far from the bustling city streets; the main public open space and front door to the tower is some three stories above the street.

On a psychological level – and bearing in mind the evolution of tall buildings and their corresponding effects on the public realm – this form of movement puts one in mind of the medieval fortress. It could be argued that it intentionally expresses a fundamental elitism, much as do suburban gated communities. One of the most pressing challenges for urban design with regard to tall buildings is thus to provide healthy alternatives to their fundamental insularity. These new self-contained islands (or archipelagos of contiguous, disconnected islands), set within dense cities but not integrated into them, create class-based, street-killing social barriers.

However, before giving up on the great potential of the urban ground plane as a social mixing palette altogether, contrast Pudong and Tokyo's Roppongi Hills with a tall building that is probably less remarkable as an artifact in its own right, but which offers a much more humane public-realm experience. Tokyo Midtown, a mere stone's throw from Roppongi Hills in Central Tokyo, does have its own subway station. However, in recognition of the huge impact of the ground plane on the human psyche, the design intent is to reinforce the feeling of entering the development's buildings from the street level. While the building entries, plazas, and

9.6 Large public green spaces such as Tokyo Midtown's Great Lawn are unusual in Japan
Image: © 2011 AECOM; photography by Dixi Carillo

indeed the streets sit atop many levels of parking and subway tunnels, the sense of Midtown is the sense of being firmly on the ground. The entire complex is steeped in positive, useful, contained open space for the development's 20,000 daily residents and countless other visitors. Buildings are pushed to the street edge, with active retail frontage, and then open up to frame a large entry plaza. While Midtown has an iconic tower (tall by Tokyo standards, at fifty stories), the green space does not simply surround the tower. The buildings embrace and frame green space in service of a public realm—the open space becomes the focus. Likewise, the design of the open spaces themselves reflects their lack of servitude to the buildings.

Paving and planting in the central plaza is a self-referential system, a contemporary abstracted reference to a Japanese tatami mat in proportion and color. The plaza paving has bold black bands that refer to tatamis' silk

9.7 The masterplan for Tokyo Midtown
Image: © 2011 AECOM

borders. These are laid out parallel to the central tower, but otherwise do not relate to the buildings' facades in any way. There is no border, threshold, or mediating element of any kind where landscape meets building, whether the facade is metal, glass, or stone. Put simply, the buildings contain the plaza, but there is no formal "picture frame" around it. In part, this is a result of the primary open space goal: an informal, contemporary interpretation of traditional Japanese landscape. However, it is also a simple function of the fact that the three buildings bounding the plaza are not aligned with each other, making it nearly impossible for the ground plane to meet each facade in the same way. The result is a landscape designed as an independent system, following its own rules, as nature does: landscape as earth, with buildings placed on or cut into it.

In order to achieve this level of independence from architecture in the landscape, when that landscape

9.8 Tokyo Midtown's plaza, with its distinctive tatami-inspired paving
Image: Steve Hanson

9.9 A detail of the plaza's paving
Image: © 2011 AECOM; photography by David Lloyd

9.10 Tokyo Midtown – an airy glass canopy soars above but interacts with pedestrians below
Image: © 2011 AECOM; photography by David Lloyd

actually rests atop architecture, agreements about structural systems must be made early in the design process. Too often architects of tall buildings turn to landscape to solve problems rather than acknowledging that the long-term success of the project will be measured by the two fundamentals mentioned early in this chapter: impact on users and economic returns. That the two are intimately interwoven is a given. All other things being equal, a development that lacks pleasant, human-scaled urban realm elements, or that tries to tack them on at a later date, will not succeed as well as one that includes them in its planning from the outset. But the requirements for open space that any landscape architect would relay to the architect and structural engineer are too often given short shrift, with landscape tending to turn into "extra" space to be filled with building infrastructure.

Architects and engineers surely understand the need for large, column-free interior spaces, but rarely acknowledge the need for equally flexible outdoor space. Seemingly trivial aspects (such as the need for two meters of unencumbered planting depth for an on-structure development) must be honored. Trees are very heavy, and while a grid of them can be a beautiful thing, it should be the result of design intent rather than a structural building grid. Such "compromises" often result in solutions that are both innovative and beautiful. At Tokyo Midtown, the downdraft solution for the central plaza became an enormous glass canopy some ten stories off the plaza. Its effect of mitigating climatic issues as well as defining space, is dramatic, as it rests atop huge, white, tree-shaped columns.

Midtown is a lively place in a vibrant neighborhood. Its mix of uses – retail, office, hotel, residential, museums, and entertainment – ensures its standalone success; it too could have been an island. However, its openness to the street and programmed network of open spaces ensures its place as an integral part of the city social fabric rather than as an island within it. The project was from the outset modeled roughly on Rockefeller Plaza – hence the "Midtown" name. Though much greener, with fully half the site devoted to open space, like Rockefeller Plaza the buildings frame public space rather than resting within it, and that public space plays host to the kind of events and day-to-day leisure activities that create urban habitat and define a city's character.

It remains to be seen whether the Tokyo Midtown model will be adopted by other new tall buildings within the public realm, although in an era of increasing infill development and burgeoning urban populations this would seem to be the smartest approach. While their predecessors (such as the Omaha State Capitol building in Nebraska, or New York City's Empire State Building) attained early iconic status largely through their height and importance as civic symbols, the new supertowers have taller orders to fill, and cast longer shadows. The continuing success of many of the first "skyscrapers" is largely dependent on the urban infrastructure and public realm that has evolved alongside their use as buildings; remove those, and it is not difficult to imagine their cultural significance diminishing to the point of historical footnote. While anomalous tall buildings in smaller communities can still function as the "steeples" and power symbols of old, most towers in newer, less-dense cities have none of the vitality or community one associates with the most iconic examples.

The reason is not hard to find. Tall buildings in numbers, in places like Manhattan, succeed in large part because the urban core around them is serendipitous, the result of decades of real estate building and demolition, services, retail, and the hundreds of other things that give sense of place, and due also to the idiosyncrasies of a community acting over time. People will cluster around and enrich the places that provide them with what they need, and part of what attracts people to any place – whatever height its buildings – is the overall quality of the urban experience.

While new developments are by their very nature and process difficult to plan, a pattern is emerging. A tall building may be a short-term success where vital and successful communities already exist, where it is a novelty that draws people in. For long-term success, however, the building must "give back," supporting the intangible needs of its users and the economics of the urban fabric that surrounds it. Whether tall buildings can develop meaningful harmony and social purpose out of whole cloth in the interstices that separate them is an experiment to be played out over the next hundred years.

References

Alexiou, Alice Sparberg. (2010). *The Flatiron: The New York Landmark and the Incomparable City that Arose with It.* New York: Thomas Dunne.

Amsterdam.info (undated) "Amsterdam Architecture: Canal Rings Houses." Online. Available HTTP: http://www.amsterdam.info/architecture/ (retrieved October 26 2010).

Buckley, Michael and Strauss, Robert. (1992). *Tibet: A Travel Survival Kit.* South Yarra, VIC: Lonely Planet.

La Boda, Sharon, Ring, Trudy and Salkin, Robert M. (1996). *International Dictionary of Historic Places: Middle East and Africa.* Chicago: Fitzroy Dearborn.

Chapter 10
Internal Environment and Planning

Jason Pomeroy

Introduction

Global population increase is having a significant impact on the built environment in which we live. The consensus amongst demographers is that by 2050 the global population will be 9.2 billion – an almost four-fold increase on 1950 figures. In 2007, for the first time in history approximately half of the world's population was living in cities, highlighting the continued trend towards inner city transmigration correlated with rapid economic progress. Such factors have seen both population and building densities intensify; and have heralded the emergence of the tall building as not only a symbol of the city or a representation of economic progress, power and prestige, but a means of optimizing land use in the wake of increasingly high land prices. As expressed eloquently in research into low- to medium-rise high-density alternatives by Leslie Martin and colleagues (Martin and March 1972), the tall building is by no means a panacea for the design of high-density environments, though it is a building typology that will be present on the city skyline until alternative, more environmentally responsive designs can be embraced. The view that the tall building has been the root cause of many socio-physiological and environmental ills has been shared by both the general public and some academics. Recurrent issues of perceived density, lack of social space, illegibility, and compromised health, well-being, noise, security and maintenance have contributed to the sense of community disconnection that has beset the tall building typology.

This chapter seeks to address these issues by firstly considering what constitutes community, before examining the impact of population increase, technology, secularism and industrialization on the city. It goes on to discuss how our sense of community and the arena in which communal activity and interaction previously existed were displaced; and how our experience of internal and external high-density environments became compromised. Through a series of built examples from different climates, it demonstrates that there has been a paradigm shift in the way tall buildings are being designed, such that they seek to balance space with built-up area through the provision of sky courts and sky gardens. It argues that these spaces bear a certain resemblance to historical building precedents that also offer social and environmental betterment. It concludes that a greater sense of community can be fostered, and more comfortable internal environments re-established, through creating sky courts and sky gardens using more objective design methodologies.

Defining Community

What is community? It can be defined as a group of people living in the same locality or having the same religion, race, profession or interests. Its various written definitions invariably reference the built environment ('commune', 'community centre', 'community home'), which tells particularly of the intrinsic relationship between social groupings and the space or building needed to foster a sense of community, be that the piazza, mosque, office, clubhouse or café. These environments provide opportunities for individuals with similar interests to engage with one another, turning *space* into *place*, fostering a sense of community and identity through casual interaction and co-presence.

Community need not be fettered by geographical location, as we see in the virtual examples of the 'chat room', 'face book', or 'transpatial community',[1] which are independent of place and part of individuals' commitment to an institution, group or association. However, urbanist Eduardo Lozano (1990) argues that such technological advancement has paradoxically contributed to the 'break-up of urban life and community … much human contact, which used to take place in streets, plazas and parks has become packaged; romance has been replaced by singles bars or computer dating agencies'. Lozano's assertion that community life has been compromised by technological advancement and the disjuncture between the professional practice of creating architecture and the traditional cultural practice of human habitats being built by the very community that lives, works and plays within them is no doubt a compelling argument. Sociologist Richard Sennett (1976) suggests that seduction by mass manufactured goods and the individual need to understand one's character and personality increasingly negated the need for public social interaction. What is apparent in both commentaries is the association of communal activity with the urban environment and, more to the point, the argument that those spaces which once encouraged the casual meeting of social groupings are being eroded. What are the socio-environmental consequences of the transformation from the city of open spaces to the city of objects?

From a City of Spaces to a City of Objects

Socio-economic change due to technology, industrial capitalism and secularism saw a correlative impact on urban morphology. Up until the eighteenth century, the city was determined from the outside in, with rationalized voids acting as outdoor rooms that dictated the shape of the city and provided a means of planned social interaction or chance meeting, whether trade and commerce, political activity, or religious and cultural events. Meanwhile, buildings' solid form accommodated urban idiosyncrasies by acting as infill elements. Such a celebration of public life reaffirmed the predominance of space over object. By the middle of the eighteenth century, however, 'public space was implicitly traded for the private object; a deal that formally represented the beginning of the end of the res publica' (Rowe 1997). Shifting patterns in society saw the public realm go into decline and the development of the freestanding private object. The need for more housing, improved public utilities and transport infrastructure determined the space from the inside out, with rationalized solids of core structure and service elements dictating the form of the

10.1 Figure-ground diagrams of (top) eighteenth-century Rome and (bottom) twentieth-century Portland, Maine illustrate the growing dominance of object over space
Image: Eduardo Moix

building within the city, and the void spaces became the habitable space left over. By the twentieth century, the transformation was complete. The freestanding private object building sat within open, undifferentiated space, and became the means of absorbing urban idiosyncrasies. The modern city of towers represented the antithesis of the traditional city, heralding the pre-eminence of object over space and the erosion of a public realm that otherwise would have nurtured civility amongst society. Such a physical transformation is lucidly summarized by urbanist Colin Rowe's description of the diametrically opposite figure-ground diagrams of the traditional and modern city: 'one is almost all white, the other almost all black … in both cases, the fundamental ground promotes an entirely different category of figure – in the one object, in the other space' (Rowe and Koetter 1978).

Such a paradigm shift in urban habitat can be largely attributed to Le Corbusier's cure for the slums and disease that lay behind the Haussmann facades of Paris, which sought paradoxically to decongest the city centre by increasing density, using high-rise structures. These buildings would contain 'perfect human cells which correspond[ed] most perfectly to our physiological and sentimental needs' (Hall 2002) while allowing the automobile to take precedence over pedestrians. Le Corbusier and his like greatly influenced a post-war generation of architects, both in Europe and around the world, spawning a legacy of high-density development that borrowed heavily from his concepts in order to address slum clearance, increasing land prices and an increasing birth rate that would lead to overcrowding. 'The big cities, many of which were not averse to keeping their own people rather than exporting them to new and expanded towns, read all this as a signal to build dense and build high' (Hall 2002).

The ability to re-house the masses in sanitized high-density environments with, in the best examples, supporting communal facilities, indoor streets and outdoor raised plazas owed much to the early visions of Fourier and then Le Corbusier; at the same time it sounded the death knell for pre-existing forms of community use of spaces. Social groupings and complete neighbourhoods, accustomed to low-rise urban environments that permitted casual interaction, were being dismantled and relocated to high-rise urban environments. The very same groups that once gathered to do their laundry or share in common activities found that the spatial mechanisms that permitted such communal activity and spontaneous meetings with neighbours were being socially and spatially engineered. In the worst cases, high-rise, high-density developers and authorities failed to understand the importance of such spaces for improving amenity, well-being, health, productivity and social interaction, and these were often omitted for economic reasons.

J.G. Ballard's novel, *High Rise*, highlighted the potential for developments to be poorly conceived enclaves divorced from their surrounding context and crudely executed by local authorities, and social and physical disjunctures were all too clear in real high-density estates such as Pruitt Igoe, Illinois. Here, a multitude of ills resulted in the development's eventual abandonment and demolition. Technological advancement paradoxically compounded such socio-physiological, psychological and environmental ills in the tall building typology. Increasing dependency on energy-consuming artificial light and air-conditioning to counteract deeper and more populated floor plates, and the vertical extrusion of the same residential uses from the ground plane, further increased the tall building's disconnection from the rich milieu of open space and greenery within the urban fabric and from people's ability to forge a sense of community through co-presence. Building-related illnesses, due to lack of natural light and ventilation; restlessness and social disorder given sparse or non-existent social and recreational facilities; illegibility and lack of diversity given uniform stacked floor plates; and soaring crime figures caused by lack of a sense of ownership and lack of surveillance – all have contributed to the tarnished image of tall buildings within the fabric of the city. This has spawned a radical rethinking of the tall building, socially and spatially.

From Outside In to Inside Out

Despite the debilitating consequences of high-density tall building developments of the past, academics and built environment professionals are increasingly considering tall building solutions, more sustainable ones that can reduce energy consumption while fostering a sense of community and enhancing the interior environment. The provision of open space within the confines of the private tall building object, in a fashion not too dissimilar to the space/object hybrid described by Rowe and Koetter in their book *Collage City* (1978), is one such approach being explored. It is based largely on an understanding of how the positive attributes of natural light and ventilation, and their very source within the building, *space*, benefit the health and social well-being of the individual, group or association.

Natural light and ventilation are essential for the survival of living organisms, and their quantitative provision within buildings can be found in the research of

Cambridge academics Nick Baker and Koen Steemers (2000) in their formulation of the LT method of calculation – a tool that considers active and passive areas of lighting and thermal performance in order to shape more comfortable environments. Space is essential for social interaction and recreation and was a critical element in Le Corbusier's high-density communities. In *Vers une architecture* (1923), a section – his 'fifth point' – was dedicated to the rooftop garden as a supplement to the open recreational spaces on the ground.

Consideration has also been given to the potential benefits of balancing natural light and ventilation, and fostering greater communal interaction through open spaces such as verandas, sky courts and terraces. The present author has postulated that sky courts and sky gardens have the potential to be alternative civic spaces, forming part of a broader multi-level open space infrastructure that seeks to replenish social space within the urban habitat and provide ease of movement (Pomeroy 2008). Engineer Kenneth Ip has considered architect Ken Yeang's works and the balance between creating a tropical regionalist identity and environmental responsiveness, with particular consideration given to the environmental performance of sky courts (Puteri and Ip 1986). Dr Joo Hwa Bay (2004) goes further, exploring the socio-environmental intricacies of vertical terraced spaces, in particular whether their ability to reinterpret the essence of the *kampung* tradition[2] is conducive to a high-density outdoor living environment fostering similar communal interaction. Roger Ulrich (1986) has considered the inclusion of planting and landscape as a means of enhancing the physiological and psychological health and well-being of the individual within the high-density built environment. The environmental benefits of planting roof tops has been aptly demonstrated by Mayor Daley's influence in Chicago, where 2.5 million square feet (232,258 square metres) of roof tops were landscaped in order to help reduce the urban heat island effect (and consequently ambient temperatures and energy loadings) within the vicinity, in addition enhancing biodiversity. Yeang's (2002) treatise on sustainable tall building design takes the city's diverse mélange and transposes it vertically in order to reconsider the tall building as a mixed-use extrusion of the city, punctuated by open spaces that help establish vertical communities.

These approaches among practitioners and academics are perhaps a welcome departure from the non-contiguous, floor-plate-stacking exercises of many twentieth-century tall buildings, and arguably they reinforce Oscar Newman's social theories about reducing crime and improving sense of community (Newman 1972). His research involving New York's projects during the 1970s concluded that tall buildings were more prone to crime and social disjuncture due to their sense of disconnection from the more integrated environments at street level – places that provided natural surveillance by those living and working there, as well as opportunities to play out one's youth, socially interacting, resting and recuperating. Such theories have created considerable debate, as evidenced, for example, by the argument of Professor Bill Hillier and Professor Julienne Hanson (1987) of University College London that co-presence and the ability to recognize who is and who is not a stranger within a legible network of more highly integrated streets creates the most defensible spaces, in contrast to overtly engineered spaces that are functionally specific and less integrated.

Collectively, these attributes offer the potential for not only more habitable but also more pleasurable tall and dense built environments, ones which encourage communal interaction. Particular historical building typologies have demonstrated that open spaces, incorporated into private development, help replenish the space lost for the community and offer environmental benefits at the same time. The eighteenth-century *hotel*, an aristocratic private residence, incorporated a semi-public court within its curtilage, which could be used as a place for meeting and greeting. It established a hierarchy of void spaces, with its larger, figurative, public-space counterpart that contributed to the footfall, amenity and interaction of civil society.

The nineteenth-century galleria, a retail object of private speculation, incorporated a semi-public thoroughfare that could be used as a means of transition between larger public spaces and, as with the hotel, provided an environment for social interaction within the broader urban fabric. Such privately managed open spaces in the form of the court and arcade accompanied the street and the square as means of balancing space with object, and formed a collection of alternative spaces that attempted to recapture elements of public life within the private curtilage. To this end, the court and arcade may be suitable models for the sky garden and sky court as alternative social spaces within the tall building typology for the twenty-first century.

Sky Courts and Sky Gardens

Sky courts can be spatially defined as interstitial spaces that balance the figurative (semi-public) void within the solid of the (private) tall building object. Just as one normally finds a proportion of open space to built-up area in groundscraping mixed-use developments, sky courts start to balance open space to built-up area ratios vertically, within the tall building. In doing so, they can help foster community through the provision of open space for the interaction of social groupings as well as provide opportunities for greater penetration of natural daylight and ventilation into deeper floor plates, thus further enhancing the internal environment and making it more habitable.

Newton Suites, Singapore (2007) aptly demonstrates the incorporation of sky courts for their socio-environmental benefits. The thirty-six storey building features a series of sky terraces that seek to offer amenity space for residents every five floors, and in doing so instil the sense of sub-communities within a vertical neighbourhood. Designated green space and vertical greenery acts as an environmental buffer to the low-angled east- and west-orientated sun, thus helping reduce solar heat gain. Such spaces reinforce the notion of a vertical reinterpretation of the tropical verandah, acting essentially as social places for meeting within the community and a critical spatial element within the urban fabric.

Socio-environmental considerations need not be confined to the realms of exterior spaces, however, as demonstrated in architect Norman Foster's Commerzbank, Frankfurt (1997). Hailed as Europe's first ecological high-rise, the office tower was conceived as three 'petals' of triangular office floor plates, grouped around a central 'stem' formed by a full-height atrium. Sealed sky courts, four storeys high, rise up through the building, rotating every four storeys to the next face. These afford the employees opportunity to view down to the sky court and cityscape beneath, or up to the sky court and sky above.

These spaces provide a social focus for the office employees, as places for meeting, events, lunches or

10.2 The WOHA-designed Newton Suites features sky terraces that reinterpret the traditional tropical verandah and its role as a community meeting place
Image: © Philip Oldfield/CTBUH

remote working. They have a social dimension, 'with coffee bars and seating tucked amongst the plants. They are thus intrinsic to Foster's vision of the tower as a community of villages with each garden as a village square/green for the 240 employees who directly overlook it' (Davey 1997).

Sky courts, in their interstitial positioning in a tall building, can also be a useful source of convenience, recreation and amenity that negates the need to travel groundwards for groceries, gymnasium visits or relaxation in open space. As tall buildings increasingly embrace a mixed-use programme, sky courts provide a forum for establishing new social relationships between building occupants. Just as spaces between a horizontal mix of uses can be places of recreation and amenity that foster a sense of community, so too can sky courts act as the socio-spatial gel that glues a mix of vertical uses and their social groups and associations together, thus fostering vertical communities.

The Shard at London Bridge Tower is a case in point. At seventy-two stories and a little over 310 metres tall, the tower is the tallest mixed-use structure in Europe. The first twenty-six floors above the public piazza provide 55,741 square metres of modern high-specification office space with winter gardens. A five-star

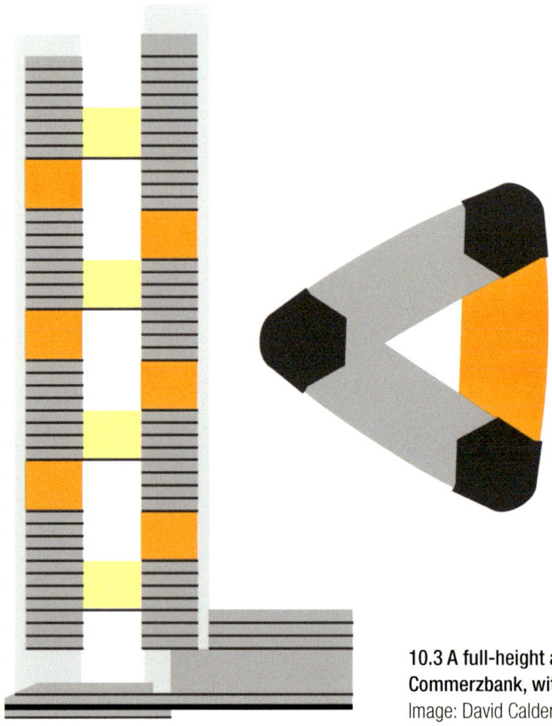

10.3 A full-height atrium forms the 'stem' of Norman Foster's Commerzbank, with floor plates forming the 'petals'
Image: David Calder

10.4 Sky courts, as here in the Commerzbank tower, act as social foci or 'village greens'
Image: Foster + Partners

10.5 London's Shard at London Bridge Tower will exemplify mixed-use development, creating a vertical community topped by a triple-height viewing gallery
Image: Stellar Property Group

hotel with 200 rooms is housed from the 37th to the 51st floors, with residential apartments from the 52nd to the 63rd. Separating the working from the living spaces is a three-storey sky court at the mid-level, acting as the community space that gels the disparate functions together. Such a space is designed not only to provide memorable views of London but also to house retail, bars, restaurants, leisure, performance and exhibition activities as well as social spaces for the tower's inhabitants and the broader community. Renzo Piano, architect of the tower, sees it 'as a small vertical town for about 7,000 people to work in and enjoy, and for hundreds of thousands more to visit. This is why we have included shops, museums, offices, restaurants and residential spaces.'

Sky courts also serve as transitional conduits within the tall building by connecting the disparate vertical circulation methods, whether ramp, stair, escalator or lift. Further circulation to higher levels within the tower is often made possible via transfer lobbies, which sky courts may serve as, thus increasing footfall through the court and encouraging activity, chance meetings and social interaction. When sky courts are integrated into broader movement strategies via skyways, podium decks or skybridges, it can lead to a greater social integration through pedestrian permeability – from the fabric of the city, through the tower and beyond. This begins to mitigate the risk of visual disconnection and separation from the activity of the street at ground level, as the horizontal and vertical means of circulation within a complex of tall buildings serve to create 'new eyes on the street in the sky'; this can aid security through the recognition of who is friend and who is foe. Furthermore, it presents an opportunity to escape from one tall building to another via skybridge, thus reducing the need for phased evacuation, which not only may be economically unviable due to the number of escape stairs required, but can compromise life safety.

The Pinnacle, Singapore (2009) demonstrates such an approach, using twelve sky gardens to interconnect its seven fifty-storey, high-density social housing blocks, comprising 1,848 family units. The sky gardens reinterpret the void decks of past social housing blocks as a series of elevated social spaces. Intermediate gardens at the 26th floor serve the residents only, whilst the 50th floor rooftop garden is accessible to the public as well as residents. It is perhaps this sense of pedestrian permeability analogous to the arcade that allows us to consider the sky court not only as a destination place of recreation and planned meeting, but also as a transitional space of movement and chance meeting.

In addition to improving the internal environment for individuals and reducing energy bills, the inclusion of sky gardens at the top of tall buildings can also provide opportunities to observe a memorable skyline (and thus be income generating), or a further setting in which social groupings with similar interests can interact, play or relax. The Marina Bay Sands, Singapore is an integrated resort and a destination for meetings, incentive travel, conventions and exhibitions (MICE) that will accommodate up to 52,000 people, making it one of the largest such facilities in Asia. Crowning the three prominent fifty-seven-storey hotel towers is a sky park 200 metres above the ground. The park, over one hectare in

10.6 Sky gardens in the Pinnacle project, Singapore, provide a series of elevated social spaces that create pedestrian permeability
Image: Arham Daoudi

area, is the world's largest public cantilever and provides a variety of amenities within a lush tropical landscape setting.

Just as the Empire State's 86th floor observation deck was able to help the development weather the storm of financial crisis by drawing visitor receipts of $2 million in its first year of operation, so the Marina Bay Sands sky park became an income generator through the levying of an entrance fee to gain panoramic views of Singapore's skyline from its observation deck. Its rooftop bars, restaurants and gardens have become a popular destination, providing an alternative social environment for locals and tourists alike during the day and into the night, while its rooftop pool and performance areas provide further recreation and amenity to fee-paying guests.

In all of these examples, time is a critical factor to be considered in fostering a community. Just as a play may have a variety of plots, subplots, and changing scenes that are enriched during the course of a narrative, so a city has a variety of changing community scenes that are created, altered, strengthened and dissolved through space and time. Sky courts and sky gardens can attract different social groupings as spectators of the visual scene of a changing skyline during the course of a day, month, year or decade. The same or similar social groupings, however, may also be actors within a localized social scene, one which may change the function of the sky court or garden over a period of time. For instance, the use of the space as a means of observing a sunrise or sunset may be very different to its use as a recreational destination or meeting place at a different hour; it may embrace a completely different use according to changing societal need.

Communal groups that form and disband can also imprint an element of territoriality on a place, thus implicitly excluding people from a particular social scene. Students may gather within sky courts outside of school hours to share notes before disbanding; office workers may meet with fellow workers from different departments for coffee or lunch breaks within a working day, before returning to their respective departments; residents may meet with neighbours and friends during the weekend or in the evenings before retiring to their homes; and tourist groups may gather to observe a panoramic view, disbanding at closing time.

What this suggests is that the uses of sky courts and sky gardens, from which social groupings can develop, are often dictated by the predominant tall building function that retains them, be that library, office, apartment or hotel. These spaces have explicit rules of exclusion – such as the operating hours of a corporation or an entrance fee – and inclusion – for example being part of

10.7 Marina Bay Sands' sky park offers a unique experience to tourists and locals alike and is a significant income generator
Image: Chloe Li

a student, office, residential or tourist community. Such societal rules may be challenged in the future, however, through greater exploration of vertical mixes that transcend conventional time, space and social structures. As in the case of the Shard at London Bridge Tower, the ability to live, work, play or visit within the same tall building could generate greater social convergence in these alternative spaces, which in turn would start to imprint sky courts with twenty-four-hour civic qualities.

Conclusion

The ability to balance the open space of sky courts and sky gardens with the built-up areas of the tall building object challenges the conventional twentieth-century tall building approach of repetition with a more hybrid form. This offers civil society greater socio-environmental benefit within the new vertical urban habitat; it can help improve internal environments and foster new vertical communities through greater social interaction and co-presence.

In their current guise, sky courts and sky gardens can be deemed semi-public social spaces with public domain characteristics. They form part of a hierarchical network of open spaces that replenishes and complements existing open space programmes on the ground, embodying the critical qualities that define the civic realm and wider community. They are spatially constrained by the tall buildings that retain them, often strongly identified with their function, and socially constrained by the implicit rules of a community or the explicit rules of the institution, company, association or group that governs them. When more integrated within the building's internal planning and its vertical circulation methods, sky courts and sky gardens can provide ease of movement as transitional space (like the arcade); and income generation as a destination space (like the court).

Increased footfall and more integrated movement patterns through these social spaces provide more opportunities for chance meetings and new relationships; this also enhances security through improved surveillance. Sky courts can also provide greater penetration of daylight and natural ventilation into the traditional floor plate. When planting is incorporated, the cooling effect in the immediate environment of evapotranspiration can improve thermal comfort levels, reduce urban heat island effect, and create an environment more conducive to social interaction, with its attendant socio-physiological benefits.

It is at this socio-environmental juncture that the design of sky courts and sky gardens can be enhanced via more objective means. Hillier and Hanson's (1986) 'space syntax' method quantifies aspects of social pattern without reference to individual motivation, origin or destination, land use or density, or other factors that may have influence. In so doing, it provides a mechanism for a predictive theory of mass movement based on individual rational choice based on spatial cognition. With mixed-use tall buildings increasingly being conceived as vertical cities, space syntax is highly applicable to the design of sky courts, and may be used to improve their socio-spatial integration and intelligibility, as well as footfall, through their spatial re-configuration within the tall building. Environments that can foster communal interaction can thus be enhanced by more quantifiable means.

Dr Ong Boon Lay's (2003) *Green Plot Ratio* seeks to address this issue by assigning values to particular plants based on the surface area of greenery. This is achieved by adapting the Leaf Area Index, a biological parameter 'used to monitor the ecological health of natural ecosystems and to mathematically model and predict metabolic processes. As such, it can be used to quantify planning metrics in biological terms' (Ong 2003). Such a method can safeguard against the loss of greenery post-development by taking into account horizontal, diagonal and vertical surfaces for planting with turf, shrubs, palms and trees, thus ensuring the same (or a greater) quantum of vegetation is retained on the site to help balance the ecosystem. Similarly, environmental modelling that considers the impact of solar path, humidity, temperature and wind flow can assist in defining areas for social interaction and optimized pedestrian movement.

When these objective measures are applied to the design of sky courts and sky gardens, perhaps more spaces can be created that help foster more successful vertical communities.

References

Baker, N. and Steemers, K. (2000) *Energy and Environment in Architecture: A Technical Guide*, London: Taylor & Francis.

Bay, J.H. (2004) 'Sustainable community and environment in tropical Singapore high rise housing: the case of Bedok court condominium', *Architectural Research Quarterly*, 8(3/4).

Davey, P. (1997) 'High expectations', *Architectural Review*, 202(1205).

Hall, P. (2002) *Cities of Tomorrow*, London: Blackwell.

Hillier, B. and Hanson, J. (1986) *Space is the Machine*, Cambridge: Cambridge University Press.

Hillier, B. and Hanson, J. (1987) 'The architecture of community: some new proposals on the social consequences of architectural and planning decisions', *Arch. Behav.*, 3(3).

Le Corbusier (1923) *Vers une architecture [Towards an Architecture]*, Paris.

Lozano, E. (1990) *Community Design and the Culture of Cities*, Cambridge: Cambridge University Press.

Martin, L. and March, L. (eds) (1972) *Urban Space and Structures*, Cambridge: Cambridge University Press.

Newman, O. (1972) *Defensible Space: Crime Prevention through Urban Design*, New York: Macmillan.

Ong, B.L. (2003) 'Green Plot Ratio: An Ecological Measure for Architecture and Urban Planning', *Landscape and Urban Planning*, 3(4).

Pomeroy, J. (2008). 'Skycourts as transitional space: using space syntax as a predictive theory' in *Tall and Green: Typology for a Sustainable Urban Future*, congress proceedings, Council on Tall Buildings and Urban Habitat 8th World Congress, Dubai, 3–5 March 2008.

Puteri, S.J. and Ip K. (1986) 'Linking bioclimatic theory and environmental performance in its climatic and cultural context: an analysis into the tropical highrises of Ken Yeang', in *PLEA 2006*, congress proceedings, 23rd Conference on Passive and Low Energy Architecture, Geneva, Switzerland, 6–8 September 2006.

Rowe, P.G. (1997) *Civic Realism*, Boston, MA: MIT Press.

Rowe, C. and Koetter, F. (1978) *Collage City*, Boston, MA: MIT Press.

Sennett, R. (1976) *The Fall of Public Man*, London: Faber and Faber.

Ulrich, R.S. (1986) 'Human responses to vegetation and landscapes', *Landscape and Urban Planning*, 13.

Yeang, K. (2002) *Reinventing the Skyscraper*, Hoboken, NJ: Wiley.

Notes

1 A 'transpatial' grouping involves a relationship independent of space. Examples are membership of a club or clan, or a visiting lecturing post at a university.

2 The *kampung* is a traditional village comprising dwellings and essential conveniences, often created using locally sourced materials, and is indigenous to South East Asia.

Chapter 11

Atria: A Vital Ingredient of Sustainable Tall Buildings

Swinal Samant

Atria, in the form of grand entrances, courtyards and sheltered semi-public areas, have been around for about 2000 years (Saxon, 1983). Iron and glass technology of the industrial revolution in the nineteenth century led to the covering of large courtyards, achieving significant improvements to the indoor climate, and this concept of the atrium as a social hub led to its widespread use in taller public buildings in the late nineteenth century. An essential feature of early tall buildings in New York and Chicago, atria enabled natural lighting and ventilation and drew away fumes from oil- and gas-fired lamps. Despite the use of atria in buildings in the late nineteenth and early twentieth centuries (albeit that this use was quite conservative), by the First World War the development of this concept had declined to a halt (Saxon, 1983).

This decline was largely due to changes in the New York building regulations. The move to taller buildings with smaller footprints and proportionally larger floor spaces eventually led to the rise of fully glazed Modernist towers that were predominantly mechanically conditioned and artificially lit. Although many pioneers pushed forward the atrium concept, this particular building feature remained largely disused until John Portman's early 1970s inclusion of atria as iconic design features in the Hyatt Regency Hotels in the USA. Since the late twentieth century, the atrium has been reconsidered, often with the rationale of climate modification and energy efficiency. The latest generation of tall buildings have made admirable progress in their response to environmental, social, cultural and economic concerns, often with the use of atria.

Atrium Functions

An atrium's ability to contribute to several aspects of a tall building makes a compelling case for its incorporation. Tall buildings are often principally concerned with making landmark public statements through iconic forms and soaring heights. Inside, a full-height atrium can create a sense of awe due to the contrast of an atrium with its surroundings, and indeed with human proportion. Many atria, such as the one in Shanghai's Jin Mao Tower (1999), have been built purely for aesthetic delight and are an expression of wealth and extravagance, power and grandeur.

The influence of these dynamic spaces permeates through to the different parts of a building, lending it social, functional and spatial coherence (Bednar, 1986). Atria can be powerful, iconic spaces that act as focal points, improving links and orientation within and between buildings. They lend a strong identity to a building and strengthen its sense of place and relationship with the urban landscape and city.

Atria serve as a sensitive medium through which a high-rise building is connected to the ground and its context – the urban fabric and infrastructure and the wider city. The 1995 Osaka World Trade Centre atrium is designed such that the complex assimilates with its context while also providing principal access to its tower. Jean-Paul Vigier's 2001 Cœur Défense building in the heart of La Défense in Paris is Europe's largest office complex, comprising one hectare of floor space. It successfully employs an atrium as a physical and symbolic unifying entity; the soaring 44-storey atrium space links 180-metre tall twin towers and three other eight-storey buildings. It sits on top of a three-level base, the top level providing connections to Charles De Gaulle esplanade and access to all buildings, restaurants, a conference centre and services.

Atria form a seamless link between the public and private realms, drawing people into the heart of a building via a warm and welcoming entrance and reception spaces. These vertical arcades can be characterised by shops, cafés, restaurants, performance or exhibition spaces and retail facilities that present opportunities for social interaction, natural surveillance and people-watching, reinforcing their commercial viability. Adopting an ecological approach, the 26-storey EDITT Tower in Singapore addresses the lack of continuity between the street level and the vertical, sealed environments of typical tall buildings characterised by a repetitive, compartmentalised stratification of floors. It achieves this through sky courts, atrium spaces, sky plazas and vertical landscaping, including wide, landscaped ramps that rise up to six floors and are lined by shops, cafés, performance spaces and viewing decks (see Figure 11.1).

Increased urban densities and inner-city living in the twentieth century led to an increase in high-rises but a decline in the quality of the public realm. This could be partly counteracted in vertical structures through use of atrium spaces. These can embrace a multitude of functions, particularly when combined with sky courts and sky gardens, to create lively, sustainable vertical communities (for more on this, see Chapter 10). More recently, atria of various sizes, shapes and orientations, including vertically stacked atria, have been used strategically in tall buildings in combination with smaller atrium spaces and sky courts and gardens of varied character, successfully extending public amenity spaces at height. For example, Norman Foster's 1997 Commerzbank building in Frankfurt incorporates a series of four-storey sky gardens on one side of the triangular plan, flanked on the other two sides by 16.5-metre deep offices which spiral around the central atrium to form a cluster. These sky courts are essentially used for meetings, social interaction and relaxation; they play an ecological role, bringing in daylight and fresh air; and they enable visual links between the interior and the exterior (Figure 11.2).

11.1 Social spaces with vertical landscaping, EDITT Tower, Singapore
Image: T.R. Hamzah & Yeang Sdn. Bhd.

Such atria play a vital role in forging meaningful links between sky courts and the different spaces at upper levels, creating a web of transitional and recreational spaces, and improving movement and legibility within the vertical urban fabric of cities. The Shard at London Bridge Tower (2012), a mixed-use development by Renzo Piano, includes two sky gardens and a series of atria at different heights of the building to support the

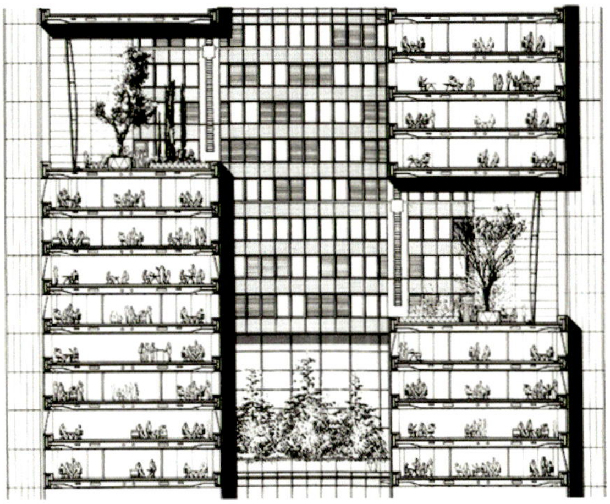

11.2 Section through the Commerzbank in Frankfurt showing the relationship between the atrium, sky gardens and offices
Image: Foster + Partners

social needs of its proposed vertical community of 7,000 residents. Given the projected rise in urban population and proposed mixed-use tall buildings in urban environments, the use of such atria is pertinent.

An atrium can aid adaptive reuse and sensitive refurbishment of historic buildings. In the Bank of Nova Scotia headquarters in Toronto by WZMH Partnership (1988), a dramatic 14-storey high glazed atrium connects the 68-storey office tower with the historic limestone bank headquarters building, and forms a striking entrance (Figure 11.3). Below ground level is the pedestrian concourse that links this development to Toronto's 28-kilometre PATH network of pedestrian subways and the largest underground retail centre in the world. Here, the atrium, in more than one sense, makes an impressive link between the past and the present, and the immediate and wider city context.

Given the large volume of an atrium, the consequent sizeable enclosing surfaces present opportunities for display of artwork. In the atrium of the Bank of Nova Scotia, adorning the north wall of the original banking headquarters is a magnificent decorative marble bas-relief sculpture; it is counterpointed on the south wall by the largest oil-painted abstract mural ever commissioned in Canada.

The use of an atrium is no way limited to any individual typology, and reasons for its use can differ greatly. Its use for creating a strong image, aesthetic appeal, day-lit environments and social hubs is established in corporate and commercial environments. To express transparency and sustainability in their head office, the Deutsche Post AG tower in Bonn, Germany (2002) uses

11.3 A striking entrance atrium – Scotia Plaza, Toronto
Image: Courtesy WZMH Architects

a glazed atrium to bridge its two towers. Atria arguably create day-lit spaces with pleasant views in deep-plan buildings, which are associated with: psychological benefits, occupant well-being and improved productivity in offices; improved student performance and behaviour in educational settings; and uplifting and healing environments contributing to improved recovery rates and the health and well-being of patients, staff and visitors in healthcare environments. In educational and healthcare settings atria can act as significant social nodes and improve wayfinding. Tokyo's 50-storey Mode Gakuen Cocoon Tower (2008) is the second tallest education building in the world. It incorporates a series of three-storey, day-lit atrium lounge spaces on intermittent levels, in the east, south-west and north-west, placed between rectangular classrooms that rotate around the central service core. Young (2009) refers to these atria as the more familiar 'schoolyards' where social interaction thrives, and which, in this case, offer enviable views of the city (Figure 11.4). The three-storey high and approximately 20-metre wide atrium glazing is supported by double-arched Vierendeel truss beams at each floor

11.4 Atrium lounge space in the Mode Gakuen Cocoon Tower, Tokyo
Image: Koji Horiuchi

level, which carry the weight of the glazing and resist wind pressure while not obstructing views (Tange and Minami, 2009).

From an economic perspective, atrium buildings have the potential to be competitive in terms of marketability, securing premium lets and generating revenue (Bednar, 1986). Shallow-plan perimeter spaces combined with atria offer natural ventilation and daylighting opportunities, reducing the operating costs typically associated with deep-plan buildings. However, higher capital costs are associated with a larger building footprint, roofing, skylights and shading devices, fire and smoke control systems, maintenance and landscaping (Bednar, 1986; Saxon, 1983). The increased costs of atria have until recently confined their use to corporate commissioned towers; elsewhere the financial outlay and reduced lettable floor area have dissuaded developers. However, with the perceived importance of quality design and sustainability increasing, developers and authorities are yielding to the notion of less area-efficient buildings creating more energy-efficient environments. An example of this is the developer-led, multi-let, village-style Heron Tower by Kohn Pedersen Fox in London (2011; see Case Study 17).

Environmental Atria

The environmental potential of an atrium is fundamental, and it is achieved mainly through passive heating, cooling and daylighting. An atrium acts as a sheltered or buffer space between the exterior and interior that may not be fully conditioned but brings in daylight while excluding wind, rain and temperature extremes. Atria act as heat sinks or stores and are used as return air plenums and ventilation chambers. The atrium has been developed particularly effectively in cooler climates at higher latitudes, where expanses of glass may not succumb to excessive solar gain and overheat, while at cooler times the atrium provides a thermal buffer to the external climate. However, despite the obvious environmental opportunities presented by atrium buildings, they are often simply utilised for their aesthetic attributes.

Atria: Ventilation and Fire and Smoke Control

An atrium can be used to enhance natural ventilation, mainly through cross-ventilation and stack effect, and is particularly effective when combined with sky courts and sky gardens in tall buildings. However, temperature stratification and unimpeded movement of air within the atrium potentially pose a significant fire risk; these can be controlled by splitting the atrium into sections, as is the case in the central full-height atrium of the Commerzbank, which has horizontal glazed screens at 12-storey intervals.

With the objectives of climate-conscious design and a 60 per cent reduction in energy consumption, the Manitoba Hydro Place building in Winnipeg (see Case Study 4), where climatic conditions are extreme, represented a huge challenge. The main body of the building, which rests on a three-storey podium, comprises two 18-storey office towers that form an 'A' shape converging to the north and splaying open to the south, with a series of three 6-storey atria spanning between them (Figure 11.5). The atria maximise passive solar gains and southerly winds; they act as buffer spaces and deliver pre-conditioned air to the adjoining spaces via adjustable vents incorporated in the raised floors. In winter, recovered heat from exhaust air and passive solar radiant energy is used to warm the fresh air. The building is designed to operate in response to Winnipeg's seasonal climatic variations, and a 24-metre tall waterfall featured in each of the atria either humidifies or dehumidifies the incoming air. In summer and natural ventilation mode, the building relies solely on outdoor fresh air, using automatic and manually operated windows. A soaring solar chimney to the north is used for passive ventilation which relies on the natural stack effect occurring within the atria; the glass louvres at the top of the solar chimney draw out stale air in summer and interim seasonal periods, while in the winter heat is recovered from the exhaust air to warm the multi-level parking and to preheat the incoming cold air (KPMB, 2009). In this case, atria form part of a larger environmental strategy, at the same time providing valuable amenity and circulation spaces and improving visual links between the different office floors.

Atrium buildings require special design considerations with respect to an effective fire safety strategy, integration of which during the early design stages is vital to reducing economic, functional and aesthetic impact. 'An atrium provides a route by which smoke and fire can spread from storey to storey much more rapidly than in the equivalent non-atrium building' (BSI, 2008). Nevertheless, 'controlling the spread of fire in atrium buildings is a lesser problem than the hazard to life safety caused by the spread of smoke' (Bastings, 1988). Safety of the occupants is greatly endangered if the smoke produced during the fire is spread into an atrium's adjacent spaces. The potential spread of fire and smoke through the atrium can have a significant effect upon the number of persons at risk, the escape procedures and firefighters' activities (BSI, 2008).

Selection of an appropriate strategy for an atrium building depends on diverse factors, including the building's function and the nature of its occupants, its physical characteristics and the degree of openness of the spaces adjacent to the atrium that have a significant impact on the design of the evacuation routes. Balconies around an atrium may not be used as escape routes unless they are properly sealed against smoke leakage. Moreover, an atrium's size, proportions and other characteristics play a major role in developing an integrated approach to both natural ventilation and its related smoke control strategies. For example, an atrium's height will impact upon stack effect, while openness within the atrium walls, sizing of the inlets and outlets and pressure within the adjacent rooms will determine smoke flow and control. Finally, a study of the relationship of the building with its surroundings is vital to prevent fire from spreading to neighbouring properties. Various other strategies are also necessary, including appropriate design of safe escape routes, secure from the effects of fire or smoke; the use of fire-resistant materials in the atrium's walls to reduce the risk of fire propagation; and the use of sprinklers to suppress fire and reduce temperature.

Atria and Daylighting

Daylighting is recognised as one of the key benefits of the atrium form, contributing to the building's aesthetics, experience and environment. Glazed atrium spaces allow for adjoining spaces to have larger windows without considerable heat losses or gains, potentially increasing the amount of occupied space that can be naturally lit and so replacing artificial lighting and its associated cooling load.

Daylight performance of an atrium building is complex and depends on the predominant sky conditions, the roof and fenestration system, atrium orientation and geometry, the design and reflectance of the

11.5 Environmental atrium, Manitoba Hydro Place, Winnipeg
Image: KPMB Architects

atrium facades and floor, and the characteristics of the adjoining spaces.

Goncalves (2007) highlights improved daylighting in office buildings as the most considered and desirable change evident in sustainable tall buildings the world over; this is often achieved through the incorporation of atria and multi-storey peripheral and internal communal spaces. The European model has been to incorporate more detailed facades in response to the local climate and the building's functions. Highly glazed facades, including double- and triple-skin facades with sun control devices between layers of glazing, retain the high-tech aesthetic so widely sought and enable views and daylight penetration while providing solar shading and ventilation, reduced noise transmission, improved thermal performance and control of infiltration of air and water from outside.

Particular roof types and atrium geometries can be exploited to bring in daylight into the lower reaches of a tall building. These include two- and three-sided atria and sky courts; higher atrium surface reflectances and variable openings in an atrium's facade, with smaller openings at the top and a progressive increase in size of openings lower down; and higher floor-to-floor heights and shallower adjoining spaces.

In London's Lloyd's Register of Shipping complex (2000) by Rogers Stirk Harbour + Partners, two glazed atria slotted between radiating 14-storey office wings act as thermal buffers, allow daylight penetration into the office spaces and provide views into and out of the building. Along with the marble floor, the atrium's glazed facades act as light reflectors, enhancing lighting conditions while limiting solar heat gains in summer and heat losses in winter. Glass balustrades and the use of light, opaque walls lend transparency to the central atrium space and create bright, daylit conditions.

Deutsche Bank Place at 126 Philip Street, Sydney (2005), a premium sustainable mixed-use/office tower by Foster + Partners and Hassell, is a result of extensive whole-building energy modelling and a meticulous and transparent process of measurement and assessment. The 160-metre tall atrium runs the full height of the building and is designed to draw daylight from all directions; this is enabled by the use of high-performance glazing and facade systems, with a high shading coefficient yet optimised daylight transmission into the atrium and its adjoining offices.

London's Heron Tower implements atria in its design by vertically subdividing the 36-storey building into three-storey 'blocks', each of which is connected by a three-storey atrium on its glazed north elevation, serving as a focal amenity space for the operation (Figure 11.6). A concrete core protects the south facade, while the east and west facades are triple glazed, with anti-glare and heat-resistant blinds. On each 'block' the two upper floors are recessed at the centre, enabling daylight penetration into the offices. The considered orientation of the building and the atria cuts out any need for solar shading. In addition to taking advantage of stack effect, the blocks are supported by localised heating and mechanical ventilation with a heat recovery system, with additional occupant control, however, for opening windows. The building offers solutions to the problems of temperature stratification and unequal daylight distribution characteristic of tall atrium spaces, reduces costs associated with large-scale centralised mechanical systems, and ensures effective management of the environment.

11.6 Three-storey atrium in the Heron Tower, London
Image: Hufton + Crow, courtesy Kohn Pedersen Fox Associates

Conclusions

Atrium buildings improved living conditions in ancient Rome, and today, following a revival, they are helping to create economically, socially and environmentally sustainable buildings. It is estimated that 61 per cent of the world's population will live in urban environments by 2030. This, combined with the expected growth in the world population – to 9.1 billion by 2050 (United Nations, 2005) – means that resources will become stretched. In a sustainable future, mixed-use tall buildings that combine life, work and leisure will be omnipresent in urban settings, and atria will play a central role in realising this vision. Architecturally, programmatically and structurally, atria form the substantive heart of a building. More recently, atria in the form of varied amenity spaces of superior aesthetic qualities have taken on a much broader, social role in tall buildings, creating vertical communities at a human scale. In their accomplished use of atria and sky courts, several contemporary buildings have set a precedent for the new, sustainable skyscrapers.

Atrium buildings are often energy inefficient due to a lack of consideration of this aspect of their design. Where developments are driven primarily by building aesthetics and pursuit of iconic architecture, their overall sustainability is often questionable. Inefficiency may also be due to the particular site characteristics, construction economics and programmatic requirements of a building. Further, due to complex structures and construction requirements, this building typology can give rise to inefficient usage of space, high embodied energy and increased construction costs. As Baker and Steemers (2000) state, "Accepting the atrium is now a very common feature in large public and commercial buildings, it becomes all the more important to ensure that it does not commit the building to a lifetime of high energy consumption."

However, atria in fact present excellent opportunities for improving the environmental sustainability of tall buildings. An atrium is an essential all-weather socialising space, offering opportunities for bringing in vegetation and connecting the outdoors or the enclosed spaces between and within buildings. It provides an intermediate environment that can filter and manipulate environmental factors to create desired conditions through passive means. Atria manifest recurrent themes of natural ventilation, daylighting, passive solar gain, solar shading and evaporative cooling, but it is the combination of these strategies that is vital.

The success of tall atrium buildings relies on thorough evaluation of various performance variables and the complex trade-offs between them to achieve an optimised solution, while retaining design integrity and inherent architectural merits. Advances in materials and structural technology, mechanical controls, shading devices, high-performance double- and triple-skin facades including advanced glazing and roofing technology, intelligent lighting systems, and sophisticated fire and smoke control and air handling equipment may be employed to develop creative environmental responses. Atrium manifestations have altered, reflecting technological developments and complex geometries realised due to advanced computational capabilities, giving rise to more innovative forms and more energy-efficient, cost-effective and responsive contemporary architecture. While this provides architects with greater freedom, integrated performance-based design approaches and increased collaboration between designers and engineers is paramount. With greater recognition and viability of sustainable tall buildings, atria will be a vital defining ingredient in this new era.

Acknowledgements

The author would like to thank her students, Miguel Juarez and Phillip Oluwatoyin, for their help in data collection.

References

Baker, N. and Steemers, K. (2000) *Energy and Environment in Architecture: A Technical Design Guide*, London: Taylor & Francis.

Bastings, D. (1988) *Fire Safety in Atrium Buildings*, BRANZ Study Report SR15, Building Research Association of New Zealand.

Bednar, M. (1986) *The New Atrium*, McGraw-Hill.

BSI (British Standards Institution) (2008) *BS 9999:2008 Code of Practice for Fire Safety in the Design, Management and Use of Buildings*, London: BSI.

Goncalves, J.C.S. (2007) 'Sustainable and Tall', *CTBUH Journal*, 18–21.

KPMB (2009) *Manitoba Hydro Place*, KPMB. Available at: http://www.kpmbarchitects.com/index.asp?navid=30&fid1=0&fid2=37#desc (accessed 9 October 2011).

Saxon, R. (1983) *Atrium Buildings: Development and Design*, New York: Van Nostrand Reinhold Company.

Tange, P. and Minami, M. (2009) 'Case Study: Mode Gakuen Cocoon Tower,' *CTBUH Journal*. Available at: http://www.ctbuh.org/LinkClick.aspx?fileticket=c2hAbvHAGUQ%3D&tabid=71&language=en-US (accessed 10 August 2010).

United Nations (2005) *Population Challenges and Development Goals*, Department of Economic and Social Affairs of the United Nations Secretariat. Available at: http://www.un.org/esa/population/publications/pop_challenges/Population_Challenges.pdf (accessed 10 August 2010).

Young, N.M. (2009) 'Tokyo Give Us an Education,' editorial, *WorldArchitectureNews.com*, 23 January. Available at: http://www.worldarchitecturenews.com/index.php?fuseaction=wanappln.projectview&upload_id=10970 (accessed 9 October 2011).

Part Three

Sustainable Responsibilities and Risk

Introduction

Dave Parker

Tall buildings—skyscrapers—have evolved a long way from their nineteenth-century origins. This evolution has rarely been smoothly linear and predictable, distorted as it has been by technological breakthroughs and economic woes alike. For a decade or two towards the end of the twentieth century, however, there was something of an evolutionary plateau. Tall buildings became a largely unremarked feature of most city skylines and their development, design and construction became as routine as they had ever been. However, all this has now changed.

During this plateau period architects felt they had the language and the understanding to translate their clients' briefs—be they for landmark, potentially award-winning designs or for buildings which combined the maximum net lettable floor area with the minimum construction cost—into towers that would both satisfy clients' expectations and attract enough tenants to make the project economically viable. Engineers had the design tools to translate the architects' designs into structurally safe and efficiently serviced buildings. True, there was much less consideration of the long-term welfare of the occupants than is the norm today, and there was an underlying assumption that energy would remain cheap and plentiful, but on the whole tall buildings constructed during these decades performed as expected.

By the turn of the twenty-first century, however, expectations were much wider, among developers, tenants, and society as a whole. As Part Two explained, there was an increasing awareness of the impact tall buildings could have on those who lived and worked inside them and nearby. Internal environment became a priority almost equal to external form. At about the same time, two other imperatives came to the fore, factors that would eventually accelerate the evolution of the tall and supertall building towards a distinctive twenty-first century form. Cosy assumptions about the safety and security of landmark buildings crumbled after 9/11. And the twin spectres of global warming and peak oil sent shockwaves through the tall building community.

Sustainability became the watchword—clients and tenants had to at least make gestures towards energy conservation and recycling. After some false starts, genuine progress has been made. The latest generation of buildings is also more resilient, better able to cope with potentially traumatic events ranging from terrorist attacks to extreme weather. All this rapid progress is down to the availability of new tools for designers and constructors alike.

In Part Three, eminent contributors consider the impact on tall building design of the advanced structural analysis software now available. Rectilinear glass-clad towers are now definitely last century; in the first decades of this century architects have experimented with an almost unbelievable variety of forms, from the elegant and exciting to the bizarre and even ridiculous—and structural engineers have made it possible to turn these imaginings into reality. As other chapters reveal, new and improved materials and technology are playing their part, recycled materials are now commonplace on many projects, and the number of towers sporting photovoltaic panels and/or wind turbines is steadily growing. Under their increasingly sophisticated facades, modern buildings now incorporate complex hazard mitigation systems. Tall building occupants are now safer, more comfortable and significantly more productive thanks to the accelerated evolution of tall buildings during the last decade or so. This evolution could help urban societies cope with the manifold challenges we are almost certain to face within our lifetimes.

Chapter 12

Sustainability and Energy Considerations

Rick Cook, Bill Browning and Chris Garvin

In 2009, humanity crossed into a new experience—it became an urban species. For the first time in history, more people now lived in cities than in the countryside. In the next three decades it is expected that 70 percent of the world's population will live in cities. At low densities, cities sprawl across the landscape consuming farmland, habitat, water, energy, and materials. As densities increase, individual dwelling units become smaller, and transportation and amenities are shared by more people, thus lowering the per capita consumption of resources. A New York City subway train achieving 540 passenger miles per US gallon (229 passenger kilometers per liter) during rush hour can be 20 times more efficient than commuting by automobile, which typically struggles to achieve better than 22 passenger miles per US gallon (10 passenger kilometres per litre) during the rush hour, thanks to most cars carrying only one commuter. Tall buildings tend to be integrated into tighter, more efficient utility infrastructures, and can rely on efficient district heating and cooling systems that are not economically feasible at lower densities. This is the context in which tall buildings have the potential to be truly sustainable.

Newer mixed-use tall buildings with residential, hotel, office, and recreational facilities integrated can reduce transportation demands and, more importantly, resources can be effectively shared between uses to improve efficiencies and balance out peak loads. For example, an office building rejects a great deal of heat, which can be harnessed by residential and hospitality uses for hot water heating. This integrated systems approach is critical to achieving truly sustainable uses within buildings.

In an urban environment, there is the opportunity to maximize resource efficiency within the building and within the larger community. Tall buildings are typically more resource intensive but are also more durable than smaller buildings. The first step is to connect building systems to the climate and maximize the natural assets of daylighting, solar radiation, natural ventilation, rainwater collection, and geothermal (ground source energy) potential. As new tall buildings are developed and existing ones are renovated, it is important to understand how the flow of resources—energy, water, waste, and materials—can be optimized and how these flows relate to adjacent buildings and infrastructure systems within the community. Utilizing distributed infrastructural systems for power generation, stormwater management, and wastewater treatment can be extended beyond the individual building to the district and neighborhood level in urban environments.

Human comfort, health, and productivity are perhaps the most important issues for the proper operation and function of any tall building, and the true purpose of the efforts of architects and engineers. Tall buildings have unique issues due to the disconnection created by their height. Buildings in the twenty-first century must be more engaged with their surroundings, resource efficient, energy exporting, and healthy for all occupants.

Energy

In smaller buildings—those under 10,000 square meters (100,000 square feet)—heat loss and gain across the skin of the building dominates energy consumption. Larger commercial buildings tend to be dominated by the internal loads created by the heat of people, lights and office equipment. Internal loads are different for residential and hospitality buildings, with higher hot water demands, but much-reduced lighting and office equipment loads. The approach to energy should be to minimize internal loads and consumption through innovative systems and opportunities for utilizing waste heat or natural ventilation as a resource. Once this is achieved, the design of the central mechanical systems and envelope can be developed as an integrated system with possible onsite generation that utilizes renewable resources. This is achievable due to the scale and complexity of tall buildings.

The exterior envelope is a critical component of a tall building's design, particularly relative to daylighting and energy demands. In the last two decades glass technology has improved dramatically. It is now possible to specify glass that lets in twice the amount of visible light to infrared light (the latter largely leading to unwanted heat gains, especially in office environments). Visual transparency is maximized by the use of low-iron glass in buildings like the Bank of America Tower at One Bryant Park in New York City. Bringing daylight through a high-performance glazing system and then bouncing that light onto the ceiling plane with light shelves, re-curved blinds, or even top-controlled Venetian blinds, helps to both light spaces naturally and control glare. Some buildings use a ceramic frit pattern on the glass to control light, others a more active system of automatic louvers and shades. A well-designed task/ambient lighting system that responds to the amount of daylight can have operating lighting loads of under 0.4 watts per square foot (4.3 watts per square meter). With reduced perimeter and lighting loads, the requirements for cooling loads are also reduced.

With the smaller floor plates of tall buildings, people spend more time near the outer walls, so the radiant surface temperature of the inner pane of glass becomes an important factor in people's perception of comfort. If the surface temperature is significantly higher or lower than the indoor air temperature, then the occupants will perceive the space as being too hot or cold, independent of the actual air temperature. Adding low-emissivity (low-E) coating, gas fillings and suspended low-E films can dramatically increase the thermal performance of glazing, to the point where, in some cases, perimeter heating can be eliminated. Including thermal breaks in the glazing frames is extremely important to improving the total glazing system performance. Added insulation in the opaque portions of the building envelope will also improve the envelope's performance.

Due to the gaps at slab edges, it is generally easier to have thermal continuity of the outer skin with a curtain wall system than with infill panels. In some cases it makes sense to move to a double-skinned building, with an outer glass layer separated from an inner glass layer by a space ranging from five inches to several

12.1 Cambridge Public Library facade interior. In a double-skin wall, dynamic climate control devices can be integrated with MEP strategies to improve building energy performance, daylighting, and exterior views
Image: © William Rawn Associates

12.2 High-performance low-iron glazing units provide natural daylighting and energy efficiency, in addition to creating strong visual connections to neighboring features —Bryant Park is shown here from One Bryant Park (COOKFOX Architects)
Image: Courtesy COOKFOX Architects

feet in thickness (12.7 centimeters to a meter or more). Moving air through the space, the glass partially tempers and helps to minimize heat gain or loss in the adjoining occupied space. In an aspirated double skin, outside air comes in at the bottom of a unit and is exhausted at the top. Behnisch Partners utilizes an open double skin facade in Norddeutsche Landesbank in Hanover to direct fresh air from the courtyard to the offices while minimizing exterior noise and pollution. In harsher climates, the air would be pulled from inside the building and then exhausted to a return air plenum.

Air distribution, critical for human comfort, is a major energy consumer in buildings. In commercial buildings, high-efficiency Variable Air Volume (VAV) systems are the standard. These systems are well known and have a good track record, although they tend to require significant interstitial space to coordinate with building structure, wiring, and plumbing. Under-floor air delivery systems using raised floors as distribution plenums are becoming increasingly popular for a number of reasons, including dramatic reduction in duct work and fan boxes, individualized airflow control through the use of small operable vents at each workstation, ease of reconfiguration of wiring and spaces, lower fan power, higher delivery temperature of air allowing more use of direct outdoor air, and better air quality as the result of pollutant displacement rather than diffusion.

As some very tall buildings have different uses at different elevations, for example broad floor plates being used for offices in the lower portions, hotels in the middle, and residential for the smaller upper floor plates, it is not unusual to see more than one MEP solution used in a tower. In extremely tall buildings it is possible to experience cooling loads at the base and heating loads at the top.

Some tall buildings are exploring the use of natural ventilation when weather conditions are appropriate. In general these buildings have relatively narrow floor plates and are located in moderate climates, with examples including the RWE headquarters in Essen, Germany, the Commerzbank headquarters in Frankfurt, the Lloyd's Register of Shipping in London, and the San Francisco Federal Building in California. Integrating the facade design, mechanical systems functioning and controls, and occupant use and controllability allows for energy savings on mechanical systems greater than 30 percent in these climates.

Vertical transportation increases its percentage of energy consumption as the number of elevators and complexity of strategy increases with height. Elevators are not usually a huge energy consumer in most tall buildings. The transference of the concept of regenerative braking technology from hybrid cars has significantly lowered their energy use. At a certain point in very tall buildings (above 60 floors), however, vertical transportation becomes a significant user of energy. Elevator shafts also create pressurization changes in buildings which can give rise to challenges for air distribution and increase energy consumption.

12.3 Retrofitting older tall buildings will be one of the most important sustainability strategies in many cities. The GSW Building in Berlin is a highly successful example of adding a new high performance envelope to an existing tall structure. The south façade features occupant adjustable sliding panels that are part of a system to modulate light and airflow
Image: © Bill Browning

Efficient envelopes, daylighting, lighting systems, and space conditioning can all reduce the energy use of a tall building. An additional strategy to lower the overall carbon emissions of a building connected to a carbon-intense electrical grid is the use of thermal storage. This can be small scale—the use of phase-change dry wall to moderate temperature swings, for example. It can also be large scale, for instance the use of ice storage tanks to lower the peak cooling demand. The Bank of America Tower uses 44 ice tanks chilled at night and then melted in the afternoon to balance out the energy demands of the building. While this strategy slightly increases the net energy usage of the building, it can substantially reduce

carbon emissions as it avoids drawing power when a utility grid is peaking and using the most carbon-intensive generators.

Some tall buildings are also being used to explore the possibilities for onsite power generation. The Condé Nast Building at 4 Times Square and the Solaire in New York City were two of the first tall buildings to incorporate photovoltaic panels into the facade. In both cases the solar cells provide a very small portion of the total building energy load; however they do help to reduce the peak demand load and therefore lower energy bills. The Bank of America Tower uses a 4.7-megawatt gas-fired turbine to generate power, and then uses the waste heat internally for heating loads and to run absorption chillers. This produces significant energy cost savings.

A further step is the concept of net-zero energy buildings. These are structures that use onsite renewable energy systems to produce energy in amounts that on an annual basis exceed the amounts they consume from the electric grid or other fossil fuel sources. The proposed 52-story mixed-use Water Street Tower in Lower Manhattan, New York City, was designed in 2007 as a tall, thin building skinned in a pixellated pattern of panels of glass and solar cells. The lower third of the building would be mostly glass and contain office space, while the upper two-thirds would be increasingly covered with photovoltaic cells and contain residential units. In this near net-zero energy scheme, the solar cells would produce enough energy to power the office space during the day, and the residential units would pull power from the nighttime grid. Currently the project is on hold.

Skidmore, Owings & Merrill's Pearl River Tower (2011) in Guangzhou, China was conceived as a very large-scale experiment in building-integrated renewable energy. The building has an internally ventilated double wall to reduce the heat gain and uses radiant cooling with displacement ventilation to reduce energy consumption. Its narrow floor plates allow maximum daylighting. Photovoltaic cells are incorporated into the roof and into horizontal sunshades on the east and west facades. The building has two areas where the facade curves inward to direct air into through-building channels housing wind turbines. While the wind turbines will produce only a small portion of the building energy, the use of the through-building channels will significantly reduce the structural wind loading and thus allow a lighter, more efficient building structure.

12.4 The mixed-use Water Street Tower, planned for Lower Manhattan, utilized BIPV (building integrated photovoltaics) to achieve almost net zero energy
Image: COOKFOX Architects

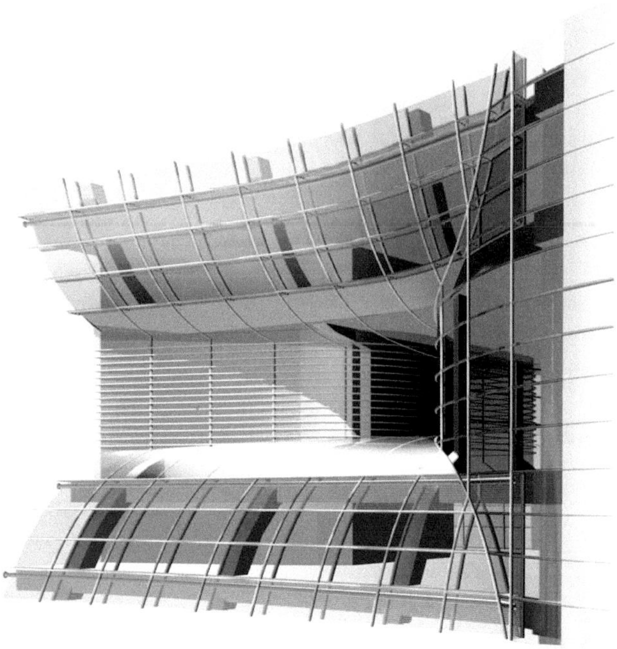

12.5 The Pearl River Tower in Guangzhou, China (Skidmore, Owings & Merrill) integrates wind turbines into two horizontal openings, harnessing the acceleration of the wind across the face of the building. The project was designed to be net-zero energy
Image: Skidmore, Owings & Merrill

Water

Frequently, water use in a tall building is thought of as a linear process: potable water comes in and is pumped to various uses, while wastewater and stormwater are discharged into drains. Potable water is essential for building occupants. While, unlike energy sources, there are no readily interchangeable substitutes, there are ways to radically rethink the water requirements of tall buildings. The best way to truly understand water in the context of any building project is to produce a dynamic water balance chart. This exercise will lay out all of the uses of water (both amounts and required qualities), sources of water, and how much can be reclaimed, captured or recaptured. Then the most efficient water-using devices can be selected to further reduce loads, and areas for non-potable use can be identified. These strategies can reduce a tall building's potable water demand by 40 to 50 percent.

In many places stormwater runoff is seen as a problem that has to be managed; instead it can be a great resource for mechanical system make-up, for toilet flushing, and for process water uses. The Bank of America Tower captures stormwater that falls on the roof and facades, and stores it in a series of staged cisterns that are placed above elevators as they drop off in the core and in the concrete foundation below the elevators. This strategy is relatively low cost, as the structure is already in place. The building also captures grey water from sinks, and groundwater from sump pumps. This groundwater, like the stormwater, is then filtered, cleaned, and stored in the cisterns.

Material and Waste Resource Efficiency

Large populations can produce large amounts of waste. Moving this stream of materials through a tall building can be a logistical nightmare. In many buildings, all of the waste is commingled. In some, recyclable materials are separated and hauled separately. Some have separate internal shafts for trash and recyclable materials while others use compactors. Selection of a system is based upon space constraints and the type of materials being managed. Typically these materials are packaged and exported from the site. One opportunity to reduce waste is to employ anaerobic digestion to turn organic and paper waste into compost, grey water, and energy in the building. Anaerobic digestion is a biological process using microorganisms to break down organic matter in the absence of oxygen.

Considerations of sustainability and materials use in tall buildings frequently focus on the carbon impacts of material production, transportation of the material to the site, energy content of the material, and its impact on indoor environmental health. Embodied energy—the amount of energy required to make a material—is an indication of its carbon footprint. Concrete is the most ubiquitous building material and accounts for 8 percent of annual global carbon dioxide emissions. Cement production is energy intensive and the curing process releases additional carbon dioxide—the production of one ton of Portland cement releases one ton of carbon dioxide. To reduce this impact, many projects are reducing the amount of Portland cement in the concrete mix by the use of pulverized fuel ash (PFA), ground granulated blast furnace slag (GGBFS), or other processed wastes, which have varying degrees of reactivity. Such materials have a lengthy track record of successful use and, if properly used, can have beneficial effects on fresh concrete properties and long term strength and/or durability. There are also many well-proven chemical admixtures that can reduce Portland cement content or improve long-term durability. By replacing part of the Portland cement in concrete with PFA, the builders of the Bank of America Tower reduced its carbon footprint by 56,250 tons.

Steel is the other major material used in tall buildings. Architectural steel frequently has a recycled material content as high as 90 percent, significantly reducing its embodied energy content. The carbon footprint of aluminum used in curtain wall construction is entirely dependent on recycled content. In a low-rise building, the weight of material has few implications; in a tall building, however, the weight has significant, compounding structural implications—heavy materials require more structure to support them. Tall buildings are typically built to higher standards and utilize more durable materials, which reduce the need for replacement during the lifetime of a building.

Thinking about the overall lifespan of buildings is important. In many buildings, the embodied energy of materials is only about 20 percent of the total energy footprint over the building lifetime; the majority of the remaining energy use is in the operation. Therefore, control systems and the operational team are critical to maintaining and improving the sustainability of the tower over time.

A final concern about materials is related to health impacts. Avoiding materials that negatively impact the health of factory workers, construction workers, and the occupants of the building is a major issue for all buildings. The off-gassing of Volatile Organic Compounds (VOCs) is usually one of the more problematic indoor health issues; this can occur both from the material itself and from the compounds used in cleaning processes. It is also important to ensure that materials and air-handling systems do not harbor microbial organisms.

Well-being

A real challenge with tall buildings is that they disconnect the occupants from the larger urban community due to their extreme height. Tall buildings can cause vertigo and sickness in some people. Occupant wellbeing is a critical component of all building projects, and tall buildings must therefore address the problem of disconnect.

The true cost of operating a commercial building is found in its people. In the United States energy costs are typically 1 percent of total salaries, benefits and personal equipment, while rent is typically 10 percent of total salaries. So a 1 percent gain in productivity is equivalent to eliminating energy bills. Starting in the 1990s, a number of case studies identified 6 to 16 percent gains in productivity in green buildings.

These gains, although financially significant, are in many ways just a placeholder for a more important issue: people's well-being. The gains in productivity emerged through implementing measures to increase energy efficiency, which had the positive side-effect of also improving the physical working environment. As part of the desire to create a healthier, more sustainable society, the question emerges of how to design buildings and places that maximize human well-being. Part of this is ensuring that toxins are prevented from entering occupied spaces. So what can be done to make the built environment the best possible place for people to live and work?

The framework for thinking about maximizing people's well-being comes from Edward O. Wilson's concept of "biophilia," or "the innately emotional affiliation of human beings to other living organisms" (Wilson 1993: 31). This emerging field of research combines work from evolutionary psychology, anthropology, archaeology, geography, neuroscience, and a number of other disciplines to assess how the built environment affects human well-being. There are now several hundred papers that document various aspects of the human need for contact with nature, and the psychological and physiological effects of this interaction. From this work, architects have realized that eliciting a biophilic response can be achieved by implementing Nature in the Space, Natural Analogs, and Nature of the Space.

Nature in the Space, or bringing plants, water, and animals into the built environment, is something that humans have done since time immemorial. There have been several tall building designs that specifically introduce natural elements: the sky gardens in Norman Foster's Commerzbank in Frankfurt, Germany; Roche Dinkerloo's Ford Foundation in New York City; and the tropical plants and hanging gardens in Ken Yeang's skyscrapers.

Natural Analogs are materials and patterns that evoke nature and can be categorized into three broad types: ornamentation, biomorphic forms, and the use of "natural materials." Cultures around the world ornament their buildings with representations of nature, such as carvings and paintings of fruits, flowers, leaves, acorns, seashells, birds, and other animals. The Finnish architect Eero Saarinen and the Spanish architect Santiago Calatrava are famous for creating explicitly biomorphic buildings that look like trees, shells, bones, and wings. Finally, natural materials, where their inherent qualities are perceptible, are another type of natural analog.

Nature of the Space deals with how people respond psychologically and physiologically to different spatial experiences. English geographer Jay Appleton was perhaps the first modern researcher to attempt to codify which elements of landscapes people found most appealing, but Nature of the Space touches on a human

12.6 Sky lobbies in Gensler's Shanghai Tower in Shanghai, China feature biophilic characteristics "nature in the space" (gardens) and "nature of the space" (prospect)
Image: Gensler

12.7 The design of the EDITT Tower combines development with ecological restoration in Singapore. This approach seeks to rehabilitate this urban site by enabling the process of ecological succession. A unique feature is the well-planned facades and vegetated terraces that extend up the entire building
Image: © 2010 T.R. Hamzah and Yeang Snd. Bhd.

legacy found in many traditional geomancy systems. The research into the spatial patterns in preferred landscapes attempts to reach a deeper, non–culturally specific understanding of spatial patterns. Today, researchers cite between seven and sixteen patterns, such as experiences of "enticement" or "peril." For example, an elevated view across a distant landscape is a highly preferred spatial pattern that Appleton called "Prospect." It is a view made still more appealing if it contains shade, trees, and water.

A contrasting but also preferred spatial pattern is what Appleton called "Refuge," in which the back is protected with shelter overhead, so that the space embraces and nurtures the occupant, as in an inglenook next to a fireplace. The biophilic experience is stronger when patterns occur together. For example as a space that has both Prospect and Refuge, as occurs when sitting under the overhanging roof on the raised porch of a bungalow. Given the nature of tall buildings, which elevate us above our surroundings, there is often a feeling of being disconnected; this may be partly mitigated through utilizing these biophilic techniques.

The Challenge Ahead

The development of tall buildings can improve the sustainability of cities as they continue to urbanize, if they maximize resource efficiency and are built adjacent to robust transportation systems, and provide healthy environments for people to work and live in the city. At the same time, existing tall buildings and other structures have to be upgraded in order to protect cultural heritage

12.8 Replacing Boston's Government Center, One Congress Street is designed to re-engage the finely grained urban fabric in order to reconnect the site and its buildings with the city's people, sense of place, and natural environment. Conceived both from the inside out and from the outside in, the project will re-knit the surrounding neighborhoods, create a recognizable icon on the skyline, and produce the highest caliber sustainable workplace in Boston
Image: COOKFOX Architects

and urban diversity while creating the sustainable city of the future. A good example of the retrofitting opportunity is the recent renovation of the iconic Empire State Building in New York City. Originally designed for natural ventilation and views, the building was drastically altered for mechanical conditioning over time. The new offices for Skanska, the Swedish construction company, are designed to return to some of the original conditions. Through a series of efforts designed by COOKFOX Architects, the project cut the energy use of the space by 49 percent, and the project achieved a US Green Building Council LEED Platinum rating. In addition, the owner of the building has recently embarked on a multi-year upgrade of the building's infrastructure and facade. The opportunity to retrofit the Empire State Building almost 80 years after its completion, together with the other examples in this chapter, gives us a sense of how tall urban buildings can become sustainable buildings.

Bibliography

Heerwagen, Judith and Orians, Gordon (1993) "Humans, Habitats and Aesthetics," in S. Kellert and E.O. Wilson (eds.) *The Biophila Hypothesis*, Washington, DC: Island Press, 141–56.

Kaplan, Rachel, Kaplan, Stephen, and Ryan, Robert (1998) *With People in Mind: Design and Management of Everyday Nature*, Washington, DC: Island Press.

Kellert, Stephen (2008) "Dimensions, Elements, and Attributes of Biophilic Design," in S. Kellert, J. Heerwagen, and M. Mador (eds.) *Biophilic Design*, New York: John Wiley & Sons, 13–15.

Wilson, E.O. (1993) "Biophilia and the Conservation Ethic," in S.R. Kellert and E.O. Wilson (eds.) *The Biophilia Hypothesis*, Washington, DC: Island Press, 31–41.

Chapter 13

Towards Energy Independence

Russell Gilchrist, Roger Frechette and Dave Parker

A supertall building is often referred to as a city unto itself. This vertical assembly of steel, concrete and glass houses many of the same activities that one might find in any other small city. For a tall building to function, energy must be provided, otherwise business cannot happen, society cannot flourish, life cannot be sustained. Currently, the world's tall buildings are, probably without exception, linked to national grids supplying gas, electricity and potable water, and thus are major, if indirect, consumers of fossil fuels. Even without the potential for climate change triggered by increasing atmospheric carbon dioxide levels, building owners are soon likely to face the impact of soaring fossil fuel prices. Architects and engineers are being pushed to deliver ever-more energy-efficient buildings, to minimise dependency on fossil fuels and to significantly reduce the carbon footprints of their projects. Achieving these goals requires a holistic approach.

For the purposes of this chapter, a high-performance tower (HPT) is defined as a structure that demands a significantly lower input of fossil fuels during its construction, operation, maintenance and final demolition. It is suggested that the quasi-mathematical formula HP = R^2AG gives a guide to the essential elements of the design process, where

- R = Reduction – *Reduce the HPT's appetite for consumption*
- R = Reclamation – *Once energy is introduced, re-use it over and over*
- A = Absorption – *Tap into the natural energy streams making contact with the HPT*
- G = Generation – *Turn the HPT into a highly efficient power plant*

Reduction

Perhaps the most straightforward and obvious approach to saving energy in a tall building design is to reduce the building's 'appetite for consumption'. There are many strategies and technologies that can now be used to achieve very significant reductions in both power and potable water demand, compared to requirements in the past. This general approach is referred to as 'reduction', and is covered in more detail in Chapter 12.

Reclamation

Tall buildings require a significant amount of energy to operate on a daily basis. The principle of reclamation is simple: once energy has been injected into the building, that energy should be captured and re-used as often as possible. Every electron or water molecule recaptured reduces the amount of overall consumption of 'new' energy. The impact on energy consumed, carbon emitted and utility costs can be substantial. A few examples of reclamation strategies follow.

Rainwater reclamation systems

In many environments, implementing rainwater reclamation systems could save a significant amount of water on an annual basis – up to 50 per cent. Without such facilities these water streams are usually combined with other, contaminated waste-stream sources prior to entering a municipal treatment facility. Treatment is then required for all of the contaminated water in order to provide potable water to consumers. However, most buildings today use potable drinking water for make-up water for cooling towers, landscaping irrigation and many other applications for which potable quality is irrelevant. Significant potable water could be saved by substituting rainwater captured from the new impervious areas on a site before it is mixed with more contaminated wastewater streams. Collected water is first processed through a stormwater management filtration system and transferred to a large storage tank.

Grey water recovery systems

Wastewater from the building's sinks, showers and laundry systems is normally known as 'grey water'. Such water can be treated and stored, then used for toilet flushing, cooling tower water make-up and site irrigation. It is not, however, recommended that biological grey water purification techniques be utilized in a healthcare facility. Due to the required high level of sanitation, healthcare facilities contribute a large amount of antibacterial soaps and cleansers to their wastewater systems.

Mixed-use tall buildings offer water consumption and grey water production profiles that can be balanced. For example, in mixed-use buildings, large quantities of grey water can be captured from the residential portion of the building, where grey water production is high. This recovered water can then be transferred to the office building cooling towers, where water consumption is very high and potable water use is not necessary.

Some buildings with deep basements rely on sump pumps to keep the basement dry. Groundwater collected in this way can also be stored after passing through the stormwater filtration system.

Cooling coil condensate recovery

Tall towers require very large quantities of ventilation air from the outside. Usually, this air must be cooled and/or dehumidified before it is suitable for use in the building. This process can yield significant quantities of high-quality condensate water, which can be extracted from the building's ventilation system cooling coils and collected in the same storage as reclaimed rainwater. There will normally be no need for the condensate water to pass through the initial water treatment filtration process and it can be discharged directly to the grey water storage volume.

Heat recovery

In the course of normal building operations, internal air and water absorb energy, and much of this energy is lost as the air and water are discharged back into the environment after use. Reclaiming a significant proportion of this normally wasted energy is relatively straightforward, proven technology is widely available, and such reclamation can be economically beneficial in the long term.

Outside air is brought into the building for ventilation proper and building pressurization, and internal 'used' air is exhausted to the outdoors to remove contaminants. Incoming air needs to be heated or cooled in order to maintain a comfortable indoor environment. An air-to-air energy recovery system takes the surplus energy from one airstream and transfers it to the other. Thus the exhaust air preconditions the incoming air, reducing the need for energy-hungry conditioning by other means. Enthalpy wheels – basically rotary air-to-air heat exchangers – can be integrated into most 100 per cent outside air handling units. Plate-type or heat-pipe energy recovery devices can be utilized when the exhaust air stream is likely to have an unusually high moisture content or high level of contaminants.

Grey water in particular usually contains significant amounts of energy, which will raise the temperature of any wastewater storage. This could encourage unwelcome bacterial growth, including the infamous legionella bacillus. Passing the grey water through a simple heat exchanger before discharging it either to waste or into store will extract useful energy that can at the very least help reduce domestic hot water energy demand.

Elevator regenerative braking systems

Traditionally, an elevator car is linked to a counterweight via a pulley system. An ascending car full of passengers obviously needs power to raise the car, usually electricity. A full car descending also needs energy to control the speed of descent and brake it at the chosen floor. In a typical non-regenerative drive, energy is dissipated as heat in a set of resistors when braking occurs, resulting in reduced efficiency and creating additional waste-heat loads in the building.

Conversely for regenerative drives, electrical power is generated when the elevator travels up with a light load, travels down with a heavy load, or during elevator deceleration. This electricity is returned to the building's power grid to be reused for such things as lighting and heating. Such a recycling system can reduce the overall elevator energy usage by up to 70 per cent compared to systems with non-regenerative drives, lowering overall building operating costs and delivering significant annual savings to building owners and tenants.

Absorption

Tall and supertall buildings generally stand well clear of their neighbours and are open to higher, more consistent wind flows and solar exposure than 'groundscrapers'. High wind loads and solar gain have usually been seen as negative factors, requiring elaborate precautions to minimise their effects on the building and its occupants. This is starting to change, however, as clients and designers begin to appreciate that wind and sun are energy streams that can be 'absorbed' by the building using a range of well-proven technologies. Aligning the building to ensure maximum beneficial exposure to these energy streams is the first and possibly most crucial step. A choice of technologies is then available.

Mixed-mode ventilation

Heating, ventilation and cooling (HVAC) typically accounts for over 50 per cent of the energy demand in buildings, thus the potential benefits of natural ventilation can be quite substantial. A dramatic reduction in the initial and lifetime costs of the mechanical cooling system can be achieved. Windcatchers and their cousins are beginning to appear on many tall building projects; however, natural ventilation is unlikely to be capable of providing all the heating and cooling required in a building. A 'mixed-mode' combination of natural and mechanical ventilation is usually the most practical alternative.

Geothermal systems

Geothermal energy was originally defined as heat from the earth's core. It is a clean and virtually limitless resource, but unfortunately there are only a few locations around the world where this energy can be tapped directly, mostly close to the edge of tectonic plates where hot volcanic rocks can be found relatively close to the surface. Deep boreholes are drilled down to the hot spots, fluid circulated and electricity and district heating generated. Few design teams will be fortunate enough to have this as an option; a more realistic alternative is the heat that may be extracted from a few metres below the surface. This is almost entirely solar energy absorbed by the earth, and temperatures are low but very stable.

As a rule of thumb, annual average soil temperature can be taken as annual average air temperature, and generally lies in the range 7 to 21 degrees Celsius. An important factor is that the shallow soil temperature lags behind air temperature to a degree that is a function of depth, and will generally be higher than the ambient air temperature in winter and lower in summer. This lag is more exploitable at depths below 3 metres or so, where peak soil temperatures occur in October or November in the northern hemisphere. Subsoil properties and the movement of groundwater influence how much energy can be extracted across the site.

Soil temperatures are normally too low for heating – although they can easily be utilised for summertime cooling – but there is still a lot of energy in the soil, which can be extracted and 'concentrated' by a heat pump. Heat pumps can be thought of as 'refrigerators in reverse', circulating a fluid through buried pipework known as 'ground loops' then extracting the energy from the fluid and transferring it to the space heating or domestic hot water system. In practice, most modern heat pumps use 1 kilowatt of energy to extract at least 3 kilowatts of energy from the ground or other sources, such as ponds, lakes or even underground aquifers.

A less aesthetically appealing if initially cheaper option is air-to-air or air-source heat pumps, which have to be mounted on exterior walls. These are most effective in milder climates where long periods of sub-zero temperatures are rare.

Heat pumps can usually be run in reverse as coolers, either discharging waste heat back to the atmosphere or, in more advanced applications, transferring it back into the ground or water to be reused later in

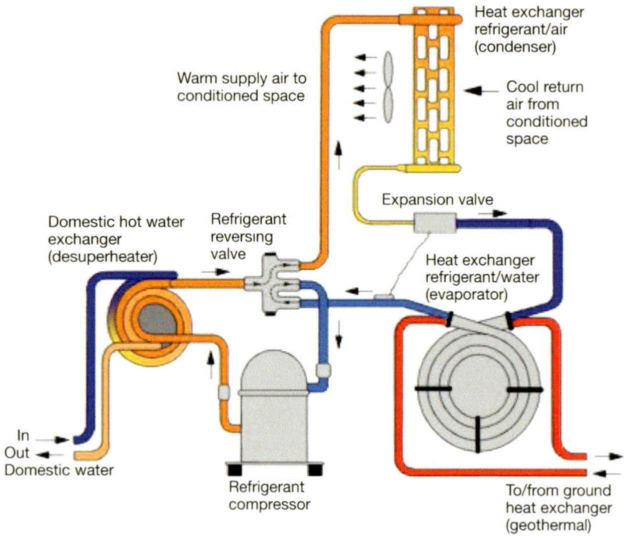

13.1 Heat pumps collect and concentrate heat from the ground, the air, or nearby water
Image: Oklahoma State University

13.2 Once famous for the energy-saving gold coating to its glazing – gold worth more than $1 million at the time was used – Toronto's Royal Bank Plaza (1979) is now cooled by the waters of Lake Ontario

the year (see the subsection on 'Energy stores'). The vast majority of heat pumps are electric powered; absorption heat pumps, by contrast, are powered by heat, which can come from the sun or from the building's own waste heat. Efficiency is lower than conventional vapour-compression heat pumps but maintenance costs tend to be lower as well.

Tall buildings usually have relatively modest footprints and sit on restricted sites. This limits opportunities for straightforward shallow ground loops, and deeper boreholes have to be considered, or perhaps the incorporation of pipework into deep building foundation piles. Opportunities may still exist, however: unshaded parking areas are one possibility, and the potential of any aquifers or other bodies of water below or close to the building are often worth investigating, if only for summertime cooling.

Solar concentrators

Thermal energy from the sun has been used to improve the internal environment of buildings since antiquity. Dedicated solar concentrators were first used for this purpose in nineteenth-century Sweden, and today there are several well-developed technologies available to collect and utilise solar thermal energy. Collected energy can be used for space and water heating, to drive absorption chillers or stored for later use (see the subsection on 'Energy stores').

A typical tall or supertall tower can make little use of some types of solar collector, due to its relative lack of otherwise unused horizontal or pitched areas. A tower's flat roof might just be able to hold enough evacuated tube or compound parabolic concentrators to produce enough high temperature water to drive an absorption chiller, but for space and hot water heating the only realistic option in most cases is the facade.

Facade-integrated solar thermal collectors are available in a wide range of sizes, finishes and levels of complexity. Some absorb the solar heat in a fluid, others produce hot air. In summer, such collectors also have an important shading function. Surplus heat may be discharged back to the atmosphere or transferred to an energy store (see the subsection on 'Energy stores). Increasingly, the pumps and fans in such systems are powered directly by photovoltaic arrays (see the subsection on 'Photovoltaics'), which again helps to minimise the building's overall grid dependency.

For maximum efficiency, all forms of solar collectors need to be aligned with the sun, and there are a

13.3 Solar thermal collectors are now available as facade-mounted systems
Image: © Sonnenkraft

number of systems on the market in which the collectors are motorised and follow the sun's path across the sky throughout the year. This greater efficiency, of course, comes at the cost of higher capital and maintenance costs and potentially lower reliability. Where the option of such alignment is available, the usual compromise north of the equator is to install the collectors at the angle of latitude; the further from the equator the steeper the angle of the panel. This poses some challenges in locations near the equator, where the collectors should be aligned close to the horizontal. Set against this, of course, is the greater solar energy potential at these latitudes, which can compensate for any loss of collector efficiency.

Generation

National and local electricity grids are convenient but not particularly efficient. Up to 67 per cent of the energy in fossil fuels burned in power stations is lost during generation and transmission. There was a time when power stations were relatively small and located close to or actually within the cities they supplied. Not only did this reduce transmission losses, it also gave the opportunity for the waste heat from the generators to be tapped for district heating schemes. London's iconic Battersea Power Station actually supplied heating water to a housing estate on the opposite bank of the River Thames. But as power stations grew much larger in the pursuit of greater thermal efficiency they moved out into the countryside, and district heating opportunities disappeared.

Producing both power and useable heat is known as 'cogeneration', and there is increasing potential for individual buildings – or groups of buildings – to go down this route. Many of the options involve the more efficient burning of fossil fuels or hydrocarbon-rich wastes. Again, some of these options are less attractive in the case of tall buildings due to the space requirements. However, with relatively large facade areas that could be used to collect solar energy and the possibility of siting wind turbines at height, tall buildings do offer many opportunities. Microgeneration of all types will naturally increase initial building costs, but with soaring demand for fossil fuels pushing up prices and anxieties about the security of supply growing, the long-term benefits of microgeneration are looking ever more attractive.

Governments throughout the world are now offering a range of financial incentives to encourage building owners to invest in microgeneration technologies. Which technology or, more likely, combination of technologies is the most appropriate for any particular project is very dependent on its location and function. In a few cases there may be the opportunity to tap into a nearby watercourse and set up a small-scale hydropower project. This has the advantage of reasonably predictable energy output unaffected by diurnal variations, although seasonal variations are likely, but few projects will be lucky enough to have this option. More widely available techniques are described here.

Cogeneration, trigeneration and beyond

Producing both usable heat and electric power from one installation is dubbed 'cogeneration'; using the heat for both space heating in winter and to drive an absorption chiller in summer is known as 'trigeneration'. Some installations also yield significant amounts of high-temperature steam which can be harnessed for industrial purposes, and the term 'quadgeneration' is coming into use for such systems.

Most installations are powered by some form of hydrocarbon fuel, most commonly natural gas. Many can also run on biogases produced from various types of hydrocarbon-rich wastes by such processes as anaerobic digestion and pyrolisation. These latter processes are unlikely to be a viable option for the majority of tall buildings due to the storage space required and the need for frequent deliveries. The same applies to the straightforward combustion of woodchips and the like, local, sustainable sources of which are often hard to find.

Whatever the fuel source, it can be burned in modern high-performance steam engines, conventional internal combustion engines, the smooth-running

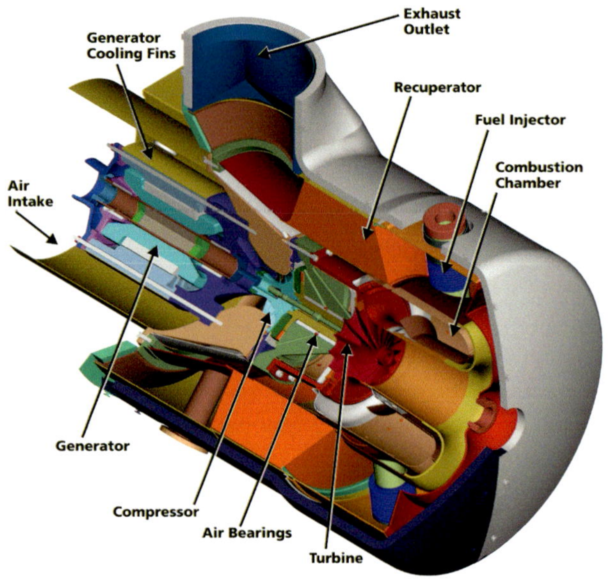

13.4 Microturbines are at the heart of several packaged CHP systems
Image: Capstone Turbine Corporation

external combustion constant-speed Stirling engine in any one of its several variants, or the new generation of high-speed microturbines. All drive conventional generators, while the waste heat from the installation is collected for re-use.

Although potential efficiencies are high, there is one intrinsic characteristic that has to be taken into account. There will be a fixed relationship between power and heat outputs, which means that there will frequently be moments when one or the other is in surplus. Unwanted electricity can be 'stored' on the grid; surplus heat can be put to other uses or transferred to an energy store (see the subsection on 'Energy stores').

A potential alternative to microturbines and the like is the fuel cell, of which several types are becoming available. These take hydrogen-rich gases and combine them with oxygen from the air in a 'reverse electrolysis' process that yields both electric power and heat. Early fuel cell designs demanded very pure hydrogen and operated at temperatures up to 1,000 degrees Celsius, but current offerings are somewhat more tolerant and user friendly. Fuel cell packages running on natural gas and biogases are now available. Historically, the impetus for fuel cell development has come from the aerospace and defence industries, and as a result the technology is probably not yet ready for practical use in a typical building.

Photovoltaics

There is a seductive simplicity about solar photovoltaic (PV) technology. A classic PV installation has no moving parts, is silent and unobtrusive and needs little maintenance. It produces low-voltage direct current electricity, normally rectified to yield high-voltage alternating current, and significant quantities of heat. In fact, electricity output is in inverse proportion to the temperature of the photovoltaic elements, so cooling is essential. If cogeneration is not the objective, passive convective cooling with the waste heat discharged to atmosphere is usually sufficient.

The vast majority of PV options currently available use silicon semiconductor technology. More efficient materials based on rare and often toxic metals such as gallium, cadmium and indium are available at a price, but questions regarding the sustainability of such materials in the long term are still not yet fully answered. An alternative is to utilise less efficient, less toxic but significantly cheaper materials which can be economically used on a larger scale – there are even PV roofing membranes now available which can transform an entire roof into a PV array.

Crystalline silicon PV cells are made up into rigid modules; modules are assembled into panels and arrays. These have a distinctive appearance that may not be to everyone's taste. An alternative is offered by semi-translucent and 'transparent' PV cells that can be bonded onto glazing, although as the heat they generate would cause problems to the building's internal environment if used on a single-skin facade their logical home would seem to be on the outer skin of a double-skin facade.

Orientation of PV arrays follows the same rules as solar thermal collectors. Efficiency likewise increases if the panels can track the sun's passage across the sky. Two alternatives to large-scale tracking are to use mirrors and lenses to concentrate the sun's rays from a wide range of angles, or to motorise individual cells to track the sun. At least two U.S. companies, SolFocus and Solaire, offer PV panels based on these technologies.

Solar PV electricity production fits best with the daytime demands of office towers, while the heat it yields is in most demand from residential accommodation at night. To balance these conflicting demands and to deal with the usual solar energy problems of intermittency, some form of energy store will usually be necessary (see below). To meet, say, 10 per cent of a typical office building's annual needs, every 1,000 square metres of floor space would need at least 80 square metres of standard silicon cells.

One of the few tall buildings to invest in large-scale PV to date is the 118-metre Co-operative Insurance

Tower in Manchester, England. Built in 1962, its original cladding was a mosaic made up of four million individual tesserae, which had begun to fall off within six months of the building's completion. In 2005 it was reclad at a cost of £5.5 million with 7,244 integrated silicon solar panels, some of which were lookalike 'dummies' for shaded areas of the facade. Each of the near 5,000 active panels is rated at 80 watts, giving a potential peak output close to 400 kilowatts, and, despite early teething troubles, it is on track to produce around 180,000 kilowatt hours annually. This is still the largest facade-integrated PV array in Europe.

London's Heron Tower also features a major PV installation (see Case Study 17). Located on the southern facade, the 3,374 square metres of silicon arrays are predicted to reduce the building's overall carbon

13.5 London's Heron Tower features a facade-mounted solar PV array
Image: Kohn Pedersen Fox

footprint by 2.2 per cent annually. Other buildings with similar concepts are beginning to roll off the drawing boards, spurred on by increasingly beneficial government subsidies.

Wind turbines

Wind energy offers many potential advantages, and is currently the fastest growing energy source in the world. Part of this popularity is down to the perception of wind energy as a clean, emission-free and sustainable technology. It is also true that, compared to other alternative energy sources, wind power is a mature technology that requires relatively low capital investment. There are several practical questions to be addressed, not least that of the intermittency of wind-derived electricity, and there are vocal groups in many countries lobbying against wind farms on the grounds of their visual impact on the environment, the risk they represent to migrating birds and bats and the need to keep fossil fuel-burning generators online to compensate for low wind conditions. A combination of a tall building and wind turbines, however, could answer many of these objections.

Relatively small horizontal axis wind turbines mounted on the roofs of low- to medium-rise buildings in an urban environment have proved to be ineffective and uneconomic. The turbulent low-level urban airflow is the main factor in this poor performance; it is also true that in wind turbine terms, size is everything. Wind farms usually locate their massive turbines – now towers reaching 70 metres with blades more than 50 metres in diameter, each producing more than 5 megawatts – on open moorland or out at sea, where wind patterns are more stable and predictable. Tall buildings experience similar wind patterns, so the potential to tap into this free, 'inexhaustible' energy stream is high. (Although doubts have recently been expressed on just how much energy can be extracted from wind globally before wind and weather patterns are adversely affected; Miller, Gans and Kleidon, 2010.)

Simply bolting a wind farm-size horizontal axis turbine onto the roof of a tall building is unlikely to be a practical option. The loads fed into the building's structure will be very high and vibration could also be a major problem. An array of smaller turbines would be a more realistic alternative, although vibration would still have to be considered.

A less well-known but almost as well-developed alternative is the vertical axis wind turbine. This has the advantage of siting the heavy generator at low level within the equipment, simplifying maintenance and reducing structural weight. Several types have been developed. Two inherent disadvantages have reduced their efficiency compared to horizontal axis designs: for part of their revolution the blades are travelling against the wind to some degree, and they are generally much lower and thus operate in the more turbulent airflows near the ground. The latter is likely to be less of a factor on the roof of a tall building.

On a positive note, loads fed into the structure will be significantly lower from vertical axis turbines, as will vibration, as vertical axis machines do not suffer from the asymmetrical blade loading that affects their horizontal axis cousins. Some manufacturers now offer a third type, basically a classic spiral-bladed vertical axis design mounted horizontally. These are mainly intended for roofs and parapets, although some are recommended for vertical mounting on building corners at high level.

Perhaps the most interesting development in recent years is the building-integrated wind turbine. Such designs can involve shaping the roof to accelerate prevailing winds through an array of relatively small turbines mounted in ducts, or shaping the entire building to produce the same effect on fewer, larger turbines. One recent example of this latter approach is the Pearl River Tower in Guangzhou in China. The building's shape optimises the pressure differential between the windward and leeward sides of the building to enhance the wind speed through the two aerodynamically shaped tunnels located at the mechanical floors of the building. Each tunnel contains two vertical axis turbines, which take advantage of the increased air velocity to generate significant power.

In London, the distinctive profile of the 43-storey residential Strata Tower is already a local landmark. Three high-level integrated wind turbines are expected to generate around 50 megawatt hours annually, around 8 per cent of the building's total energy needs. Architects BFLS say independent analysis by environmental consultants IES has confirmed that the building will eventually achieve more than a 70 per cent reduction in CO_2 emissions compared to current UK building regulations.

Wind power's most publicised downside, its intermittent and unpredictable output, is less of a problem for buildings connected to national or local energy grids. Short term surpluses can, with the cooperation of the utility supplier, be 'stored' on the grid, and any shortfall made up from the grid. Various metering arrangements of varying generosity exist throughout the world.

A more practical objection is the sheer size of installation needed to generate only a fraction of the building's energy demands. Estimates vary, but one rule of thumb is that a modern high-specification office space will consume around 200 kilowatt hours per annum per square metre. To supply, say, 20 per cent of that figure from wind energy requires an installation with an annual output of 40 kilowatt hours per square metre, which

13.6 Integrated wind turbines are expected to generate 8 per cent of the energy needs of London's Strata SE1
Image: Yui Law/BFLS

implies that in practice every 1,000 square metres of office space would require something like a 20-metre diameter horizontal axis wind turbine mounted on a 30-metre high tower. This would be rated at 100 kilowatts. Building-integrated turbines would have to be even larger due to their fixed orientation.

Energy stores

Wind and solar power, in particular, suffer from intermittency and unpredictability of supply. Electricity may be 'stored' on the grid, but if the grid goes down this stored electricity is unavailable. Heat is often in embarrassing surplus during summer and in deficit during the cold winter months. Storing both electricity and heat within the building is now becoming a much more practical option, especially if the storage facilities are designed and constructed as part of the normal building programme.

Fossil-fuel backup generators are often installed as insurance against grid failures. These still have their place, but new generations of flow batteries, cryogenic and flywheel energy storage devices and unitised regenerative fuel cells fed by solar PV panels and wind turbines should one day offer a greater degree of independence from both fossil fuels and electricity grids. Such independence will come at a price, however: where the electricity surpluses are expected to be modest and storage on the grid is impractical or financially unattractive, the simplest option is to convert the surplus into heat and store it in a thermal store.

Many types of thermal store are in use all over the world, ranging from high-temperature water tanks to low-temperature underground pits, and including ever more efficient use of phase change materials. Some applications are basically ground- or water-source heat pumps running in reverse, transferring heat out of the building to be stored for later retrieval by reversing the heat pumps. Diurnal stores take in heat energy during the day and release it at night, intermediate stores hold enough energy to compensate for several windless and sunless days, and seasonal storage conserves summertime heat for winter use.

Adding effective thermal storage to a HVAC system increases the efficiency of most alternative energy systems significantly. One option is to increase the building's thermal mass, either by the use of concrete structural elements or phase change materials. This is only diurnal storage: adding a basement-located low-temperature intermediate water store constructed from waterproof, lightweight concrete will increase performance. This could sit alongside stormwater or grey water storage. High-temperature water storage or phase change storage are other options. Such a store could receive inputs from solar thermal panels, PV panel cooling, grey-water heat exchangers and the like, and help to both minimise the building's overall energy demand and smooth out the variations in input and output.

Conclusion

Few buildings, even high-performance towers, will ever be totally free from dependence on national electricity and potable water grids, and hence they will be significant consumers of fossil fuels for the foreseeable future. By their very nature, however, tall buildings offer designers unique opportunities for minimising this dependency and achieving a significant and reassuring degree of self-sufficiency. Yes, some projects much lauded for their pioneering 'sustainable' designs are in practice only examples of cynical 'greenwashing', and making more than a token gesture towards lower carbon footprints is a challenging task, but the route is well charted. Tall buildings sit on readily available energy and are bathed in readily available energy, and relatively well-developed technologies are available to exploit this energy. In addition, due to the greater financial and professional investment in a tall building, more experimental sustainable technologies can perhaps be justified, which could have benefits for the entire built environment. Ignoring these opportunities is no longer a sensible option.

References

Miller, L.M., Gans, F. and Kleidon, A. (2010) 'Estimating maximum global land surface wind power extractability and associated climatic consequences', *Earth Syst. Dynam. Discuss.*, 1, 169–89.

Chapter 14
Recent Advances in Technologies, Techniques and Materials

Richard Marshall and Graham Knapp

Introduction

Tall buildings have traditionally been associated with innovation, the introduction of elevators sparking a change in what was possible in terms of the number of occupied building floors. This then led to progressive challenges to the prevailing methods of construction, and a move away from load-bearing masonry to steel-framed buildings.

In recent years there has been a fundamental shift in the way in which projects are designed, the use of more sophisticated analytical and drawing methods meaning that ever more complex structures – which previously would have been impractical to design, let alone construct – have been built, including inclined and twisting structures.

Changes in design approaches to tall buildings have been mirrored by changes in construction technology. In particular, more complex 3D forms can be fabricated through the extensive use of 3D modelling and detailing software plus automated manufacturing methods. As the number of super-high-rise buildings that have been constructed increases, our collective knowledge of what works and of the performance of the buildings is changing. With an increased focus on sustainable design, it is clear that understanding the performance of constructed projects from an energy or material use standpoint is essential in order to provide a basis for meaningful targets and standards for future designs.

Such innovations in design technology are one aspect of the changing situation; another is the availability of a variety of materials – higher strength concretes, composite materials, different modular systems and 'new materials'.

It seems certain that the future will be mediated by a more efficient utilisation of materials in construction, achieved through greater understanding on the 'demand' side and also greater understanding of building behaviour and the improved (or variable) properties of materials. The most ambitious currently conceptualised 'tall building' must be the so-called 'space elevator' being studied by NASA, which would be constructed from carbon nanotubes and which, at approximately 36,000,000 metres in height, would be around 45,000 times as high as the Burj Khalifa, the current 'world's tallest'. This confirms that there is still significant scope for continuing innovation and development for tall buildings.

This chapter looks at three different aspects of new materials and innovations in the design of tall buildings:

14.1 Improved 3D modelling enables complex forms to be realised – GCC Bank, King Abdullah Financial District, Riyadh
Images: Courtesy Gensler and Buro Happold

first, those innovations related to design technology (such as new analysis methods); second, building technology (effectively changes in design approaches); and third, construction technology and new materials (changes in choice of materials and how to erect and construct tall buildings). Finally, some consideration is given to what the future may hold for tall buildings.

Design Technology

The information revolution that has characterised changes in society in general in recent decades has also had an (admittedly slow and cautious) impact on how tall buildings are designed. Changes in design of the structures of tall buildings can be differentiated into changes which influence the 'demand' side of the equation – for example designing for different or lower forces, or the use of better understanding of stiffness and mass requirements to achieve acceptable levels of human comfort – and those which influence the 'capacity' side – such as more sophisticated tools of analysis and concepts to deal with nonlinear behaviour (elastic design being a key underlying presumption in much of the earlier work carried out by structural engineers for tall buildings). Some specific examples follow.

Nonlinear analysis, including staged construction and pushover

Traditionally when engineers have analysed tall buildings they have relied upon linear elastic analysis models – effectively 'wished-into-place' structures. It has been recognised for decades that this has an impact on the accuracy of analyses of particularly differential movements in structures. This has been considered mainly a 'serviceability' issue, classically in the case of columns shortening more than cores, causing floors to be 'out of tolerance'. Approaches to calculating vertical shortening (particularly with concrete structures, which are of course subject to temporal nonlinearity, that is, shrinkage and creep effects, which continue to change the geometry of the structure over time), have been developed and used in projects. There are two particular instances in which the previously used methods have proven to be inadequate. First, where different 'vertical' elements are connected together (outriggers in particular, but also link beams, transfer beams, etc.) the forces induced in elements due to 'internal restraint' become significant. The second instance occurs where there is a significant 'out-of-balance' net force acting on the building (twisting or leaning buildings, or 'cut-backs', for example), with this being the more challenging of the two to predict and compensate for.

A recent advance has been the introduction of a 'staged construction' function in one of the commonly used analytical programmes (ETABS), which means that at least part of the time variation in terms of building displacements and consequently forces can be studied with greater ease. This does not allow for the longer-term displacements related to shrinkage and creep, therefore for concrete structures there is a need for engineering judgement to be applied to the problem formulation part of the analysis.

In addition to these challenging 'static' issues there is also the need to consider the dynamic analysis of buildings, in particular seismic design. Recent earthquakes in New Zealand and Japan emphasise that even in seismically active locations with long-established seismic engineering traditions and codes which explicitly allow for seismic loads it is still difficult to design buildings which can resist higher levels of seismic accelerations than have been envisaged based on past experience. One method that has been developed to deal with this is the concept of a 'pushover' analysis – effectively an iterative analysis that progressively adds lateral load to a structure until it forms a 'mechanism' and 'collapses'.

More detailed loading information, such as wind tunnel tests and site-specific wind speed, and site-specific seismic hazard studies

Traditional approaches to determining the magnitude of wind and seismic forces acting on tall buildings have been based upon relatively limited, small-scale modelling of wind effects, which have been used to develop codes of practice and standards for 'typical' buildings. It has been recognised for decades that in order to design efficient and safe tall building structures it is necessary to understand the level of wind loads to a greater degree of accuracy than is achievable with simple code approaches.

The traditional approach, informed by aeronautical and marine engineering, has been to use wind tunnel tests to determine the 'drag' on a building, assuming that it is effectively a 'rigid' body. A dynamic portion of loads has then been added numerically, based upon the dynamic properties determined by the analysis of the building structure by the engineer. Alternative approaches have been to use pressure tap models to determine the static pressures and then add the dynamic portions numerically; to develop aero-elastic models, which effectively model, in a simplified manner, the dynamic behaviour of buildings so that the dynamic component of the overall load can be recorded directly; or to use numeric methods. These have typically not been considered reliable or fast enough for real design purposes. The accuracy of the wind loads determined is being improved through the use of better analytical methods to 'add' the dynamic portions of the wind loads and by taking a more detailed look at the wind speed applicable to a particular site.

A huge increase in the number of bespoke wind tunnel tests carried out has been seen over the last two decades. This has allowed the major wind tunnel centres to invest in more sophisticated measurement equipment and model-building facilities. Rapid prototyping of models is becoming routine, linked to the increase in 3D modelling described previously, and processes are becoming increasingly streamlined.

Increases in computer power allow these centres to process pressure tap results from hundreds or thousands of pressure taps, making full allowance for the time-varying pressures across the building. This is also likely to lead in time to more sophisticated time-domain analysis of structural performance. As buildings get taller, scaling effects become more significant and larger scale models are needed to validate the results of small-scale tests.

Another trend is to look at a site-specific seismic hazard assessment, particularly for super-high-rise buildings and for buildings in locations where the local codes are 'older' and typically either inaccurate or overly conservative. Because in areas of high seismic risk the seismic loads dominate the design of building structures, in particular the detailed requirements for structures designed to behave in a ductile manner, it is efficient to look at what the appropriate site design spectra are, rather than use simplified peak ground accelerations.

Performance-based design based upon broader understanding of comfort and displacement criteria

The observation has been made that projects such as the twin towers of the World Trade Center in New York were designed using a performance-based approach to wind engineering, since at the time the entirety of the project lay outside of realms of empirical methods and standards and codes – it was necessary to consider what the actual limiting criteria (accelerations, sway, interstorey drift, elevator rope movements) were. However this approach has not been codified.

Performance-based design under seismic loading has been developed successfully over the last twenty years. Essentially it involves setting performance criteria (such as no perceivable damage, no damage to structural elements and no collapse) and then return periods for which an event is expected to be able to meet the performance target. This is in essence a development of the limit state concept of both serviceability and ultimate to cover further 'limit states', which was in its own way a development of the early static, elastic methods of considering design. This is currently being extended to the development of performance-based wind design. Rather than working to a 50-year return period, wind speed is used as a basis for the development of both serviceability and (factored) ultimate state stresses in the structure, meaning a variety of different return periods are used for different wind speeds (and hence wind loads) in order to assess the behaviour of the structure against other criteria.

This is coupled with advances in understanding the basis of perception of motion and the limits for building motion which should be prescribed by code. Typically, the aim was to limit the lateral sways under loads to a level at which there is no damage to non-structural building elements (such as cladding, partitions and finishes), whether achieved via limiting overall drift (to total building height divided by 500, for example) or via limiting the inter-storey drift (a less conservative limit). This then changed to limiting the raking component of inter-storey drift (recognising that the cladding and partitions are damaged by local displacements and not those which have occurred at levels below). This has effectively meant a relaxation of the deflection limits, which has made the limits on acceleration and torsional velocity – which are intended to limit movements to a level where an acceptable percentage of a population will object for a certain return period (for example 2 per cent of people objecting in a ten-year return period). Recent developments include a clearer understanding of how we perceive motion when we are sitting, lying down or working, as well as an appreciation of how we react to different frequencies of motion. Research is ongoing, with, for example, Rowan Williams Davies and Irwin (RWDI) carrying out motion studies on a testing array in Newfoundland that was developed for naval architecture in order to demonstrate the variation in the ability of individuals to perceive motion. Code limits are currently based on limiting accelerations, but research suggests that the rate of change of acceleration – 'jerk' – is more relevant; however limits for and practical approaches for estimating jerk are not currently in place.

These two avenues are certain to lead to more efficient designs, as they provide a more 'tuneable' basis for tall building structural design.

Distributed damping for ULS design

The traditional approach to the design of tall buildings was to limit the overall sway under wind loads to a particular ratio of height (overall height divided by 500, for example). This has changed with the advent of taller, lighter forms of construction and the realisation that perception of motion is more critical, and more related to accelerations under lateral induced building motions due to wind loading.

The criteria by which buildings are designed have been adjusted over time, moving away from, say, a ten-year return period acceleration limit in milli-g for different usages to a number of different acceleration criteria at different return periods. In order to control

14.2 Visco-elastic damping distributed across multiple locations enabling redundancy and hence reliability for ULS design
Images: Courtesy Buro Happold

building accelerations it has generally been a case of allowing for a form of applied damping, typically a Tuned Mass Damper (TMD), although various other forms of damping devices (including Tuned Fluid Dampers and Slosh Dampers) have been used. These devices have tended to be in one location on the building – since the first type has generally dominated, a mass at the top of the building 'swinging' to oppose the building sway is most effective. Due to the lack of redundancy and reliance on maintenance to ensure the devices work, it has been common practice to design these devices to reduce accelerations and amplitude of motion at the serviceability limit state (SLS) but still to design for the strength of Ultimate Limit State (ULS) without considering the influence of the additional applied damping.

More recently, consideration has been given to various forms of distributed damping systems, for example incorporating visco-elastic dampers in diagrid braces and beam elements, and fluid-viscous damped outriggers (see Chapter 16). In order to use this sort of approach it is important to consider redundancy, reliability, maintenance, differential shortening, performance under seismic loading and fire protection. This appears to be a very useful conceptual approach to the 'demand' side of the equation for tall building design. It is likely that the recent proliferation of different systems will be followed by some refinement and rationalisation in coming years.

Computational wind engineering

Computational Fluid Dynamics (CFD) methods are increasingly used in the early stages of building design to test ideas, illustrate key effects and improve understanding of wind flow around tall buildings. These techniques have come from the aerospace and motor industries, being adapted for the larger geometries and more turbulent conditions typically found in the built environment. Although it is tempting to think in terms of 'wind tunnel versus CFD', the best results often come from combining CFD and wind tunnel studies, taking advantage of the strengths of both. CFD techniques have now developed to the point where environmental wind conditions across large sites can be examined within realistic project timescales and at lower cost than wind tunnel testing. However, this technology cannot yet reliably be used as the sole means of determining structural loads on a complex building.

As with wind tunnel testing, good quality CFD simulation is not sufficient on its own to understand wind effects on a development. CFD must be combined with a statistical study of site wind speeds, the effect of surrounding terrain and the impact on human comfort, pollutant spread, structural loading and other aspects of design.

The key benefits of computational modelling are:

- rapid prototyping and testing of options
- can be lower cost than wind tunnel testing
- flexible geometry definition
- good understanding of average wind conditions
- full 3D wind pattern, not just point values.

Disadvantages are:

- not currently recommended for structural loading
- requires specialist staff, software and hardware
- frequently less efficient than wind tunnel testing for gust effects.

Building Technology

More extensive use of diagrid and exoskeleton structures

Both steel and concrete diagrid and exoskeleton structures have been designed and constructed. These represent an alternative to the 'outrigger and core' structure as a currently popular form of construction (compared to the earlier tube, bundled tube and other 'tube' variants

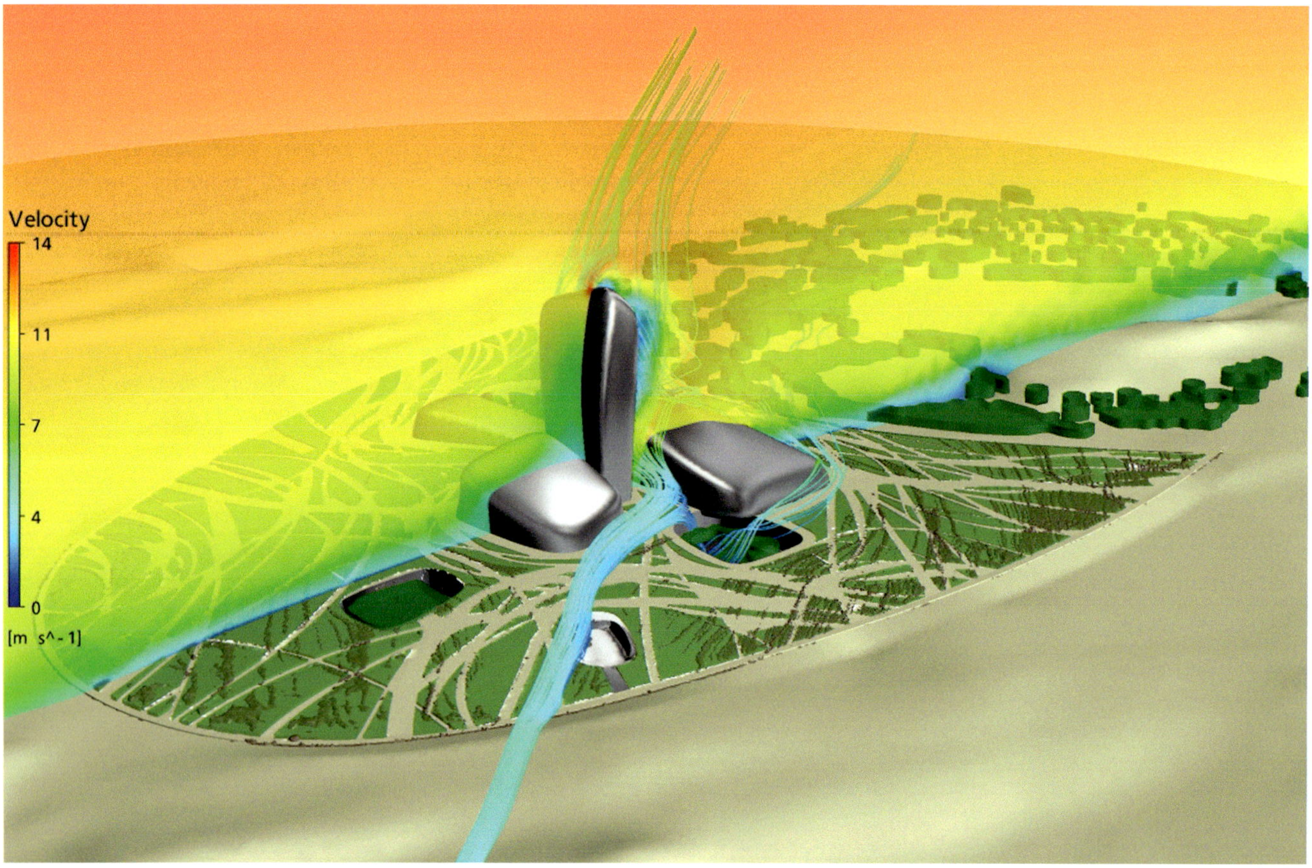

14.3 Environmental modelling using CFD simulation – King Abdulaziz Centre for World Culture and Knowledge, Damman, Kingdom of Saudi Arabia
Image: Courtesy Buro Happold

suggested by Fazlur Khan in the late 1960s and early 1970s – see Chapter 16).

Examples of steel diagrid structures range from the more 'conventional' 30 St Mary Axe (the Swiss Re Building) and Hearst headquarters, to the recent Guangzhou International Finance Center, to more 'extreme' versions such as the Capital Gate project in Abu Dhabi, UAE, which has a complex 3D warped diagrid working in conjunction with a vertically pre-stressed core in order to control the lateral displacements due to the asymmetry of the building.

These forms of 'external' structure are interesting as they provide both relative freedom of architectural form and a relatively efficient structure for resisting vertical and lateral loads. Another advantage is that the 'pattern' of the diagrid can be modified, either while remaining concentric or as an alternative when becoming eccentrically braced – there are currently unexplored opportunities for 'tuning' the form of building response (shear versus flexural/frame bending) in the lateral load resisting system, and for incorporating more intrinsically ductile detailing for areas within seismic zones.

Monitoring performance of buildings and using data for new projects

A frequently voiced frustration among designers of tall buildings is the difficultly of benchmarking building performance from design to construction and against other buildings for a variety of criteria.

In order to design efficiently there is a need to be able to verify whether the theoretical predictions that have been used as the basis of design are matched in the constructed building. Specifically for structural engineering, this would be the intrinsic damping coefficient of the built structure (at realistic amplitudes), the building sway under a defined wind load (at a specified return period) and building accelerations and velocities. The trend has been to specify various forms of monitoring in tall buildings, particularly super- and megatall towers, incorporating either active or passive damping systems.

The other area in which progress is being made is the use of instrumentation during construction in order to verify and refine the assessments of the building movements made during design. This is particularly

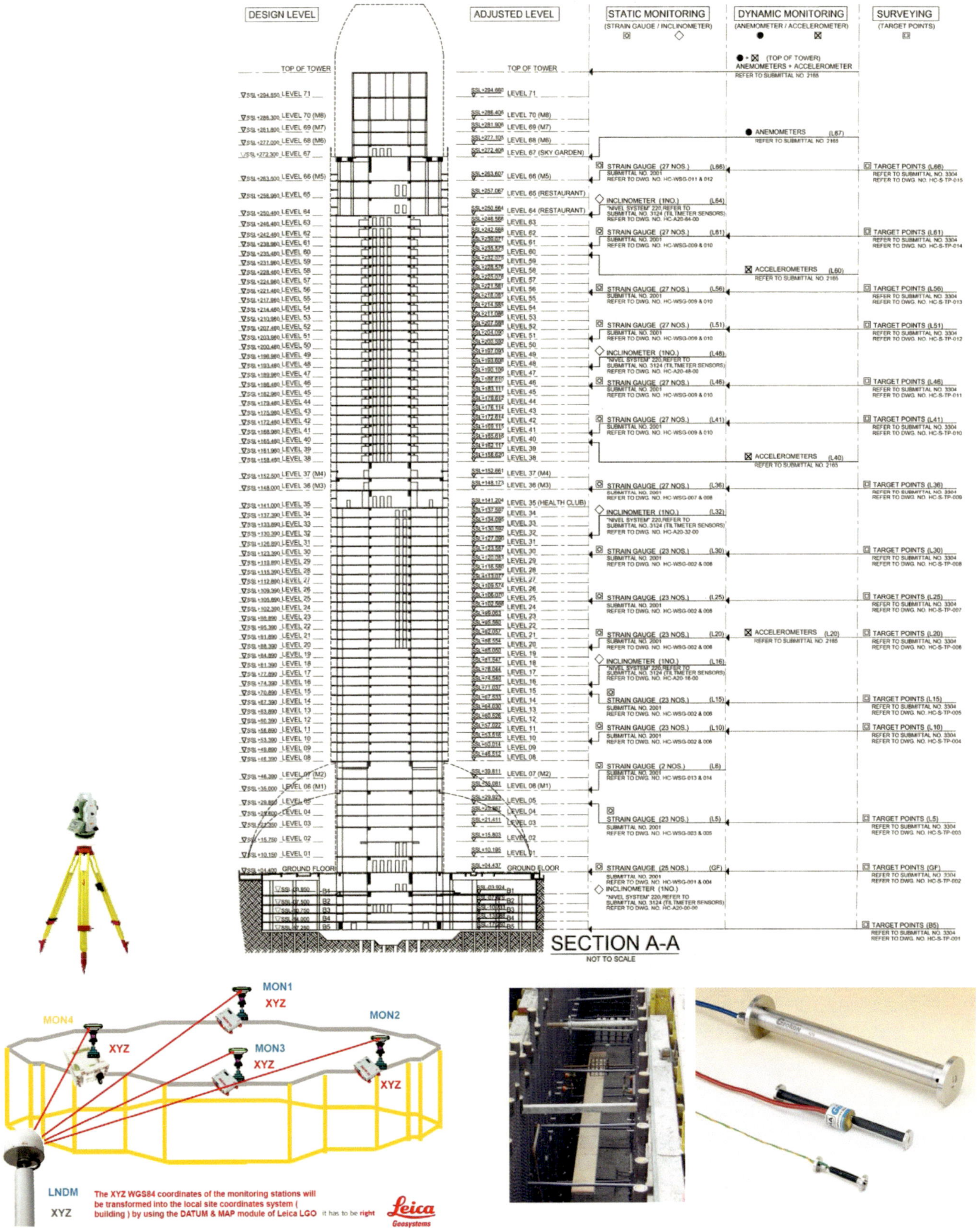

14.4 Building instrumentation and monitoring to inform future design – Landmark Tower, Abu Dhabi

Images: Courtesy Buro Happold and Leica Geosystems

important in asymmetric buildings where, in order to construct the cores within verticality tolerances, it is necessary to build to a modified geometry so that it may deform into its intended final geometry.

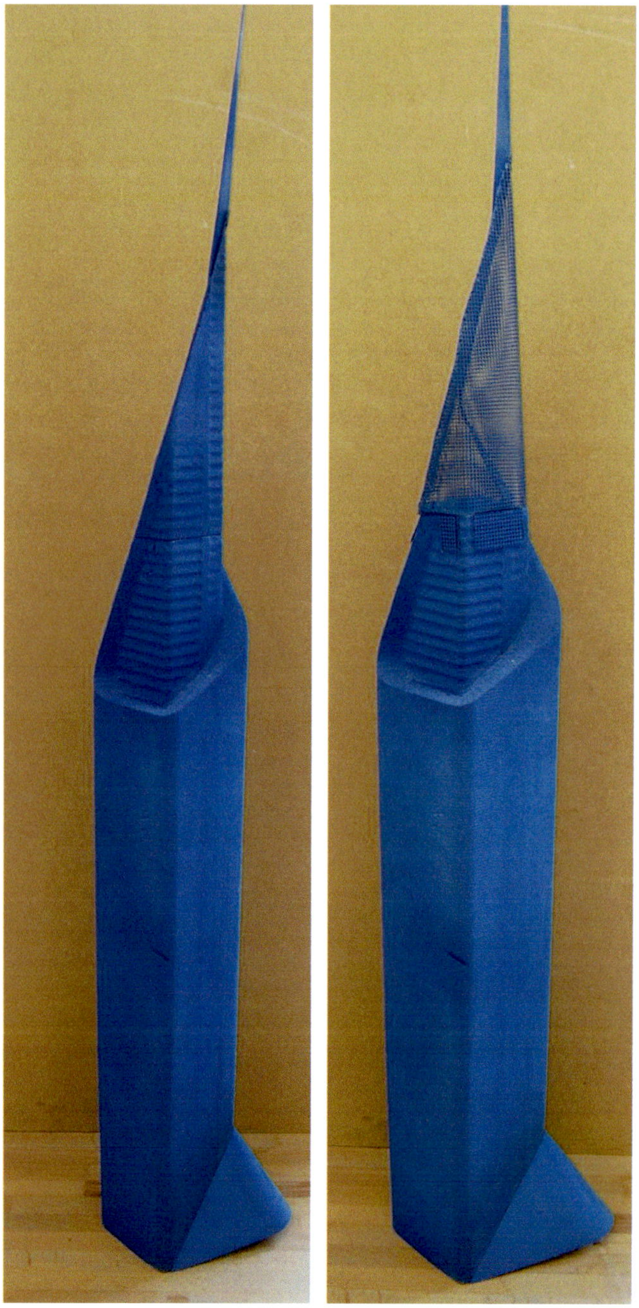

Porous buildings, tops of towers and varying geometries

One easily observed fact about tall buildings is that the wind 'loads' at the top are often the driving factor in the determination of, first, the wind shears acting on the building (that is, the magnitude of the wind load) and, second, the bending moments induced (allowing for the dynamic and static components of the wind loading).

One approach to creating architectural forms while reducing wind loads is to incorporate porosity into the fabric of the building. This can be observed in the Sports City Tower (Aspire Tower) in Doha, Qatar, effectively a volume created in mesh with a central solid core and suspended 'occupation' modules, and also in the design of the Arriyadh tower project in Riyadh, Saudi Arabia. Here a similar concept has been used at the top of the tower to create a passive damping system for the vortex shedding behaviour, therefore reducing the overall wind loads and wind-induced lateral accelerations. Given that many projects incorporate architectural features at the top, above the top inhabited floor, this seems to be a useful method of reducing overall wind loads while maintaining the desired architectural appearance. This could be considered alongside the various tall buildings that move with the prevailing wind, that is, they respond to applied wind loads with changing geometry.

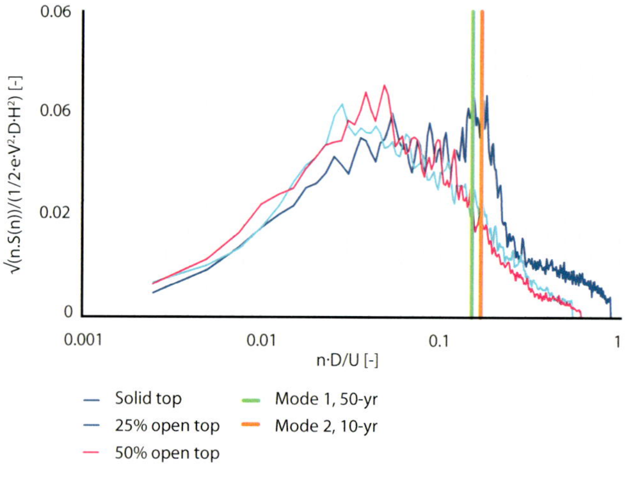

14.5 Investigating Architectural Form to reduce structural loading – Ritz Carlton Hotel, Riyadh. (left) Solid crown (middle) Porous crown (right) Significant load reduction through increased porosity
Images: Courtesy Buro Happold and BMT Fluid Mechanics

Construction Technology and Materials

Concrete technology

A large percentage of the tall buildings constructed in North America in the twentieth century had principal structural systems built in steel. However, internationally there are regions where concrete has been very much more used; generally this is due to a combination of factors, principally economics and the familiarity of local contracting organisations with the use of concrete. The use of high-strength concrete, in particular, has been routine throughout the world for decades.

There have been some more recent advances in concrete technology that are of interest to tall building designers. Very High Strength Concrete (VHSC), defined as having characteristic compressive strengths of f'c = 240 MPa, has been developed. This concrete has steel fibre reinforcement incorporated in the mix and achieves a flexural tensile strength of 40 MPa. Typical high-strength concrete mixes have been noted as having ductility and tensile capacities which are not enhanced as much as the compressive strengths – VHSC is an attempt to overcome these issues. Given the premium paid for useable floor area in tall buildings, VHSC provides a potential route to smaller cores and columns in taller buildings.

Self-Compacting Concrete (SCC) has been widely used, and a particular application of use to the designers of tall buildings is in large area mass concrete pours, such as for the foundation rafts of tall buildings. Typical examples are the Al Faisaliah tower in Riyadh (6000 m^3 at 375 m^3/hr) and the Landmark tower in Abu Dhabi (16,000 m^3 at 400 m^3/hr), where the use of SCC enabled the full pours to be carried out without construction joints (and associated delays, risk and potentially large

14.6 Use of SCC for large area raft pours with congested, multi-layer reinforcement – Landmark Tower, Abu Dhabi, one of the world's largest continuous concrete placement operations
Images: Courtesy Buro Happold

amount of interface shear reinforcement). Equally important, SCC reduces the risk of problems with concrete placement and compaction where multiple layers of large diameter bars are used.

Lightweight concrete has obvious benefits for use in a tall building context. The typically utilised 'structural' lightweight concretes have a density of around 14 to 19 kN/m^3 and have been used advantageously, for example, for the One Shell Plaza project in Houston (1971), which at 52 storeys in height is the world's tallest lightweight concrete building. An example of a potential alternative form of lightweight concrete is Fibre-reinforced Aerated Lightweight Concrete (FALC). This is an attempt to combine the advantages of fibre-reinforced concrete (increased flexural and tensile strength, hence ductility) and lightweight concrete. A study on mechanical properties has been carried out which indicates a density of between 11 kN/m^3 and 15 kN/m^3 and strengths of between 12 and 40 MPa, with polypropylene and carbon-fibre reinforced concretes having properties which seem most appropriate for use as structural (particularly slab) elements.

Steel composite floor panels

The weight of floor construction is always an issue in tall building design, with lightweight concrete and composite steel solutions having been adopted to reduce the loads that the gravity and seismic load resisting systems need to deal with. Another form of floor structure which has been used in stadia projects and which has potential benefits for high-rise construction is the SPS floor system. This consists of two steel plates bonded with a polyurethane elastomer core – effectively a steel composite floor panel. This has the advantages associated with off-site fabrication and also reduced weight compared to conventional reinforced concrete floor slabs, particularly for locations where steel buildings are economic and where the self-weight is particularly critical. For example, its use was considered by Snohetta and Buro Happold for the Ras Al Khaimah gateway tower, which leans and twists, to reduce the movements which need to be 'built out' during construction, and of course the magnitude of the *P*-delta effects.

Piezoelectric materials, smart materials and smart buildings

Piezoelectric materials are those that convert mechanical energy into electrical energy when the material is strained. Piezoelectric dampers have been developed as an example of 'controllable materials' which can be incorporated into buildings. Effectively the dampers are used as part of a 'smart' structural system – they respond to the structural motion without sensors and actuators.

This form of damping has been tested in Japan on a 30-storey apartment building, where it was found to be effective in reducing the accelerations experienced under particular levels of wind loading by up to 50 per cent, compared to the undamped case.

Forms of applied damping which are distributed and which do not rely upon centralised sensor and control arrays are important steps towards being able to confidently design to a particular (higher) level of damping in order to reduce the 'demand' side of the equation of engineering an efficient tall building structural solution.

There are other forms of 'controllable' materials, including Shape Memory Alloys (SMAs), which have been used in the aeronautical industry. Conceptually, 'controllable' materials enable a building to respond locally to particular environmental demands without a requirement for an overall control system. An analogy is the use of optimisation analyses – iterative procedures designed to optimise a structural system according to set criteria – in the design process, which ideally leads to the structure being modified in response to its (theoretical) environment. Used together the efficiency of buildings should be enhanced.

High-strength steel, large diameter bars

When designing tall building structural systems there are two driving criteria: first, strength and, second, serviceability (which often comes down to stiffness). There is pressure on designers to minimise the 'wasted' space in buildings occupied by vertical circulation of the building services, and the structure and associated finishes. This pushes designers towards options for reducing the sizes of particular vertical load-resisting structures.

One option is to use high-strength concrete as described previously; another is to look at increasing the yield strength of the reinforcement. There have been advances made in steel reinforcement production, and there are, for example, grade S670 reinforcing bars produced by SAH which have been used in tall building projects in the US (Epic Tower, Miami) and Germany (Opernturm) with diameters of up to 75 mm. There are advantages in reducing the congestion of reinforcement, in particular in very heavily loaded and highly reinforced columns towards the base of towers. These would include increasing steel reinforcement by up to 20 per cent, which, coupled with the higher yield strength, would obviously enable higher load-carrying capacity

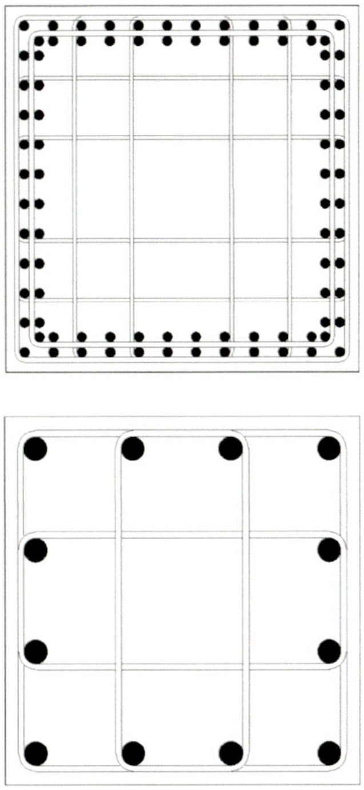

14.7 (left) High-strength, large diameter bars to reduce reinforcement congestion in compression-controlled sections. (top) Standard reinforcement, 6% (bottom) High-strength, large diameter reinforcement, 6%
Images: Courtesy Buro Happold

for a given section size. There are some drawbacks – for example the bar couplers are compression contact splice couplers and not tension couplers – which would preclude their general use, but as the number of super- and megatall structures increases the need for 'larger and stronger' is also increasing.

Steel castings and solid steel columns

With very tall buildings such as the Chicago Spire project (610 metres in height), the use of solid steel columns made from steel plate welded together has been proposed in order to minimise the size of columns and maximise lettable area, also to ensure that the architectural appearance of 'lightness' is maintained. This approach was successful in the New York Times Building in New York, in which solid sections of up to 762 millimetres (30 inches) square were utilised. The same concept was applied in the design of the Chicago Spire but with larger section sizes due to the greater height.

One issue with the use of very large steel sections (including solid sections) is the practicality of making connections, particularly where inclined members (truss web members, transfers, etc.) intersect. One answer is the use of steel castings rather than built-up sections; this was the solution proposed for the base of the Chicago Spire. The castings can be manufactured to complex geometries, allowing the members to be simple 'sticks' with machined ends. Obviously the key to success is the production of the casting without any imperfections (this requires costly manufacturing processes and testing to prove integrity) as well as tolerances sufficiently fine for members to be placed 'bearing' against the casting (since for very large steel-to-steel connections the main force transfer is through direct bearing rather than through welds or bolts).

Modular construction

Modular construction is effectively factory production of individual building components which enables a reduction in the number of on-site trades, particularly 'wet' trades used in finishes. The concept has mostly been advocated for locations with high labour costs and a lack of skilled or semi-skilled site labour, as well as for reasons of health and safety. Bathroom 'pods' have been used for decades, but the concepts of modular construction are now becoming more widely utilised, for example in the construction of hotels. The approach can be used for elements such as bathrooms or entire hotel or hospital rooms, or for components such as electrical rooms.

A modular construction concept such as the Verbus system can be utilised to speed up construction considerably, but has generally been used in low- and medium-rise construction. Currently crane capacities and speed of lifting, as well as the need for significant vertical and lateral load-resisting structure in addition to modules' own structural properties, make it more difficult to apply such a concept to super- or mega-high-rise projects.

Heavy lifting and strand jacking: lessons from bridge building

Increasingly in recent years, the geometry of tall buildings has become significantly different from the traditional, 'single spire' form. For example, in the Marina Bay Sands project in Singapore, the upper SkyPark links the three towers below; the Dubai Pearl project in the United Arab Emirates has a similar concept. Given that the 'premium' space in a tall building is at the top, the attraction of building horizontally at height is apparent – a similar driver is illustrated by the 'Walkie-Talkie' (20 Fenchurch Street) in London, which widens as it goes up.

In order to construct such large sections of structure at height, a recent trend has been to consider the use of heavy lifting technology from bridge building. For example, on the Marina Bay Sands project the SkyPark was lifted into place in segments using a 'strand-jacking' technique by VSL.

Phase-change materials

An example of a non-structural application of newer technology is phase-change materials (PCMs), which are simply substances (such as water) that liquefy and absorb and store heat when the temperature reaches a certain point, and do the reverse once the temperature drops.

PCMs have been developed (by BASF, for example) to be incorporated into construction materials such as drywall or blockwork. Their purpose is to reduce the daytime peak temperature and increase the nighttime low temperature – that is, to moderate the temperature within a building.

Bi-Steel and steel plate shear walls

In the past, the types of structural systems typically used for steel buildings were framed or braced systems in which the individual members were hot-rolled sections. For concrete structures, a traditional approach has been to use the walls around the lift and stair shafts of buildings as 'shear walls'. A more recent structural approach is the use of steel plate shear walls.

One example of this is the Corus 'Bi-Steel' system – modular steel plate panels connected together by transverse steel bars to form hollow sections. These sections, typically floor-to-floor height, are erected and bolted or site welded together, with each lift successively being filled with concrete, making the final structure a 'composite' structural steel/concrete shear wall. Advantages include curved panels which would be more difficult to achieve in concrete, thinner sections for the same stiffness, reduced construction time (due to off-site manufacture) and panels which are easier to lift than equivalent precast panels (although obviously for a tall building the lifting is significant to the project construction planning).

Another, similar example is the use of steel plate shear walls (SPSW) in the Tianjin Jinta tower in China. This project employed a dual structure of perimeter composite sections (columns being concrete infilled steel tubes and beams hot rolled steel H sections), moment-resisting frames and a central core structure of moment-resisting frames infilled with steel plate. This changed to braced frames and then moment-resisting frames at the top, the core and perimeter systems being linked via outriggers at various plant levels. This structure has a number of advantages and disadvantages in common with a typical concrete-to-steel comparison (for example the need for fire protection and the impact on cranes), but has a key advantage for a tall building with irregular gravity load distribution (that is, a leaning or twisting tower) in that, being steel, its movements can be calculated and allowed for more accurately than in a similar concrete structure, in which, in particular, it is extremely difficult to predict the long-term vertical and lateral displacements and consequential changes to loads in outriggers, and hence to determine mitigation measures.

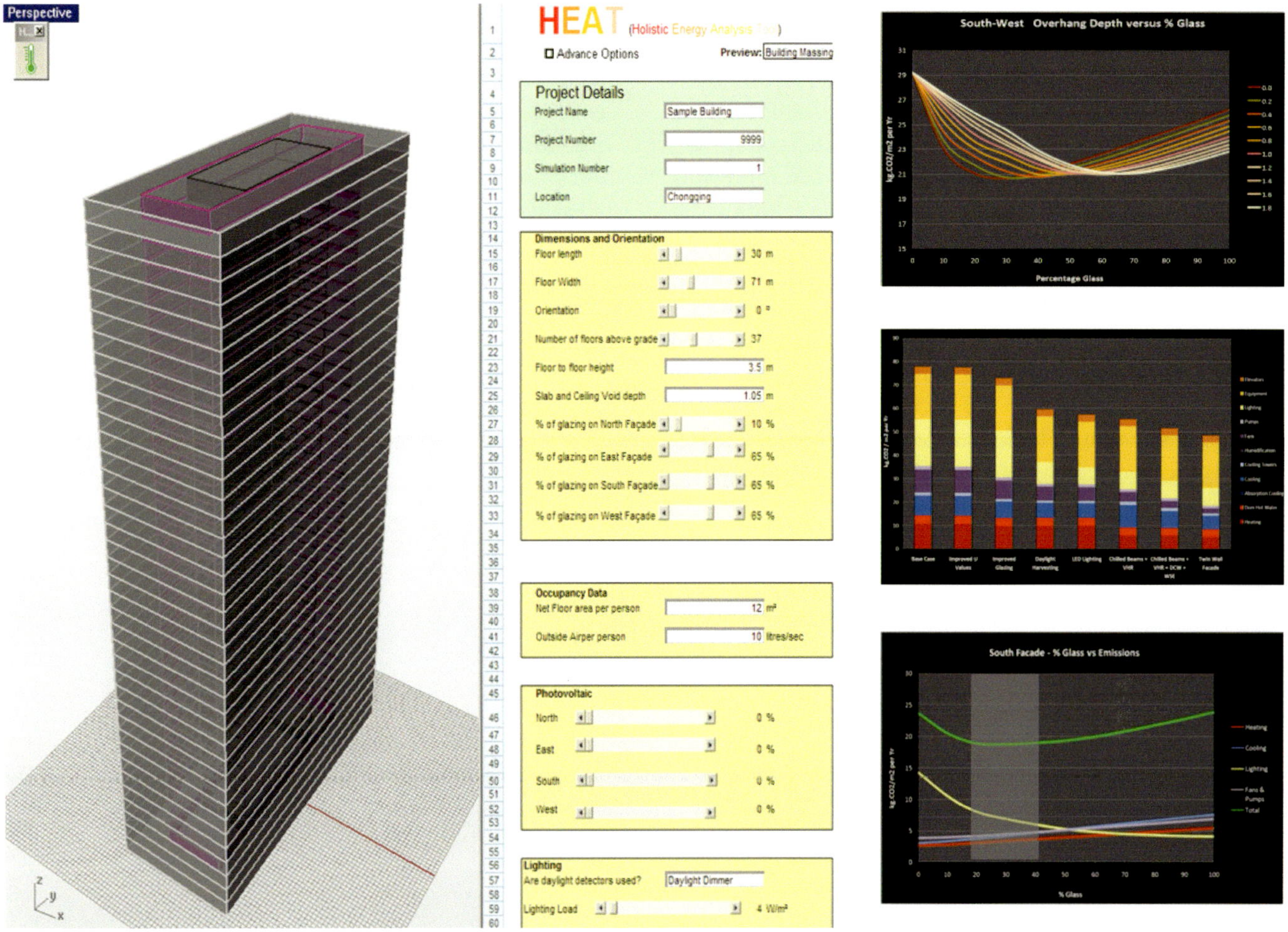

14.8 ZERO-E – a Woods Bagot/Buro Happold JV interactive carbon analysis tool
Image: Courtesy Buro Happold

The Future

The future seems to be dominated by a few key challenges for humanity (including engineers):

1. scarcity of resources
2. globalisation
3. climate change.

It seems clear that the trend towards tall buildings will continue, principally in emerging markets such as China, India and Brazil, but also in established markets, as illustrated by recent projects such as The Shard in London, Tour Phare in Paris and the Freedom Tower in New York. Legislation regarding energy use, sustainability in general and efficiency in particular will have an increasing impact. This can be viewed on a city scale, where large efficiencies can be achieved through considering the tall building as one key element in a sustainable community or master plan.

Various schemes have been prepared with greater or lesser effectiveness to produce electricity via photovoltaics, wind turbines and moving buildings. There is a trend towards 'ZERO-E' design, effectively designing buildings that have a net zero energy impact – this is expected to continue.

Smart buildings are effectively buildings which respond to the environment (a seismic event, wind, night and day) by changing their physical characteristics – changing geometry in response to wind loads, changing thermal properties to respond to the ambient temperature (using phase change materials). Such buildings seem to be a logical trend, as the technologies and supporting design methods required are developed.

There has long been an expectation that construction will follow a maturity curve into becoming more of a 'manufacturing' process than a 'trade' process – virtual prototyping, modular design and off-site manufacturing processes are likely to converge, enabling significantly more challenging heights, configurations and geometries to be constructed.

Mirroring the 'skipping' of technological development which has been observed in the telecommunications sector, it is likely that in the new 'BRIC' states the developing world's interest in technology and innovation from an economic standpoint will drive early adoption of concepts and materials in a less structured and legislated environment than in Europe. In the same way that the boom in tall buildings in the Middle East in the early part of the century has led to significant advances in understanding of the design and construction of tall buildings which has then fed back into the developed economies, it would be logical to see the BRIC states as being places where significant developments will occur quickly.

At the time of writing (2011), much of the current world financial climate is poor, but there is still a global need for tall buildings. This will continue, albeit in a less extreme form than pre-recession, and the coming decades should see rapid changes in our understanding and abilities in respect of the design and construction of tall buildings.

Chapter 15
Risk Considerations in Tall Building Design

S. Shyam Sunder, Jason D. Averill and Fahim Sadek

Introduction

The public expects a certain level of safety when using commercial, institutional, or residential buildings. Public safety is managed through the systematic assessment and mitigation of risk. The concept of risk is used to characterize the likelihood of an unfavorable outcome or event. Risk is based on the probabilities of hazard occurrence and resulting consequences, and is often separated into three components (as shown in Figure 15.1):

- *Hazards* are events with the potential to cause unwanted consequences.
- *Vulnerabilities* are characteristics of the building that render the building susceptible to damage due to hazard occurrence.
- *Consequences* are the results of the hazard and vulnerabilities that reflect the level, duration, and nature of the loss, such as building damage or collapse, personal injury or loss of life, economic and property losses, interrupted operations, and damage to the environment. The consequences of an event can be measured in terms of a value system or metric. Consequences of the failure of a tall building can be catastrophic and are generally categorized as human impacts (public health and safety), economic impacts (direct and indirect effects), psychological impact (public confidence), and functional impact (continuation of operations; see FEMA 426/BIPS 06 2011). As a result, extra care should be given to managing the risk and mitigating the vulnerabilities of tall buildings.

In a probabilistic form, risk can be expressed as the probability of occurrence of loss greater than a certain loss metric, LM, and can be evaluated as:

$$P[\text{Loss} > \text{LM}] = P[\text{Hazard}]\ P[\text{Damage} \mid \text{Hazard}]\ P[\text{Loss} > \text{LM} \mid \text{Damage}]$$

where $P[\text{Hazard}]$ is the probability of hazard occurrence, $P[\text{Damage} \mid \text{Hazard}]$ is the conditional probability of damage, given the occurrence of the hazard (a measure of vulnerability), and $P[\text{Loss} > \text{LM} \mid \text{Damage}]$ is the conditional probability of the loss or consequence given the occurrence of the damage.

Elms (1992) introduced the *context* as an additional element for defining the risk. The context, which provides a frame of reference for the risk analysis and assessment, can differ between various stakeholders,

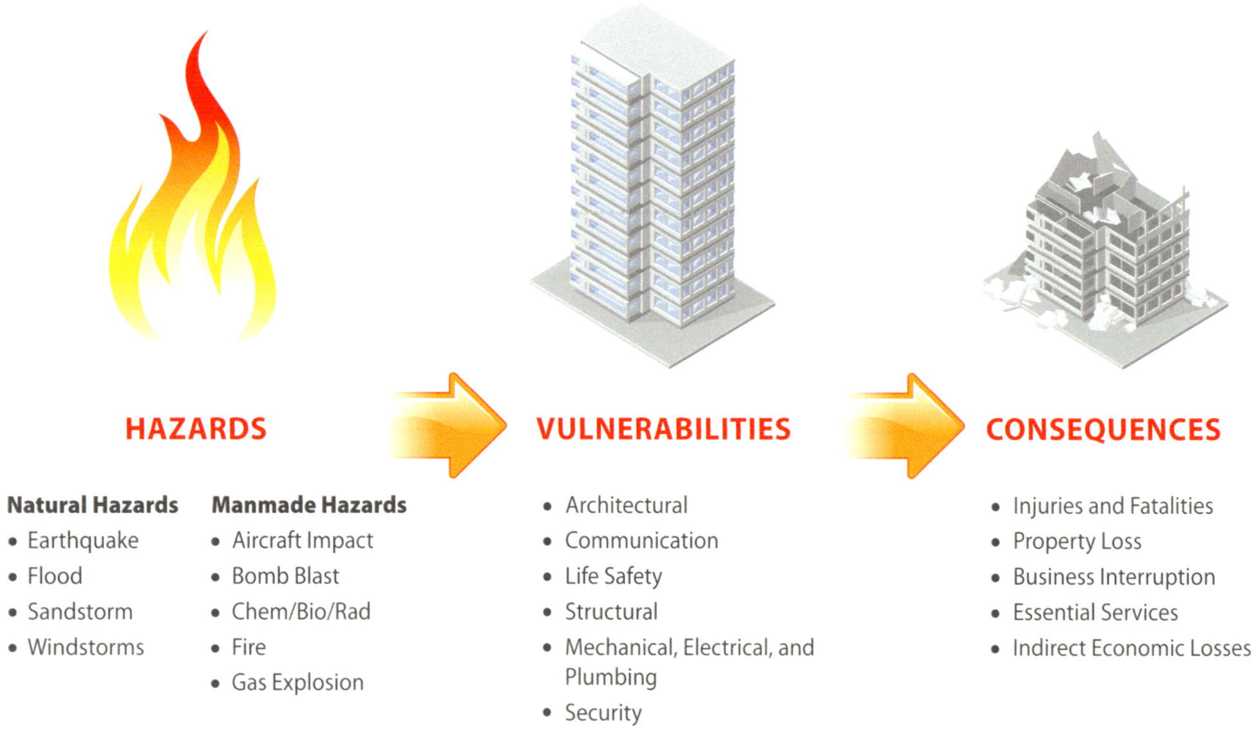

15.1 The three components of risk: hazards, vulnerabilities, and consequences

including building occupants, management groups, government agencies, and other decision makers. Individuals, in general, are risk averse and expect a substantial benefit in return for accepting marginal increases in risk. Governments and large, self-insured organizations tend to be risk neutral (see Ellingwood 2001, NISTIR 7396 2007, CIB 2001).

Furthermore, the public does not "rationally" consider all risks equally: low probability, high consequence events are considered less tolerable to the public than higher probability, lower consequence events. The public also views incidents involving large numbers of people differently from incidents involving individuals, for example airline versus automobile accidents. Similarly, the public is averse to multiple fatality fires in tall buildings, but is generally less concerned with fewer per-incident fatalities in residential buildings. Finally, the context is determined by the need for risk management, where investment in risk mitigation needs to be weighed against available resources.

In the design of tall buildings, there are multiple sources of uncertainty. Some of these uncertainties are inherent (aleatory), for example hazard occurrence and material strength, while others are knowledge based (epistemic), for example design assumptions and approximations in analytical models. For structural design, uncertainties in hazards and strengths have historically been accounted for using load and resistance factors within a reliability framework (ASCE 2010). Risk is the natural consequence of uncertainty. Thus, risk cannot be eliminated, but must rather be managed or mitigated. Stakeholders and society must accept a certain level of risk in order to achieve objectives. What the designers and owners of tall buildings need is a quantification of a socially acceptable risk level.

Implicit in building codes and standards are the socially acceptable consequences, or acceptable risks, for building hazards; see Chapters C1 and C2 of the American Society of Civil Engineers' ASCE 7-10 (ASCE 2010). However, "acceptable risk" for tall buildings is not explicitly quantified and is a relative term that can be determined only in the context of what is commonly used in other fields (for example cancer risk or risks in the nuclear, chemical, and offshore industries), the investment required to marginally reduce the risk, and the losses due to increase in risk (Ellingwood 2001). According to Paté-Cornell (1994), the risk of failure below which society normally does not mandate any regulatory guidance, is on the order of 10^{-7} annually. As stated earlier, the acceptable level of risk depends on the point of view of different stakeholders. Risk management procedures are used to evaluate the economic and social trade-offs between increased occupant safety, decreased damage, repair cost, downtime reduction, and higher construction and other costs.

Hazards and Vulnerabilities

Due to its complexity, a tall building is considered as a system of systems with many interactions and interdependencies. These include (see Figure 15.1):

- *the architectural system,* including architectural layout, interior space plan, partition walls, and glazing and cladding (facade) systems
- *communication and information technology systems*
- *the life safety system,* including active and passive fire protection systems, the fire alarm system, and the egress and emergency responder access system
- *the structural system,* including the gravity- and lateral load–resisting systems
- *mechanical, electrical, and plumbing (MEP) systems,* including water supply and distribution, gas distribution, electric power generation and distribution, and the air-handling system (heating and cooling)
- *the security system,* including physical barriers and security plan, equipment, operations, and procedures

A single hazard can affect, damage, or disrupt the operation of more than one system. For example, the performance of the structural system during a given hazard can have a direct impact on the performance of other systems, including architectural and life safety systems, in addition to utility infrastructure that can be housed inside, atop, or below the building. Ultimately, where the hazard results in the collapse of the structural system, all other systems would also fail.

Hazards can be natural or manmade. 'Natural hazards' typically refers to natural events such as a windstorm or an earthquake. In contrast, manmade hazards originate from human activity and include accidental (technological) hazards and deliberate attacks such as arson or terrorism. Accidental hazards such as accidental fire or accidental vehicular impact have unintended consequences. Deliberate attacks are malicious acts of force and violence against persons or property such as a bomb explosion or a chemical, biological, or radiological (CBR) attack.

It is desirable to quantify a hazard by its annual probability or mean rate of occurrence. The Poisson probability model has been used to model events that occur randomly in time, such as natural and accidental hazards that are not altered by deliberate human intervention (Melchers 1999, Ang and Tang 1975). The Poisson model, however, does not apply to incidents involving terrorist or arson events, as these are deliberate events directed at a specific target to maximize social disruption and impact.

Building codes and standards consider natural hazards such as seismic and wind loads in designing tall buildings by explicitly imposing a minimum design basis. In general, tall buildings are designed to comply with the prescriptive requirements, and thus may be vulnerable to rare natural hazards beyond those required by building codes and standards (such as tornadoes or earthquakes larger than those specified for design).

Consideration of multiple hazards when assessing risks to tall buildings that are exposed to a variety of natural and manmade hazards may result in cost savings, efficiency, and improved performance. A multi-hazard design should carefully consider the synergies, or lack thereof, among the mitigation strategies for various hazards. For instance, designs to resist disproportionate (progressive) collapse may be beneficial for seismic resistance and vice versa, due to enhanced structural integrity and redundancy. On the other hand, physical barriers around the building perimeter are effective against vehicle bombs, but can cause difficulties for fire engine access in the event of fire. Another example is the conflict between earthquake and blast loading, where a reduction in mass can improve the seismic performance but decrease blast resistance. This section provides an overview of major hazards that may have significant adverse consequences for tall buildings and highlights building systems' vulnerabilities.

Natural hazards

There is considerable statistical and scientific information that enables the incidence of future natural hazards to be estimated using mathematical probabilities. For example, earthquake severity can be specified based on a level of shaking (an acceleration value) that corresponds to a given probability of exceedance (say 2 percent) in a given number of years (say 50) or a probability that the level of shaking would be exceeded in a given period (say 2500 years), or, more recently, the shaking associated with a quantitative risk. The probability of exceedance of earthquakes, severe windstorms, and flooding is commonly expressed by use of the terms "return period" or "mean recurrence interval" (MRI). These are defined as the average time, in years, between the expected occurrences of an event with a specified intensity. Recent research on structures subjected to non-simultaneous multiple hazards has shown that MRIs of limit states may be shorter under multiple hazards than under any of the individual hazards (Duthinh and Simiu 2010).

Climate change may affect the frequency and intensity of natural hazards in the future. According to the Intergovernmental Panel on Climate Change (IPCC), an increase of greenhouse gases in the atmosphere will likely boost temperatures over most land surfaces, though the exact change will vary regionally. Possible impacts of climate change on natural hazards may include (IPCC 2007):

1. Increased frequency of heavy precipitation events resulting in increased frequency of flash floods and large-area floods in many regions, especially at high altitudes
2. Increased frequency and intensity of wildfires, especially in forests and peatland
3. Increased intensity of tropical cyclone activity, including hurricanes and typhoons
4. More frequent and intense storm surge due to the combination of rising sea level and more intense storms.

Earthquakes

The main parameters defining the seismic hazard are the peak ground acceleration, the duration of strong ground motion, and its frequency characteristics, which can be described by the response spectrum of the ground motion. Earthquakes can cause catastrophic damage, including collapse of buildings and massive loss of life (Figure 15.2). The structural systems of tall buildings can be particularly vulnerable to earthquakes. Structural components such as beams, columns, walls, bracings, and connections can suffer damage due to seismic excitations. Structural damage and/or severe deformations of the structure can damage other non-structural systems including architectural, life safety, and MEP systems. Earthquakes can also rupture gas lines or cause other hazardous conditions that may result in post-earthquake fires. Additional information can be found in Chapter 19 of this book.

Windstorms (hurricanes, typhoons, cyclones, and tornadoes)

The parameters defining wind hazards are wind speed, direction, and the aerodynamic response of the building (depending upon building shape, geometry, height, and exposure), including possible aeroelastic effects. Novel pressure measurement technologies used in conjunction

15.2 Partial collapse and severe damage to the 21-story Torre O'Higgins office building in Concepción caused during the February 27, 2010 Maule, Chile earthquake (magnitude 8.8)
Images: Jay Harris, NIST

15.3 Damage to the building envelope of the 26-story Hyatt Regency Hotel in New Orleans Business District caused during the 2005 hurricane Katrina. Broken windows allowed rain and debris into the building resulting in additional damage

Image: Keith Porter, University of Colorado at Boulder and SPA Risk LLC

with computer-intensive methods allow the substantially improved estimation of wind effects on tall buildings. Recent studies show that for tall buildings that respond dynamically to wind, larger load factors may need to be implemented to account for additional uncertainties associated with dynamic characteristics such as damping ratios and natural frequencies (Gabbai et al. 2008). The building envelope (glazing and cladding systems; Figure 15.3) and the structural system can be vulnerable to severe windstorms. Structural damage and/or severe deformations can damage other non-structural systems. Inter-story drift can damage glazing and cladding systems, and large accelerations can cause occupant discomfort. In addition, wind-borne debris and water penetration during windstorms are a concern for tall buildings.

Coastal and riverine flooding

Coastal flooding (inundation of land areas along the ocean coastline caused by sea waters above normal tidal levels) includes storm surge during tropical cyclone weather systems (hurricanes, tropical storms, and typhoons) and tsunamis (surge induced by seismic activity) after earthquakes. Riverine flooding includes inundation of floodplains due to the accumulation of runoff from rainfall or snowmelt. The parameters defining the hazard are the water depth, including wave height in the case of storm surge, the water velocity for hydrodynamic effects, and flood duration. For tall buildings, flooding results in inundation and water damage to the lower floors and to the building systems housed therein, including emergency power generators, fire protection systems, and elevator pits, but does not generally jeopardize the structural safety of the building. However, the flooding could be extensive enough to seriously affect people evacuating the building in either normal or emergency modes.

Sandstorms

A sandstorm is a windstorm that carries sand through the air, forming a cloud near the ground. The phenomenon is common in arid and semiarid regions. The primary sources of airborne dust are the Sahara and the drylands around the Arabian Peninsula, with some contributions from Iran, Pakistan, India, and China. When wind reaches a critical velocity, grains of sand begin to roll forward along the ground surface. At higher wind speeds, sand particles move by "saltation," a process in which particles are temporarily lifted and then bounced along the surface in a hopping/jumping motion. Sandstorms represent a minor risk to tall buildings compared with other hazards, but could result in infiltration of sand into the building, affecting cladding, occupants, electronics and mechanical equipment, food, and drinking water.

Manmade hazards

In comparison with natural hazards, probabilities of occurrence of manmade hazards can be more difficult to estimate. For instance, deliberate attacks are functions of sociopolitical trends and it may be hard to assign a probability distribution to their occurrence. Statistics and trends for manmade hazards can be obtained from various worldwide organizations, for example, for terrorist attacks worldwide, data may be available from the U.S. National Counterterrorism Center (2009). Following are several manmade hazards that may be considered in the design and risk assessment of tall buildings.

Aircraft and vehicular impact

The parameters defining the hazard of aircraft impact include aircraft mass, impact speed and orientation, and the amount of jet fuel. Buildings are not specifically designed to withstand the impact of fuel-laden commercial aircraft, and building codes do not require buildings to consider aircraft impact. However, after the crash of a B-25 bomber into the Empire State Building in 1945, designers of tall buildings became aware of the potential for aircraft collision with buildings. Organizations such as the National Transportation Safety Board (NTSB) maintain statistics on civilian aviation accidents, including those involving buildings. According to NTSB, more than half of all accidents occur at airport sites and only 30 percent occur at distances greater than 8 kilometers from the airport. The building collision rate for military aircraft is lower than that for civilian accidents. Thus, for buildings some distance away from airports, the impact probability is less than 10^{-7} annually (NISTIR 7396 2007). However, certain tall buildings might require a site-specific analysis. All building systems in the path of the aircraft and/or its debris could be vulnerable to aircraft impact. Damage to structural components can result in progressive failure of the structural system, which may also damage non-structural systems. Fire resulting from an impact due to the dispersion of jet fuel is another hazard that can cause substantial damage to many building systems and contribute to progressive failure. Some designs may also consider the possibility of vehicular impact where structural elements and other systems exposed to vehicles could be vulnerable.

Fire

The parameters defining fire hazard include the combustible load, nature of the combustibles, compartment ventilation, and bounding surfaces. Compartment areas larger than 1,000 square meters or the presence of materials that pose a high fire hazard, such as large fuel loads, gas storage tanks, flammable liquids, or explosive materials, represents a higher threat. Between 2003 and 2006, tall buildings in the U.S. experienced an average 13,400 fires each year, causing 62 deaths, 490 injuries, and $179 million in property loss (Hall 2009). In addition, fires in tall buildings are more difficult for the fire service to extinguish. The leading sources of ignition for tall building fires are electrical and gas-fueled equipment, cigarettes, and short circuits (USFA 2002). The ignition of a fire has a mean rate of occurrence on the order of 10^{-6} per square meter per year for offices and commercial facilities (CIB W14 1983, SFPE/SEI 2003).

Following ignition, the likelihood of flashover (a fully developed fire capable of inflicting significant structural damage) is affected by the presence and reliability of fire detection and mitigation systems. Given a typical flashover probability on the order of 0.01 to 0.02 (assuming an available sprinkler system), the mean occurrence rate of a structurally significant fire is thus on the order of 10^{-8} per square meter per year (NIST TN 1681 2010). Examples of tall building fires include One New York Plaza in New York (1970), First Interstate Bank in Los Angeles (1988), One Meridian Plaza in Philadelphia (1991), the World Trade Center (WTC) building 7 in New York (2001), the Windsor building in Madrid (2005), and the Mandarin Oriental Hotel in Beijing (2009). Figure 15.4 shows examples of column and floor damage due to fire. Building systems exposed to fire, including the structural, architectural, and MEP systems, could be vulnerable to the damaging effects of fire. The primary effects of fire on the structural system are thermal expansion and diminished strength and stiffness that can result in component and connection damage or failure, which if widespread could potentially lead to fire-induced disproportionate collapse.

Fire also poses danger to building occupants. Occupants may be exposed to heat, smoke, and toxic gases. Fractional effective dose (FED) and fractional effective concentration (FEC) are the methods often used to assess the accrual of various toxic gases an occupant may be exposed to during evacuation (ISO 13571 2007). Additional information can be found in Chapter 23.

Bomb blast and gas explosion

The parameters defining bomb blast hazard include the bomb size, measured in equivalent pounds of trinitrotoluene (TNT), and the standoff distance, measured from the center of the charge to the component of interest. Tall buildings are often located in dense urban environments and contain underground parking and loading docks that make them attractive to vehicle bombs. Examples include the 1993 bombing of the WTC north tower, New York; the 1993 Bishopsgate bombing of the NatWest Tower, London; and the 1996 bombing of the Khobar tower, Saudi Arabia. In 1998 alone, there were more than 225 bombing attacks against buildings in the U.S., not including residential properties (FBI 1998). This suggests an annual occurrence rate on the order of 2×10^{-6} per building. However, the frequency of such events can vary substantially from year to year.

Gas explosions pose a potential hazard to tall buildings that house natural gas pipelines. Leyendecker and Burnett (1976) indicate that the mean rate of occurrence of gas explosions in dwellings in the United States was about 2×10^{-5} per dwelling per year. That rate is about an order of magnitude lower than in the United Kingdom. In buildings with a large number of dwelling units the rate of occurrence of explosions is higher than in single-tenant buildings.

15.4 (right) Buckling of a column and beam flange in the 9-story WTC 5, and (above) sagging of a floor system in the 38-story One Meridian Plaza office building, due to fire
Images: (right) FEMA 403 (2002) (above) USFA-TR-049 (1991)

Bomb blast and gas explosions may have grave consequences for occupants near the source. Additionally, all exposed building systems can be vulnerable to blast or explosion. Loss of critical structural components of the building (such as columns or transfer girders) due to blast can result in additional structural damage or even disproportionate collapse. Secondary fires due to blast are another hazard that can cause substantial damage.

Chemical, biological, or radiological (CBR) attack

Three parameters defining the hazard of CBR attack include the exposure, the duration, and the concentration of the CBR agent. Each of the CBR agents may have different human effects and methods of attack. CBR attacks are an emerging threat and are of great concern because of the large area contaminated, the numbers of people affected, and the high economic cost of response and recovery. CBR threats may be delivered externally or internally to a building. External threats may be released at a standoff distance from the building or may be delivered directly through an air intake or other opening. Interior threats may be delivered to accessible areas such as the lobby, mailroom, loading dock, or egress route. Of particular concern is the use of a radiological dispersion device – known as a "dirty nuke" or "dirty bomb" – which combines conventional explosives with radioactive material with the purpose of scattering dangerous amounts of radioactive material over a large area. The air-handling system and the water supply of a building are particularly vulnerable to CBR attacks.

Risk Mitigation

When considering threats to tall buildings, there are two primary approaches to mitigate the risk: (1) prevent the hazard from occurring, or (2) manage the effects of the hazard once it has occurred. Prevention is nearly always the preferred strategy – though generally not possible for natural hazards – as prevention obviates the need for consequence management and may be far more cost effective. For example, if aircraft impact were considered a potential hazard, prevention would likely be the most reasonable approach as mitigation may be cost prohibitive. For other threats, a blend of prevention and mitigation strategies may be most appropriate. Fire safety in tall buildings, for example, would emphasize strategies to both prevent an unwanted interior fire (for example ignition resistant furnishings) and mitigate the effects of a fire that does ignite (for example sprinklers).

Risk assessment methods

In order to evaluate mitigation strategies, a risk assessment framework is necessary. Explicit techniques for risk assessment have their origins in the 1930s and 1940s. There are several methods for risk assessment, varying in complexity: checklist, narrative, index, and probabilistic (Watts 1981). Checklists enumerate vulnerabilities and risk mitigation considerations specific to a particular threat and target. More complex than checklists, narrative methods qualitatively describe best practices to consider for a threat and/or target. An index is a quantitative evaluation of consequences and risks for a building relative to a designated set of predetermined hazard and vulnerability criteria and values. Finally, the most sophisticated techniques are probabilistic methods. Probabilistic methods quantify components of risk using techniques such as event trees, fault trees, or influence diagrams.

Specific to protection of buildings, FEMA 426/ BIPS 06 (2011) and FEMA 452 (2005) formalize the risk assessment process based on an index method, utilizing a four-step risk assessment methodology:

1. Assess the hazards for the building
2. Assess the vulnerabilities (site and building systems design evaluated against the hazards and protections)
3. Identify the consequences of the hazard using asset value (including criticality of asset and number of people) as a metric
4. Assess the risk (likelihood and impact for each threat in the context of mitigation measures).

Hazard, vulnerability, and consequence ratings (in the range 1 to 10, from low to high) are defined for each potential hazard and building function and system. The risk matrix is calculated as:

$$\text{risk rating} = \text{hazard rating} \times \text{vulnerability rating} \times \text{consequence rating}$$

Based on the risk rating, mitigation strategies are developed for the matrix elements with the highest scores using a risk-based framework. Other risk assessment and mitigation resources include *Risk Assessment and Risk Communication in Civil Engineering* from the International Council for Building (CIB 2001), the Building Safety Guidebooks from the Council on Tall Buildings and Urban Habitat (CTBUH 2002), the American Society for Heating, Refrigeration, and Air-Conditioning Engineering's *Guideline for the Risk Management of Public Health and Safety in Buildings* (ASHRAE 2009), and *Risk Assessment in Engineering* from the Joint Committee on Structural Safety (JCSS 2008).

Design options for tall buildings are driven by economics and performance considerations. Life-cycle cost assessment methods have been standardized (see ASTM E917 – 05 (2010), *Standard Practice for Measuring Life-Cycle Costs of Buildings and Building Systems*). Use of consensus economic standards ensures full consideration of both up-front and long-term costs (or savings) for building design options. When cost-benefit analyses are combined with risk assessment tools to evaluate design and mitigation options, stakeholders are able to optimize design decisions.

Mitigation strategies

Based on the identification of specific hazards and vulnerabilities, mitigation strategies can be implemented for each building system to minimize unwanted consequences. The effectiveness of mitigation strategies will generally depend on their reliability, capacity, and redundancy. The following is a brief overview of common mitigation strategies for primary building systems.

Architectural system

Architectural design plays a major role in determining the performance of tall buildings during natural and manmade hazards. The building architecture should permit an effective design against hazards, while the structure must support the function and aesthetics of the building. Therefore, close coordination between the architect, engineering team, and security specialists is necessary to ensure that architectural features do not result in undesirable performance. The building configuration, including size, shape, and proportions, influences the stiffness and mass distribution in the building. For lateral resistance to seismic and wind loading, desirable configurations may include reduced height-to-base ratio, equal floor heights, no abrupt changes in stiffness or mass, symmetrical plan shape, balanced resistance in all directions, maximum torsional resistance, and direct load path without cantilevers.

Certain building configurations, such as U or L shapes, and buildings with re-entrant corners tend to trap and reflect shock waves, amplifying blast effects. Long, uninterrupted spans and large compartments may render the building vulnerable to fire (NIST NCSTAR 1A 2008), requiring special precautions for fire protection and evaluation of the ability of structural components and connections to resist fire effects. Partition walls, cladding, and glazing systems need to accommodate interstory drift under lateral loads and to be properly anchored to the primary structural system. Strategic material selection (windows, drapes, or cladding) may be implemented to resist blast pressure or projectiles originating outside the building envelope and also severe winds and windborne debris. More details can be found in FEMA 426/BIPS 06 (2011) and Naeim (2001).

Security system

Security is the primary defense for manmade threats and is tailored to consider individual threat profiles. Security measures need to be integrated into the architectural layout of the building. Essential to the security plan and design is the implementation of appropriate countermeasures to deter, delay, detect, and deny attacks. Often the countermeasures include the layered defense concept (the "onion" philosophy), which provides for increasing levels of security from the outer areas of the building site towards the inner, more protected areas. Setbacks and physical barriers may be implemented to prevent proximity of vehicular or other explosive devices.

Screening of entrance points using machine scanning, video, animal, or trained personnel methods may be implemented to detect more portable threats. Secured access may be provided at locations of unusually heavy loads such as tuned mass dampers, generators, vaults, water tanks, and swimming pools, and to sensitive areas such as mailrooms and air-handling systems. Furthermore, lock-down or other isolation methods may help mitigate the potential for armed attacks. For additional details, the reader is referred to Conrath et al. (1999), FEMA 426/BIPS 06 (2011), and UFC (2003).

Structural system

A large body of research and design experience has been developed for design and mitigation for natural hazards, especially seismic and wind effects (See Chapters 19 and 20). Recently, best practice guidelines have been developed for mitigation of blast effects (Conrath et al. 1999), disproportionate structural collapse (NISTIR 7396 2007) and fire effects on the structural system (NIST TN 1681 2010).

Structural systems that provide a continuous load path for all vertical and lateral loads, wherein structural components are tied together with connections capable of developing the full capacity of the members, are highly desirable. Structural redundancy enhances building robustness and mitigates disproportionate collapse. Redundancy is based on designing alternate load paths for the vertical and lateral load-resisting systems of the building. Ductile design and detailing also improve damage tolerance since structural members and their connections need to maintain their strength while undergoing large deformations. Increased reserve capacity of the structural system allows the building to absorb damage without collapse; the exterior walls of the WTC towers that remained standing after extensive damage due to aircraft impact are a case in point. In addition, hardening key structural elements to resist abnormal loads can be an effective mitigation measure.

Life safety systems

Life safety systems include active and passive fire protection systems, the fire alarm system, and egress and emergency response access systems. Fire protection systems – including sprinklers and other specialized suppression, fire detection or alarm, compartmentation, smoke control, and structural fireproofing – are critical to tall building safety due to the high likelihood of fire occurring during the service lifetime. Combustibility and ignitability of contents, use of smoke barriers or fire rated assemblies to contain the heat and smoke, and fire alarm systems to notify occupants of fire are primary means of limiting the fire threat to occupants. The effectiveness of fire protection systems depends on their reliability, capacity, and redundancy.

Tall buildings present unique challenges for evacuation and emergency response planning due to their height. As such, tall building designs should consider the overall occupant evacuation time relative to the onset of a hazardous condition, including travel distance, the number and location of exits, and the width of the stairwells and access doors. Recent innovations in the design of elevators for use by occupants and first responders during emergencies may significantly improve tall building evacuation and emergency responder access. Note that shelter-in-place could be the preferred response to some threats. In addition, the availability of the egress/access system during an emergency may depend upon the resistance of the components to specific threats (for example the use of fire-rated partitions or hardening shaft enclosures against overpressures).

MEP systems

Mechanical heating, ventilation, and air conditioning (HVAC) systems can constitute a key vulnerability since they connect large portions of the building through penetrations in the floor, wall, and ceiling assemblies. In particular, the ductwork provides means to disperse toxic gases and particles throughout a building. There are several defenses typically employed to protect the air-handling systems: source protection, zoning, sensing, segmentation, and filtration. Source protection protects the supply air components from inadvertently or intentionally drawing hazardous gases or particles into the system. Locating the intakes away from ground level minimizes the intake of inadvertent pollution (from idling vehicles, for example) and limiting access to the intakes minimizes the potential for intentional introduction of hazardous substances. Zoning involves segmenting the HVAC system into independent sections that can limit the distribution of airborne substances between different zones of a building.

Sensors can be located within the air handling system to detect the presence of smoke or other hazardous particles, but this approach is not widely used at this time. Presumably as sensor and control technology advances, the sensors may be integrated into the building alarm system and may be designed to produce specific actions, such as segmentation. Segmentation includes opening or closing dampers within the system in order to either stop the flow of air or redirect the airflow towards an exhaust outlet.

Finally, filters can be utilized to ensure that hazardous particles are trapped within the system, thereby reducing the potential consequences. Gaseous air cleaning is increasingly being applied to remove harmful gases. For sandstorms, use of air intake with sand provisions and use of washable filters can provide a measure of protection. For additional information, see CDC/NIOSH (2002, 2003).

Communication and information technology systems

Communication systems are critical to emergency management and response. First, fire alarm systems, public address systems, and other mass notification technology should be designed to produce the correct protective actions for occupants during a range of hazards. Specific actions can be communicated with audible or text-based instructions. Second, communication systems should be established and tested by emergency responders to ensure effectiveness throughout the building; specific attention should be paid to radio frequency transmissions due to the significant quantities of concrete and steel typically used in tall buildings. In addition, extra security and protection should be provided for critical information technology equipment and data.

Role of Building Codes and Standards

Building codes and standards are essential for managing risk in the interest of public health, safety, and welfare. Traditional, prescriptive building codes and standards provide a minimum threshold level of safety. However, a performance-based approach enables innovative and cost-effective design solutions, and may be necessary to account for hazards and owner requirements beyond the scope of the prescriptive codes. Performance-based design is a formal method for predicting the performance of a building subject to a specific hazard or range of hazards. Over the past two decades, substantial improvements have been achieved in the accuracy and efficiency of the calculation methods and software tools necessary to support performance-based design assessments for natural and manmade hazards.

As an example of a performance-based approach document, the *2009 ICC Performance Code* (ICC [International Code Council] 2009) captures the concept of the asset value of the structure (a measure of consequence) by specifying performance requirements over a range of event magnitudes. The *2009 ICC Performance Code* quantifies the design event magnitude based on the mean return period for a range of natural hazards. Performance level (for example maximum acceptable damage) is a function of building importance and event magnitude. Tall buildings would generally be classified as Performance Group III, unless they contain tenants critical to the continuity of life safety services (such as fire, police, or government officials), contain hazardous materials, or critical infrastructure (such as power generation or water distribution), in which case the building would be designed to deliver Performance Group IV outcomes. As an example, the expected performance of a tall office building would be immediate reoccupancy after a small-magnitude event, but a short repair period may be allowed before reoccupancy following a large-magnitude event.

Another example of a performance-based document is the *NFPA 5000: Building Construction and Safety Code* (NFPA [National Fire Protection Association] 2009), which provides performance objectives for building design in lieu of prescriptive requirements. In addition, there are national and international bodies worldwide that are developing performance codes and standards for natural and manmade hazards. National examples are the British Standards Institute (UK), AFNOR (France), DIN (Germany), and Standards Australia.

The European Union (EU) created CEN (Comité Européen de Normalisation) to represent the collective EU view (though individual national bodies continue to function), and member countries may not develop national standards where CEN standards are under development. Likewise, per the Vienna Agreement, CEN defers to international standards, including ISO (International Organization for Standardization); see Bukowski (2002).

Building codes and standards are subject to modifications in response to large-loss events. Examples include the 1911 Triangle Shirtwaist fire in New York City, the 1968 Ronan Point apartment collapse, the 1994 Northridge earthquake, and the 2001 WTC building collapses. New code requirements may reflect new insight gained from analysis of the event, new understanding based upon research subsequent to the event, or may simply reflect a shift in public support for a previously known issue.

Based on its investigation of the collapse of the WTC buildings, the National Institute of Standards and Technology (NIST) issued thirty recommendations for improvements to tall building codes, standards, and practices (NIST NCSTAR 1 2005, NIST NCSTAR 1A 2008). The premise behind these recommendations is that consequences of failure or collapse of tall buildings are tremendous when compared to those for the same hazard or set of hazards in a low-rise building. As a result, tall building owners and designers, as well as public officials, must determine appropriate performance requirements for buildings that are at risk due to their locations, use, occupancy, historic or iconic status, and so on. These NIST recommendations resulted in substantial changes to U.S. building codes and standards, including:

- An additional exit stairway or occupant evacuation elevators for buildings more than 128 meters in height
- A minimum of one fire service elevator for buildings more than 37 meters in height
- Increased bond strength for fireproofing
- Field installation requirements for fireproofing
- Inspections of fireproofing thickness, density, and bond strength after the rough installation of mechanical, electrical, plumbing, sprinkler, and ceiling systems
- Increasing by one hour the fire-resistance rating of structural components and assemblies in buildings 128 meters and higher
- Adoption of the "structural frame" approach to fire-resistance ratings, which requires all members of the primary structural frame to have the higher fire-resistance rating commonly required for columns
- Luminous markings delineating the exit path (including vertical exit enclosures and passageways) in buildings more than 23 meters in height
- New structural integrity requirements for improved resistance to disproportionate structural collapse.

Summary

Consequences of the failure or collapse of a tall building are catastrophic. Therefore, extra care should be given in assessing and managing the risk to, and mitigating the vulnerabilities of tall buildings. This chapter defined the concept of risk as the product of the probability of the hazard occurrence, the building vulnerability, and the resulting consequences. The chapter also identified the most common natural and manmade hazards that affect tall buildings and suggested mitigation strategies for managing or reducing the risk to a variety of tall building systems. Furthermore, social expectations including the role of building codes and standards in managing the risk to tall buildings were discussed.

Acknowledgments

The authors benefited from the review comments of the following experts from NIST: Stephen Cauffman, Robert Chapman, Richard Gann, John (Jay) Harris, Eric Letvin, H.S. Lew, Joseph Main, Therese McAllister, Nancy McNabb, Andrew Persily, and Emil Simiu. This chapter, however, is the responsibility of the authors.

Bibliography

Ang, A. H-S. and Tang W. H. (1975) *Probability Concepts in Engineering Planning and Design: Basic Principles*, John Wiley, New York.

ASCE (2010) *Minimum Design Loads for Buildings and Other Structures*, SEI/ASCE 7-10, American Society of Civil Engineers, Reston, VA.

ASHRAE (2009) *Guideline for the Risk Management of Public Health and Safety in Buildings*, ASHRAE Guideline 29-2009, American Society for Heating, Refrigeration, and Air-Conditioning Engineering, Atlanta, GA.

ASTM E917 – 05 (2010) *Standard Practice for Measuring Life-Cycle Costs of Buildings and Building Systems*, ASTM International, West Conshohocken, PA.

Bukowski, R. W. (2002) "The Role of Standards in a Performance-Based Building Regulatory System," in Almand, K., Coate, C., England, P., and Gordon, J. (eds.) *Performance-Based Codes and Fire Safety Design Methods*, 4th International Conference Proceedings, March 20–22, 2002, Melbourne, Australia, pp. 85–94.

Bukowski, R. W., Budnick, E. K., and Schemel, C. F. (1999) "Estimates of the Operational Reliability of Fire Protection Systems," in *International Association of Fire Safety Science (IAFSS)*, October 4–8, 1999, Chicago, IL, pp. 87–98.

CDC/NIOSH (2002) *Guidance for Protecting Building Environments from Airborne Chemical, Biological, or Radiological Attacks*, Publication No. 2002-139, Center for Disease Control and Prevention/National Institute for Occupational Safety and Health, Cincinnati, OH.

CDC/NIOSH (2003) *Guidance for Filtration and Air-Cleaning Systems to Protect Building Environments from Airborne Chemical, Biological, or Radiological Attacks*, Publication No. 2003-136, Center for Disease Control and Prevention/National Institute for Occupational Safety and Health, Cincinnati, OH.

CIB (2001) *Risk Assessment and Risk Communication in Civil Engineering*, CIB No. 259, Conseil International du Bâtiment, Rotterdam.

CIB W14 (1983) "A Conceptual Approach Towards a Probability-Based Design Guide on Structural Fire Safety," *Fire Safety Journal* 6(1), 1–79.

Conrath, E., Krauthammer, T., Marchand, K. A., and Mlakar, P. F. (1999) *Structural Design for Physical Security: State of the Practice*, American Society of Civil Engineers, ASCE/SEI, Reston, VA.

CTBUH (2002) *Building Safety Guidebooks: Assessment Guidebook/Enhancement Guidebook*, Council on Tall Buildings and Urban Habitat, Chicago, IL.

Duthinh, D. and Simiu E. (2010) "Safety of Structures in Strong Winds and Earthquakes: Multi-Hazard Considerations," *Journal of Structural Engineering* 136, 230–3.

Ellingwood, B. R. (2001) "Acceptable Risk Bases for Design of Structures," *Progress in Structural Engineering and Materials* 3(2), 170–9.

Elms, D. G. (1992) "Risk Assessment," in D. Blockley (ed.), *Engineering Safety*, McGraw-Hill International, Berkshire, UK, pp. 28–46.

FBI (1998) *1998 Bomb Incidents*, FBI Bomb Data Center General, Bulletin 98-1, Federal Bureau of Investigation, Washington, DC.

FEMA 403 (2002) *World Trade Center Building Performance Study: Data Collection, Preliminary Observations, and Recommendations*, Federal Emergency Management Agency, Washington, DC.

FEMA 426/BIPS 06 (2011) *Reference Manual to Mitigate Potential Terrorist Attacks Against Buildings*, Department of Homeland Security, Washington, DC.

FEMA 452 (2005) *Risk Assessment: A How-To Guide to Mitigate Potential Terrorist Attacks Against Buildings*, Federal Emergency Management Agency, Washington, DC.

Gabbai, R. D., Fritz, W. P., Wright A. P., and Simiu, E. (2008) "Assessment of ASCE 7 Standard Wind Load Factors for Tall Building Response Estimates," *Journal of Structural Engineering* 134, 842–5.

Hall, J. R. (2009) *High-Rise Building Fires*, National Fire Protection Association, Quincy, MA.

ICC (2009) 2009 *ICC Performance Code for Buildings and Facilities*, International Code Council, Falls Church, VA.

IPCC (2007) *Climate Change 2007: Synthesis Report: Contribution of Working Groups I, II and III to the Fourth Assessment Report of the Intergovernmental Panel on Climate Change*, IPCC, Geneva, Switzerland.

ISO 13571 (2007) *Life-Threatening Components of Fire: Guidelines for the Estimation of Time Available for Escape Using Fire Data*, International Organization for Standardization (ISO), Geneva.

JCSS (2008) *Risk Assessment in Engineering: Principles, System Representation and Risk Criteria*, Joint Committee on Structural Safety, Zurich, Switzerland. Available at: http://www.jcss.ethz.ch/publications/JCSS_RiskAssessment.pdf

Leyendecker, E. V. and Burnett, E. (1976) *The Incidence of Abnormal Loading in Residential Buildings*, Building Science Series No. 89, National Bureau of Standards, Washington, DC.

Melchers, R. E. (1999) *Structural Reliability: Analysis and Prediction*, John Wiley, New York.

Naeim, F. (ed) (2001) *The Seismic Design Handbook*, 2nd edition, Kluwer Academic Publishers, Boston, MA.

National Counterterrorism Center (2009) *2009 Report on Terrorism*, Washington, DC. Available at: http://www.nctc.gov/witsbanner/docs/2009_report_on_terrorism.pdf

NFPA (2009) *NFPA 5000: Building Construction and Safety Code*, National Fire Protection Association, Quincy, MA.

NISTIR 7396 (2007) *Best Practices for Reducing the Potential for Progressive Collapse in Buildings*, National Institute of Standards and Technology, Gaithersburg, MD.

NIST NCSTAR 1 (2005) *Federal Building and Fire Safety Investigation of the World Trade Center Disaster: Final Report of the National Construction Safety Team on the Collapses of the World Trade Center Towers*, National Institute of Standards and Technology, Gaithersburg, MD.

NIST NCSTAR 1A (2008) *Federal Building and Fire Safety Investigation of the World Trade Center Disaster: Final Report of the National Construction Safety Team on the Collapse of World Trade Center Building 7*, National Institute of Standards and Technology, Gaithersburg, MD.

NIST TN 1681 (2010) *Best Practice Guidelines for Structural Fire Resistance Design of Concrete and Steel Buildings*, National Institute of Standards and Technology, Gaithersburg, MD.

Paté-Cornell, E. (1994) "Quantitative Safety Goals for Risk Management of Industrial Facilities," *Structural Safety* 13(3), 145–57.

SFPE/SEI (2003) *Designing Structures for Fire*, Society of Fire Protection Engineers, DesTech Publications.

UFC (2003) *DoD Minimum Antiterrorism Standards for Buildings, Unified Facilities Criteria (UFC) 4-010-01*, Department of Defense, Washington, DC.

USFA (2002) *Highrise Fires*, Topical Fire Research Series 2(18), U.S. Fire Administration.

USFA-TR-049 (1991) *Highrise Office Building Fire, One Meridian Plaza (Philadelphia, Pennsylvania)*, Technical Report Series, U.S. Fire Administration.

Watts, J. (1981) "Systematic Methods of Evaluating Fire Safety: A Review," *Hazard Prevention* 18(2), 24–7.

Part Four

Engineering the Tall Building

Introduction

Dave Parker

In 1984, Nobel Prize-winning biologist Sir Peter Medawar wrote:

> Considering how much civic pride in America is invested in having a skyscraper taller than anyone else's, people may wonder why a skyscraper may not grow as high as a city's citizens may wish. The answer is obvious when pointed out: unless the upper stories are to remain uninhabited, the proportion of floor space allocated to elevators in the lower stories soon becomes grossly uneconomic – a truth of which we are vividly reminded by those tall buildings on which the elevators crawl up the outside.

This was in a book entitled *The Limits of Science* and serves as a warning of the dangers of crystal-ball gazing, especially the risks involved in venturing outside one's own specialism. Sir Peter died in 1987: what he would have made of the Burj Khalifa or the Petronas Towers is unknowable. Yet history is littered with examples of the pace of scientific and technological development leaving respected pundits looking foolish. Over the last two or three decades the knowledge base of tall building professionals has broadened dramatically, while at the same time new tools and materials have become widely available. Part Three described how tall buildings are being shaped by these new possibilities, and how in the twenty-first century the iconic skyscraper is evolving way beyond anything that could have been imagined in 1984. Here in Part Four the focus is on how the visions and concepts of the client and architect are realized as stable, efficient, safe and durable structures.

As the first two chapters of Part Four confirm, the structural limits of tall building construction are still out of reach, and the height that any project may aspire to is determined more by the optimism—or ego—of the client than any inadequacies in the strength of materials or ingenuity of the structural engineer. What will motivate the construction of a tower that tops the 828 metres of the Burj Khalifa is unlikely to be a major technological breakthrough; the necessary technology and expertise already exist. Politics and nationalism are likely to be the driving force, backed largely by sovereign wealth funds. Symbolism will be far more relevant than economics.

One trend that seems to be emerging as humankind becomes a primarily urban species and cities become ever larger and denser is the development of sites for tall buildings that would have been seen as impractical or too risky even a few years ago. The benign ground conditions and negligible seismicity of Manhattan and Chicago encouraged the development of the modern skyscraper; now supertall buildings are rising on foundations more than 100 metres deep in seismically active locations open to hurricanes or typhoons. As the latter three chapters of Part Four demonstrate, the skills and technology are available to deal realistically with such extreme conditions. Occupant comfort can be maintained even in extreme wind conditions; occupant safety during even major earthquakes is also deliverable.

In another quarter of a century these thoughts may generate the same wry amusement as those of Sir Peter Medawar above. It is to be hoped that the actual future will turn out to be far more exciting than anyone can possibly imagine right now.

Chapter 16
The Limits of Materials and Structures

Richard Tomasetti, Joseph Burns and Dennis Poon

Introduction

Limitations are best perceived as a function of time. With the accelerating pace of technological change, today's limits are often yesterday's science fiction, and probably tomorrow's routine standards. Nobel laureate Sir Peter Medawar (see the introduction to Part Four) did not anticipate the ways double-deck elevators, shuttles and sky lobbies would overcome the "obvious" limit of vertical transportation for super tall towers. So it would be wise to respond to the frequently asked question "How high can we go?" with humility: tomorrow's technology is sure to overtake today's answer. But even today, there is no theoretical structural limit until building heights are considered in miles or kilometers.

How do the limits of material properties affect the answer to the above question? And are there limits on structural systems that affect the answer?

Consider, for example, that a uniform prismatic column of normal-weight concrete with 12,000 psi (80 MPa) crushing strength (a reasonable value today) will reach its nominal strength limit just due to its own weight at a height of approximately 8,500 feet (2,590 meters). A uniform steel column with 50,000 psi (345 MPa) yield strength will reach its nominal strength limit at 14,700 feet (4,480 meters). Taper either column uniformly and its maximum height triples; tapered wire is used on winches to perform deep sea sampling for the same reason. Looking to the future, NASA predicts that using carbon/epoxy composites will extend limiting factors of self-weight theoretically to as high as 114 km (71 miles).

These heights, however, would require that sufficient lateral bracing is present, a load factor of 1.0 (no safety factor) is used rather than 1.4, and no other load is supported by the column. Essentially, the building would be a forest of columns and no floors. Obviously this is unrealistic since the purpose of a column is to support live and dead loads applied by occupied floor areas. Wind and seismic loads will generate overturning forces that will increase axial loads on some columns, as well as flexural stresses that add to axial stresses. All these effects reduce practical maximum heights. Limiting sway and accelerations due to lateral loads will further affect the practical height limit for a specific building design.

Since strength alone is a poor predictor of high-rise design limits, other ways that material properties interact need to be considered for governing performance factors for high-rise design. These interactions differ from those for low-rise building design. Governing performance factors for high-rise building structures usually include limiting lateral deflections (sway) and accelerations (occupant comfort) under dynamic wind and seismic loads, and speeding construction when working on the "island in the sky" that is the rising structural frame of a high-rise building. In contrast, governing performance factors in low-rise buildings are typically strength and local stability (buckling) due to vertical and lateral loads.

These two statements are not mutually exclusive. For low-rise buildings, deflections can govern the design of individual members, and for high-rise buildings the design of individual members – such as columns at lower floors – is usually governed by strength. For high-rise buildings, overall stability may be a governing factor as lateral loads and vertical loads act on the deflecting structure (the *P*-Delta effect).Which material properties interact with these governing performance factors?

1. *Strength*, where structural load-resisting capacity governs.
2. *Stiffness* (modulus of elasticity, or MOE), where deflection governs. Stiffness relates to material selection in subtle ways; while structural steel stiffness is independent of strength, concrete stiffness (MOE) varies approximately in proportion to the square root of its strength.
3. *Ductility*, where seismic criteria govern. Ductile materials accept large deformations beyond their proportional limit before failure and absorb energy in the process. To develop beneficial ductility also requires proper member proportions (compact sections), member details (confinement) and connection details.
4. *Inherent damping*, where acceleration criteria govern. Damping reduces accelerations from cyclic excitation under both wind and seismic conditions. It also helps reduce lateral dynamic forces and resulting building deflections.
5. *Mass*, where seismic forces, occupant comfort in windy conditions or dynamic wind loads from cross-wind excitation govern.
6. *Strength-to-weight ratio*, where both strength and mass are considerations.
7. *Stiffness-to-weight ratio*, where both stiffness and mass are considerations.
8. *Long-term behaviors*, including creep and shrinkage.

Taller buildings require a larger fraction of the construction budget to control dynamic properties, resist larger forces and deal with the special construction challenges of placing and working with materials high in the sky. Innovation and creativity based on material properties result in a cost-effective structure compatible with the architectural design, budget and overall acceptable building behavior.

Strength and Strength-to-Weight Ratio

The start of this chapter pointed out that if only self-weight is considered, a high-strength concrete column could hypothetically extend to a height of 8,500 feet (2,590 meters) if prismatic or 25,500 feet (7,770 meters) if tapered, while a steel column could hypothetically extend 14,700 to 44,000 feet (4,480 to 13,410 meters) for the same conditions. The different heights reflect a strength-to-weight ratio of high-strength concrete roughly half that of structural steel. That does not mean steel is automatically preferable for columns of super-tall construction, as other considerations are involved: stiffness, mass, damping and floor construction, for example.

Buildings include floor slabs for occupiable space, which add weight and accumulate live loads and additional dead loads. If the need for application of load factors in design, uniform tower floors and gradually reducing column sizes is considered, but no lateral loads or stiffness requirements, height limits for some hypothetical strength-based examples can be determined:

- For all-concrete construction with columns of 12,000 psi (80 MPa) compressive strength occupying 10 percent of the ground floor area, a building can reach about 3,500 feet (1,070 meters).
- For all-steel construction with columns of 50,000 psi (345 MPa) yield strength covering 5 percent of the ground floor area, a maximum building height would be about 10,500 feet (3,200 meters).
- For a tower of mixed construction with concrete-filled metal deck floors on steel beams, and concrete columns occupying 10 percent of the ground floor area, maximum height is about 4,000 feet (1,220 meters). This demonstrates that using lighter floor slabs permits a taller building than the all-concrete case.

For the design of actual supertall buildings that must address lateral loads and stiffness requirements, the maximum height would have to be reduced by about one-third for concrete column systems, and by about a half to two-thirds for steel column systems.

The above examples are limited as they consider prismatic building shapes even though the columns are tapered. Additional height can be achieved with a nonprismatic building shape, such as that of the more than 1,000-meter Kingdom Tower under construction in Jeddah, Saudi Arabia, which will be the tallest building in the world upon its completion (see Figure 16.1). The Kingdom Tower design concept is a synthesis of wind engineering, architecture and structural engineering. The structural system contains the architectural spaces and

16.1 Kingdom Tower, Jeddah (rendering)
Image: © Jeddah Economic Company/Adrian Smith + Gordon Gill Architecture

is formed by vertical and inclined reinforced concrete bearing walls and coupling beams.

Gravity loads on sub-grade conditions strongly influence the cost effectiveness and practicality of going higher. When a skyscraper is built on shallow rock of good quality (say 100 tons per square foot or 10 MPa allowable bearing), the foundations may be simple, economical footings that require only a small fraction of the total building construction budget. But when good quality rock is far below the basement excavation or is so deep it cannot be found using conventional borings, as is the case in many parts of the world, foundations may require construction of very deep structural elements such as large piles or caissons in friction or bearing, often capped by a large, thick foundation mat.

As an example, for the 1,483-foot (452-meter) Petronas towers in Malaysia, the foundation depth is almost 30 percent of the building height (see Figure 16.2). This affects the relative attractiveness of different strength-to-weight ratios: for a building sited on poor soil, any opportunity to reduce total building weight helps reduce the extent and cost of comparatively expensive foundations.

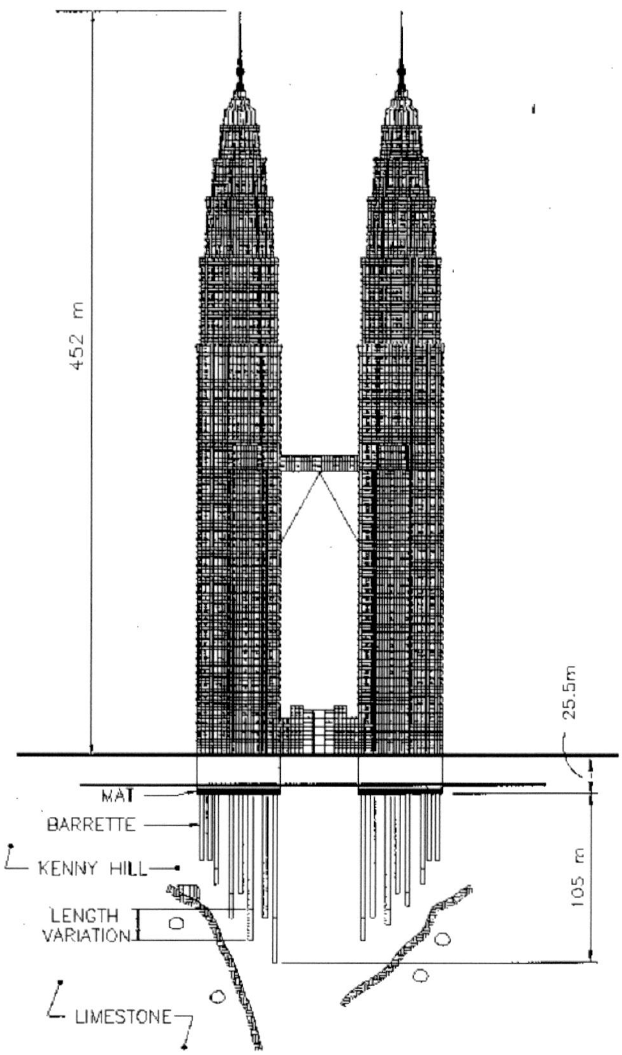

16.2 Petronas Towers, Kuala Lumpur, including foundations
Image: Thornton Tomasetti

Stiffness

Satisfactory building designs meet both strength and serviceability requirements. For a supertall building, strength concerns the safety of both occupants and neighbors, while serviceability – behavior of non-structural elements and comfort in windy conditions – concerns occupants. Both strength and serviceability are related to structural stiffness.

Stiffness relates to strength through stability and wind loads. An important component of overall building lateral stability is the *P*-Delta effect: when gravity load (*P*) is displaced laterally by a distance (Delta) due to wind, seismic or unbalanced gravity loads, it generates increased forces and additional deflections throughout the structure. Added deflections generate further *P*-Delta forces and deflections in an iterative cycle. A certain level of structural stiffness is required to ensure stability, by having the structure reach an equilibrium deflected shape before members are overloaded. If a structure is too flexible the deflections keep increasing with subsequent cycles until a critical member buckles or is overloaded. Note that controlling *P*-Delta effects is not a matter of simply adding structural material; that would both reduce Delta and add to *P*, making for only a small deflection reduction. Distributing structural material more efficiently is the better way to maximize stiffness.

Another aspect of stiffness related to strength is story drift under seismic loading. Building codes limit story drift under anticipated seismic conditions. Limits differ based on the drift determination method. For example, amplified design-level values are used in ASCE7-10 (ASCE 2010), while elastic response to a service-level (63-year) event and inelastic response to a rare (2,475-year) event are used in the China code (2010). These limits are applied for four reasons:

- Limiting the extent of damage during an earthquake, although the codes anticipate that damage will be extensive in any large event.
- Avoiding concentration of seismic energy and damage at "softer" stories.

- Avoiding instability failure from "ratcheting" deflections, where story drift is too large for recovery so it increases in subsequent cycles of motion.
- Avoiding pounding against adjacent structures across gaps or seismic joints.

In ASCE7-10, amplified seismic story drift is limited to 1.5 percent of story height for tall commercial buildings with high occupancy. See Chapter 19 for more information on seismic design.

Wind loading is a more complex issue. Drag forces and resulting downwind deflections are directly related to building shape (streamlining) and the square of the wind speed, addressed in building code reference documents such as ASCE7-10. Drag is relatively insensitive to changes in building period (seconds per first-mode sway cycle, or 1/frequency) in the period range of conventional buildings. Wind also creates lift, or crosswind forces. The resulting crosswind deflections are purely dynamic: cyclic swaying related to the building period, building shape and building dimensions. Crosswind effects are due to vortex shedding (see Figure 16.3). Discrete swirls of wind separate from the building faces, causing alternating crosswind suction forces, which cause dynamic forces resulting in both lateral and torsional deflections and accelerations. The resultant motion is dubbed vortex-induced oscillation (VIO).

It is the perception of acceleration and of torsional rotation that causes occupant discomfort, one of the most important serviceability considerations, and therefore wind-induced acceleration and rotation at upper floors must be minimized in tall building design. Vortices are shed at a rate determined by building plan dimensions and wind speed. As the frequency of vortex shedding approaches the natural frequency of building sway, crosswind accelerations become more problematic. Increasing building stiffness increases building sway frequency, which in turn increases the wind velocity required for resonant vortex shedding. Stiffness high enough to avoid vortex shedding resonance in the likely range of wind speeds would be the goal.

There are other dynamic effects to consider, however:

- Lateral acceleration is related to building deformation. If one could increase building stiffness without changing any other dynamic properties, the reduced deformations would reduce acceleration.
- Lateral acceleration is inversely related to building period. If one could stiffen the building by 21 percent without changing building mass, the period would shorten by 10 percent. The shorter period would increase acceleration, partially counteracting the benefit of reduced deformation.

16.3 Vortex shedding
Image: Thornton Tomasetti

- Lateral acceleration is inversely related to building mass. If one could increase building mass by 21 percent without changing stiffness – by having thicker floors, for example – building period would lengthen by 10 percent without changing deformation. Acceleration would decrease, as long as the period change does not lead to VIO resonance.
- If building stiffness and building mass were both increased – a realistic situation when added stiffness comes from added material – building period may not change. Deformation and acceleration would be reduced, but this approach does not address VIO resonance, and adding significant weight to the building is an expensive way to improve occupant comfort.
- Stiffening the building to avoid a VIO peak from resonance in the range of anticipated service level storms, perhaps 10-year storms or less, may not avoid VIO for strength design, where the wind speed is 50 to 70 percent faster at the "ultimate" (700-year+) wind speed associated with strength design requirements. VIO resonance at strength-level wind speeds may result in forces too large for practical designs.
- When increased stiffness (and mass) may not achieve occupant comfort from wind-induced acceleration, or strength-level VIO is a concern, increasing building damping with supplementary devices or aerodynamic shaping become more practical approaches. For very tall buildings that have both low frequencies (long periods, say above eight seconds) and are subjected to high wind velocities, additional recently developed design strategies which integrate architecture and structure have proven effective. Shaping and damping are further discussed in the section in this chapter on dynamic behavior.

Serviceability must also consider the interaction of the structural frame with other building elements, including the facade, nonstructural partitions, elevators (with regard to building sway inducing cable sway), building mechanical, electrical, plumbing and fire protection services and architectural finishes. Typically this interaction is controlled by a combination of service-level drift limits – a preliminary overall drift target of height/500 to height/400 under the 10-year service wind load is often used – and appropriate detailing to permit differential movements where they are anticipated, such as at internal partition top tracks. Torsional deformations that can add to interstory racking, particularly along the facade, must be identified early in design and mitigated where this is practical. Torsion can result from asymmetrical building shape, surroundings affecting wind loading, eccentricity between center of stiffness and center of wind loading, eccentricity between center of stiffness and center of mass, and offsets in the lateral stiffness system. Torsional stiffness can be improved by placing lateral load-resisting elements as far out from the building center as possible, for example by locating braced frames or moment frames along the facade.

All these issues relate to material stiffness in structural systems that place "non-structural" limits on building height. Efficient use of structural materials to control deflections starts with considering the nature of tall building behavior.

Overall sway at the top of a tall building is most important when considering occupant comfort in windy conditions and, for very tall buildings, when determining elevator performance. Sway results from flexural behavior (Figure 16.4[a]) and shear behavior (Figure 16.4[b]) in combination (Figure 16.4[c]). Interstory drift from flexure increases at upper floors. Interstory drift from shear relates to story shear stiffness and may decrease at upper floors where the story shear force is less. Attention is usually focused on interstory drift, but interstory racking is most important when considering nonstructural facade and partition interaction with the primary structural system (Figure 16.4[d]).

Outriggers reduce overall and interstory drift from flexure by using the vertical stiffness of perimeter columns to generate a moment in the braced core opposing core tilt (Figure 16.4[e]). Outriggers at multiple levels reduce overall and story drift more efficiently (Figure 16.4[f]). Where all perimeter columns are engaged by outriggers and belt trusses, as core and perimeter move together racking is greatly reduced (Figure 16.4[g]). Shanghai Tower, presented in Case Study 14, is a good example: outriggers at multiple levels tie the concrete core to concrete mega-columns that carry the perimeter through belt trusses. All vertical building elements work together to minimize drift and racking.

As a building gets taller but not wider, its aspect ratio, or slenderness, increases. Meeting strength requirements, drift limits and acceleration limits becomes more difficult. Simply increasing column areas to resist the greater vertical load will not be sufficient to handle the increasing lateral loads and deflections. The most practical and efficient way to control building deflection and average drift on supertall buildings is to use an exaggerated taper; it simultaneously reduces shear and overturning from upper level winds and greatly increases the overturning moment of inertia at lower levels. Gustave Eiffel used the concept to establish the profile of his eponymous tower, and it appears in modified form in the record-breaking Burj Khalifa tower, a buttressed core design (Figure 16.5 overleaf).

Because lateral stiffness and strength requirements are affected by building aspect ratio, material MOE, member areas, floor size and column layout, as well as local wind and seismic conditions, there is no way to determine a theoretical maximum tower height that considers lateral effects.

Dynamic Behavior

As discussed, building wind behavior includes downwind and crosswind effects. Vortex shedding (Figure 16.3) causes crosswind dynamic forces, which result in lateral accelerations that can cause occupant discomfort. Tall buildings are designed to control lateral acceleration through one or more control strategies. Minimizing VIO by avoiding resonance between vortex shedding and building sway rates is an important strategy. As discussed, avoiding resonance through structural stiffening has practical limits. But because the vortex shedding rate is related to wind speed and building plan dimension and shape, other strategies can be applied. To control VIO, designers can:

- Stiffen the building – a potentially costly and limited strategy (as discussed).
- Taper the building. The vortex shedding rate depends on both wind speed and building plan dimension, so changing plan dimensions with height means vortex shedding will occur at different rates on different portions of the tower for the same wind speed. This reduces overall crosswind forces on the building at all wind speeds.
- Vary building shape so it "looks different to the wind" at different elevations and from different

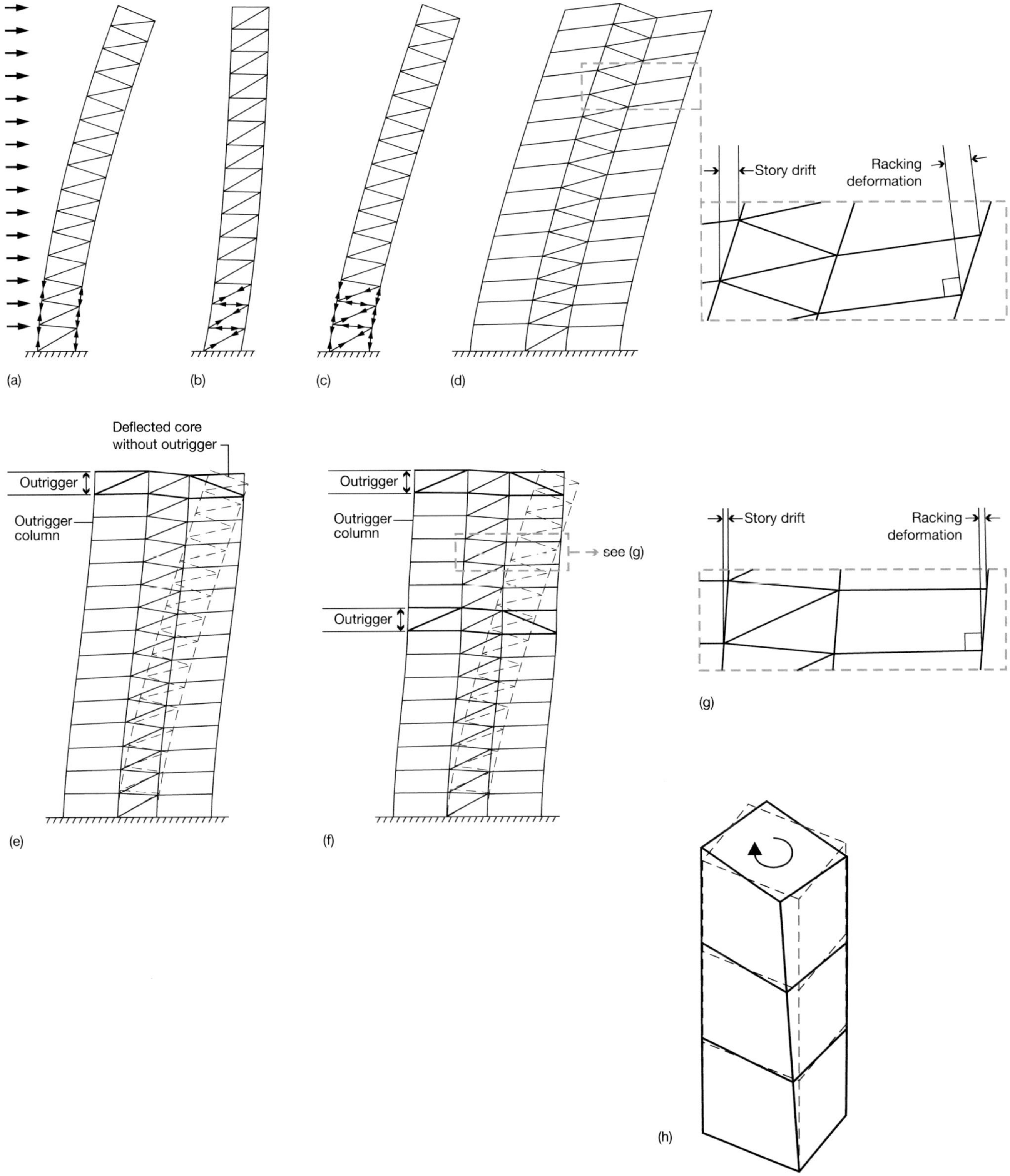

16.4 (a) Tube or flexural sway (b) Frame or shear sway (c) Combined sway (d) Interstory drift and racking if core acts alone – racking larger than story drift (e) Top outrigger (f) Two outriggers (g) Interstory drift and racking at core with outriggers – both racking and story drift greatly reduced (h) Torsion

Image: © Thornton Tomasetti

16.5 (left) Burj Khalifa (right) Eiffel Tower
Images: (left) SOM/Nick Merrick © Hedrich Blessing
(right) © Lucavanzolini/Dreamstime.com

16.6 Shanghai Tower: non-uniform, twisted and tapered, shape (left), building structure (centre), skin support (right)
Images: (left) Gensler (centre and right) Thornton Tomasetti

directions, to inhibit synchronized formation of vortices. The tapering, twisting outer skin of the 2,073-foot (632-meter) Shanghai Tower reduced crosswind overturning moment by 30 percent compared to an equivalent prismatic box tower design (see Figure 16.6). When the twist angle was changed from 100 degrees to 120 degrees the crosswind overturning moment was reduced by 10 percent. Even so, the design cross-wind moment is 170 percent of the along-wind moment because the along-wind tower motion is automatically damped when its cyclic deflection moves upwind.

- Articulate the surface. Much like the spoiler on a racing car or flaps on an airplane wing, building protrusions or irregularities disrupt the smooth airflow necessary to form strong vortices that

16.7 Taipei 101 stair-step corners
Image: Thornton Tomasetti

generate large crosswind excitation. The double stair-step corners of the 1,666-foot (508-meter) Taipei 101 dramatically reduced VIO effects at both service and ultimate wind speeds in this typhoon-prone location (see Figure 16.7). Other projects have employed more modest chamfers and notches.

- Bleed air. Holes through building corners deliver pressurized windward-face air to sidewall suction zones, disrupting vortices. This strategy has been used on several existing and planned tall buildings.
- Supplement damping, after shaping has been incorporated into the design. Damping directly reduces building accelerations by converting a fraction of the energy embodied in building sway into heat. Less energy means smaller deflections and less acceleration. Draining off a little energy in each cycle is especially effective in reducing the accumulation of vortex energy causing crosswind excitation.

Buildings have some inherent damping from friction between structural and nonstructural elements such as partitions and cladding, and within structural materials, for example micro-cracks in concrete. Tall buildings with concrete lateral load-resisting systems are often considered to have inherent damping that is 1.5 to 2 percent of critical (100 percent of critical would stop a displaced building from swaying through the neutral position). Tall steel-framed buildings are often considered to have inherent damping that is 1 percent of critical. The greater deformations under severe events such as earthquakes are considered to generate up to 5 percent of critical damping. But that would not apply to serviceability conditions such as occupant comfort in windy conditions. Where inherent damping combined with reasonable adjustments to stiffness and mass are insufficient to achieve desired acceleration limits, supplementary damping has been successfully used on many large buildings to achieve total damping of 4 percent, 6 percent, or more, dramatically reducing service-level accelerations. While supplementary damping also reduces building forces under ultimate wind loads, unless it can be considered fail-safe it is not used in determining strength design forces. In high seismic regions building mass is minimized to reduce seismic forces while the structure is stiff to control interstory drift. Where the resulting relatively short period creates high service-level wind accelerations, supplementary damping offers a cost-effective solution. Supplementary damping lifts one potential limitation on building height and reduces the significance of different inherent damping values of concrete and steel framing.

Supplementary damping for service (wind comfort) or ultimate (seismic) benefits has been successfully provided by several systems:

- Visco-elastic damping was distributed as numerous small dampers throughout the World Trade Center framing in New York. It used thin layers of a polymer designed to strain in shear, and then relax just in time for the strain direction to reverse.
- Viscous (velocity-related) dampers are used both as distributed elements and concentrated in a single location. Like automotive shock absorbers, they are efficient operating through a long travel distance. Distributed dampers are used with links and levers or clever bracing geometry to amplify local structural movements.
- Concentrated dampers in a tuned mass damper (TMD) are driven through large relative movements by the inertia of a large moving mass "tuned" to sway at the same rate as and in the opposite

16.8 Taipei 101 passive tuned mass damper
Image: Motioneering

direction to the building (see Chapter 19). The mass size depends on the intended damper energy and tuning accuracy; a TMD with a larger mass is more tolerant of slight mistuning.

- A passive TMD uses gravity pendulums or links and requires no external power supply or control system, so it is theoretically always available (see Figure 16.8).
- An active TMD is similar in concept but relies on electronic monitoring and control systems that activate and provide feedback to the device. With more accurate tuning a somewhat smaller mass provides the same energy absorption capacity.
- The water-based tuned liquid column damper (TLCD) is a long, narrow tank with upturned ends, sized so water rises and falls alternately at the ends in synch with building sway. Water flowing turbulently through metal grates absorbs energy.
- The water-based tuned sloshing damper (TSD) is sized to generate shallow waves moving across a tank and through one or more metal grates. Water turbulence provides less efficient damping than manufactured hydraulic dampers, so the weight of tanks and fill water is considerably greater than the weight of a passive or active TMD, but the construction costs may be less.

To maintain occupant comfort, accelerations should generally be below 15 to 18 milli-g for residential buildings and 20 to 25 milli-g for office buildings for the 10-year design return period. For supertall buildings with very long periods, accelerations are also checked at wind speeds associated with the one-month to one-year return periods. For the one-year return period wind, recommended accelerations are below 10 milli-g for residences and 15 milli-g for offices.

Ductility Effects on Seismic Performance

In regions with high seismic activity, structural system ductility is essential for practical and economical construction. During an earthquake a highly ductile system can accept large inelastic distortions that absorb energy without failing. Building codes reflect this capability by permitting ductile systems to be designed for smaller forces. To have a reasonable assurance of ductility, codes provide guidelines and requirements for member proportions, connection designs and structural system geometries. For taller buildings many codes include a requirement for a dual system, in which a stiff core or core-and-outrigger structural system is augmented by a ductile moment frame capable of resisting at least 25 percent of the design story shear forces even if it attracts less force by relative stiffness. In return, the design forces may be somewhat lower, although a minimum base shear, intended to cover forces from higher mode effects, may govern anyway.

The degree of benefit offered by the dual system approach can now be studied and compared to other, non-dual designs by "performance-based design," or PBD, in which computer models reflecting material ductility and geometric nonlinearity (*P*-Delta effect) are subjected to time histories of multiple seismic events, scaled to local seismic hazard levels. By tracking model behavior in small time steps, instantaneous structural deflections and member strains can be mapped on material nonlinear behavior curves to determine if the element is still elastic, has yielded or has gone into its work hardening range with limited ability to resist future load cycles or future seismic events. PBD can help identify the most critical elements and load paths to aid proper sizing of structural elements and maximize overall structural ductility and seismic performance.

Building codes are tailored to address the most typical, smaller projects. PBD offers a way to support and justify the innovative structural designs not anticipated by building codes that will be necessary to achieve the next generation of supertall buildings. Designers have used the PBD approach to justify non-dual systems in some jurisdictions, and both San Francisco and Los Angeles, California, have PBD guidelines for this purpose. Turkey is developing its own PBD guidelines. In China, PBD is used to confirm performance of dual system designs under the code.

For information on seismic performance, see also Chapter 19.

Structural Systems

The chart in Figure 16.9, developed by Fazlur Khan and published by CTBUH in 1980 (Khan and Moore 1980), assigned optimal height ranges for different structural systems. It still applies for many designs, although for the tallest towers newer building technology has overtaken it:

- Outrigger systems are used for super tall towers much higher than the 60 stories indicated on the chart. This was made possible by stiffness being provided through the use of high-strength concrete shear walls, made economically attractive by the lower relative construction costs of simple steel framing to shear walls compared to moment frames. This approach was made popular by architectural and leasing preferences for wider, column-free spaces with open, glassy facades. Examples appear in the case studies at the end of this book (see for instance Case Studies 1, 8 and 14). These designs typically include belt trusses which transfer all perimeter gravity loads to a few massive outrigger columns to minimize their net uplift and extract maximum strength benefits from members used for lateral stiffness.
- Fin walls are being extended from the core to the perimeter of tall residential towers to provide outrigger benefits within demising walls. The 2,716-foot (828-meter) Burj Khalifa is an extreme example, with the central core of a "Y" floor plan buttressed by corridor walls along each leg and demising walls across each leg.
- Exterior diagrids replace exterior vertical columns and sloping braces with diagonal members that resist both gravity and lateral loads. This differs from the exterior diagonalized tube in the old chart. Dual-function diagrid members make more efficient use of material but require that members remain elastic in earthquakes; otherwise, both lateral and gravity support would be lost upon member yielding or buckling. The original design for Tameer Tower in Abu Dhabi is a good example (Figure 16.10).

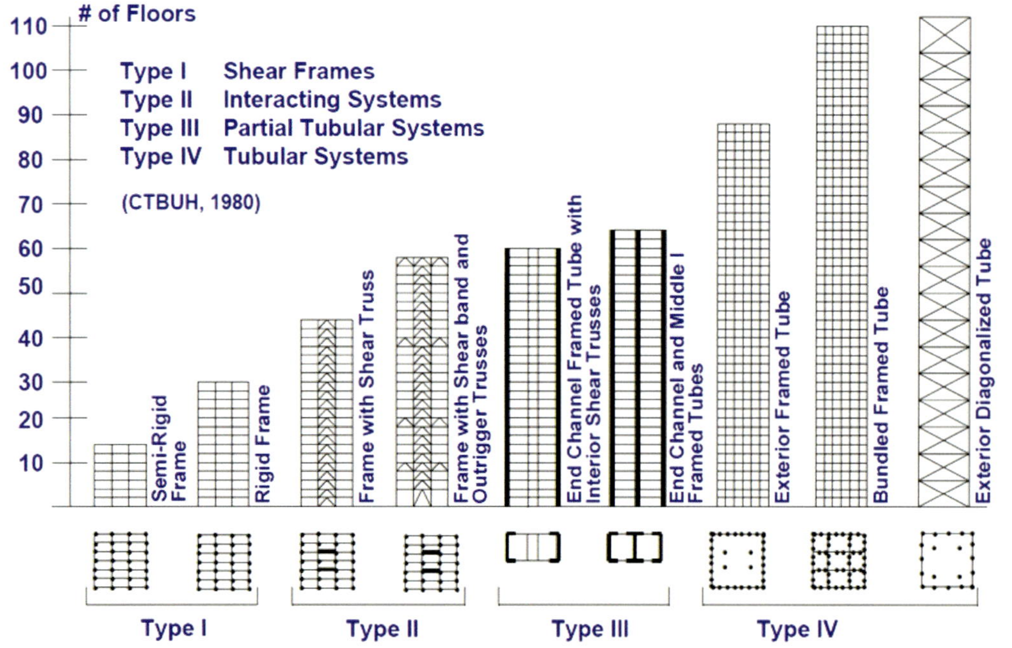

16.9 General limits of structural systems
Graphic: CTBUH

16.10 The original Tameer Tower design featured a diagrid exterior with sloping members acting as both columns and braces. Rendering
Image: Gensler

Long-Term Properties

Concrete, rather than steel, has increasingly been employed in recent supertall building structures. High-strength concrete provides strength and stiffness economically, provides greater inherent damping and simplifies construction by being placed in self-climbing forms at great heights using ground-based pumps. However, creep and shrinkage are long-term volume change properties of this material. Cumulative shortening in tall structures can grow to significant values, making prediction and control of differential shortening important design and construction requirements.

- "Creep" refers to the continued reduction of member length over time under constant compression stress, as typically exists in high-rise building columns and walls. Creep diminishes over time, with most creep taking place during the first several years of a building's life, but some creep continues for several decades.
- "Shrinkage" refers to the gradual reduction in concrete member size after the initial hydration phase when fresh concrete is placed. Shrinkage reflects the fact that concrete is placed with excess "water of convenience" beyond that required for the chemical process of hydration. Evaporation of that water creates internal stresses that contract the concrete. Even without evaporation, concrete chemistry creates some autogenous (nondrying) shrinkage, but this is a smaller effect.
- The potential amount of creep and shrinkage varies with the concrete mix design and admixtures used.
- The actual amount of creep and shrinkage is also affected by the quantity of steel reinforcement present (more reinforcement restrains and reduces both creep and shrinkage) and the ambient relative humidity that the concrete will eventually reach in equilibrium.
- The timing of creep and shrinkage is affected by the time and level of compressive stress imposed on a member (earlier loads cause more creep, larger loads cause proportionally more creep) and the member surface-to-volume ratio (a higher ratio permits faster drying and correspondingly faster shrinkage and creep).

While these effects are time dependent, several methodologies provide good estimates of creep and shrinkage, permitting structural engineers and contractors to implement methods to mitigate the effects on the completed structure.

For tall towers the main concern is a difference in the amount and timing of creep and shrinkage between core columns or walls and perimeter columns. Even if both sets of elements use the same concrete mixture, they generally have different stress levels, reinforcement ratios and surface-to-volume ratios, leading to different creep and shrinkage values and rates. If a structure has a concrete core and steel perimeter columns the difference over time is much larger, since steel columns do not creep or shrink. Framing within a tall building will not undergo the full effects of differential shortening among different vertical structural elements because they are cast incrementally over a long time. Along the way some degree of compensation is automatic, for example floor cast levels will not undergo effects from differential shortening occurring below that level. Intentional compensation may include "column camber," lengthening selected columns or walls to allow for future shortening. Such a strategy is coordinated between the engineer and the contractor with owner input, since floor levelness will vary with time.

Fire Protection

Fire safety is an important part of tall building design, both for occupant egress and for resistance to structural collapse (see also Chapter 23). Steel and concrete have time- and temperature-related properties that are known from extensive testing since the beginnings of tall building construction in the late nineteenth century. Steel may receive any of a number of fire protection materials, from archaic, heavy and costly concrete or terra cotta encasement, to contemporary sprayed-on fireproofing, to more impact-resistant cementitious spray-on, lath-reinforced spray-on and intumescent coatings at higher cost. Steel can even be left exposed, if it is used only for occupant comfort drift control rather than system strength and stability. The appropriate balance point for durability, cost and performance for supertall buildings is under continuous discussion.

Concrete is inherently fire-resistant but it is not fireproof. Its resistance comes from the combined action of rebar cover, thermal mass, steaming of free water and breaking chemical bonds of hydration. Concrete cover requirements over steel reinforcement are intended to delay the temperature rise of steel to avoid its loss of strength. More cover thickness is specified for higher fire ratings. Thermal mass explains why a concrete slab provides fire resistance: it takes time for the slab to heat up enough to reach a critical ignition temperature on the opposite, protected side. Steam from free water is both a benefit and a risk: when the temperature rise is gradual, steam can form below the concrete surface and then escape through naturally occurring pores.

When the temperature rise is fast, as in a petroleum fire, the limited pores in dense, high-strength concrete may not allow easy venting and the internal pressures can build up to the point where areas of the surface burst, causing progressive spalling that reduces member size and exposes reinforcing steel to higher temperatures. Adding polypropylene microfibers to the concrete can help by creating additional venting channels once the fibers melt, minimizing explosive spalling. Once a fire begins breaking hydration bonds, the concrete is no longer of structural value so this process is not sustainable. The extent of concrete breakdown can be determined after a fire by a simple chemical indicator. As buildings get taller, fire resistance requirements will likely be reexamined to see if increases in required hourly ratings are in order, given the time it takes to reach and attack a fire and the consequences of failing to extinguish a fire within a reasonable time.

Any new structural material used for strength and stability will undoubtedly be required to meet at least the same fire resistance levels as concrete and protected structural steel. Current high-strength composites rely on temperature-sensitive organic resin binders, making fire protection more difficult. New materials are likely to provide a way around this limitation.

Acknowledgments

Thanks to Len Joseph, P.E. and James Kent for their careful review and commentary on the text for this chapter.

References

ASCE (2005) *Minimum Design Loads for Buildings and Other Structures*. Reston: American Society of Civil Engineers.

Code of China (2010) *Technical Specification for Concrete Structures of Tall Buildings*, JGJ 3-2010. Beijing: Code of China.

Khan, Fazlur R. and Moore, Walter P. (eds.) (1980) *Tall Building Systems and Concepts*, Council on Tall Buildings and Urban Habitat Monograph on Tall Buildings Series. Reston, VA: American Society of Civil Engineers.

Chapter 17
Structural Possibilities

William F. Baker

Introduction

The very existence of skyscrapers is dependent upon structure. Along with the invention of the elevator, the creation of tall building structural systems has enabled the era of the skyscraper. As the structural engineering of tall buildings advances, so do the opportunities to create architecture that had not previously existed. Structure is the armature upon which all other parts of the building reside, so it must respond to, and work with, the forces of nature and the architectural functions of the tower. The possibilities of structure form the basis for the possibilities of skyscrapers.

Structural solutions, or "structural possibilities," have greatly expanded in the decades following the Second World War. Today there exists a broad range of structural solutions that have been successfully realized, although it should be noted that structural possibilities are not limited only to current existing solutions. In nature, new species are created through the evolution of existing life. Tall building designers can also create (invent) new systems by evolving the existing systems. Unlike nature, designers are not limited to evolution; they also have an opportunity to create something completely new. Solutions are influenced by the height of the tower, the location in which it resides, and the desires of the owner and the design team, and solutions can range from highly rational to somewhat capricious. All structural systems must satisfy the demands of nature and the requirements of building codes, hence all of these creations must embrace some overriding principles. Furthermore, because economics are often the determinant factor in whether a building is realized, successful creations usually embrace structural solutions that are both rational and efficient. These engineering-based solutions are often the starting point for extending the range of structural possibilities. This chapter will discuss the challenge of creating rational, efficient, and economical solutions, since these attributes form the basis of most realized designs; it will also address designs that ultimately wander from these initial goals.

In addition to evolution and invention, there are creations that result from the development of new and evolving materials, construction technologies, and computational tools. Along with a heightened understanding of structural behavior, these advances provide opportunities that did not exist in the past. Engineers and designers will find new solutions to the challenges presented by tall building design as technologies evolve and new insights occur.

"The Idea": The Structure as a Cantilever Beam

The fundamental conceptual simplification of the tall building is that it is a vertical cantilever beam. As such, it is globally statically determinant, and the approximate total forces on the tower are known *a priori*. Furthermore, the forces on the tower can be conceptualized as overturning moments and shears that result in flexural and shear deformations of this conceptual beam.

By taking something as large and complex as a huge skyscraper, something formed by thousands of individual beams, columns, walls and slabs, and imagining it as one simple entity, the designer can design these giant structures rationally. All buildings cantilever from the ground in the same way as a tree or a lamppost does. In a certain sense, then, the tall building is the simplest of all structural problems. The support conditions dictate that there is only one direction for the loads to go: from the sky to the ground. However, the art and science of tall building design lies in selecting the appropriate system to carry these forces from where they occur in the air down to the structure's foundations.

It is a structural irony that the taller the building, the more pure the beam needs to be and, in some ways, the simpler the solution. A tree, then, is a good analogy for a slender tall building. Trees are generally very slender and singular structures, with one trunk emerging from the ground until it devolves into a series of branches that cantilever from the trunk. It is this purity of system and hierarchy that is often emulated by successful tall building designs.

A quick look at the Willis Tower (née Sears Tower) in Chicago, Illinois, demonstrates this point. Although the system is often described as a bundled tube, it is formed by the rigid interconnection of a myriad of beams and closely spaced columns, but acts as if it were a giant steel box beam cantilevering from the ground, a beam with four webs and four flanges. Openings in the frames, however, reduce the structural efficiency of this beam to about 70 percent of the efficiency of a steel beam with no openings. This reduced efficiency is caused by shear deformations in the frame. Such deformations do not negate the beam concept, but do make the conceptual evaluation more sophisticated than a beam with no openings, where shear deformations can be neglected. As it rises out of the ground, the structural frame of the Willis Tower becomes lighter and setbacks in the building further reduce its stiffness. The tower is essentially a beam with a variable cross section. The manner in which the original design team conceptually modeled the tower as a beam with a variable stiffness (moment of inertia) is shown in Figure 17.3.

17.1 The Willis Tower, Chicago—a giant beam
Image: Skidmore, Owings & Merrill LLP/© Timothy Hursley

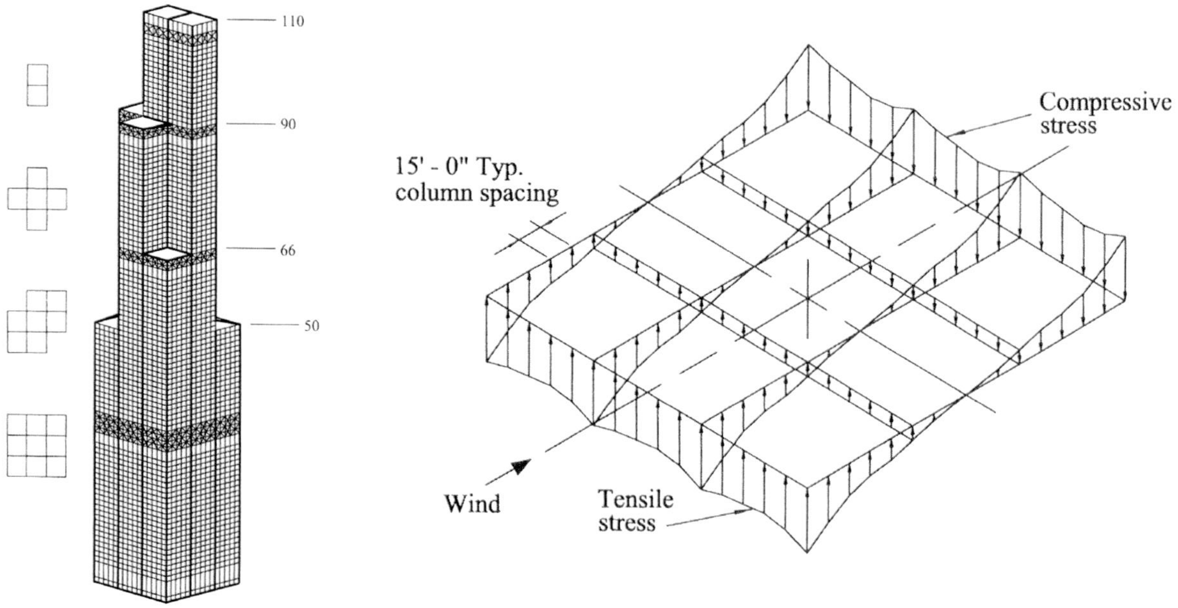

17.2 The Willis Tower as a box beam with four webs and four flanges
Graphic: Skidmore, Owings & Merrill LLP

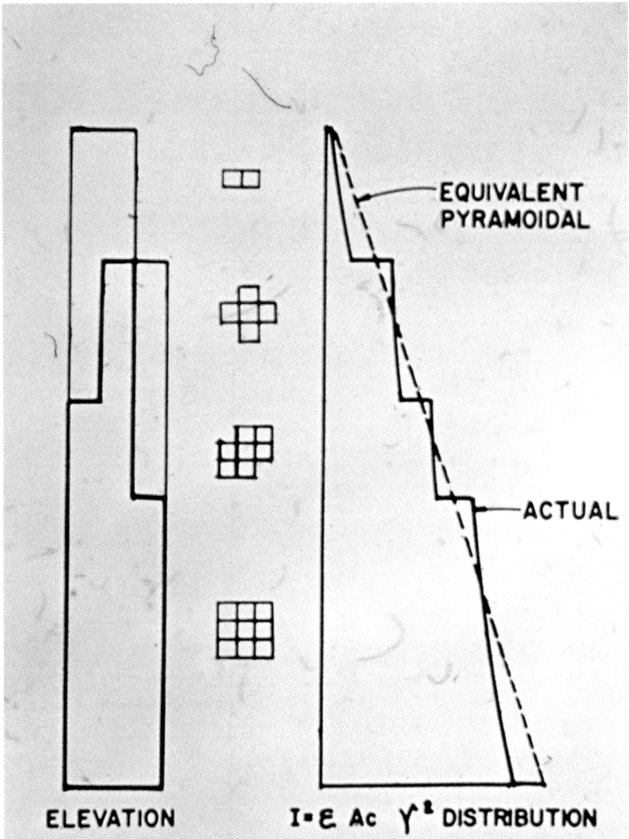

17.3 The Willis Tower as a beam with variable stiffness
Graphic: Skidmore, Owings & Merrill LLP

Shorter, squatter buildings can be more conceptually complex. While they still cantilever from the ground, the structural system is often a series of parallel sub-assemblages or individual elements that behave in a complicated joint action.

As noted earlier, engineers typically refer to cantilever structures as "statically determinant." This means that at any height of the building, the total forces are generally known. Gravity loads are the summation of everything above a given elevation, the wind shears and overturning moments are integrated from the top down, and even the seismic forces can be approximated in this manner. Before the design begins, the weight of the building's gravity and seismic loads can be estimated based on a density in the order of 1.6 kN/m^3 for an all-steel office building and up to 3.2 kN/m^3 for a residential building composed entirely of concrete. Not only are the forces known in a global sense, but the behavior is generally also known; the magnitude of the tower's movement will be limited to values determined by building codes or good practice. For example, the period of the building can be first approximated by simple rules, such as the number of stories divided by 10, or $T = C(h)^x$ (American Society of Civil Engineers, 2010), where C varies from 0.070 to 0.085. (In this equation, T equals the building's fundamental period, while C and x are period coefficients dependent upon the building's lateral system; h equals the height of the structure.) This predetermination of global forces and movements is conceptually important because it enables the design team to understand the range of stability and overturning concerns in the initial stages of design.

Once the initial loads on the building have been estimated, the design team must choose or create a structural system to resist them. In general, lateral forces, particularly wind loads, are the most dominant considerations, and the actual magnitude of the wind forces is greatly affected by the shape and harmonics of the tower. Any structure that carries wind forces to the ground (be it a tree, a lamppost, or a skyscraper) must resist two major structural phenomena: shear and flexure.

The taller and more slender the building, the more efficient the shear resisting system must be. Whether it consists of diagonal bracing, moment frames, walls, or individual elements, the shear resisting system is essential for carrying the lateral loads to the vertical elements that, in turn, resist the overturning forces on the cantilever. For example, in the Willis Tower, the four frames that are parallel to the lateral forces form the shear (sliding) resisting system that delivers the accompanying overturning forces to the cross frames. Another example would be the John Hancock Center, also in Chicago. Its structure delivers the wind shears to the corner columns when the X-bracing is parallel to the wind forces. The braced frames on the windward and leeward faces then share the wind load amongst the various columns in order to resist the overturning wind forces.

A stiff shear system is necessary for the entire building to act as a single giant beam rather than an aggregation of individual elements or subsystems. Because it is not possible to create a completely rigid shear system, there exists a phenomenon called "shear lag." This occurs when the overturning strains and stresses are not

17.4 The John Hancock Center, Chicago
Image: Skidmore, Owings & Merrill LLP/Ezra Stoller © Esto

distributed linearly from the center of the tower, resulting in a less effective use of the tower's vertical elements to resist the overturning moments on the structure. This is evident in Figure 17.2, where not all of the columns in the "flanges" are uniformly stressed.

Although efficient and stiff shear resisting systems can reduce shear deformations to a small portion of the target deflections, it is not practical to do the same for the flexural deformations. The deflection of a steel I-beam without holes is almost totally attributable to the flexural bending. For a given layout, this deflection can generally only be reduced at the cost of increasing the size of the columns and/or walls. Beam deflection can be halved by doubling the area of the cross section, for example. The great expense of reducing deflections by increasing the cross-sectional area of the vertical element places very practical limits on reducing flexural deflections.

Structural Systems

Skyscrapers have existed for over a century. As materials became more refined, new construction technologies emerged and the engineering problem became better understood. New structural systems were created which allowed towers to become taller and more slender. The list of systems will never be complete; the evolutionary process will continue as long as skyscrapers are designed and built.

There have been several attempts over the years to classify structural systems that are appropriate for buildings of various heights and slenderness. Lateral systems for buildings that are either short or wide do not have to be as efficient or as disciplined as those of very tall, slender towers. If a structure of normal floor size is in the range of 20 to 30 stories, simple moment frames are often sufficient to resist the forces. Simple concrete walls or steel bracing around an elevaor bank is often all that a building of modest height or slenderness needs to provide stability and resist lateral loads.

Fazlur Khan, the lead structural designer of the John Hancock Center and the Willis Tower, developed a series of charts that illustrate the potential structural systems for towers of various height and material (Khan 1973: 127). His initial work concerned buildings that were either all concrete or all steel, although he later developed composite systems that used both steel and concrete. Figure 17.5 is an extension of Khan's earlier charts.

The differentiating characteristics of these structural systems correlate with their specific shear resisting systems and the manner in which the shear system delivers the forces to the system that resists the overturning moments (the flexural system). For both steel and concrete structures, the list of shear systems includes moment (rigid) frames, diagonal bracing, and walls. Some of these systems are more common in one material than the other. Steel structures typically utilize moment frames and diagonal bracing, for example, while moment frames and walls are common in concrete structures. Examples of systems realized in either material do exist, however.

The stiffer the shear system, the more the shear deformations can be suppressed, but this only works if it delivers the shear to an efficient flexural system.

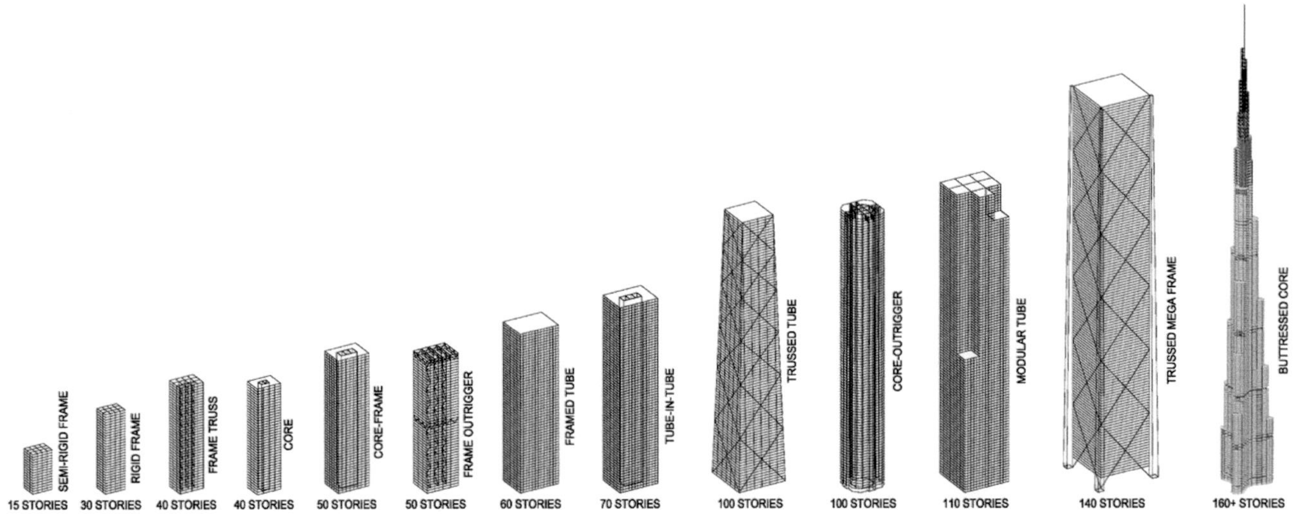

17.5 Tall building systems
Graphic: Skidmore, Owings & Merrill LLP

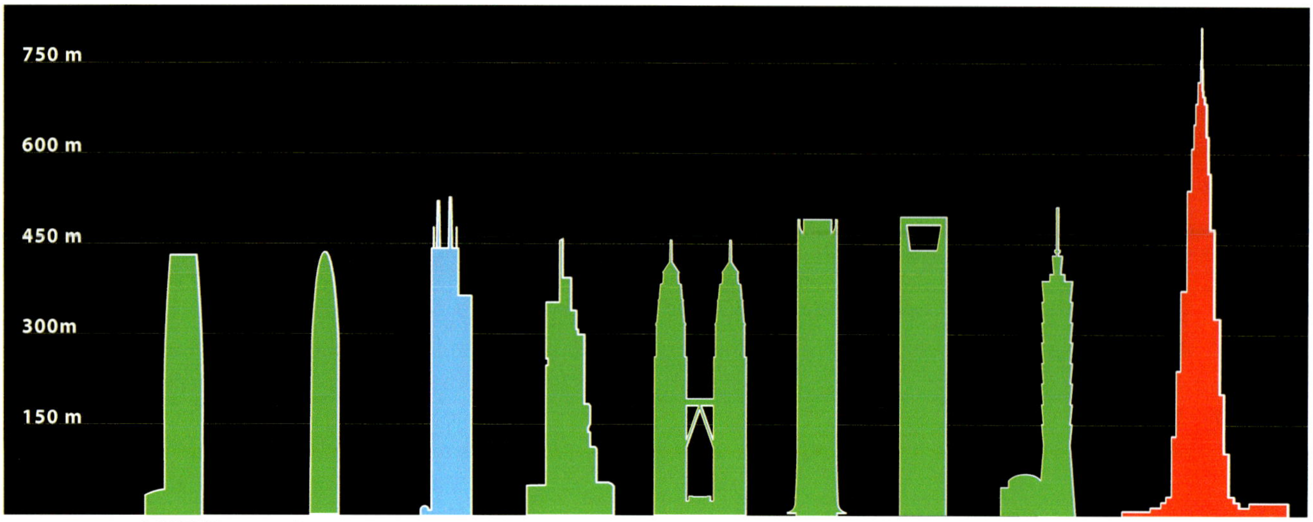

	GUANGZHOU INTERNATIONAL FINANCE CENTER	KK100 DEVELOPMENT	WILLIS TOWER (FORMERLY SEARS TOWER)	ZIFENG TOWER	PETRONAS TOWER 1 AND 2	INTERNATIONAL COMMERCE CENTRE	SHANGHAI WORLD FINANCIAL CENTER	TAIPEI 101	BURJ KHALIFA
Location	Guangzhou	Shenzhen	Chicago	Nanjing	Kuala Lumpur	Hong Kong	Shanghai	Taipei	Dubai
Year Built	2010	2011	1974	2010	1998	2010	2008	2004	2010
Height	439 m	442 m	*442 m	450 m	452 m	484 m	492 m	508 m	828 m

STEEL CONCRETE COMPOSITE

* BUILDING HEIGHT DOES NOT INCLUDE BROADCAST ANTENNAS

17.6 Structural materials of the world's tallest buildings
Graphic: Skidmore, Owings & Merrill LLP

The flexural system is essentially the vertical elements (columns and walls) that resist the overturning moments. As for a beam, the further these vertical elements are from the center, the more efficient they become. An element that is twice as far from the center can resist double the overturning moment and is four times stiffer.

Lateral loads on a tall, slender tower will generate very large tension forces. From a practical point of view, the tension forces and the tension splices are difficult and expensive to deal with, whereas, although compression forces require attention, the splices and connections are more easily achieved. Gravity is reliable and can be very helpful in economically resisting the tension forces. As buildings must resist both wind and gravity, the most economical solution is to create an integrated system that can accomplish both, with the major gravity resisting columns and walls also acting as the major wind resisting columns and walls. Because maximum gravity and lateral loads are unlikely to occur at the same time, it is economical to use the excess capacity of one structural system to resist the forces of the other. If the elements resisting these tension forces are pre-compressed with the weight of the building, the structure is much more easily designed and executed, and at a lower cost than a structure that must resist giant tension forces.

Structural Materials: Steel, Concrete, or Both?

A review of the ten tallest buildings completed or near completion by the end of 2012 reveals that structural systems fall into a limited number of categories. Figure 17.6 shows that the systems consist of structures that are primarily steel, primarily concrete, or a composite mixture of the two. For most of the skyscraper era, structural steel has been the dominant material, but structural frames built entirely of concrete or a mixture of steel and concrete have become more common in recent years. Structural frame material choice is dependent upon many things, including usage, mechanical (HVAC) systems, local construction expertise, cost of materials, and foundations.

Structures are often divided into two subsystems: horizontal framing and vertical framing. Usage and the perceived expectations of the anticipated users of the building often determine the choice of horizontal framing or floor framing. Although these are not hard and fast rules, residential buildings with limited need for horizontal ducts, etc., are often constructed with thin concrete floors that permit a minimum floor-to-floor height and high utilization of the building volume. Office floors, which require long spans and extensive horizontal mechanical distribution systems, are often steel framed.

Vertical elements can also be steel or concrete. Concrete is very economical in compression when compared to structural steel and, in many tall buildings, the vertical elements are dominated by compression. The negative attribute of concrete vertical elements is that they tend to be larger and heavier than the same element produced in steel, reducing the usefulness of floors and causing foundation problems. Furthermore, a large part of the structure in a tall building is there to hold the weight of the structure itself. A heavy structural system can give rise to a cascade effect, requiring larger columns and walls that in turn cause even heavier loads and a larger structure. The larger size of concrete structural members, compared to those in steel, tends to also lead to a loss of floor area.

Concrete buildings tend to be more massive, which leads to more expensive foundations and higher seismic forces; however, the lower cost of material often overrides these considerations. Furthermore, while structural steel has always been a highly industrialized and mechanized technology, concrete has recently become comparable to structural steel in speed of construction. Given a building geometry reasonably consistent over a large number of stories, modern self-jacking formwork systems can lead to fast construction, where the tower is essentially built as if it were a vertical factory. Complex and varied high-rise towers are possible in concrete, but the speed of construction often suffers because of the need to constantly adjust the formwork or introduce special forms. From a wind engineering perspective, concrete frames tend to have more mass and inherent dampening, both of which help decrease wind motions (see Chapter 20 on designing for wind).

Structural steel works well in tension, whereas concrete must be reinforced with steel in the presence of tension. Steel beams and columns are generally much lighter than concrete elements performing the same function. Furthermore, the behavior of structural systems made of steel tends to be more predictable than the behaviour of those made of concrete. Steel is also more dimensionally stable; because of creep and shrinkage effects, concrete elements continue to deform over time, which can lead to unintended movements and redistribution of forces. Steel structures are often more suitable for complex and varying geometries, as it is often easier to customize a shape in steel. From a seismic point of view, steel has less mass and often results in lower seismic forces. Steel's greater ductility often has an advantage in resisting seismic forces.

Composite systems, also known as mixed systems, try to use steel and concrete in ways which use each material to its advantage. These systems are quite frequently used as structural systems for tall buildings.

17.7 Self-climbing formwork and tower cranes
Image: Skidmore, Owings & Merrill LLP

Although there are numerous ways that concrete and steel systems are integrated, a common solution begins with concrete walls surrounding the central core of the building. These walls are then encompassed by structural steel floor framing, which, in turn, is supported at the perimeter by either steel or concrete columns. If the core itself is too slender to go to great heights, outriggers composed of either steel or concrete will connect the core to the perimeter columns. In doing so, the outriggers transfer the overturning moment from the narrower core to the wider perimeter.

The two systems are frequently combined for fiscal reasons. Structural steel is often the preferred material for long span floor planning, as one might find in an office building. However, concrete is much less expensive when carrying compressive loads. It is not unusual to find that concrete is only one fifth to one eighth the cost of steel when used in compression for either strength or stiffness. Therefore, even though concrete columns are generally larger than corresponding steel elements, the loss of floor area is often offset by the financial gains. The most viable solution often lies in the combination of both materials, with steel floor framing and concrete vertical elements.

Slenderness

Slenderness is a major factor in the design of tall buildings, and the slenderness of a tower is usually defined as the ratio of its height to its smallest whole-base dimension. With new advances in technology and construction techniques, buildings have become more slender. The all-steel towers of the 1960s and 1970s were in a slenderness range of 6.0:1 to 6.7:1. Today's concrete and composite towers are sometimes as slender as 8:1 or 9:1. There are some instances of towers even more slender, but the cost of the structure rises exponentially with extreme slenderness.

The importance of slenderness cannot be overemphasized, as it is often more relevant than height considerations. A very slender 40-story tower can have stability and movement issues that are more difficult to solve than an 80-story tower with a large base. Depending on the relative importance of movement versus strength, the structural quantities normalized over the supported floor area (kilograms of steel per square meter or cubic meters of concrete per square meter) can often vary with the cube of the slenderness ratio for very slender towers.

One way to reduce the slenderness is to utilize the entire width of the building in resisting the overturning forces (flexural forces). In the towers of the 1960s through the 1980s, the lateral systems were primarily located on the perimeter. This approach helped to minimize the slenderness of the tower. However, the shear resisting system that interconnected the vertical elements often resulted in dense structures that limited the openness of the building and made shaping the towers difficult.

An interesting development in the last 30 years is the move toward interior shear resisting systems that are then connected to the perimeter at discrete locations, rather than exoskeletons. There are many reasons for this, but in order to understand the evolution in systems, one must first understand the science of the systems themselves. Generally, as in the case of an exterior system, the perimeter is regarded as the most efficient place to resist the overturning moments on the tower, because the vertical elements are farthest from the center of the building. The tension and compression from the wind decreases in proportion to the distance from the center, and the stiffness of the vertical elements increases by the square of the distance from the core. However, the wind shear is most efficiently resisted by systems that are continuous and uninterrupted, such as solid walls or diagonal bracing. It is often difficult to place these elements on the perimeter of a building because they limit the view out of the tower and restrict the ability to manipulate the shape of the perimeter. An interior system may be employed in such cases, one where walls and diagonal bracing are placed within the core of the building by carefully arranging the elevators, stairs, and mechanical shafts, all of which need to run vertically. In these instances, one can achieve systems that are very stiff in shear. A slender core, however, may be limited in height by excessive flexural deformations. This predicament results in the great prominence of core and outrigger systems, where the core resists the shear and the outriggers deliver the overturning moments to the vertical elements at the perimeter.

Figure 17.8 shows the lateral system for the ten tallest buildings in the world as of the end of 2012. A review of the summary shows that the more recent towers tend to utilize an interior shear resisting system with the overturning forces resisted on the perimeter.

Scale

As towers become ever taller and more slender, the issue of scale must be understood.

Just as in nature, systems that exist on one scale may not be possible on another. System proportions will also change with scale, for the same reasons that the proportions of the bones of a mouse are so much different than the proportions of the bones of an elephant. A structure's appropriate scale is determinant upon three major factors: height, area, and distance from the core to the perimeter.

In general, the taller the building, the wider the base must be. Human beings spread their feet apart to increase their stability; similarly, as a building becomes taller, the base becomes proportionally wider and thicker. Point towers such as the Willis Tower, Taipei 101, the Shanghai World Financial Center, and most of the other existing world's tallest buildings, scale geometrically by the cube. Therefore, if the Willis Tower were to double in height, the floor area would grow cubically

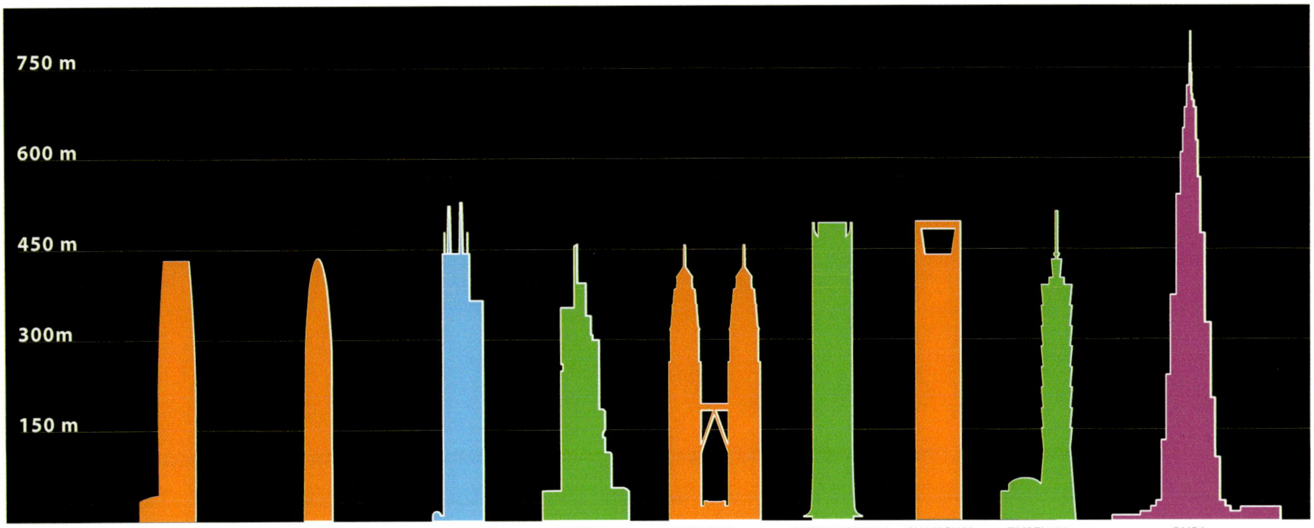

	GUANGZHOU INTERNATIONAL FINANCE CENTER	KK100 DEVELOPMENT	WILLIS TOWER (FORMERLY SEARS TOWER)	ZIFENG TOWER	PETRONAS TOWER 1 AND 2	INTERNATIONAL COMMERCE CENTRE	SHANGHAI WORLD FINANCIAL CENTER	TAIPEI 101	BURJ KHALIFA
Location	Guangzhou	Shenzhen	Chicago	Nanjing	Kuala Lumpur	Hong Kong	Shanghai	Taipei	Dubai
Year Built	2010	2011	1974	2010	1998	2010	2008	2004	2010
Height	439 m	442 m	*442 m	450 m	452 m	484 m	492 m	508 m	828 m

FRAME – OUTRIGGER
CORE PLUS PERIMETER **
CORE – OUTRIGGER
BUNDLED TUBE
BUTTRESSED CORE

* BUILDING HEIGHT DOES NOT INCLUDE BROADCAST ANTENNAS
** MAY ALSO INCLUDE OUTRIGGER

17.8 Structural systems of the world's tallest buildings
Graphic: Skidmore, Owings & Merrill LLP

(2^3) and increase by a factor of 8. As it stands today, the structure's base measures 69 meters × 69 meters, and the distance from the glass to the core is 23 meters. The steel-framed system of the Willis Tower is at the limits of its scale. If the roof of this steel tower were to be rescaled to 610 meters, the base would grow to 95 meters and the total area in the building would increase from a very large 410,000 square meters to an astronomical 1,100,000 square meters. At this height, the structural system of the Willis Tower would be inappropriate, and a different structural system would have to be used.

Total built area is a major factor in the design of tall buildings, as there are practical limits on the total floor area of a tower. In addition to the great cost of finishing and fitting out an enormous building area, there are also limits on the economic ability of a given real estate market to absorb this much area. Many recently proposed supertall towers remain unrealized due to the problem of excessive floor area.

A further consideration is the practical limit concerning the distance from the core to the windows, in order to allow for external views and for light to enter the interior space. This is in direct conflict with the desire for a wide footprint on a tall building. For these reasons, new structural systems are needed as towers become taller and more slender.

Architectural Massing

Perhaps the most important structural parameter in a tall building is the architectural massing, or shape, of the building. Not only does it define the complexity of the tower, it usually has a significant influence on the wind forces on the building (see Chapter 20).

Wind is the dominant structural consideration in the design of tall buildings. Wind flows around a tall building are often unstable. As the wind passes by, the flow separates from the side of the tower, creating a series of vortices that cause pressure differentials across the building. Although the drag effects of wind on a tower are important, these forces are generally not the controlling concern. It is the crosswind forces caused by the vortex shedding, in conjunction with the natural harmonics of the tower, which can lead to very large forces that are perpendicular to the direction of the wind. Because of the speed of commonly occurring wind events and the normal range of building width, there will be certain wind events where the vortex shedding matches the natural harmonics of the tower.

The rate at which these vortices occur is related to the wind speed, as well as the width and shape of the floor plate. Whereas the speed of the wind is set by nature, the geometry of the tower is set by the design

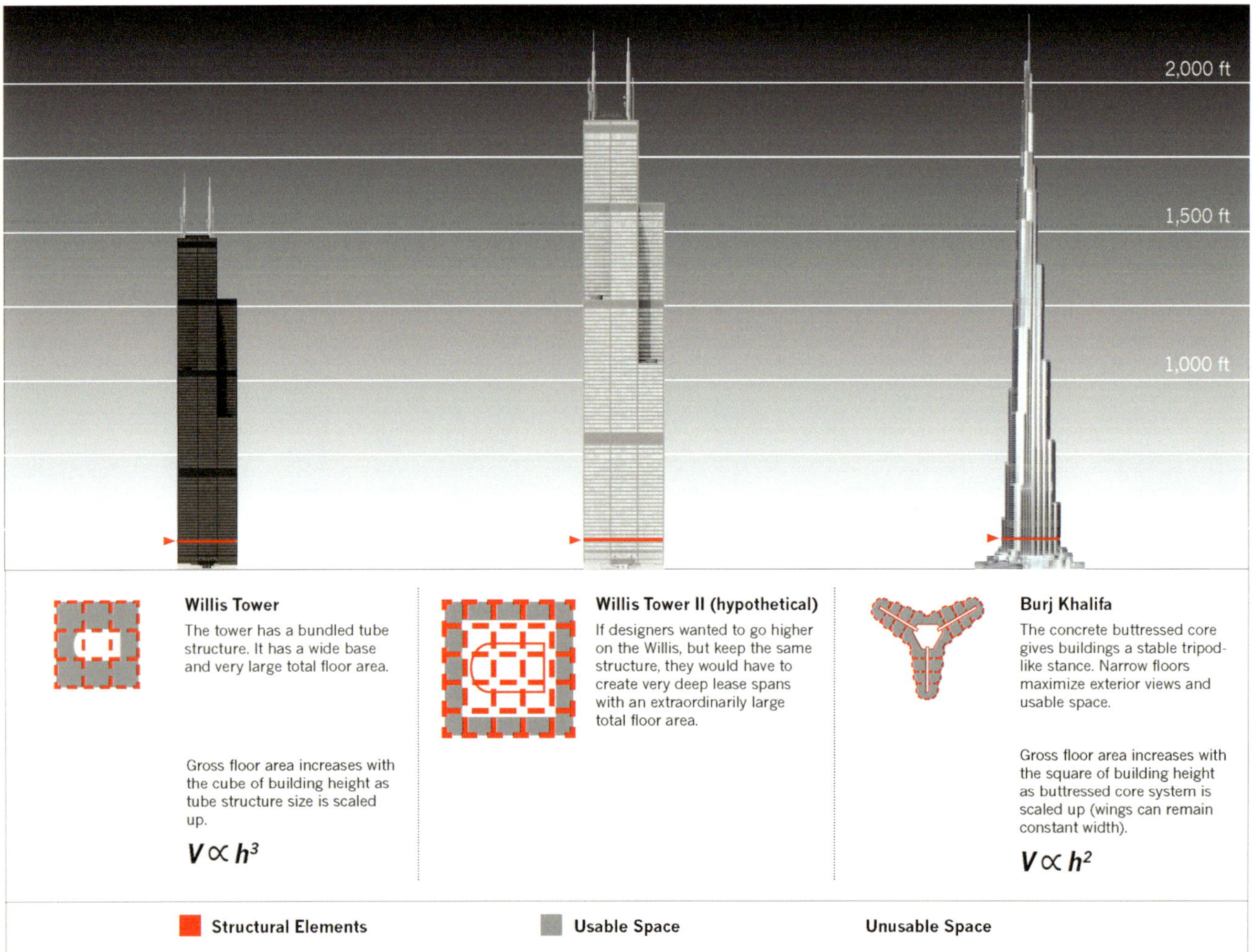

17.9 The issue of scale
Graphic: Skidmore, Owings & Merrill LLP

team. Extremely large forces can be generated if the building's profile encourages the development of highly organized vortex shedding at frequencies that harmonize with the natural frequency of the structural system. Varying the shape of the building from top to bottom, however, can 'disorganize' the vortices, so that only a small portion of the tower will be generating vortices with harmonic frequencies at any one time. Building corners are often where the vortex is initiated by flow separation. Vortex intensity here can be greatly reduced by having a corner that is somewhat canted or curved. It is also true that other geometric conditions, such as notches or open floors, can allow air to bleed through the building and disrupt the formation of strong vortices. Forces on a building whose shape leads to the formation of highly organized vortex shedding can be many times larger than those on one shaped to avoid such effects.

For a building of a given shape, the dynamic behavior of the tower can also have a dramatic effect on the scale of the forces. A building can be tuned by adjusting its stiffness and mass distribution in order to minimize the resonant behavior for storms likely to occur within the life of the building.

Design Philosophy

It should be noted that many proposed tall towers and most proposed supertall towers are not ultimately realized. In order to increase the chances of success, an overarching design philosophy can be helpful. Many of the fully realized tall towers have the following traits in common:

17.10 Burj Khalifa—the buttressed core
Graphic: Skidmore, Owings & Merrill LLP

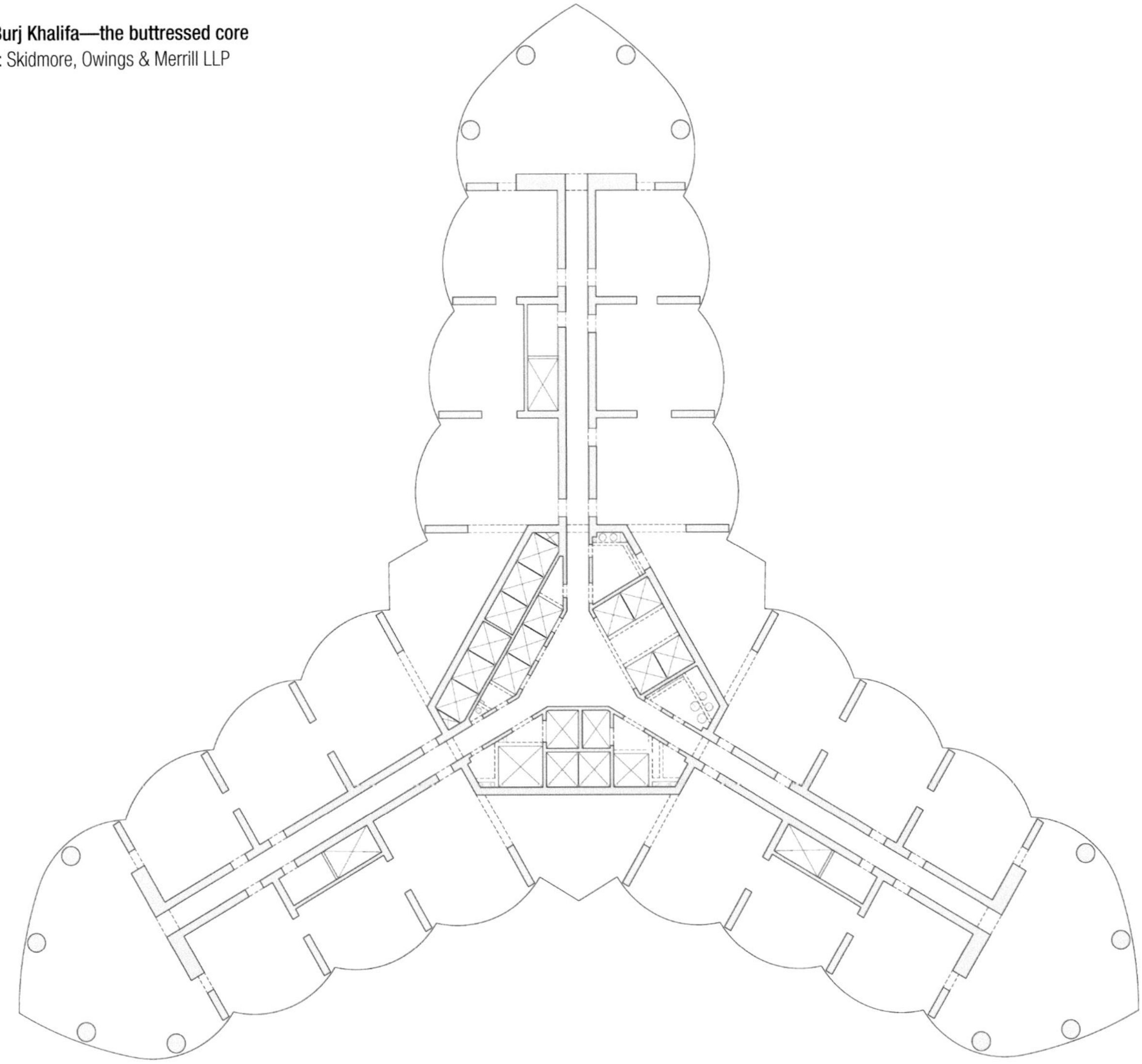

- The towers are based on a simple, rational structural system. This system is the driving, central organizing device on which all other systems (architectural, MEP, vertical transportation) are arranged.
- The structural ideas behind the structural system are clear. If one begins with a clear structural idea, the building will have an inherent logic, and in turn will likely be easier to build.
- The structures are efficient and economical.
- The structures are easily built.
- The structures are capable of being built quickly.
- The structures and building systems have a clear hierarchy. Not every component of a tall building is of equal value. By setting a hierarchy of importance, the design team can more adeptly solve the problems that may arise. By setting a hierarchy, the structure is ultimately more organized and more thoughtfully designed.

Future Towers

Although they only represent a small portion of all tall buildings, supertall towers test the ingenuity of the design team and capture the imagination of society. These designs are at the limit of what is possible and, as such, are testing grounds for new ideas and new insights into the tall building problem. As discussed earlier, as buildings become taller, the size of the base needs to become larger, but the depth of the floor plate may not. Also, the total built area needs to be within limits; often, proposed supertall buildings have extremely large total floor areas. In and of itself, too much floor area can render a proposal unviable. The Burj Khalifa may provide clues to the development of future supertall towers.

The buttressed core structural system of the Burj Khalifa was able to separate the issues of base size from lease span. The base structure extends approximately 49 meters from the center but the lease span ranges within a very normal 8 to 11 meters. This particular structural system works well with a residential layout. Because the lease span is generally independent from the base size, the total floor area is also reasonable. Other layouts would be necessary for an office building. It will be the creative solutions to these issues that lead to successful supertall towers of the future.

Structural possibilities are limited only by our ability to create new solutions for the demands of nature, height, material, and dreams. Whether a new structural system evolves from the refinement of its predecessor or is created as an entirely new device, designers and engineers should reexamine the "tall building problem" using modern technological and material advancements to create the next generation of supertall. As manifested in the structures themselves, the sky truly is the limit.

References

American Society of Civil Engineers (2010) *Minimum Design Loads for Buildings and Other Structures.* Reston: American Society of Civil Engineers.

Khan, Fazlur (1973) "Structural Design of Tall Buildings" in P.K. Thomas (ed.) *Proceedings of the National Conference on Tall Buildings.* New Delhi: Indian National Group of the International Association for Bridge and Structural Engineering.

Chapter 18
Foundations and Basements

Tony A. Kiefer and Clyde N. Baker, Jr.

Introduction

Foundations and basements of major high-rise buildings can account for significant relative cost (and risk) for the structure, even though the below-ground improvements are rarely seen or appreciated. The cost and risk of installation of the substructure can be greatly magnified if the site geology is not considered early on in the project planning. In spite of this fact, it appears that too often buildings are planned with little consideration of the difficulties that may exist below ground.

The recent worldwide building boom has seen tall buildings being constructed at an unprecedented rate. Unfortunately there have been some recent spectacular failures related to subsurface construction. Many of these failures have been captured on video phones and posted on the internet, where literally millions of viewers have watched catastrophes in progress or seen their immediate aftermaths within hours or days of the events. These examples serve as stark reminders that lives, reputations, personal freedom, millions of dollars, and construction schedules are routinely at risk when construction occurs below the ground's surface. Such recorded failures have included:

- The collapse of a retaining wall and complete inundation of a multi-level basement excavation that occurred in seconds in the Middle East in 2007. Fortunately no one was killed but the project was delayed for more than a year. This dramatic video should convince architects and engineers of the dangers of deep excavations adjacent to a waterway. Nevertheless, because most of our cities have grown up as port cities at the edges of rivers or oceans, excavations are constantly being extended through soft landfills often only metres away from open bodies of water.
- The collapse of a multi-story residential building under construction in China in 2009. Undoubtedly the building was designed for lateral loading conditions from wind and seismic effects but it was not designed for the lateral load resulting from excavating a deep pit on one side of the building while filling the excavated soil on the opposite side of the building. As a result the building fell over, killing one of the workmen in the pit. Those found responsible are serving prison terms. Failure to consider the construction sequencing during design can lead to load cases far in excess of those required by code.

- The 2009 implosion of a 31-story condominium building in Texas that experienced over 30 centimeters of differential settlement during construction and could not reliably be repaired. This project is the subject of a US$145,000,000 lawsuit. The building was of unprecedented height in the area, thus specific experience on how the proposed foundation systems would perform was not available. A peer review might have prevented this failure.
- The collapse of a four-story high pile of concrete blocks used as a reaction for a pile load test in Singapore in 2011. Fortunately no one was injured in this accident, which was caught on video and posted to the internet within hours. Modern procedures for conducting load tests exist which completely eliminate the risk and expense of reaction frames and massive weight stacks. Were the older, more time consuming, dangerous, and expensive procedures used truly needed, or were they selected due to inertia, lack of knowledge on the current practice, and because "this is how we have always done it?"

Close Team Collaboration

One key tenet in successful building design and construction is that the project structural engineer, architect, and geotechnical engineer need to be a closely cooperating team throughout the project. The geotechnical engineer should ideally be involved at the planning and proposal stage of the project to provide an idea of the subsurface conditions and likely foundation and basement requirements that should be considered even before the project begins. This cooperation will assist the team in developing an exploration program appropriate for the specific site conditions. During the design phases of the project, the geotechnical engineer should be involved in project meetings to review proposed design changes, and should also assist in preparing or reviewing load test specifications and testing specifications for foundation construction. A representative of the geotechnical engineer should be on site during foundation construction to act as the eyes and ears of the architect and structural engineer, to check that construction is performed in accordance with the design intent and to raise the "red flag" if any below-grade construction procedures occur which may not have been considered in the design.

Peer Review

Peer review is a very important aspect of foundation and basement design, particularly where unprecedented building heights or excavation depths are considered. There is a risk in such situations of the local geotechnical engineer becoming very conservative or nonconservative due to a lack of experience with projects of a similar magnitude. Input from an engineer with relevant international experience is recommended in such situations. In the best scenario, the peer reviewer should be included in the early stage of a project to provide input into the exploration program and proposed design before a local report is written. This results in teaming and collaboration rather than possible disagreement, which could result in project paralysis. Top architects and structural engineers routinely rely on their most trusted geotechnical engineers to perform limited peer review of every project, even if the owner is not aware of or directly paying for this service.

Subsurface Exploration

Significantly different foundation design and construction practices exist throughout the world. These differences relate to local precedents and experience, local codes versus international codes, differences in labor costs versus material costs, and variations in legal and contracting procedures. Of course, probably the greatest factor affecting local practices and experience is geology. The geology in every city of the world varies, and even within a city there is likely to be major differences in subsurface conditions. Obvious differences should be expected between land reclaimed from the sea, versus a valley floor at the foot of mountains, versus highlands, for example. Because of these differences, the exploration program will be expected to vary greatly from site to site. For illustrative purposes, it is possible to categorize sites into one of four site types, as summarized in Table 18.1.

As can be seen in the table, for each successive site type, the soil or rock strength can be expected to increase by a factor of about five and the soil or rock stiffness to increase by an order of magnitude. Thus, the same foundation supported on Site Type 1 soils should be expected to settle ten times as much as the same foundation supported on Site Type 2 soils. This 100-fold difference in strength and 1000-fold difference in stiffness is in stark contrast to normal building materials such as concrete, which might vary by a factor of only two in stiffness and four in strength. This large difference explains some of the risk of the underground. Measure or interpret the ground incorrectly or change it during construction and the building response could be an order of magnitude different from that expected.

Site type	Description	Typical value of UCS (MPa)	Typical value of E (MPa)	Example cities
1	Soft deep soil, no bedrock	0.1	2.5	Las Vegas, Mexico City
2	Competent soil or reachable bedrock	0.5	25	Chicago, Boston, London, Moscow
3	Intermediate geomaterial (IGM)	1–5	250	Dubai, Doha, Abu Dhabi, Jeddah, Manila
4	Soft to hard shallow rock	10+	1000+	New York, Seoul, Busan, Riyadh

Table 18.1 Soil profile site types and their likely major properties

UCS: unconfined compressive strength
E: elastic modulus

Boring and Sampling

The typical high-rise project may consist of a tower and low-rise podium. For such a structure a minimum of five to nine borings should be considered for the tower. Boring spacings vary greatly based on local practice, but typical values of 30–45 meters are common. For the Petronas Towers, an example of Site Type 2 conditions, the boring spacing was only 8 meters, to search for solution cavities in the very deep, karstic limestone bedrock.

Boring depths will vary based on expected foundation type and site type. For a mat or mat on piles foundation in Site Types 1 and 2, boring depths should equal twice the mat width. For Site Type 3, depths of one to one-and-a-half times the mat width would likely suffice. For hard rock sites with no concern for cavities, borings extended as little as 6 meters into rock (twice the diameter of the pile) may be adequate. The deepest borings performed for a building structure have been on the order of 150 to 200 meters. Minimum boring depths of 30 meters are needed for seismic site classification per the International Building Code (International Code Council 2009).

A typical podium may be several stories high, possibly with several basements. Thus loads in the podium will be lighter than the tower and may result in uplift due to hydrostatic pressure. The depth of borings in the podium should extend at least to the expected retention system or pile depth plus an additional 5 to 10 meters. Borings should be along the property alignment for retention wall construction, and not just at the corners. If water inflow below the retention wall is an issue, boring depths may need to extend through permeable sand or gravel layers to an impermeable layer (clay or mudstone) so that meaningful permeability tests can be performed for design of construction dewatering or basement drainage.

In Situ Testing

In situ testing should be part of any exploration for a high rise. In situ testing allows the soil or rock to be tested in the ground with relatively limited disturbance in comparison to extracting a sample and testing in the laboratory. Many materials, including sands, gravels, and weak, weathered and highly fractured rocks or IGMs, cannot be sampled intact and tested in the laboratory.

The most common in situ testing methods are summarized in Table 18.2.

The most important in situ test that has been used on most major high-rise buildings is the pressuremeter (PMT). This is a probe about 300 millimeters long which is lowered into a borehole, typically between samples, where it is expanded against the walls, affecting a ring of soil which may be 1 meter or more in diameter. Increments of pressure versus hole diameter are measured to provide a stress–strain curve of the soil or rock. A pressuremeter of the proper capacity should be selected so that the soil or rock can be loaded to failure and the ultimate material strength determined. Thus the advantage of this device is that a much larger sample of soil or rock is tested than a 50-by-100-millimeter sample returned to the laboratory. Also, the PMT measures the soil or rock under the in situ stress state, including whatever fissures or fractures may exist. A graph of elastic moduli measured by PMT at major project sites around the world is shown in Figure 18.1.

Another key soil or rock property that should be measured for most projects is the shear wave velocity. This is needed to accurately determine the seismic site classification per IBC. Standard penetration test (SPT) and unconfined strength measurements can also be used, but these methods may result in a less accurate determination. Shear wave velocities need to be measured to a depth of 30 meters. It is not uncommon for these measurements to allow the Site Class to change from D to C or B to A, for example, which would result

Testing method	Site type	Depth limit (m)*	Measures
Pressuremeter (Menard)	1, 2	100	Soil strength and stiffness
Pressuremeter (Elastmeter/TriMod/Probex)	2, 3, 4	100	Hard soil to soft rock strength and stiffness
Pressuremeter (high pressure dilatometer/Goodman jack)	4	100	Moderate to hard rock strength and stiffness
Standard Penetration Test (SPT)	1, 2, 3	150	Granular soil relative density and strength
Cone Penetration Test (CPT)	1, 2	60	Soft soil strength and water pressure
Vane Shear Test (VST)	1	25	Soft clay strength
Packer Test	1, 2, 3, 4	25	Soil or rock permeability locally
Pump Test	1, 2, 3, 4	25	Soil or rock permeability globally
PS Logging	1, 2, 3, 4	150	Down-hole compression and shear wave velocities
Cross-Hole Logging	1, 2, 3, 4	30	Cross-hole shear wave velocity

Table 18.2 Essential in situ tests for most projects

*Depth limit based on typically available equipment, physical limitations, or likely depth required for buildings

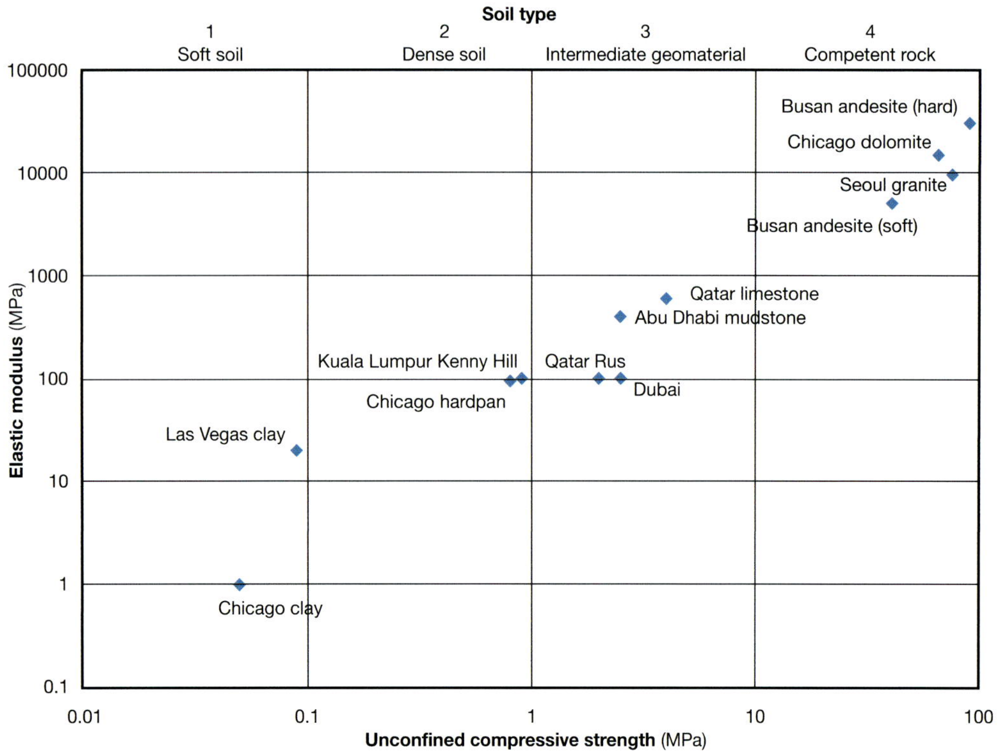

18.1 Elastic modulus of soil or rock for different site types, measured by pressuremeter
Graphic: AECOM

in significant reductions in seismic base shear. These reductions would more than pay for the cost of the testing. Geophysical testing is also used to obtain the low-strain modulus of the soil or rock and provides a macro view of the variation of ground stiffness.

Load Tests

For major structures, load tests of deep foundations can often be the most useful and cost-effective method of testing to help fine tune and pare a foundation design to the minimum size while defining the true factor of safety that exists. Thus, sacrificial load tests taken to geotechnical failure should be considered part of the exploration process to provide maximum benefit during the foundation design. In practice, however, load tests are performed for many other reasons, including:

- Because they are required by code or local practice
- As a proof test on a production pile to check construction compliance
- To exceed local code-defined allowable capacities
- To prove acceptable performance (in settlement or deflection)
- To check the contractor's ability to construct the pile and achieve the design load, particularly where new procedures are proposed.

Often, the traditional approach to pile design includes limited geotechnical exploration and in situ testing resulting in very conservative design values (and large, expensive piles). Then, during construction, production piles are randomly selected and subjected to load tests, a typical criterion being that perhaps 1 to 2 percent of the piles should be tested statically and an additional 5 percent tested dynamically. These load tests cannot be taken to failure; the pile is loaded to perhaps only 125 to 150 percent of the design load. If the design has been very conservative it is likely that the pile will easily pass this test with very limited movement—typically only a few millimeters. The designer and contractor can then pat themselves on the back and present the favorable results to the owner.

But is this really the best way to extract value and gain information from the expensive and time-consuming load tests? A better method, proven in many case histories, is to perform advance load tests on sacrificial piles before final foundation design. The purpose here is not to reach an arbitrary or conservative assumed design load, but to test the pile to geotechnical failure so that the ultimate friction capacity and end bearing capacity can be measured. By measuring the achievable side friction and end bearing, an appropriate factor of safety (perhaps 2.0) can be applied and a much smaller pile results than in the first scenario. At a factor of safety of 2.0, this pile may move 15 to 25 millimeters at the design load, which is acceptable. In this case, the owner is not paying for an embedded high factor of safety or performance that is not needed. Even for the tallest towers in the world, two or three such advance load tests have proven adequate.

Modern Methods

In recent decades significant improvements and innovations in load testing have evolved which make the time-consuming construction of load frames, reaction piles, tie downs, or kentledge stacks largely unnecessary. The best methods available today that the architect and engineer need to be aware of are:

- The bi-directional (O-cell) method
- Large-strain dynamic testing
- Statnamic testing.

Figure 18.2 presents a schematic of the bi-directional load test method using an Osterberg load cell (O-cell). In this ingenious method, the hydraulic jack, which is normally sandwiched between the top of a pile and a load frame held down by reaction piles or enormous weights is embedded instead into the bored pile itself. The position of the load cell will typically be within the lower half of the pile. If the pile is expected to support load in friction without end bearing, the load cell might be placed at the shaft midpoint. If the pile is expected to develop both friction and end bearing, the load cell might be placed around the lower third of the pile. Conversely, if the pile is to develop primarily end bearing the load cell might be placed at the base of the pile.

In addition to the cost and space savings of the bi-directional method, another great advantage is the load capacity that can be reached. For conventional kentledge tests, a maximum practical load limit may be about 30 MN. Figure 18.3 shows the 30 MN kentledge load test conducted for the Petronas Towers. For load tests performed with hold down piles, the practical limit is about 50 MN. In the bi-directional method; however, multiple load cells (two to five) can be combined at a single level to act in concert. Further, load cells can be placed at multiple levels so that the single test pile can be loaded in stages and sections. The bi-directional

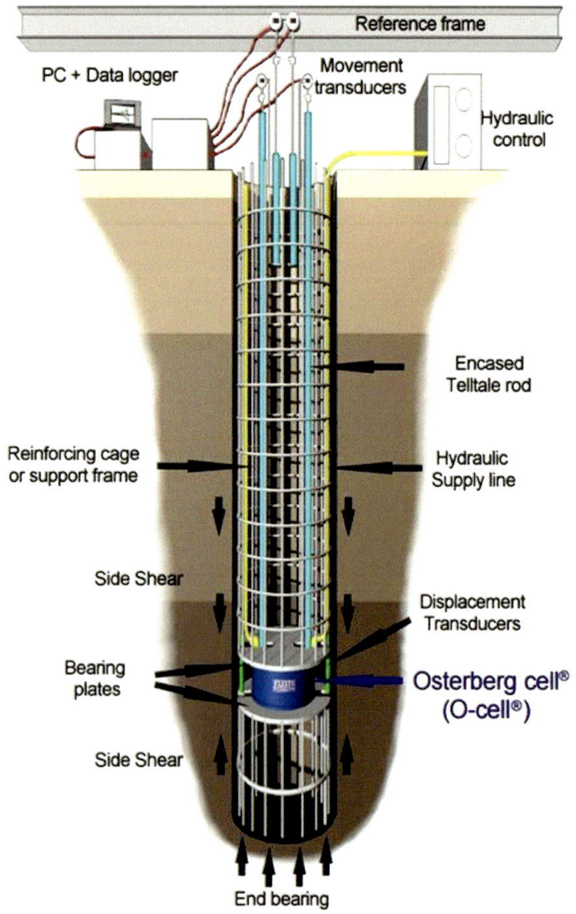

18.2 The bi-directional load test method in which the jack is embedded within the pile
Graphic: Loadtest

18.3 A 30 MN kentledge load test conducted for the Petronas Towers in Kuala Lumpur, Malaysia
Image: AECOM

loads applied are additive at each level such that record load tests as high as 350 MN have been conducted by this method (for the Incheon Bridge and Incheon Tower, South Korea). The completed reinforcing cage with multiple load cells embedded at two levels for a 96 MN load test is shown in Figure 18.4.

Another newer and innovative method of pile testing is the large-strain dynamic method. This was adapted from pile monitoring performed on piles driven by an impact hammer. In this test the pile hammer is replaced by a large weight, which is dropped onto the pile to create a dynamic impact that lasts only milliseconds. The impact and resulting stress wave in the pile are measured by instrumentation attached to the pile and recorded by a computer. Analyses of the results are needed to interpret the equivalent static load. Dynamic tests with weights as heavy as 40 tonnes dropped from heights of one meter have been performed to test large piles to capacities as high as 40 MN. The test setup for a 40 MN, 2-meter bored pile is shown in Figure 18.5.

Statnamic testing is similar to dynamic testing, but rather than dropping a heavy weight, a reaction chamber with propellant is placed on top of the bored pile, and a large cylinder filled with gravel is constructed above the pile as a reaction weight. The propellant is ignited and reacts against the weight of the gravel cylinder, throwing it upward while pushing down on the pile. Duration of loading in the Statnamic test is about one second. One advantage of Statnamic tests over dynamic tests is that the relatively slow push does not force the pile into tension and is more similar to a static load test. The Statnamic tests are useful in difficult access conditions such as over water, and have also been adapted to performing lateral load tests. Statnamic tests with capacities as large as 40 MN have been achieved.

18.4 Reinforcing cage with four bi-directional load cells embedded at two levels for a 96 MN load test at the Central Market project in Abu Dhabi, United Arab Emirates
Image: AECOM

18.5 Large-strain dynamic load test (Apple 4) on a 2-meter diameter bored pile in Chicago
Image: AECOM

Basements

Site type, water table depth, soil permeability, basement depth, intended basement use, foundation type, earth retention system versus open cut, conventional versus top-down construction, drained versus undrained design, and neighboring structures and features are all interrelated and dependent factors that mesh in a very complex dance. Planning from the earliest stages is essential to consider feasible options and discard methods or materials which could result in severe costs or delays and which might be mitigated by excavating one fewer basement levels or simply extending more structure above grade.

A single basement level is routine construction almost everywhere, and chances are good that limited water would be encountered at this depth. A single level could likely be open cut, or relatively inexpensive sheeting or soldier pile and lagging could be used for earth support, possibly as a cantilever without the need for significant bracing or tiebacks. However, extend below one level and construction difficulties increase exponentially. Because soil and water pressures increase linearly with depth, a two-level basement retention system will need to withstand four times the lateral load of a single level. Add a third level and the load will be nine times greater. Add hydrostatic pressure to the soil pressure and the load on the retention system or basement wall doubles again. Thus if basements or temporary earth support systems are designed for drained conditions, the drainage system must work, otherwise design loads will be twice those intended and failure or excessive movement will be likely, particularly for temporary systems where the design factor of safety may only be 1.5.

The importance of construction staging and considering unbalanced lateral loads is well illustrated by the collapse of the building in China in 2009 (as discussed in the introduction to this chapter). If we assume an average design wind pressure is around 2 kPa, the lateral force on a 15-story, 30-meter wide high rise might be about 3 MN. In contrast, the lateral soil pressure on a basement might be around 7 kPa or 14 kPa for soil above and below the water table level, respectively. These pressure values are not constant however, but increase *linearly* for every meter of depth. Thus, for a 30-meter wide basement, the lateral load for a 6-meter deep basement would be about 4 to 8 MN, and greatly exceeds the wind loading. The lateral soil load is also constant and always acting without pause. It is not an intermittent load of short duration occurring briefly at the peak of an infrequent storm.

Deep basements under towers may provide an important benefit that could allow building height to be increased, especially in Site Type 1 or 2 conditions. Every meter depth of soil removed in a basement excavation could be equal to the dead load of three building stories. Thus, excavating a two-level basement to 6 meters could result in an unloading of the soil equal to the weight of a 20-story building. Even if the structure were constructed with a mat foundation with no piles, the long-term settlement would be expected to be zero if there is no net pressure increase on the underlying soil. Such a structure is called "fully compensated," and essentially floats within the soil profile. Major high rises in Mexico City and Houston have been constructed using this principle.

While this load removal is beneficial for settlement performance of a major high rise, it is usually a major detriment to the adjoining podium if the excavation extends deep below the water table level. If the exterior podium walls do not extend to an impermeable soil layer to provide a cut-off, it may not be possible (or desirable) to pump the water and provide a drained slab design. In this situation, a thick pressure slab or mat would be needed and net uplift from hydrostatic pressure would likely result. Thus, a mat alone might not be suitable, and hold-down anchors or tension piles would be needed.

Sometimes it appears that waterproofing and designing for uplift is the common procedure in an area. However, if a cut-off can be provided or if the soil is generally clayey, the quantity of water inflow will likely be low enough to be pumped. If a drained design can be accomplished, it could have significant cost savings for a project by eliminating a thick mat or pressure slab, hold-down piles, and waterproofing. Significant cost savings were achieved for the podium of the Petronas Towers by providing a drained slab, even though this was not the common practice in the area.

The designer must consider the construction dewatering aspects carefully in the design of the building foundations. The building must be constructed to a height (and weight) sufficient to balance the hydrostatic pressure before the pumps are turned off for an undrained design. Conversely, hold-down piles need to be designed for the maximum possible uplift occurring during construction, if the pumps are to be turned off before completion of the structure. The load cases used in the piling design must be clearly spelled out in the construction documents.

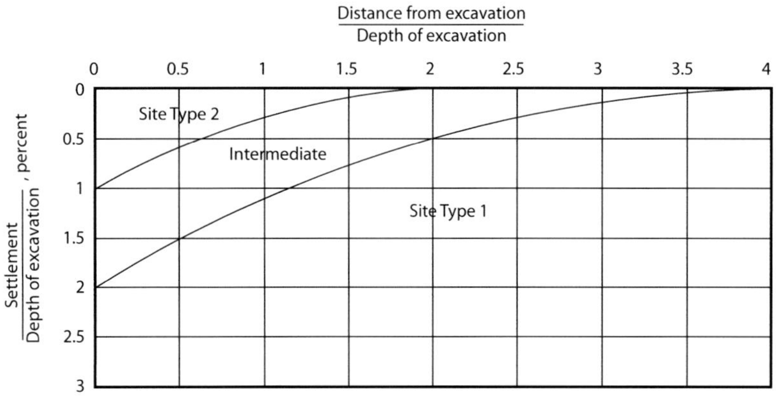

18.6 Expected settlement near soldier pile and lagging or steel sheeting walls versus excavation depth for Site Types 1 and 2
Graphic: Adapted from Peck (1969)

Adjacent Movements

One of the most important considerations in basement excavation is its effect on neighboring structures. Most lawsuits in geotechnical engineering involve damage to structures resulting from ground movements (lateral and vertical) caused by adjacent excavations. Possible ground movements could occur to structures even at lateral distances equal to the excavation depth. Movement is dependent on the major factors of site type, wall stiffness, distance between bracing levels, and excavation staging. Even the most properly designed earth retention system could be subject to severe movements if prescribed staging is not followed in the field, with overexcavation or failure to install required bracing or tie-backs the most common cause of failures. For this reason it is often wise to have the earth retention system installer and excavator tied contractually. Expected potential settlements adjacent to soldier pile or steel sheeting retention walls installed in Site Type 1 and 2 conditions are shown in Figure 18.6.

Another aspect of excavation that can greatly affect neighboring structures is dewatering. If an earth retention system is used which extends to a cut-off level and the system is relatively impermeable, dewatering will be limited to removing water within the walls of the excavation. However, if an open cut excavation is considered below the water table, significant dewatering could be needed, and the depression of the water table could extend a significant distance (five to ten times the water table drop) beyond the excavation limits. If the water table is lowered below a structure supported on shallow foundations, the building may settle as a result of the increased effective stress in the soil resulting from the drop in water pressure. Every one meter drop in water level below an existing structure would increase the soil stress by 10 kPa. The other danger of dewatering is the potential for causing the migration of fines from outside the excavation limits, which could undermine neighboring structures. Dewatering within excavations should occur with properly filtered wells and sediment tanks, which should be monitored to check that water is pumping clear.

Top-Down Construction

"Top-down" (or "up-down") construction techniques, where the ground floor slab is poured first and basement and superstructure construction proceeds simultaneously, have several benefits over conventional construction for deep basements. The main benefit is that of speeding up the construction of the superstructure, because it is not necessary for the entire excavation to be completed first. This suggests that top-down methods would likely not be applicable for buildings with only one or two basement levels. As basement depth increases, construction time savings increase with top-down; however, for more than four to five basement levels top-down methods begin to pose serious quality control issues for foundations. Another significant advantage of the top-down method for deep excavations is that the basement floor levels become the lateral bracing for the retention system, thus the cost of bracing or tiebacks is removed. However, this system also requires that the earth retention system become the permanent structural basement wall. This is likely to mean that concrete slurry walls, secant walls, or even heavy steel sheets would be the only feasible retention walls that would need to be inherently waterproof. Where steel sheets are used, the interlocks would be welded to prevent water inflow. On many projects where the basements have been used for car parking, the retention walls have been utilised as the final architectural walls with only cleaning, minor patching, and painting.

For top-down construction, foundation types are limited to those where a single foundation element can support a single column. Thus the most typical foundation type, due to its high load-carrying capacity, would be a bored pile. This would be constructed from near the original street grade and would have a deep cut-off within a casing, or could be poured up to grade within a liner. In the latter case, the pile itself would become the structural column in the basements as the excavation proceeds downward. Where the pile is cut off below the lowest basement level, a temporary steel casing would be required. Steel columns can then be plunged and embedded into the wet concrete of the bored pile, or the concrete can set and workers enter the casing to clean off the top of the pile and install a base plate. Steel columns or even precast concrete columns have been installed within the casing. This construction method is risky due to the potential concrete problems within the bored pile if it is installed by tremie methods, where the concrete is placed below water level through a pipe. Also, alignment and verticality of the piles are critical, so that columns line up without causing unacceptable eccentricity within the piles. Designers should expect eccentricities of at least one percent of the depth to the cut-off level and contingency plans should be in place if greater eccentricity occurs. For a major project with a five-level basement, an eccentricity of more than 300 millimeters occurred.

Foundations

The vast majority of all major high-rise structures are supported on bored pile foundations. In different parts of the world, bored piles are also called "drilled shafts," "drilled piers," or "caissons." Bored piles may have straight shafts or may be underreamed. A mat is often cast on top of the bored piles, particularly in seismic areas or where a deep basement extends below the water table in a permeable material. The remaining high rises may be founded on barrettes (rectangular "bored" piles constructed with a slurry panel excavator), driven piles, a mat alone, or even footings. The anticipated new One World Trade Center tower in New York sits on footing foundations in hard rock, and two of the Zenith towers (of 75 and 80 stories) in Busan are supported on mat foundations on rock.

A mat alone can be used to support a major tower where the foundation rests on competent rock. Where conditions degrade to Site Type 2 or 3, a mat alone may still be feasible, or a mat could include piles. In these soils, proper soil structure interaction analysis can result in sharing load between the mat and piles at considerable cost savings. Savings can increase even further if the mat alone can support the structure from a bearing capacity point of view, and the piles are added simply to stiffen the ground and act as settlement reducers. In this approach, the piles can be considered ground improvement and may be designed with geotechnical and structural factors of safety as low as 1.2. Even in Site Type 1 soils, mats alone can support a high rise if settlement is controlled by excavating basements deep enough to compensate for the weight of the structure.

Where extremely long piles are required, barrettes constructed with slurry wall panel machines may be needed. The Petronas Towers are supported on barrette foundations that extend to a maximum 135 meters, thought to be the deepest building foundations in the world. These lengths were necessary to limit differential settlements due to Site Type 2 conditions and sloping bedrock (Baker et al. 1998).

Design Responsibility

Surprisingly, there are no set rules across the world regarding who should be responsible for foundation design. In the UK and Middle East, for example, it is most common for piling contractors to design and construct a building foundation. In the process, the contractor takes the entire risk for the foundation performance. In the US and most other portions of the world, the structural engineer designs the foundations with input from the geotechnical engineer. The contractor is responsible for building what is shown on the plans and specifications, and thus takes only the risk of constructing the foundation properly. Design and performance risks rest with the geotechnical engineer and the structural engineer.

Allowable Settlement

What is the allowable settlement for a major high rise? Even this question is difficult to answer and varies based on local site conditions and practices. While a fast answer for most might be 25 millimeters (Petronas Towers), major high rises have been designed in anticipation of 60 millimeters (Burj Khalifa) and even up to 200 millimeters (Las Vegas hotels) of settlement. Some major modern high rises have experienced differential settlements as great as 500 millimeters and yet are in use today. The acceptable total and differential settlement depends on building configuration and site type. Projects supported on Site Type 4 are unlikely to settle more than a few millimeters. Elastic compression of the foundation piles would likely be greater than the rock compression if piles are used to reach the competent rock. Within Site Type 1 profiles, larger total and differential settlements may need to be tolerated in a design if a major structure is to have a constructible and economic foundation. Differential settlements between tall towers and the attached low-rise podiums are the cause of greatest distress. The final attachment of the surrounding podium to a tower is routinely delayed until the last possible moment to allow the tower to undergo as much settlement as possible without distressing the podium.

Pile Design

Bored piles are used most often for high-rise buildings because of their large carrying capacity, constructability in almost any site type, and because concrete is cheaper than steel. The Chicago Spire project was designed on end bearing bored piles 3 meters in diameter, which extended 2 meters into dolomite bedrock at a 30 MPa design bearing pressure using 80 MPa concrete. Maximum design load on a single pile was 200 MN.

Pile foundations can be designed for end bearing, side friction, or a combination of the two. In Site Type 3 and 4 conditions end bearing only may be considered where shallower material provides little side friction. End bearing piles are also possible in Site Type 2 conditions. In Chicago, buildings up to 80 stories are supported on underreamed bored piles on hard glacial clay at bearing pressures up to 3 MPa. The 60-story Blue Cross Blue Shield building was supported on underreamed piles with shafts as large as 3 meters and underreams as large as 8 meters in diameter.

Prediction of bearing capacity is usually performed using unconfined compression tests and in situ pressuremeter tests along with semiempirical formulas. A minimum factor of safety of 3.0 is usually applied to the bearing capacity estimated by these methods.

Where piles extend long distances through dense soils or intermediate geomaterials, significant side friction will be developed. In soils or weak rocks, which cannot be sampled easily, empirical formulas can be used to crudely estimate the ultimate side friction based on SPT blow counts, unconfined strengths, and qualitative rock quality classifications. Better estimates of side friction can also be made from in situ pressuremeter data. Where no harder layer can be reached, friction piles will need to be extended longer to control settlement and provide adequate capacity. Long, skinny bored piles are more efficient in friction than short, large-diameter piles. A 1-meter diameter pile 40 meters long would have the same side friction capacity as a 2-meter diameter pile 20 meters long. However, the smaller diameter pile would use one half the concrete and result in much less settlement, since twice the depth of the profile would be reinforced with concrete.

Combination friction and end bearing designs are also possible; however these designs must consider strain compatibility. Ultimate side friction is developed at very low settlements, typically less than 5–10 millimeters. End bearing, however, in dense soil or IGMs may require 50–75 millimeters or more settlement to obtain the ultimate capacity. Also, where sockets are extended through IGMs or rock, studies show that very little load will reach the base of the pile where the socket depth exceeds 1–4 pile diameters. The allowable end bearing is also heavily dependent on the quality of the cleanup job done at the base of the pile.

When friction piles are used, group effects must also be considered. If pile spacing decreases to about 6–8 pile diameters or less, interaction effects increase and the capacity of the pile group will be less than the sum of the individual pile capacities. Generally, minimum pile spacing depends on the quantity of piles in the group, but usually should not be less than 2.5–3 pile diameters. Group capacity is checked by comparing the perimeter area of the entire group to the sum of the perimeter areas of the individual piles. Thus pile lengths may need to be longer where close pile spacing results in poor group efficiency.

Uplift Design

Uplift and lateral loads must also be considered in design. Often uplift in the podium is more critical than uplift in the tower. Uplift in the podium results from hydrostatic pressure and is essentially a constant, never-ending upward-acting dead load. In towers, uplift results from intermittent wind load (or possibly seismic load), which may only be present for brief pulses during the design storm occurring perhaps every 50–300 years. In podiums, uplift may be more critical than the gravity case and thus may control the design length of the piles. Piles must be checked both for individual side shear capacity and global uplift. As pile spacing decreases, interaction effects increase and global uplift becomes controlling. Thus adding short piles between column locations may not be as efficient as gaining uplift resistance from longer piles at the columns where dead load is concentrated.

For tall towers, net uplift does not appear to be common, since most designers are favoring heavier concrete construction. However, the current trend to ever taller structures in combination with the concentration of load within relatively few super-columns seems to be resulting in greater stress swings, such that the foundation may experience 0–100 percent loading conditions pulsing at the building period. Very little real information exists regarding full-scale cyclic behavior of building foundations; however it is possible to perform cyclic loading tests on large piles using the bi-directional load test method. Unfortunately this method is limited to cycling periods no shorter than about ten minutes, and cannot apply a true tension.

The Electric Power Research Institute has performed cyclic loading evaluations in model pile studies (McManus and Kulhawy, 1991). Its conclusion is that, provided actual stress reversal does not occur (that is, there is no tension), cyclic creep will be equal to the static creep that would normally occur over the same duration of static loading. But when stress reversal occurs, creep can become large and unpredictable. So, where foundations are founded in Site Type 1 to 3 conditions, cycling between tension and compression should be avoided wherever possible, and load changes in piles ideally should not exceed 20–100 percent of the maximum compression load. In competent rock conditions (Site Type 4), creep is unlikely either statically or cyclically.

Estimating Settlement

Experience has shown that most high rises settle as a block. This is particularly true for structures supported on large groups of friction piles where the pile spacing is less than about 6–8 pile diameters. It is important to understand that in this situation, even if individual full-scale pile load tests are performed, settlement of the pile group will generally not be the same as the measured settlement of the individual pile. The group will settle more due to group effects.

A simple, powerful technique that has been used to estimate the settlement of even the tallest buildings in the world is the equivalent footing method. While there are many variations on this theme, the technique involves modeling the pile group as a solid block with a length between two-thirds of the full pile length, and the full pile length. The building load is supported by the ultimate friction acting on the perimeter of the pile block reduced by a factor of safety—typically 1.5 or even less. Any load not supported by the perimeter friction is supported by the entire block base in end bearing. The soil or rock at the base of the block must have an allowable bearing capacity that exceeds this remaining pressure. Settlement of the block is then largely controlled by the soil or rock below the base of the block. Elastic compression of the block itself is added to the settlement estimated below the base.

Ultimately, estimating the settlement is dependent on how accurately the modulus of the soil and rock can be determined. In ascending order of probable accuracy, settlements may be estimated by:

- Empirical charts and field index testing (SPT, rock quality)
- Laboratory testing of samples (consolidation or triaxial) and theoretical formulas
- In situ modulus testing (pressuremeter) and empirical formulas
- Full-scale load testing of foundation elements
- Comparison to case histories.

Finite element modeling can be an excellent tool for predicting settlement, however it is a tool that seems to be more abused than most. An expensive study is no guarantee at all of greater accuracy. Selection of the soil modulus is again the key. An excellent way to use finite elements would be to back-figure a proper modulus from full-size case histories or from full-scale pile load tests. Modeling the actual loads or stresses applied to a structure can then provide insight into differential settlements and stress concentrations that cannot be achieved by any hand methods.

Ideally, where finite element studies are planned, field and laboratory testing must be performed to provide modulus data for the foundation material. This should include geophysical tests to obtain low strain moduli (10^{-4} strain), pressuremeter tests (0.5–2 percent strain), and possibly laboratory consolidation or triaxial tests (2–15 percent strain). Graphs of strain-dependent elastic modulus data for Site Types 3 and 4 is shown in Figure 18.7. Where foundations are in Site Type 1 clayey soils, time-dependent behavior must be considered.

The problem should be modeled with at least 10 to 15 stories of the shear walls and structure in place, since the superstructure stiffness has a major effect on the prediction of differential settlements. Loads should be applied at the top of the structural elements and not simply as average distributed loads on the mat. Piles should be modeled individually. Settlements predicted for a major structure with variable length piles below the core and super-columns in Site Type 3 conditions are shown in Figure 18.8.

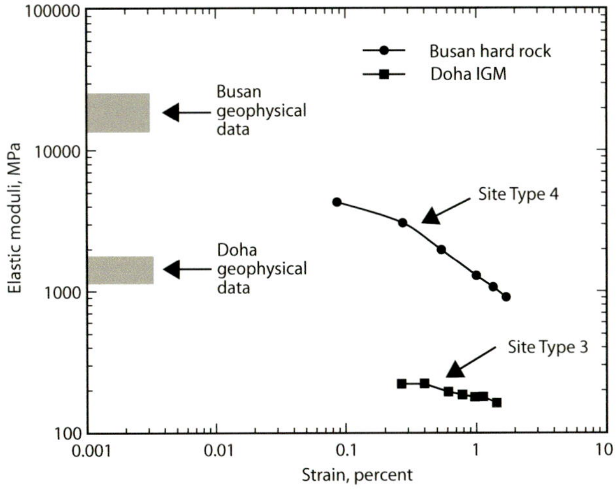

18.7 Elastic modulus versus strain curves for Site Types 3 and 4
Graphic: AECOM

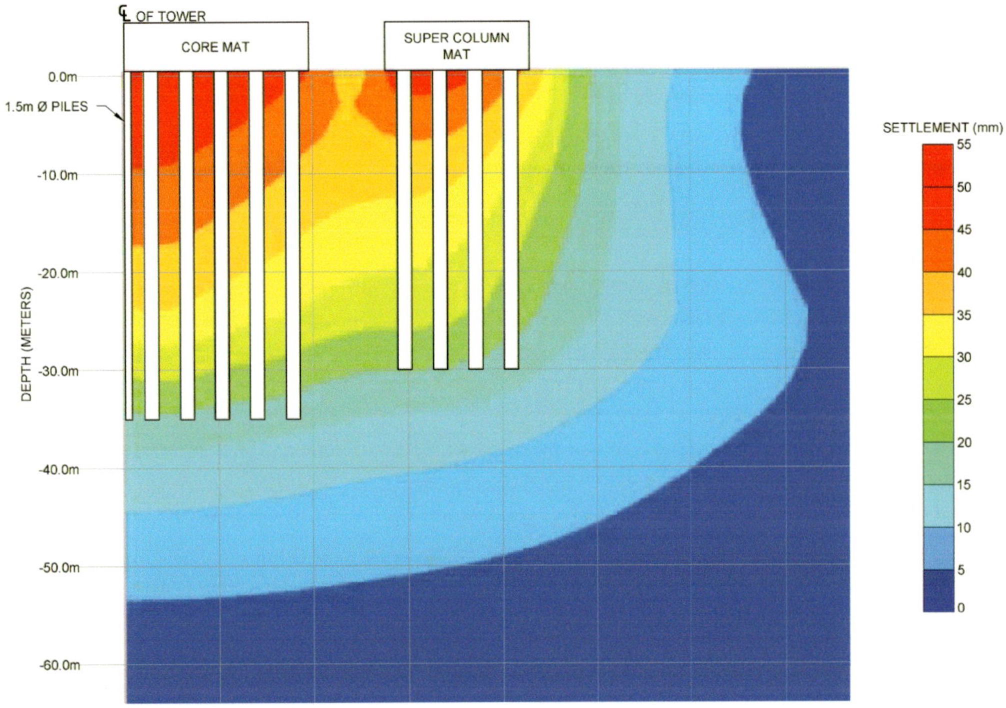

18.8 Finite element settlement prediction for a core and super-column structure on variable length piles in Site Type 3 ground
Graphic: AECOM

Bored Pile Construction

Typical bored pile diameters range from 0.8 to 3 meters, with piles as large as 5 meters in diameter possible. Bored piles may be extended one to two pile diameters into hard rock or 40 meters into IGM, as was done for the Burj Khalifa. Practicable bored pile lengths depend on equipment type, with 50 meters being a typical limit using conventional buckets or augers, which must trip into and out of the hole. For greater depths, reverse circulation drilling (RCD) equipment with down-hole hammers or cutting bits would be needed. In the RCD method, bentonite drilling fluid is used to carry the soil or rock cuttings up to the ground surface rather than tripping the cutting bit in and out of the hole. Bored piles can be excavated in the dry in clays or mudstones

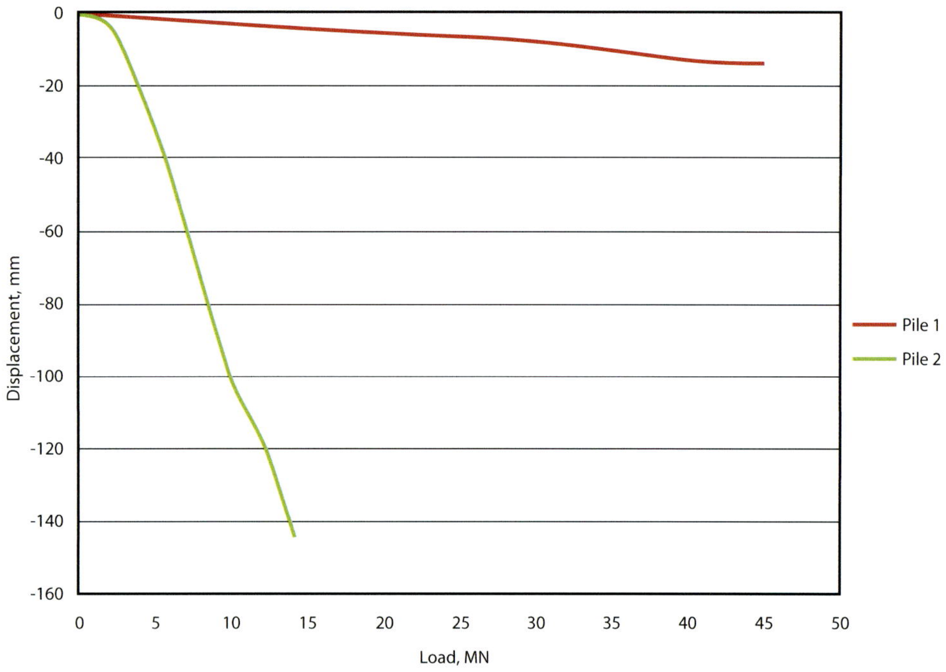

18.9 Two pile load tests on 1.5-meter diameter bored piles in Abu Dhabi mudstone
Graphic: AECOM

and can be filled with concrete by the free-fall method, provided that the base is clean and dry. In Chicago, concrete has been placed by free-fall methods to depths of over 40 meters (Kiefer and Baker 1994).

Where bored piles extend into caving soils or granular soils above or below the water table level, temporary casing, a stabilizing drilling fluid like bentonite, or polymer slurry would be needed to keep the hole open. Today, polymer slurry drilling methods are largely preferred due to reduced environmental disposal issues, easier slurry cleaning, and better shaft-side friction, since polymer slurry does not form a "cake" on the excavation walls. While polymer slurry use may still be relatively new in some locations, it does appear to be readily available in all parts of the world.

Figure 18.9 presents the results of two load tests performed on sacrificial bored piles for a project in Abu Dhabi mudstone. Even though the soil and rock conditions are the same at both locations, the ultimate loads achieved differ by a factor of three and the settlement differs by a factor of ten! The piles were constructed by the same crew, to the same lengths and diameters, in the same geology, in both cases using tremie methods with polymer slurry. So what could cause such a major difference? The good pile represents the expected capacity. During construction of the second pile, the tremie pipe plugged and was pulled out of the concrete. The obstruction was cleared and the pipe was reinserted into the concrete without a proper bottom separator between slurry and concrete. This procedure compromised the bottom half of the pile, which generated virtually no end bearing or side friction.

This example illustrates the critical importance of proper construction procedures and proper inspection. It does not matter if the design resulted from the best engineering with exotic in situ testing and sophisticated finite element procedures. Construction methods can make or break any design, if the difference in capacity is a factor of five between good and bad procedures. Suddenly, a design factor of safety of 2.0 does not look quite so large.

Construction problems can be avoided by selecting contractors with demonstrated experience on projects of similar difficulty. However, even with the best contractor and a long history of mutually successful projects, the recommended attitude to take is "trust, but verify." Testing requirements should reflect the redundancy of the design, the experience of the contractor, the construction method, geologic difficulty, and the factor of safety. Where a design attempts to "push the envelope," an inexperienced contractor has been selected on the basis of cost, the design is based on end bearing, and tremie procedures are to be used, detailed independent inspection and nondestructive testing along with production pile proof tests would be indicated. Where individual piles are to be end bearing at high capacity, proper bottom clean-up procedures are much more critical than for friction piles in a large redundant group.

Nondestructive testing is a valuable tool that is routinely specified for checking the integrity of bored piles during construction. Generally, nondestructive testing is required when tremie pouring procedures are used. Where concrete can be poured into clean and "dry" (less than 50 millimeters of water) shafts, nondestructive integrity testing is not needed. Where piles carry heavy loads or high concrete stresses, testing is more critical. Typically, where the concrete stress level exceeds 0.15 f'c, cross-hole sonic logging (CSL) is the preferred method. For lower stress levels, impulse response spectrum (IRS) methods can suffice. Planned coring of representative piles should be done to check concrete quality and provide a comparison with the nondestructive testing results.

References

Baker, C.N., Jr., Drumright, E.E., Joseph, L.M. and Tarique Azam, Ir. (1998) "Foundation Design and Performance of the World's Tallest Building: Petronas Towers," Fourth International Conference on Case Histories, St Louis, MO.

International Code Council (2009) *2009 International Building Code*, Country Club Hills, IL: International Code Council.

Kiefer, T.A. and Baker, C.N., Jr. (1994) "The Effects of Free Fall Concrete in Drilled Shafts," ADSC International Association of Foundation Drilling, Dallas, TX.

McManus, K.J. and Kulhawy, F.H. (1991) "Cyclic Axial Loading of Drilled Shaft Foundations in Cohesive Soil for Transmission Line Structures," Rpt. EPRI-EL-7161, Palo Alto, CA: Electric Power Research Institute.

Peck, Ralph B. (1969) "Deep excavations and tunneling in soft ground" in *Proc. 7th International Conference on Soil Mechanics and Foundation Engineering*, Mexico City: Sociedad Mexicana de Mecanica de Suelos, 225–90.

Chapter 19

Seismic Design

Ron Klemencic

Introduction

Seismic ground shaking has been a factor in the structural design of buildings for only slightly more than one hundred years. It is indeed a "young" science, with room for further discoveries, advancements, and new technologies. Similarly, the design and construction of tall buildings is also only slightly more than one hundred years old. It follows, then, that tall buildings in areas prone to significant earthquakes are arguably an untested building type. While significant earthquakes in Mexico City (1986), San Francisco (1989), Northridge (1994), Kobe (1995), Chile (2010), and Japan (2011) have provided insight to the performance of towers, much remains to be learned.

Buildings are growing taller and taller in response to urban density and economic pressures, leaving ample room for new ideas, research, advancements, and design technologies in areas of high seismicity.

Historical Perspective

Prior to the 1906 San Francisco earthquake, no formal building regulations in the U.S. included any earthquake design provisions. It was not until 1927 that the word "earthquake" first appeared in a building code, when the Palo Alto, California Building Code gave "optional" earthquake provisions (McClure 2006).

Early approaches to seismic engineering were largely based on observations and anecdotal evidence suggesting that structures designed with consideration for modest lateral loading (15–30 pounds per square foot) performed well. For example, in 1923 it was observed that buildings in Japan designed for a lateral force equal to 10 percent of their dead weight performed particularly well during seismic events (McClure 2006). This approach—basing seismic provisions on modest lateral forces—lasted until the 1950s.

In that decade structural engineering for earthquake demands began to consider the dynamic response of each building, as well as the inherent ductility associated with each type of structural element and its general arrangement in a building. A commentary was published by the Structural Engineers Association of California in 1967 which laid the groundwork for the philosophy that has guided seismic engineering ever since. This commentary stated that buildings designed according to code provisions should:

- resist minor earthquakes without damage
- resist moderate earthquakes without structural damage, but with some non-structural damage
- resist major earthquakes of the intensity or severity of the strongest experienced in California without collapse, but with some structural as well as non-structural, damage (Structural Engineers Association of California Seismology Committee 1967).

Building code provisions ever since have been developed around these basic philosophical statements. Even today, with the advancements in the seismic engineering of tall buildings, these ideas form the basis of state-of-the-art performance-based seismic design (PBSD).

International Practice

Seismic design provisions were being developed around the world at the same time code provisions were being developed in California. While the scientific principles are the same from country to country, there are differences in the application of those principles that should not be overlooked.

Japan and New Zealand have developed some of the most advanced thinking in seismic design. On the one hand, Japan's approach to designing and constructing tall buildings, based largely on PBSD principles, aims to create structures that often contain a high degree of structural redundancy. In New Zealand, seismic engineering is deeply rooted in a capacity-based design approach aimed at isolating locations of structural damage to predetermined elements or areas.

On the other hand, seismic design provisions in China tend to be much more prescriptive in nature. They are still based on PBSD principles, however: for example, the design of a tall building must meet specific performance objectives when subjected to site-specific levels of ground motion input.

Other seismically active countries such as the Philippines and Indonesia largely follow U.S. code provisions, which remain largely prescriptive as they relate to the seismic design of structures.

Applicability to Tall Buildings

It is intended that buildings designed using code provisions *all* have safe and acceptable performance when subjected to various levels of ground shaking. However, there are a number of unique features of tall buildings that fall outside of the general principles embodied in most code provisions. As such, it is debatable whether current provisions adequately address the performance of tall buildings. Some of the issues unique to tall buildings include their complex dynamic behavior, axial forces, size effects, and damping.

Complex dynamic behavior

It is common for the response of a tall building to strong ground shaking to be heavily influenced by complex dynamic behavior, including the impacts of higher modes of vibration. Traditional engineering practice has focused on only the first translational mode of vibration when setting strength requirements and lateral force distributions. However, for tall buildings, the second or even third mode of vibration can be equally important, if not more important, to the overall design.

The influence of these higher modes of vibration can result in significantly higher flexural demands well above a building's base (Figure 19.1), as well as shear demands three to four times greater than those anticipated by a typical prescriptive design (Figure 19.2). Failing to recognize and incorporate these demands into a tower's design can lead to undesirable and potentially unsafe results. Excessive damage, large residual deformations, and in the worst cases partial or total collapse of a tall building would result in massive financial loss and possible loss of life.

Axial forces

Due to the significant number of levels in tall buildings, axial forces in columns typically increase to very high values. Columns and walls can grow quite large in cross-sectional area as a result. As these elements grow larger and larger, they tend to attract additional axial forces due to their interaction with the floor framing and/or bracing systems. The accumulation of the additional forces caused by these interaction effects can overwhelm a column or wall's strength, potentially leading to unsafe conditions. Great care must be taken in the structural design of the tower to adequately address this possibility and to protect columns against axial failure, which could prove catastrophic.

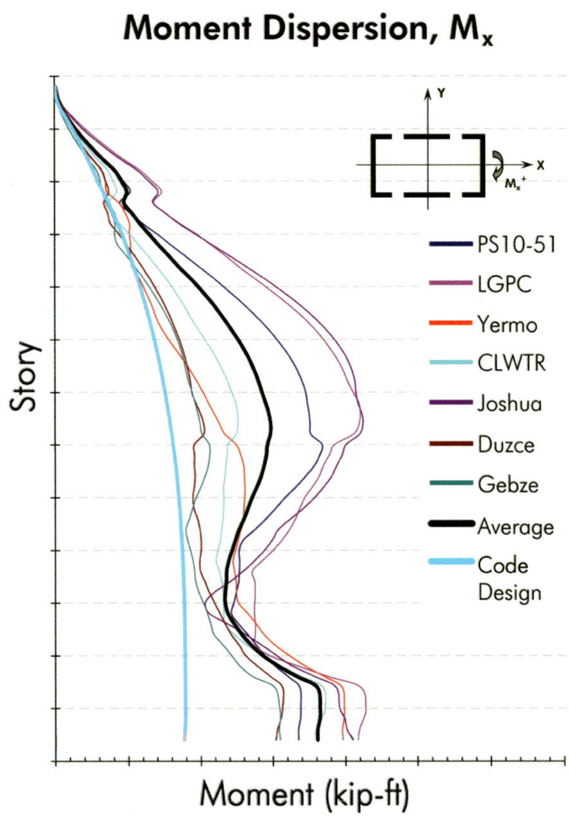

19.1 Flexural demands related to higher modes of vibration
Graphic: Magnusson Klemencic Associates

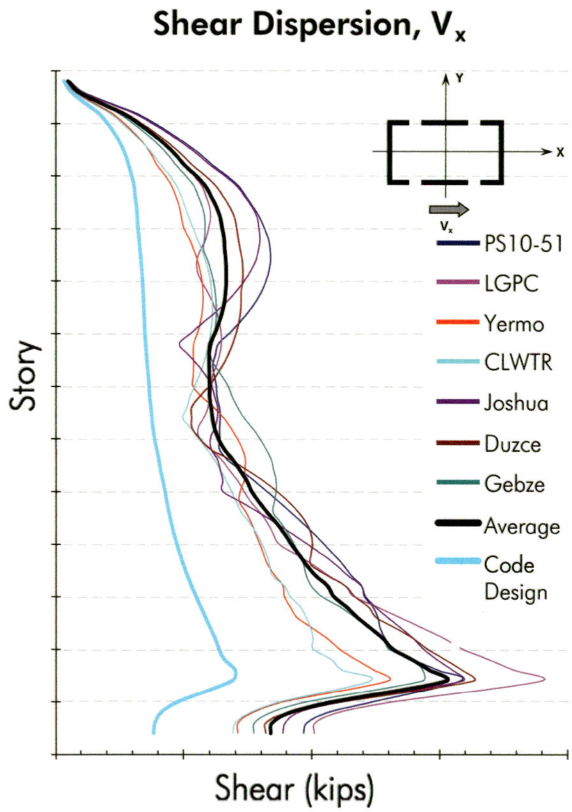

19.2 Shear demands related to higher modes of vibration
Graphic: Magnusson Klemencic Associates

Size effects

Most of the existing research forming the basis of design provisions in building codes for steel and concrete is the result of small-scale testing. As buildings grow taller, the size of columns, walls, and foundations tends to grow proportionally, and to reach sizes far in excess of anything that has been previously tested. Current engineering practice relies heavily on the extrapolation of small-scale test results to justify the design of these massive elements. However, as these elements grow to enormous size, extrapolating the research results and associated code provisions becomes questionable. Consider, for instance, a column at the base of a tall building: at 3.0 meters square and with a story height of 4.0 meters, is this still a column in the traditional sense? Most of the columns previously tested are no larger than 600 millimeters by 600 millimeters in size, with height to width ratios of 4 or more. Do building code provisions for traditional column design and the confinement of concrete and reinforcing steel remain applicable? Surely more research is required in this area and others.

Damping

A modest amount of data exists regarding the natural damping inherent in a tall building. Unfortunately, there is not a great deal of instrumentation of tall buildings, especially those subject to strong ground shaking. Available data suggests that the amount of natural damping inherent in tall buildings is quite modest: perhaps 2 to 3 percent, much lower than the traditional value of 5 percent normally considered in seismic design (Council on Tall Buildings and Urban Habitat Seismic Working Group 2008). Lower amounts of natural damping can lead to higher seismic demand levels throughout the height of the tower. This is another area where additional research will provide greater reliability in the performance of tall buildings.

State-of-the-Art Seismic Design

The seismic design of tall buildings has advanced significantly in the last ten years. Driven by the desire to create safe and reliable designs that are economical to construct, structural engineers have developed a broad spectrum of structural systems. Implementation of these designs has been facilitated by significant advances in computational power, making complex, nonlinear time history analysis more accessible. In addition, advances in wind engineering and wind tunnel testing have provided greater confidence in the wind response of tall towers.

There is growing recognition that the prescriptive provisions of most building codes do not adequately address the unique aspects of tall building behavior and performance. In general, building codes have been developed for low- and medium-rise buildings, in which the seismic response is typically dominated by the first translational mode of vibration of the structure. PBSD has provided structural engineers with an approach by which to achieve more appropriately designed tall buildings with unique framing systems.

The principles of PBSD are based on designing buildings to meet specific performance objectives when subjected to various site-specific demand levels. Traditional, qualitative objectives of resisting minor, moderate, and major earthquakes are shown to be met quantitatively, through design and analysis. This approach to design scientifically identifies and quantifies the earthquake hazards and ground shaking intensities specific to a particular building site. Then, with consideration of these site-specific demand levels, a structural framing system is conceived, designed, and detailed. Advanced computational modeling confirms the predicted performance of the design. The result is a higher degree of safety and reliability.

Guidelines on Performance-Based Seismic Design

Guidelines and recommendations for the appropriate and consistent application of PBSD have been published by a number of structural engineering groups, including:

- the Structural Engineers Association of Northern California
- the Department of Building Inspection, City and County of San Francisco
- the Los Angeles Tall Buildings Structural Design Council
- the Pacific Earthquake Engineering Research Center
- the Council on Tall Buildings and Urban Habitat.

While there are some differences in specific recommendations offered by each of these groups, the consensus view is that PBSD allows structural engineers to more appropriately and directly address the unique aspects of the seismic design of tall buildings.

Application to Recent Buildings

Some examples of recent buildings that have included unique structural framing systems and have taken advantage of PBSD methodology include:

- The Paramount Tower, San Francisco, California
- St. Francis Shangri-La Place, Mandaluyong City, Philippines
- One Rincon Hill, San Francisco, California.

These and many others represent a new direction in the seismic design of tall buildings, incorporating unique structural framing systems and using advanced computational modeling to predict and confirm their performance with respect to various levels of seismic ground shaking.

The Paramount Tower

At 39 stories tall, the Paramount Tower (Figure 19.3) remains the tallest precast concrete frame building in an area of high seismicity. Conceived through a collaboration between Charles Pankow Builders and Robert Englekirk Consulting Structural Engineers, the structural frame includes a perimeter precast hybrid moment-resisting frame. The most unique aspect of this frame is the inclusion of post-tensioning tendons, which provide the building with a "self-centering" capability in the aftermath of a major earthquake.

During an earthquake, the tower (like all towers) will sway from side to side as the ground shaking imparts energy into the building. Extreme ground shaking will trigger cracking in the concrete and the reinforcing steel will begin to yield. This damage will result in a permanent deformation or "lean" in most buildings after a major earthquake. However, in the Paramount Tower, the post-tensioning tendons in the tower's frame act to pull the tower back to center, reducing any residual deformations and thus improving the overall performance of the building.

19.3 The Paramount Tower
Image: © David Wakely, engineering by Robert Englekirk Consulting Structural Engineers

19.4 St. Francis Shangri-La Place
Image: © David Wakely, Arup

19.5 One Rincon Hill
Image: Magnusson Klemencic Associates

St. Francis Shangri-La Place

As the tallest residential tower in the earthquake- and typhoon-prone country of the Philippines, demands on the structural design of the 60-floor, 217-meter tall St. Francis Shangri-La Place (Figure 19.4) are significant. A traditional seismic design would have required many large concrete walls and stiff perimeter concrete frames.

Instead, an inventive use of outrigger walls with sixteen viscoelastic dampers allows for a more efficient and reliable bracing system. Designed by an engineering team at Arup, the structural system allows for reduction in wall thickness and column sizes, and consequently a significant reduction in the quantities of concrete and reinforcing steel compared to those normally required for a building of this height. The dampers act as shock absorbers, effectively reducing the sway of the tower during earthquakes and large windstorms.

One Rincon Hill

One of the most recent contributions to the world-famous skyline of San Francisco is the 60-story One Rincon Hill tower (Figure 19.5). The structural design created by Magnusson Klemencic Associates includes a combination of ductile reinforced concrete core walls and sets of buckling-restrained braces (BRBs), which serve as outriggers for the core (Figure 19.6). These outriggers effectively widen the structural "stance" of the tower, thus providing additional strength and stability.

The behavior of the BRBs is well understood and provides a reliable approach to effectively stiffening the tower. A more traditional outrigger system utilizing steel trusses or concrete walls could create a condition where supporting columns become overstressed, since the upper-bound strength of these outriggers is difficult to control and predict. The BRBs offer predictable lower- and upper-bound strength, making the overall building performance more reliable.

As one of the first tall towers to be designed using the PBSD methodology in California, the design was subject to extensive reviews by an independent peer review panel as well as the City of San Francisco. The resulting design and methodology has helped inspire the development of several advancements in the practice of seismic engineering.

Future Trends

While many recently constructed buildings include reinforced concrete core walls and outrigger systems, planning has begun for even taller buildings (350 meters high or more) which are likely to include more exotic structural systems. Composite construction, passive damping systems, and ultra-high-strength concrete and steel will be employed to address the growing complexities in architectural forms. PBSD methodologies will provide structural engineers with the framework to pursue these exciting new frontiers.

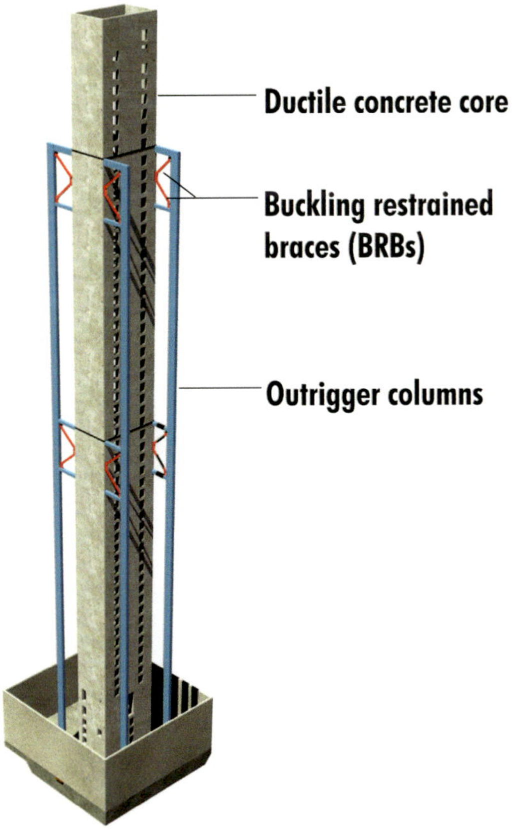

19.6 One Rincon Hill's core and BRBs
Graphic: Magnusson Klemencic Associates

HEIGHT LIMITATIONS AND CODE FOLKLORE

While the absoluteness of code-prescriptive height limits suggests some "step function" in structural performance, the reality is that the limits were set arbitrarily and without scientific basis:

- *Late 1940s* – Efforts began to study the effects of earthquakes on buildings and propose specific code language to guide design.
- *1951* – ASCE publishes Lateral Forces of Earthquake and Wind, one of the first documents to address seismic design of high-rise buildings. The document's recommendation for moment-resisting frames for buildings taller than 135 feet (41 meters) was anecdotal rather than scientifically justified: "buildings with moment-resisting frames … have had a very good record."
- *1959* – Los Angeles removes its zoning height limit of thirteen stories and 150 feet (46 meters). The limit, set in the early 1920s by the city council, was intended to keep buildings short and encourage spread-out development.
- *1960* – The Structural Engineers Association of California (SEAOC) publishes the first commentary on Recommended Lateral Force Requirements, stating "The limitations of thirteen stories and one hundred sixty feet have been established arbitrarily and are subject to further study." (Reportedly this 160 feet (49 meters) limit was a typographical error and was originally intended to be consistent with L.A.'s earlier limitation of 150 feet (46 meters).)
- *1961* – The Uniform Building Code incorporates a new height limitation for structural systems, stating "Buildings more than thirteen stories or one hundred and sixty feet … in height shall have a complete moment resisting space frame."
- *1988* – The UBC extends the allowable height for shear wall building frame systems to 240 feet (73 meters). No specific technical justification for this increase can be found.

References

American Society of Civil Engineering. (1951). *Lateral Forces of Earthquake and Wind,* American Society of Civil Engineering.
International Conference of Building Officials. (1961). *Uniform Building Code,* University of California.
International Conference of Building Officials. (1988). *Uniform Building Code,* University of California.
Structural Engineers Association of California, Seismology. (1960). *Recommended Lateral Force Requirements and Commentary,* Seismology Committee, Structural Engineers Association of California.

Chapter 20
Designing for Wind

Peter Irwin

Introduction

For tall buildings, wind forces and pressures are often the governing factor for the design of the structural system and the strength of the building envelope, being even more dominant than the forces due to earthquakes. This chapter describes the different types of wind that nature produces, how the wind behaves aerodynamically when it meets a tall building, the forces and pressures that it exerts, and the building's response to those forces and pressures. This chapter is focussed primarily on the mitigation of wind forces in tall buildings. Wind as a potential generator of energy is covered in Chapter 13.

One of the consequences of the interaction between a tall building and wind is that, besides the building being deflected by the sustained force of the wind, it is also set into swaying and twisting motions. The moving mass of the building results in large inertial forces on the structural system that effectively magnify the effect of the wind forces. These motions also add to the building's deflections and can be felt by the building occupants, especially by those in the upper floors. All these effects influence the building's design, since it must have sufficient strength to maintain structural integrity under extreme wind conditions, its floor-to-floor deflections must be small enough to avoid overstressing the building envelope and internal partitions, and the motions must be kept to within acceptable limits from a human comfort point of view.

While wind loading is in the first instance important, addressing other wind issues is also essential in arriving at a successful, problem-free design. These issues can be very varied but include: minimizing the impact of wind on pedestrians and occupants in plazas, terraces and balconies; dealing with the combination of wind and stack effect pressures on entrance doors; ensuring problematic exhausts are kept away from sensitive areas such as air intakes and operable windows; avoiding wind-induced noise problems; avoiding lift-off of roof pavers; preventing flexible appurtenances such as spires, sunshades, and other similar features from going into wind-induced vibrations; and capitalizing on the ability of wind to improve energy efficiency through use of natural ventilation or ventilated wall systems. The designer needs to be aware of these effects but space prevents a full description of all of them in this chapter.

Wind Climate and Characteristics

Windstorms occur in a wide range of sizes, with larger-scale storms generally lasting longer than small storms. At the equator, winds tend to be light, and the storms that produce the strongest winds are likely to be smaller-scale disturbances such as thunderstorms, which can occur in virtually all regions of the globe. Beyond about five degrees latitude north or south of the equator, tropical cyclones form over the oceans, and the strongest of these (called hurricanes in the Atlantic region and typhoons in the Pacific) provide the most extreme winds along the coasts that lie in their path. Since these storms require warm water to sustain their energy they die out rapidly once they pass over land. Further north or south than about 30 degrees, strong winds are created by very large-scale frontal and cyclonic systems, called here synoptic systems. These can be several thousand kilometres across and form as a result of instability at the intersection between cold polar air and warmer air from the lower latitudes. Often thunderstorms are embedded within these synoptic systems. There are also a variety of more localized patterns of windstorm, such as monsoons in southern Asia, shamals in the Middle East, and downslope winds in mountainous regions such as the Rocky Mountains of North America and the Alps of Europe.

Regardless of the type of storm, the wind near the earth's surface is slowed by friction and forms a planetary boundary layer within which the mean speed increases with height and the air is quite turbulent. In large-scale synoptic storms with strong winds this boundary layer can grow to several kilometres thick. In hurricanes and typhoons the scale of the storm is somewhat smaller and the boundary layer in the strongest wind zone, at the eye wall, is more typically five to six hundred meters thick. In thunderstorm downdrafts or in downslope winds, which extend over distances of only a few kilometers or tens of kilometers, the boundary layer depth may be only about one hundred meters. In tornadoes, which are very violent but are usually smaller-scale phenomena embedded within thunderstorms, there is almost no boundary layer, since the tornado diameter is typically no more than a few hundred meters. Thus the profile of wind speed and height depends on the type of storm being considered. Figure 20.1 illustrates the profiles of mean wind speed in several types of storm.

For the structural design of tall buildings, large-scale synoptic storms, hurricanes and typhoons are the most important causes of wind. Since the planetary boundary layer in these storms is generally between five hundred to several thousand meters thick, the buildings are typically wholly immersed in the boundary layer, and experience a wind in which the mean wind speed increases with height and which is also turbulent over their entire height. Therefore, to obtain meaningful results from wind tunnel tests on these towers, the mean wind profile and turbulence of the boundary layer need to be properly simulated in the wind tunnel. This is done by using special "boundary layer" wind tunnels with long working sections, rough working section floors that replicate the earth's surface roughness and special turbulence generators (ASCE 1999). The same boundary layer simulation is typically used for determining local loads on the building envelope and for examining wind speeds in pedestrian areas.

Since the design of buildings against natural forces is essentially an exercise in reducing the risk of failure to an acceptably small level, the statistics of the wind are important. For the determination of extreme wind loads

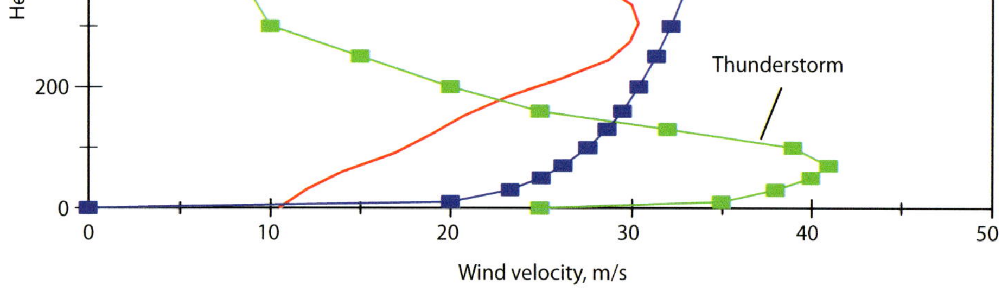

20.1 Examples of wind velocity profiles in different types of storm
Graphic: Peter Irwin, RWDI

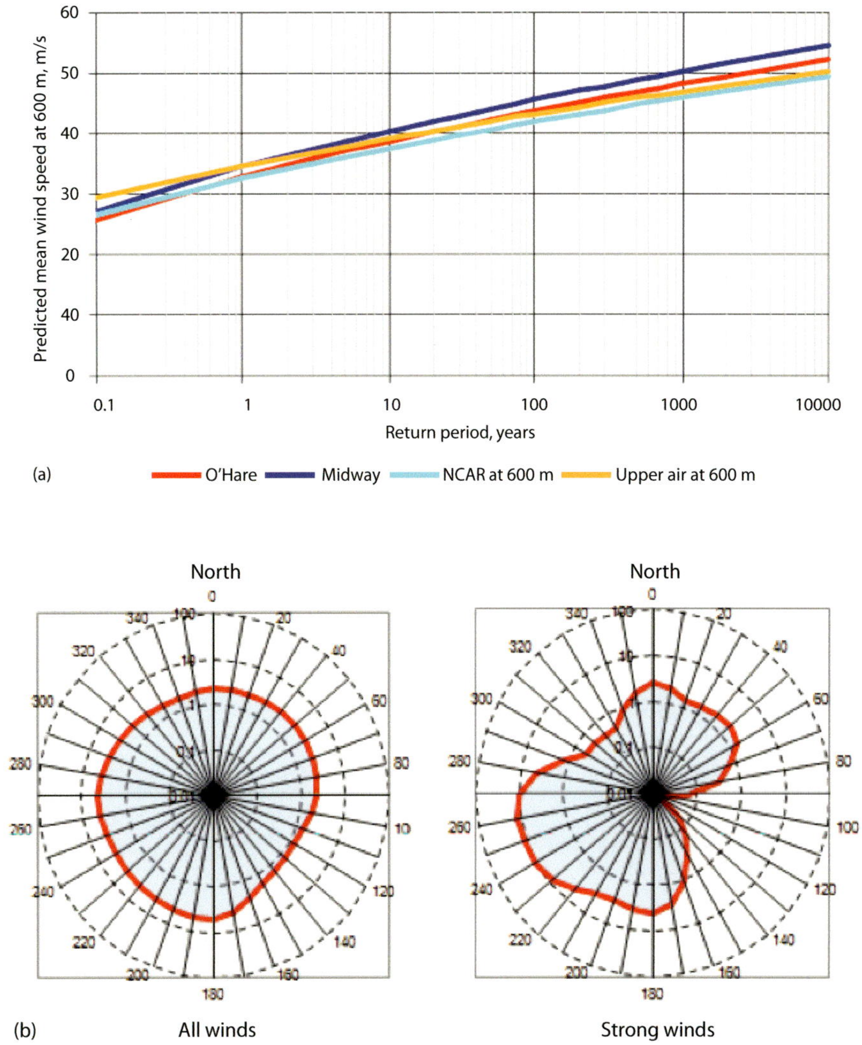

20.2 (a) Illustration of wind speed versus return period at 600 m over Chicago, as predicted from various data sources (ground-based records from O'Hare and Midway Airports, National Center for Atmospheric Research re-analysis data, and upper air balloon data) (b) Percentage of winds from within each 10-degree sector of wind direction for Chicago. The radial scale is in percent and is logarithmic to show the full range of percentages. The left plot is for all winds; the right plot is for only those winds above the five-year speed
Graphics: Peter Irwin, RWDI

the statistical analysis needs to focus on the strongest winds, including both speed and direction. The causes of these strong winds are often different from those producing everyday winds and may well result in different preferred directions. For example, in coastal areas there are often diurnal variations of onshore and offshore winds that dominate the more common wind events but have little to do with large-scale synoptic systems that cause the very high winds. For evaluation of extreme loads more meaningful statistical predictions can be obtained by separating the wind records into different categories based on the meteorological phenomena involved. Figure 20.2(a) shows an example of wind speed versus return period at six hundred meters above Chicago, as predicted from various sources of data. The spread of the predictions is an illustration of the uncertainties in wind statistics. Figure 20.2(b) illustrates the directional behaviour of all winds and strong winds in the Chicago area. This type of information can be important because the predicted extreme wind loads and motions of a tall building are often sensitive to its alignment relative to the most probable directions for strong winds.

Building Aerodynamics

Wind flow patterns around a tall building are the result of the building shape, the wind profile and turbulence of the planetary boundary layer, and the aerodynamic effects of adjacent buildings. For the simplest situation, where there is no other tall building nearby, Figure 20.3 illustrates the typical wind flows that occur. Because wind speeds in the boundary layer are higher at greater elevations, air that impacts the windward face of the building generates higher positive pressures near the top than at the bottom of the building. On the windward face there is therefore a downwards flow from the point

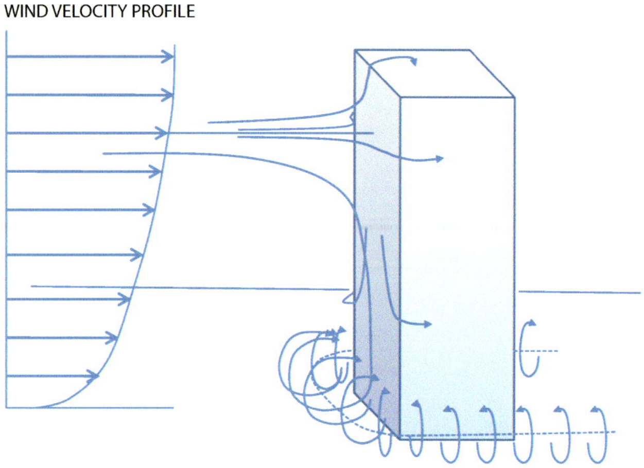

20.3 Wind flow patterns around a tall building
Graphic: Peter Irwin, RWDI

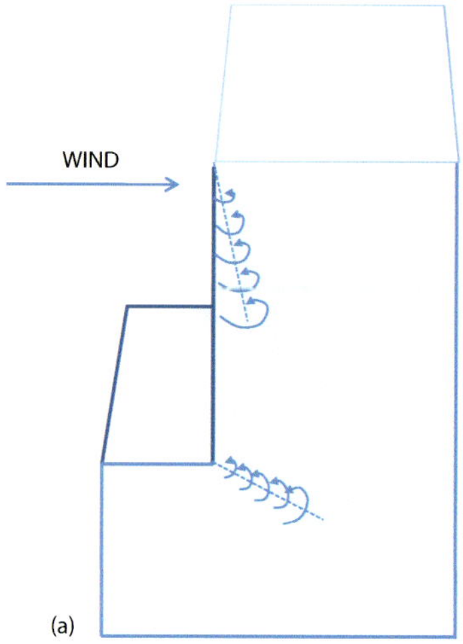

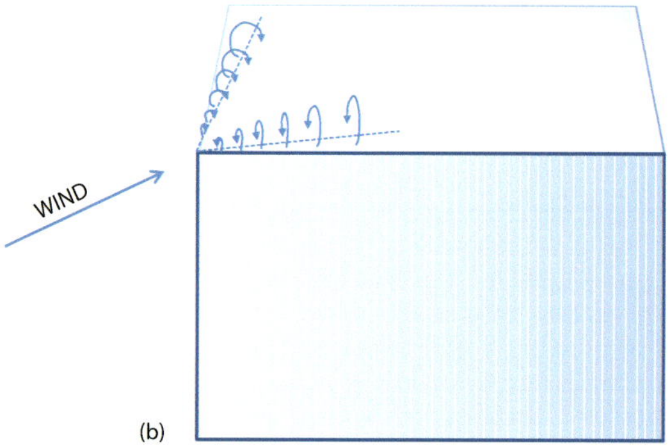

20.4 (a) Corner vortices forming at the windward edge of a sidewall and downwind of a setback (b) Formation of roof corner vortices
Graphics: Peter Irwin, RWDI

of highest pressure at roughly the three-quarter height, through a stagnation point in the center of the face, towards the base where the pressure is lower.

At the very top of the windward face the positive pressure is relieved by the ability of the wind to escape over the roof. Therefore there is an upwards flow from the three-quarter height towards the top. The flow down the windward face curls up into a vortex when it nears ground level and this vortex wraps itself around the base of the building in a horseshoe shape, as illustrated in Figure 20.3. At ground level the vortex tends to cause strong backflows just upwind of the building, highly accelerated winds around the building's upwind corners, and strong winds under the vortex on both sides of the building. The accelerated winds around the base of tall buildings caused by this vortex pattern can cause uncomfortable and sometimes hazardous conditions for pedestrians.

Generally, the side walls of the building are areas of suction as the wind curves around these faces. Near the upwind corners of the sidewalls strong local vortices can form, particularly near to discontinuities such as at the top corner or at setbacks, as illustrated in Figure 20.4(a). These are the cause of local "hotspots" of extremely high suction that have been known to suck windows out of buildings. Similar vortices can form over the roof when the wind is at quartering angles, as shown in Figure 20.4(b), leading to scouring of roof gravel and lift-off of pavers.

On tall buildings exposed to winds with low turbulence levels, large coherent vortices, extending over most of the building height, may form alternately on one sidewall and then on the opposite one (Figure 20.5).

20.5 Top view of a square cross-section building shedding large vortices that form on the sidewalls, leading to crosswind forces
Graphic: Peter Irwin, RWDI

These vortices are shed into the wake of the building and march off downwind in a regular pattern called a Kármán vortex street (named after engineer Theodore von Kármán). As the vortices form they give rise to oscillating crosswind forces on the building at a very distinct frequency, called the Strouhal frequency, f_s, expressed by the formula

$$f_s = S\frac{U}{b} \qquad (1)$$

where S = Strouhal number, U = wind speed, and b = building width normal to the wind flow. The Strouhal number is nominally a constant with a value typically in the range 0.10 to 0.30 (for rounded buildings S is strictly a function of another parameter called the Reynolds number, but for engineering purposes it can be treated as reasonably constant over certain ranges of Reynolds number). For a square cross-section it is around 0.14 and for a rough circular cylinder it is about 0.20. When f_s matches one of the natural frequencies, f_r, of the building, resonance occurs, which results in amplified crosswind response. From Equation 1 this will happen when the wind speed is at a critical value given by

$$U_{CRIT} = \frac{f_r b}{S} \qquad (2)$$

The consequences of resonance are important for the structural design of the building and will be discussed in the section on resonant loads and vortex shedding.

Wind Loads and Effects: Mean and Fluctuating Loads

Wind pressures on a tall building fluctuate not only because the oncoming wind is turbulent, but also because the building creates its own signature turbulence, including the vortices referred to in the previous section. Fluctuations due to both sources of turbulence occur much more rapidly than the changes in wind velocity due to the passage of meteorological systems over the site. Changes in velocity due to the passage of large-scale meteorological systems occur over hours whereas turbulence fluctuations occur over seconds. Observers sense the latter as gusts, whereas the former are sensed as gradual changes in general magnitude of the wind speed. The dividing line between the durations of turbulence events and large-scale meteorological events is usually set at about 10 minutes to an hour. Thus it is conventional to describe as mean loads those obtained by averaging over 10 minutes to an hour, and they are associated with the mean wind velocity and direction averaged over that time. To characterize the fluctuations in load about the mean, statistical descriptions are used such as standard deviation of load and expected peak load for the averaging period.

Along-Wind, Crosswind, and Torsional Aerodynamic Loading

It is important to note that buildings experience aerodynamic loads not only in the direction of the wind but also at right angles to it, i.e. in the crosswind direction. Both mean and fluctuating crosswind loads will occur for buildings that lack symmetry, or where the surroundings cause asymmetrical flows. However, even for a perfectly symmetrical building in surroundings that do not disturb the symmetry, while the mean crosswind loads are indeed zero, there are still substantial crosswind load fluctuations. These are due to fluctuations in both the lateral component of turbulence velocity in the approaching wind and the building's own signature turbulence, i.e. vortex shedding. As with mean crosswind loading, mean torsion loading can occur when there is asymmetry in the building and/or surroundings, but even in a perfectly symmetrical case torsion loading fluctuations will occur due to turbulence effects.

By integrating wind pressures over the whole building at any given instant a corresponding instantaneous base aerodynamic shear or moment can be derived. These overall integrated aerodynamic shears and moments are termed the background loading. The shape of the power spectrum of the aerodynamic base moment, which shows the contribution to the fluctuations of moment from different frequency ranges, is particularly important when computing the response of the building. Figure 20.6 shows typical examples of the spectrum for a tall building. In the figure, $S_M(f)$ is the power spectrum of moment and is expressed in the nondimensional form

$$\frac{f\,S_M(f)}{\sigma_M^2}$$

where f = frequency and σ_M = standard deviation of moment. By using the nondimensional form it is much easier to compare the aerodynamic characteristics of different building shapes and to build on experience with previous buildings. The frequency is also expressed in the nondimensional form,

$$\frac{fb}{U}$$

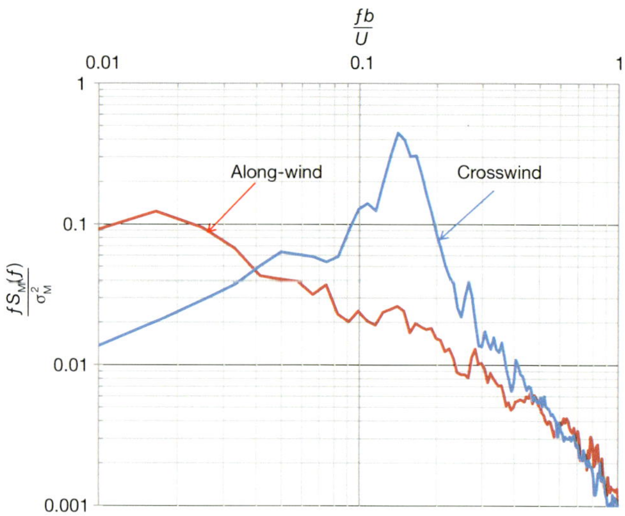

20.6 Examples of the power spectrum of along-wind and crosswind aerodynamic base moment
Graphic: Peter Irwin, RWDI

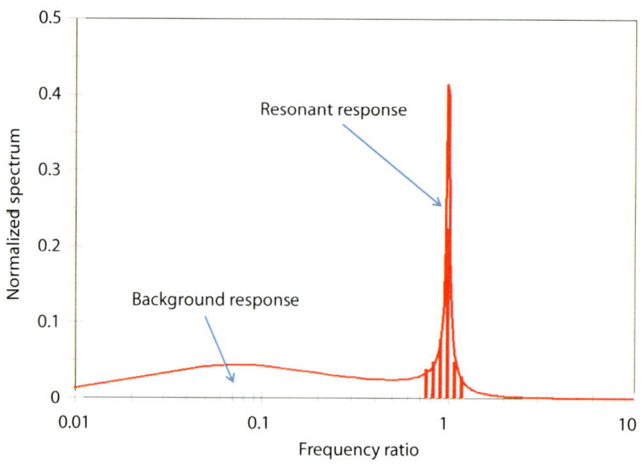

20.7 Effect of resonant response on the power spectrum of base moment
Graphic: Peter Irwin, RWDI

What is notable in Figure 20.6 is the difference between along-wind and crosswind spectra. For the example shown, the crosswind spectrum exhibits a distinct peak at a value of nondimensional frequency of about 0.14, corresponding to the frequency of vortex shedding. This is typical for a tall, slender building of roughly rectangular cross-section. The peak in the along-wind spectrum usually occurs at much lower frequency than in the crosswind spectrum and coincides with the peak of the energy spectrum of the turbulence in the oncoming wind.

Total wind loading and the corresponding load spectra are highly sensitive to the building shape and surroundings. This sensitivity makes it difficult for the simplified wind loading formulae of building codes to provide reliable estimates of the wind loads and the building response. Therefore project-specific wind tunnel tests are typically needed to provide the required accuracy and reliability. The wind tunnel tests provide not only the spectra of overall loads but also the localized peak wind loads on the cladding.

Resonant Loads and Vortex Shedding

The sustained effect of the background load fluctuations on the building is to cause it to move in its natural modes of vibration. Once the building moves, the acceleration of its mass, which reaches a peak at the extremity of the motion, results in inertial forces on the structure as a consequence of Newton's law (force equals mass times acceleration). Inertial forces due to excitation of each mode of vibration are perfectly correlated over the height of the building and are a function of the natural frequency of the mode, the mass distribution and the modal deflection shape. Although the inertial loads have as their origin the background loads, there is little correlation on an instant-by-instant basis between the two. On very tall, slender buildings it is often the inertial loads that dominate over the background loads.

Resonant response in a particular mode of vibration has the effect of greatly magnifying the power spectrum of base moment at the natural frequency of the mode. This is illustrated in Figure 20.7, in which the horizontal axis is the ratio of frequency to natural frequency of the building. The magnification effect occurs in both the along-wind and crosswind directions, and in torsion also. If the peak resonant magnification in the crosswind direction coincides with the vortex-shedding peak in the background spectrum illustrated in Figure 20.6 then very high crosswind loads can result, several times the loads calculated by building code formulae that only address along-wind loads. The effect of vortex excitation combined with resonant response on the building's crosswind response is illustrated in Figure 20.8 as a function of wind velocity.

Wind Load Distributions, Deflections, and Building Motions

For structural design, knowledge of the variation of wind load with height is needed. As inferred above, there are three contributors to the loads: the static or mean loads

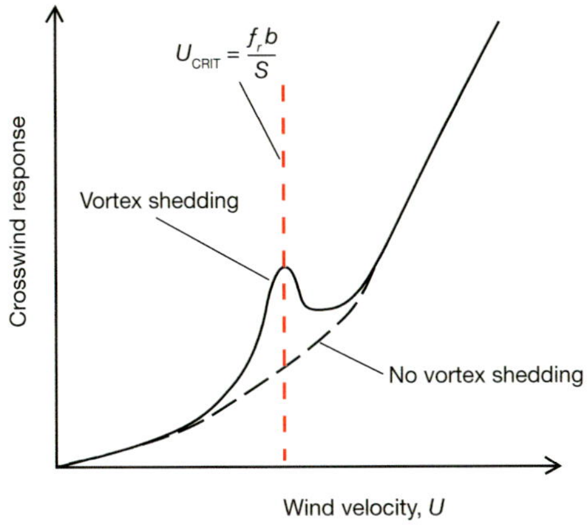

20.8 Effect of vortex shedding on crosswind response versus wind velocity
Graphic: Peter Irwin, RWDI

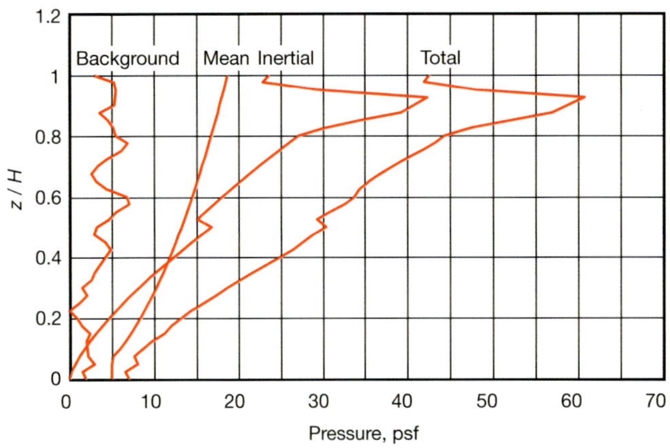

20.9 Variation of mean, background, resonant and total loads with height
Graphic: Adapted from Peck (1969)

and the background fluctuating loads, which are both purely aerodynamic, and the resonant loading which results from the movement of the building's mass. Contributions from each source vary from building to building, but on tall, slender buildings it is often found that the background loading contributes the least of the three and in the crosswind direction the resonant loads dominate. Typical variations of these loads with height are illustrated in Figure 20.9. The background and resonant loads combine in a root-sum-of-squares manner to form the total fluctuation in load. This is then added algebraically to the mean load to obtain the total wind load.

Calculated peak deflections based on the peak loads are important for the design of the curtain wall system and internal partitions. Likewise, the building acceleration and velocity can also be determined from the power spectra of loading and the building's structural properties, such as mass, natural frequencies, and damping. The accelerations and velocities should satisfy appropriate criteria for occupant comfort.

Cladding Loads

Cladding is affected primarily by the exterior local wind pressure acting on a small area of the building envelope, such as the area of a single lite or curtain wall panel. Over such a small area the exterior wind pressure is highly correlated and can be strongly influenced by local flow phenomena, such as vortices peeling off a building corner, as illustrated in Figures 20.4(a) and (b). Wind tunnel studies to determine cladding loads therefore involve models instrumented with many hundreds of pressure taps in order to measure the detailed local pressure patterns, both positive (i.e. acting into the building) and negative (i.e. acting outward and usually referred to as suctions).

Since cladding responds to the net difference in pressure between its inner and outer surfaces, the internal as well as the external pressure needs to be provided for design. Internal pressure is a function of leakage paths through the building envelope and the exterior pressures at the points of leakage, as well as any additional pressures created by the building's HVAC systems and stack effect. For buildings with significant openings such as operable windows left open or windows broken by flying debris, the internal pressures will tend to be magnified relative to the condition where all windows or other potential openings are closed. Appropriate allowance for internal pressure effects needs to be made when determining design cladding loads.

Criteria for Strength, Deflection and Motions

The structural system of a tall building needs to be designed to maintain its integrity in winds that would have very little chance of ever being exceeded. Traditionally, the approach has been to evaluate wind loads that on average are exceeded only once every 50 to 100 years. These are then multiplied by a load factor between about 1.4 and 2.0, depending on geographic region and governing code, before being applied to the structural model by the structural engineer to check its strength. After applying the load factor these loads are often referred to as Ultimate Loads. The effect of the load factor is like bringing the recurrence interval from 50 to 100 years up to an "Ultimate" recurrence interval of somewhere in the range of 500 to 2000 years. An alternative approach, which is being used more and more frequently, is to determine the Ultimate Loads directly and have a load factor of 1.0. To evaluate deflections, typically loads with a recurrence interval between 20 and 100 years are applied to the structural model, with the aim that the floor-to-floor deflections at these recurrence intervals will not cause malfunction of the curtain wall system or internal partitions. Building motions, particularly the accelerations and torsional velocities in the upper floors, are usually determined for recurrence intervals in the range of six months to 10 years and compared with specific criteria such as those published by CTBUH and ISO.

Mitigation of Wind Forces and Building Motions

Traditionally the approach of structural engineers to mitigating wind loading and associated deflections and motions on tall buildings was to stiffen the building with the aim of increasing the natural frequency. This has the effect of moving the resonant response of the building at the design recurrence interval further to the right of the spectra shown in Figure 20.6, where it can be seen that, as long as one keeps to the right of the spectral peak, the power in the aerodynamic force spectrum decreases with increasing frequency. Another approach, often used in combination with stiffening, is to add mass to the building in the upper floors. Adding mass while maintaining or increasing the frequency is effective in reducing accelerations in particular, as long as one is on the right-hand side of the spectral peak in Figure 20.6. However, for very tall buildings the natural frequencies may well put the building right in the vicinity of the spectral peak, and the traditional approaches of stiffening or increasing mass may become cost prohibitive, impractical, and even counterproductive.

With the huge increase in construction of supertall buildings in the late twentieth and early twenty-first century, other approaches to mitigation have increasingly been used. One such approach has been to add special damping systems. These effectively supplement the building's ability to dissipate the energy of vibration, which is fed by the wind. Applications of supplementary damping systems have so far been focused on mitigating

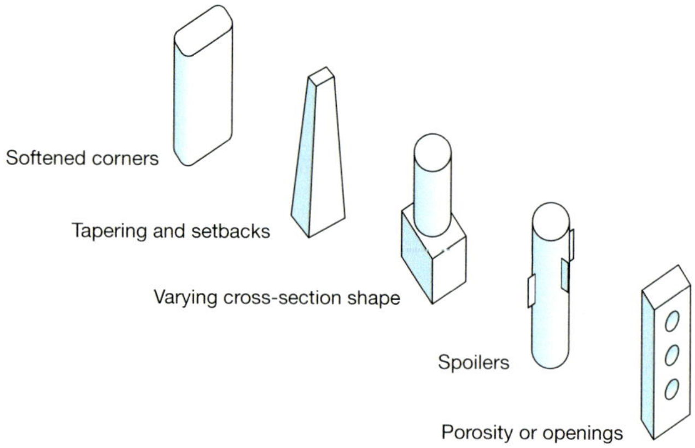

20.10 Shaping strategies for reducing vortex excitation
Graphic: Peter Irwin, RWDI

the building's motions for comfort reasons rather than reducing the loads considered for the building's strength. To rely on damping systems to reduce loads for strength design requires a higher level of reliability of the damping system than for comfort, but there appears to be no other obstacle to using this approach.

Since the need for mitigation is usually triggered by the presence of vortex excitation, rather than trying to mitigate the effects of vortex shedding through stiffness, mass, or damping measures, another strategy is to tackle the problem at its origin, that is to prevent the vortices from forming in the first place or at least weaken them. This can be achieved through designing the building shape aerodynamically. Various shaping strategies are summarized below and illustrated in Figure 20.10.

- *Softened corners* Square or rectangular shapes are very common for buildings and experience relatively strong vortex-shedding forces. However, it is found that if the corners can be "softened" through chamfering, rounding, or stepping them inwards, the excitation forces can be substantially reduced. Such softening should extend about 10 percent of the building width from the corner. Corners on the 509-meter tall Taipei 101 were stepped in order to reduce crosswind response and drag, resulting in a 25 percent reduction in base moment (Irwin 2005).
- *Tapering and setbacks* As indicated in Equation 1, at a given wind speed, the vortex shedding frequency varies depending on the Strouhal number S and width b. If the width b can be varied up the height of the building through tapering or setbacks, then the vortices will try to shed at different frequencies at different heights. As a result they become "confused" and incoherent, which can dramatically reduce the associated fluctuating forces. An example

of this strategy is the shape of the 828-meter high Burj Khalifa, Dubai (Figure 20.11), which was developed in the course of extensive wind tunnel tests.

- *Varying cross-section shape* A similar effect can be achieved by varying the cross-section shape with height, for example from square to round. In this case the Strouhal number *S* varies with height, which again, in accordance with Equation 1, causes the shedding frequency to be different at different heights. A good example is the 632-meter high Shanghai Tower (Figure 20.12). Its rounded triangular cross-section, besides tapering, rotates through about 130 degrees from base to top, thus presenting a different shape and width to the wind at different heights.
- *Spoilers* One can also reduce vortex shedding by adding spoilers to the outside of the building. The most well-known form of spoilers is the spiral Scruton strake used on circular chimneystacks. Architecturally and practically, the Scruton strake leaves something to be desired on buildings, but other types of spoiler could be used that might be

20.11 Burj Khalifa, Dubai and wind tunnel model
Image: Peter Irwin, RWDI

20.12 Shanghai Tower and wind tunnel model
Image: Peter Irwin, RWDI

20.13 Proposed tower in Inchon, Korea and wind tunnel model
Image: Peter Irwin, RWDI

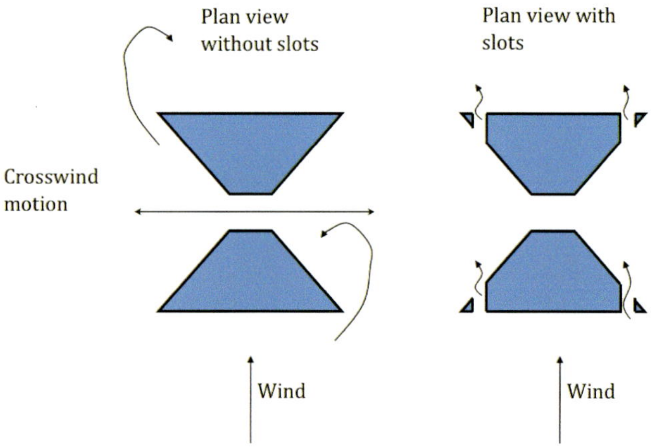

20.14 Slots to reduce vortex shedding on proposed tower in Inchon, Korea
Image: Peter Irwin, RWDI

more acceptable, such as vertical fins at intervals up the height. This approach was used to prevent vibrations on the cylindrical spire at the very top of Burj Khalifa.

- *Porosity or openings* Another approach is to allow air to bleed through the building via openings or porous sections. The formation of the vortices becomes weakened and disrupted by the flow of air through the structure. A proposed tower in Korea (Figures 20.13 and 20.14) is an example of using openings to suppress vortex shedding.

Pedestrian Environment and Wind-induced Noise

Downdrafts and vortices formed at the base of tall buildings lead to zones of accelerated winds. To illustrate how this can affect people it is worth noting that the mean wind speed approaching the top of a 200-meter high building in suburban surroundings would typically be about 100 to 200 percent higher than at ground level. The building deflects these stronger winds down to ground level. Bearing in mind that wind forces are proportional to speed squared, the force felt by a pedestrian at street level could be magnified by a factor of four to nine as they near the building.

There are a number of ways of reducing this type of effect. Changes to the building massing are the most effective, for example setting the building back from the surrounding streets on a podium, thus keeping the accelerated winds up at podium level, leaving the street level protected; or creating a series of setbacks that face the prevailing wind direction and break up the downwards flows. If massing changes are not possible then landscaping, screens, colonnades, and overhead trellises or canopies can be effective in creating local sheltered areas.

Criteria for assessing the appropriateness of wind conditions for pedestrians are divided into two categories: safety and comfort. The criteria and methods of assessment have been reviewed by the American Society of Civil Engineers in the document *Outdoor Human Comfort and Its Assessment* (ASCE 2003). Wind tunnel tests are usually needed for precise studies of pedestrian level winds but computational methods are increasingly in use, especially for initial massing studies. With respect to safety, the ASCE (2003) suggests that gust speeds above 25 meters per second, enough to blow some people over, should be limited to no more than two or three occurrences per year, corresponding to a probability of occurrence per hour of less than 0.1 percent.

For comfort it is important to consider the type of activity occurring at each location in a building and to vary the criteria depending on the activity. The ASCE document suggests, for example, that in areas where pedestrians are expected primarily to be walking from one location to another the mean wind speed should be kept to less than 5.4 meters per second for at least 80 percent of the time. At a location where people are expected to stand, such as at a bus stop or waiting at an entrance, the criterion is reduced to 3.9 meters per second, and where they will be sitting, for example an outdoor cafe or a swimming pool, it is further reduced to 2.6 meters per second. These should be regarded as guidelines only, since people's tolerance of wind varies quite widely and they are likely to be more tolerant in geographic regions that are inherently windy. Some local calibration is therefore desirable for each city.

Wind-induced noise occurs when the wind flows past building elements that cast off small vortices at frequencies that fall within the audible range, this being called Aeolian noise (the humming of telephone wires is an example). On buildings this can happen at small recesses and slots in the curtain wall system or off balcony railings, louvers, and suchlike. In the case of railings and louvers, they may also vibrate due to the vortex shedding, which can magnify the noise by further strengthening the vortices. Another mechanism for noise is the blowing of wind over a slot or hole that connects to an enclosed inner volume. The classic example of this mechanism, known by the scientific name Helmholtz resonance, is the humming noise generated by blowing over the open end of an empty bottle. The air of the internal volume acts like a spring for the mass of air around the mouth, which oscillates in and out, causing sound waves to radiate at the natural frequency of the "mass-spring" system. Another source of noise is the flow of air through small cracks at elevator doors within the building or through other small slots in the envelope. These flows result from stack effect as well as wind, creating pressure differences both between the inside and the outside of the building, and between different parts of the building's interior. The stack effect is most severe when there are large temperature differences between the inside and outside, which make the inside air either positively buoyant for cold exterior temperatures or negatively buoyant for warm exterior temperatures.

References

ASCE (1999) *Wind Tunnel Studies of Buildings and Structures*, Manuals and Reports on Engineering Practice No.67. Reston, Virginia: American Society of Civil Engineers.

ASCE (2003) *Outdoor Human Comfort and Its Assessment*, prepared by the Task Committee on Outdoor Human Comfort, Aerodynamics Committee, Aerospace Division. Reston, Virginia: American Society of Civil Engineers.

Part Five

Enclosure, Servicing and Construction

Introduction

Dave Parker

One of Ludwig Mies van der Rohe's most quoted aphorisms is "God is in the detail," although the actual author of the phrase has never been convincingly identified. In the longer run, it is the performance and durability of the cladding, the smooth functioning of the services and elevators, and the building's safety and resilience should a fire break out that determines whether a particular project can be deemed a success. Similarly, a building whose construction is marred by cost and time overruns, and subsequent expensive and time-consuming lawsuits, is unlikely to win many plaudits. In Part Five, eminent contributors consider the final phases in the genesis of a tall building and the detailed decisions that have to be taken by designers and construction managers alike.

Sometime in the middle of the last century architects fell in love with glass. Advances in glazing technology made the Holy Grail of a totally transparent building almost realizable. Hermetically sealed towers rose, clad entirely in glass, dependent on complex building services to create a tolerable internal environment all the year round. This came at the price of massive energy demands, although the development of fritting and similar techniques helped keep this under some sort of control. But those who worked on the vast open-plan floors, even those without significant acrophobic tendencies, were less at ease than the designers anticipated. In this century, however, the factors that lead to occupant wellbeing are better understood and given much higher priority, not least because of the significant improvements in productivity that can result.

In the opening two chapters of Part Five the complex imperatives that now shape the design of high-performance cladding are discussed in detail. Facades are still evolving, influenced in particular by the increasing emphasis on sustainability and carbon neutrality. Traditional single-skin facades are slowly giving way to double- and even triple-skin solutions. Natural ventilation is now seen as desirable, however difficult it may be to achieve in practice. And the threats of terrorism and increasingly extreme weather events are also shaping tall buildings in subtle but significant ways.

In practice, the biggest risk to the safety of a tall building and its occupants will always be fire, however that fire may be triggered. Part Five's third chapter is a comprehensive review of modern fire protection and safety techniques. Such measures also impinge on the increasingly complex and potentially more vulnerable services that a modern tall building demands. As Part Five's fourth chapter details, service design is also responding to new imperatives, not least ever increasing height.

Height and occupant densities impact on lift/elevator design as well. Intelligent lifts, be they double deck or shuttle, are now capable of moving more people faster and to greater heights than ever before. In Part Five's fifth chapter can be found details of the latest technological developments in the field of vertical transportation and their effect on building design.

However sophisticated and ingenious the conception of a tall building might be, the point comes when ground must be broken, concrete poured, steel erected. Modern tall building construction leaves little room for even minor errors, as Part Five's final chapter explains. Project managers carry a heavy burden of responsibility. Luckily, the skills, experience and techniques essential to the successful completion of a major tall building project are available.

Chapter 21
Curtain Walls

Nicholas Holt

The Curtain Wall

In the broadest definition, a curtain wall is any non-bearing exterior wall system attached to, or hung on, a building's exterior structural frame. It supports itself and resists external forces, such as wind or seismic loads, but is independent of the primary structural system of the building. This type of wall is usually made up of a series of framing elements, most often aluminum, infilled with a combination of glass, metal, or stone panels. These elements are then anchored back to the building's structural system. This construction method was invented in the late nineteenth century, when advances in the mass production of iron, and later steel, eliminated the need for heavy load-bearing facades, which until that point had been required as an integral part of the building's overall structural system.

The jump from load-bearing masonry walls with punched windows to curtain walls is significant in terms of production efficiencies, and also precipitated monumental changes in buildings' interior experience and exterior perception. Beyond new technological advances, the advent of curtain walls addressed ideas about transparency and repetition as a social value. Office floor plans became simpler, allowing more natural light to penetrate, and making floor plates more efficient and more egalitarian. An early and significant example of this, Mies van der Rohe's 1921 competition entry for a tower on Berlin's Friedrichstrasse, was a crystalline honeycomb completely devoid of all ornament. Its universal expressionism marked the beginning of a new era in building design.

More recently, the modern curtain wall can be traced to Mies' influence on Gordon Bunshaft of Skidmore, Owings & Merrill. For Lever Brothers' new headquarters (Figure 21.1 overleaf), Bunshaft employed a metal and glass curtain wall widely considered the first of its kind, which kicked off a new design aesthetic the world over. In a June 1952 review, *Architectural Record* marveled at the building's slim profile, commending it for bringing daylight to the open-plan offices, a new concept at the time, but one that reflected the client's vision for personal hygiene products in the twentieth century. This project, more than any other, ushered in the era of the modern curtain wall.

Establishing Initial Design Parameters for a Tall Building Curtain Wall

Tall buildings are inevitably made up of networks of interdependent systems, which in the end must function as a unified whole incorporating a wide range of performance parameters. As such, the approach to designing a tall building is multidirectional; it is a research and decision-making process that reinforces and informs design intent at many scales in a series of feedback loops.

How a building responds to its environment, how it meets its performance goals, and how it can be economically constructed are all factors that must be considered in concert. In the final realization of a building, these considerations are expressed primarily by the image and performance of the curtain wall.

The almost infinite possibilities for form-making make it necessary for architects to research the relationships between architectural forms and the large numbers of forces, energy streams, and performance requirements that interact with those forms. Following the analysis of the requisite zoning and urban parameters, designing a curtain wall facade is about how fundamental environmental forces such as wind, solar, thermal, and, most importantly, moisture are addressed, and addressed beautifully. Initially, this is where the curtain wall design must begin, by determining which parameters will drive the fundamental aspects of form and function for the facade and, ultimately, the entire building.

Although the list varies from project to project, some of the initial questions a curtain wall designer might ask include:

- *Climate* What is the climate being designed for?
  - Which climate forces can be leveraged for benefit?
  - Which climate forces require mitigation?
- *Energy performance goals:* What kind of overall energy performance is being sought and what is the expectation for the facade to participate in achieving that performance?
  - Geometry relative to sun, views and wind?
  - Shading devices?
  - Percentage of glazing allowed or required?
  - Building-integrated photovoltaics?
- *Orientation:*
  - Within the confines of the zoning regulations and urban plan, how can building orientation be leveraged for solar performance and views?
- *Real estate performance goals:* What are the real estate performance issues?
  - Desired planning module?
  - Full height windows?
  - Views?
- *Natural ventilation:*
  - Are operable windows desired or required?
  - Are other approaches to natural ventilation required?
- *Security:* Are there any security or threat considerations (blast, ballistics)?
- *Regulatory:* Are there any local code or zoning issues that might affect the overall design of the facade?
- *Other regional or local considerations:* Are there any regional factors that may fundamentally affect the wall (hurricanes, seismic factors, cultural factors)?

With the answers to these questions, the architect can develop a fundamental understanding of how the design of the wall *wants* to develop. A great curtain wall design concept will develop in an iterative fashion, with each design parameter resolving itself within the system. The trick, of course, is aligning the often disparate performance criteria within an inherently flexible system, while also effectively expressing design intent and being economical to fabricate and install.

Illustration: Harnessing climate conditions

The first step is to analze the desire for energy savings and natural ventilation that has emerged as a priority in new buildings in moderate climates. This emphasis is also starting to appear in the United States and around the world. A good example is a project where the Santa Monica–based architectural firm, Morphosis, introduced natural ventilation via a dynamic curtain wall into the San Francisco Federal Building (Figure 21.2 overleaf), aided by the University of California at San Diego, ARUP, and the Lawrence Berkeley National Laboratory,

FACING PAGE

21.1 Lever House, 1952, Skidmore, Owings & Merrill—in *Architecture: From Prehistory to Post-modernism*, the authors singled out Lever House for its precedent-setting design: "Its thin, 20-story tower was small by New York standards (sacrificing potential real estate profits to beauty and urban values), but for an International Style building it was unusually large. The mastery of proportions and especially of detailing was and remains exceptional, a harbinger of the technological excellence of the metal-glass curtain wall that was to become the special province of Late Modernist American builders." (Trachtenberg and Hyman 1996: 545)

Image: © Ezra Stoller/ESTO

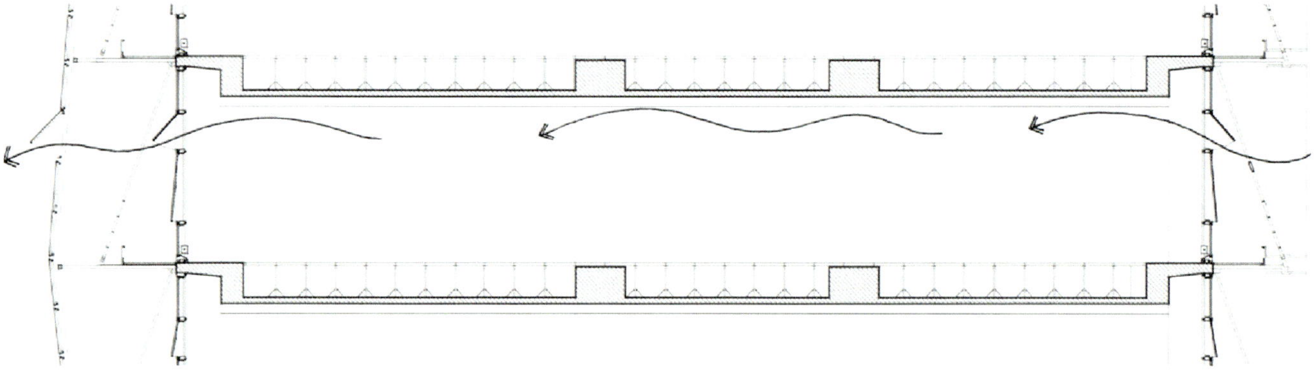

21.2 San Francisco Federal Building, 2007, Morphosis—ventilation
Graphic: © Morphosis Image: © Roland Halbe

part of the U.S. Department of Energy, which provided computer modeling for the ventilation. The goal was to take advantage of the local climate characteristics and replace mechanical cooling on certain floors with cross-ventilation at night from automated operable windows. Awning windows in the northwest and southeast glass curtain walls facilitate cross ventilation of 70 percent of the floorplate in these areas. Perforated metal scrims protect the curtain wall on the southeast from solar heat gain. In order to control solar gains enough to allow natural cooling to occur, perforated metal scrims protect the curtain wall on the southeast from solar heat gain, while frosted vertical fins provide effective shading for the forthwestern facade.

Selecting the system

Once the fundamental analysis and direction is complete, the next step is to evaluate and select the basic curtain wall system that is best suited to complement the design intent. This decision integrates issues of constructability, budget, and performance with design intent.

The most basic curtain wall systems utilized today are often referred to as "stick" and "unitized":

- *Stick systems:* Stick systems were the first generation of modern curtain wall systems and consist of premanufactured parts, or "sticks," assembled and glazed in the field. They rely heavily on field labor, which impacts both cost and quality considerations. Stick systems are utilized heavily in regions that have inexpensive field labor, but are falling out of favor in many advanced markets due to quality control considerations and the cost of labor.
- *Unitized systems:* Unitized systems, on the other hand, are fabricated at the factory and installed as finished, glazed panels in the field. This means that the quality control efforts primarily occur in a more easily controlled environment, at the factory. It also limits the size of the field team required to install the units and the number of parts they have to handle. One consideration is sizing the unitized panel relative to the number of crane picks versus using the elevator hoists to get the units to the floors for installation, as this is a site logistics issue and impacts cost.

21.3 The two most basic curtain wall systems, stick and unitized, differ in the ways they are manufactured, assembled, and installed. A stick system was used for Lever House (left), while One World Trade Center (right) has a unitized curtain wall system

Images: Left courtesy Skidmore, Owings & Merrill Right © 2011 Port Authority of New York and New Jersey

Variations on these two primary types of curtain wall systems do exist, and may be advantageous for a particular project with a specific wall configuration or set of site constraints. Examples of these variations include window wall systems (where units rest on top of the slab edge), column and spandrel systems (where vision units infill a system of opaque exterior column and spandrel panels), and hybrid systems (including a vision unit resting on the slab and a spandrel panel hung from the slab edge).

Illustration: Multi-unit panelized system

A 23-story condominium near the Hudson River in Manhattan's Chelsea neighborhood, designed by Ateliers Jean Nouvel with New York–based Beyer Blinder Belle Architects & Planners, provides an innovative example of one of these variations. In this project, 1,700 unique oversized glazing units create a dynamic pattern over the surface of the highly visible exterior wall. In this instance, the wall fabricator combined these lights of various sizes into large multi-unit panels, allowing the installer to limit the number of pieces which had to be handled on-site. The resulting wall appears to be a random array of windows of varying sizes, colors, and operation.

Developing and Refining the System

Once the basic curtain wall system has been selected, there are a host of follow-up issues that a designer needs to consider as he develops the design, details, and specifications:

- *Glass composition:* Aesthetic and performance properties (low iron, clear, tints), coatings (Low-E), frits (opaque, colored, and/or translucent ceramic printing), and the assembly of the insulated glazing unit (double or triple glazed, gas filled, sealant type and color) must be considered. The balance between solar heat gain coefficients, visible light transmittance, and U-value needs to be carefully considered in terms of overall cost-benefit of each option, relative to the interior daylighting quality and system energy performance.
- *Selection of finish materials:* Stone, terra cotta, metal, and other materials must be selected appropriately for the region and local environment, as well as verified with the specific structural requirements.
- *Coatings and paint systems:* The most commonly used metal finishings are anodized, painted, or powder coated. Each has its pros and cons that

21.4 100 Eleventh Avenue, Ateliers Jean Nouvel with Beyer Blinder Belle Architects & Planners—multi-unit panelized curtain wall system
Images: © Will Femia

must be evaluated for each specific market. Paint systems are numerous and can be selected based on interior or exterior performance (in some system types, coating types may differ from interior to exterior), environmental conditions, and maintenance schedules.

- *Coordination with building maintenance systems:* No curtain wall can avoid the need to maintain the wall, replace broken windows, and do routine inspections. Provisions for facade maintenance and inspection must be integrated into the design. The most common systems are intermittent anchor systems that include "button" tie-backs for facade maintenance rigs, or track systems that integrate a track into the facade that receives a torpedo from the maintenance rig or platform rig. Different jurisdictions prescribe different maximum heights at which these systems may be used (the integrated track system is typically allowed to go higher). This has a significant effect on the mullion profiles and configuration and should be determined early in the process of design.
- *Coordination with wind tunnel results:* Performance criteria (positive and negative pressures) determined in a wind tunnel are usually the most appropriate for tall buildings, as code values often do not effectively take into account the heights and geometric considerations of high rises. As a result, wind tunnel testing often generates a more efficient and cost-effective wall than would be the case using code-prescribed calculations.
- *Coordination with building movements and structural considerations:* Coordination of curtain wall design and overall building structural performance is a critical early decision and can affect cost directly, particularly if the curtain wall criteria require a stiffer or more robust structure in order to meet the movement or inter-story drift criteria specified for the facade or vice versa.
- *Condensation resistance and thermal performance:* Determining the desired thermal performance of the wall based on prescriptive energy code values and/or calculations will determine the necessity of providing thermal breaks in the wall (in most cases they are required). This can also affect condensation resistance, which can be calculated, but is best tested via advanced simulation software and confirmed in performance mock-ups.
- *Air and vapor resistance:* There are multiple strategies employed to prevent the infiltration of air and moisture through a sealed curtain wall. Typically, multiple lines of defense (primary and secondary seals) are required to create proper air and moisture barriers. A rain-screen system represents another strategy, in which the outermost joints are left open while the inner sealed wall acts as the air and moisture barrier. A further refinement is a "pressure-equalized" rain-screen system, where an optimized, open cavity internal to the wall between the outer open joints and the inner sealed joints equalizes air pressure differentials and external forces that can lead to infiltration.
- *Acoustic criteria:* Consideration must be given to the required Sound Transmission Class (STC) ratings, which measure the frequencies specific to the transmission of sound between floors and tenants, and Outdoor–Indoor Transmission Class (OITC), which was developed to measure the frequencies specific to exterior-based noise (focusing on a lower range of frequencies such as truck motors). Achieving the appropriate performance levels, typically more stringent in residential applications, may require adding mass or density to the wall in terms of the glass assembly, slab edge details, and mullions.
- *Window treatments:* Provision for window treatments (shades and/or blinds), particularly in wall systems dealing with glare issues, needs to be made and coordinated with future tenant ceiling options. One issue worth considering is the appearance of the window treatments from the exterior of the building—does it make sense to mask the window treatments with dark or reflective glass? Or is it better to prescribe the color and type of window treatment for tenants via a series of documented tenant standards? In speculative projects, the owner's willingness to enforce these items during lease negotiations should be considered.

Illustration: Detailing the facade

A prime example of integrating these criteria is 7 World Trade Center (Figure 21.5 overleaf) in lower Manhattan, designed by Skidmore, Owings & Merrill LLP (SOM). The 52-story tower of clear, floor-to-ceiling glass establishes a visual ambiguity between transparency and reflectivity—an almost liquid appearance achieved with a carefully chosen reflective coating on ultra-clear low-iron glass (which removes much of the typical green tint present in standard "clear" glass). Being the largest production IGU (Insulated Glass Unit) made in the United States at the time (12 feet (3.7 meters) tall by 5 feet (1.5 meters) wide), SOM specified a thicker-than-required outer light (10 millimeters versus 6 millimeters) to enhance the flatness of the glass. At the upper portions of each IGU, a ceramic frit pattern, applied in a graduated dot pattern, reduced solar heat gain while providing for excellent light transmittance

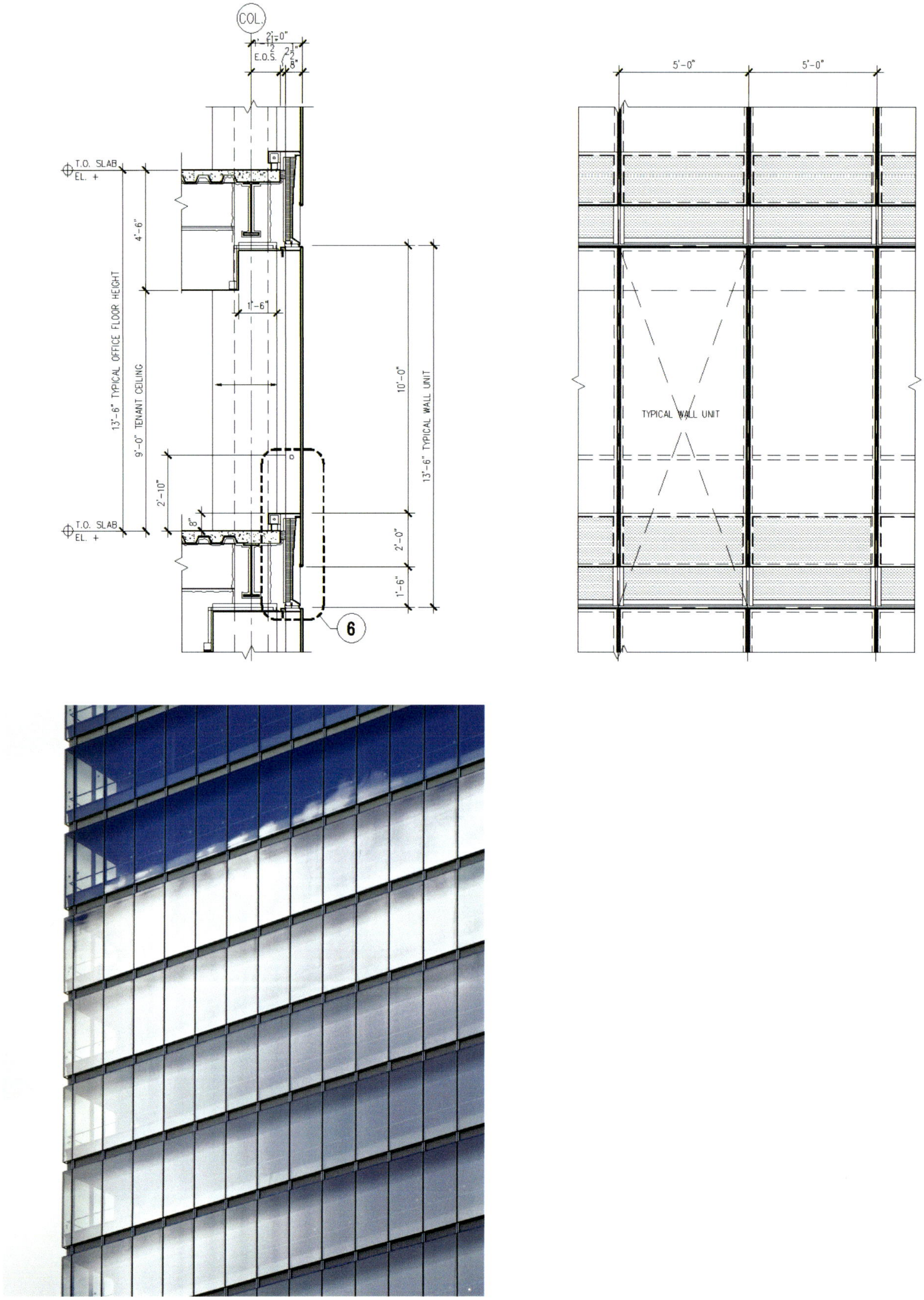

21.5 7 World Trade Center, 2006, Skidmore, Owings & Merrill
Drawings: © SOM Image: © David Sundberg/ESTO

in the interior spaces of the building. From the interior, the full height clear glass panels and transparent corners provide a unique sense of openness, as panoramic views extend from all angles. The specific glass selections and materials were tested and confirmed in numerous small-scale mock-ups and a final visual mock-up during the design and documentation phases of the project. This visual mock-up process was critical to ensuring the outcome met the design intent.

Illustration: Integrating the facade with interior function and energy performance

At the Memorial Sloan-Kettering Mortimer B. Zuckerman Research Center in New York (Figure 21.6), also designed by SOM, the team developed an array of environmentally sensitive design strategies tested via extensive whole-building simulations. Computer modeling helped evaluate energy consumption and cost matrices that ultimately led to a LEED Silver rating from the U.S. Green Building Council.

While the organization of the building's program components created a varied external profile presenting four distinct faces to the neighborhood, the development of the facade required careful study of the urban context and environmental concerns. Transparent, translucent, and opaque scrims of glass etched with graduated densities of ceramic frit (patterns of baked enamel) enclose the laboratories.

The intended transparency was achieved while also reducing solar heat gain and allowing the building to surpass city code requirements for energy efficiency. The decision to combine Low-E coating with ceramic frit came relatively early in the design process and necessitated a significant research and development effort in order to achieve the desired effect while also meeting the needs of building occupants.

In order to achieve this, the design team developed a custom computational tool that linked AutoCAD instructions placing frit dots in a graphic model. With this new tool, designers could easily modify and control the slight variations of size and spacing of the frit dot pattern and work to avoid moiré patterns, optimize solar heat gain performance, and create areas of high light transmittance where this was most effective. In turn, these AutoCAD files were provided to the manufacturer to accurately place the frit patterns on the flass.

21.6 Memorial Sloan-Kettering Mortimer B. Zuckerman Research Center, 2008, Skidmore, Owings & Merrill
Images: © David Sundberg/ESTO

Coordinating the location and density of the ceramic frit with the functions of the building occupants resulted in an alternating, checkerboard-type pattern covering the entire surface of the laboratory facades. Areas of greater transparency are located at the laboratory aisles, while areas of greater opacity reduce glare and solar heat gain at the lab benches without overly limiting required daylight. The weaving pattern contributes to a 40 percent reduction in building energy use while achieving a 76 percent daylighting factor in all occupied spaces.

Illustration: Structural integration

The New Beijing Poly Plaza (Figure 21.7) is a potent example of how facade design can be fully integrated with structural design to respond to environmental issues rather than be compromised by them. At 90 meters tall and 60 meters wide, the building required innovative approaches to resolving issues of design, engineering, and construction unique to its scale and the seismic zone of the building's site. The integration of architecture and engineering is expressed most strongly in the dramatic cable-net wall enclosing the main atrium, which is currently the largest such enclosure in the world. A cable-net wall consists of vertical and horizontal cables arranged in a rectilinear grid, used to support a glass curtain wall enclosing an atrium. Conventional thinking would have sought to support the glass wall as a single plane, using bulky trusses that would have disrupted the sense of transparency. Instead, a cable-net wall that offers a heightened sense of connection between the atrium and the city is folded and stabilized by a formidable V-cable. The cable is counterweighted by a suspended lantern, actually a small building unto itself, by a specially designed rocker pulley that allows the main cable to compensate for movement during a seismic event. Folding the cable-net wall over the V-cable reduces the deflection caused by high winds—and therefore the heft of the supporting cable—while adding a faceted dimension to the enclosure.

21.7 The New Beijing Poly Plaza, 2007, Skidmore, Owings & Merrill
Images: © Tim Griffith

Visual and Performance Mock-ups

Computer simulation and modeling tools are vital to predicting a building's performance prior to construction. However, real-world testing is still required to test solutions and study alternatives. Full-scale mock-ups provide insight that simulation alone cannot. Different types of mock-ups might be appropriate at each stage of the design process, and depend in part on client budget and special requirements—visual, wind, blast, performance—as well as the complexity of the individual systems and how they affect the performance of other systems. Mock-ups are also useful for studying human factors including tenant experience, operational and maintenance issues, and construction installation issues. The data collected from mock-ups provide confirmation on design assumptions and generate invaluable information for the architects, clients, contractors, and facade manufacturers.

Visual mock-ups demonstrate the visual aspects of the wall. These can range from small desktop assemblies and scale models to larger mock-ups built at full scale. Typically the smaller mock-ups are used to inform early design decisions which are then confirmed in the full-scale visual mock-up. Full-scale mock-ups are critical to both finalize material selection and demonstrate the design intent to the client for final approval on the design. Visual mock-ups are usually constructed of closely approximated components (though some elements, such as the glass and paint finishes, must be exact) and structured with simple interior spaces behind them to approximate a realistic background for viewing. Often these mock-ups will be hoisted on a crane to simulate the most likely viewing angles for the final project.

Architects specify the construction of a full-scale performance mock-up of a portion or portions of the curtain wall using the actual materials and methods proposed for the job. The assembly is constructed on a testing chamber at an independent testing laboratory where it is subjected to a prescribed battery of tests (according to recognized standards such as AAMA) that simulate thermal and condensation performance, dynamic wind forces, simulated wind pressures, basic air pressure differentials, seismic forces, and rain. This is done not only to ensure that the facade meets the specified performance criteria but also to provide an opportunity for the installers to test installation methods and for maintenance personnel to review the design and test glass replacement procedures. The performance mock-up must be performed using the actual extrusions, detailing, and installation methods proposed for the project, and is usually included as one of the early stages of a curtain wall fabricator's or installer's contract. The timing is important, since a failed test may require

21.8 Visual curtain wall mock-ups test the visual aspects of the wall and are often hoisted onto a crane to simulate final viewing angles; performance mock-ups test the system's weather tightness and structural integrity
Images: © SOM

modification of details or assembly sequences and will require revisions to the design before components for the wall are fabricated.

Another and oftentimes complementary approach to performance mock-up testing is in situ or field testing (also according to recognized testing standards). There are several levels of field testing, including simple water tests, where a hose is sprayed onto a defined area of the facade to determine if water enters the building, and more involved tests that may include the construction of physical testing chambers on-site. In this case, a sealed testing chamber is constructed against a portion of the installed wall to test for water tightness and air infiltration. Field tests have the benefit of testing the quality of the field installation, but occur too late in the process to economically develop meaningful corrections to the wall's design or engineering. Field testing is therefore most often used in conjunction with the standalone performance mock-up; the former for confirming quality in the field and tweaking installation methods and the latter primarily for confirming design, engineering, and fabrication techniques prior to substantial fabrication and on-site installation.

The Next Evolutionary Cycle

The modern curtain wall is no longer a passive building wrapper that keeps out the elements and brings in daylight. Decades of refinement in energy-efficient glazing, structural sealants, composite building materials, high-strength concrete, and photovoltaics have created the expectation that even standard curtain walls will be high-performance machines capable of interacting with all building systems at optimal levels.

Currently, the approaches to developing the next generation of facades fall within two primary schools of thought, though most systems borrow liberally from each. The first school of thought focuses on barrier technology, in which buildings are sealed as tightly as possible so that mechanical systems are in minimal demand. This approach is perhaps best exemplified by the Passivhaus standard that is taking hold in many European countries. Building envelopes under the Passivhaus standard are required to perform beyond the capabilities of conventional construction. Air barriers, careful sealing of every construction joint in the building envelope, additional insulation, limited glazed openings, solar shades, and sealing of all service penetrations throughout are the tools of the trade. These ideas have been applied primarily on small-scale residential projects, but hold a great deal of potential for future use in high-rise buildings.

Sealed cavity facades are another example of barrier technology. These were developed as a next generation of the double-skin facade, a typology addressed in more detail in Chapter 22

Similar to a double-skin system, a sealed cavity system (Figure 21.9) consists of a monolithic glass outer layer and an inner insulated glass layer separated by a sealed air space. A sealed cavity system relies on careful balancing of the performance of the solar-reflective single-pane outer layer, motorized blinds in the cavity, and a highly insulated inner layer. This is necessary in order to keep the temperature inside the cavity within acceptable limits and to reduce ambient energy gains and losses. To deal with temperature-related pressure changes in the cavity, a very small amount of air leakage to the outside is allowed from the "sealed" cavity, while the cavity maintains a positive internal pressure with a dried and filtered air supply. As the cavity is conceptually sealed, there is no need to size the cavity to maximize airflow, so the width of the cavity can be decreased to almost half that of a typical ventilated double-skin wall, saving space and reducing cost.

A second approach in developing the next generation of facades focuses on leveraging energy streams from the environment and applying them where needed. This is perhaps best exemplified by the work of the Center for Architecture Science and Ecology (CASE), a multi-institutional and professional research collaboration co-founded by Rensselaer Polytechnic Institute and SOM. CASE researches, tests, and prototypes next generation building technologies with the goal of realizing high-performance systems in building projects, and works with fabricators to make these systems readily available to designers and building projects worldwide. CASE's research is currently focused on next generation technologies that promise to dramatically enhance a building's performance by harnessing the surrounding energy streams, including solar and wind, and drawing them into the building in a controlled manner (see Figure 21.10 overleaf).

It is anticipated the facades of the future will likely depend on dynamic elements in order to meet ever-evolving standards for energy, wind and seismic, and structural performance. Toward this end, Permasteelisa North America, SOM, and the Adaptive Building Initiative (itself a collaboration between Hoberman Associates and Buro Happold) recently embarked upon a collaborative study to investigate next-generation, dynamically responsive facades. The goal is to design an advanced building enclosure prototype that borrows from the barrier school of thought and also harnesses energy inputs from the environment. Given that these environmental inputs are themselves kinetic (seasonal and diurnal), the team is looking beyond static facade solutions and is attempting to tackle the problem of developing a workable solution to actively adaptive facades. HelioTrace, a prototypical facade system, has been developed, and improves the wall's performance relative to daylighting and glare, reducing solar heat gain by as much as 81 percent (Figure 21.11 overleaf). This in turn facilitates the use of advanced mechanical systems which together deliver total theoretical energy savings of nearly 63 percent over the baseline existing building performance in New York City (calculated per ASHRAE 90.1 2007). With potential returns of that magnitude, it seems this line of thinking is worth further investigation.

There is a responsibility within the design community to actively develop innovative ideas to help pursue the sustainability goals required to protect our environment and our clients' investments. The world over, municipalities and the market are mandating energy, carbon, and water-use targets that are set to reach net zero or near net zero in the next two decades (2020 in Europe, 2030 in the United States). While facade performance has been improved with current technologies, as net zero designs are approached dramatic improvements in performance are going to be required that will necessitate the next great evolution in current wall design and technology. In short, the future of curtain wall is bright, and full of wonderful technical challenges and design opportunities.

21.9 Sealed cavity system testing mock-up
Image: © SOM

References

Architectural Record (1952) "Lever House," *Architectural Record*, June: 130–5.
Trachtenberg, Marvin and Hyman, Isabelle (1996) *Architecture: From Prehistory to Post-Modernism*. Upper Saddle River, NJ: Prentice Hall.

Contributor

Benjamin Reich, Enclosure Specialist, Skidmore, Owings & Merrill.

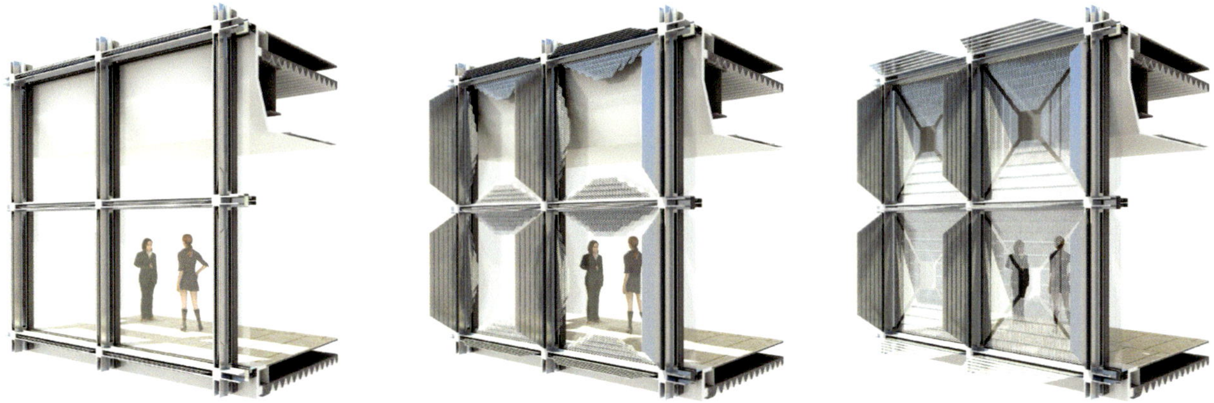

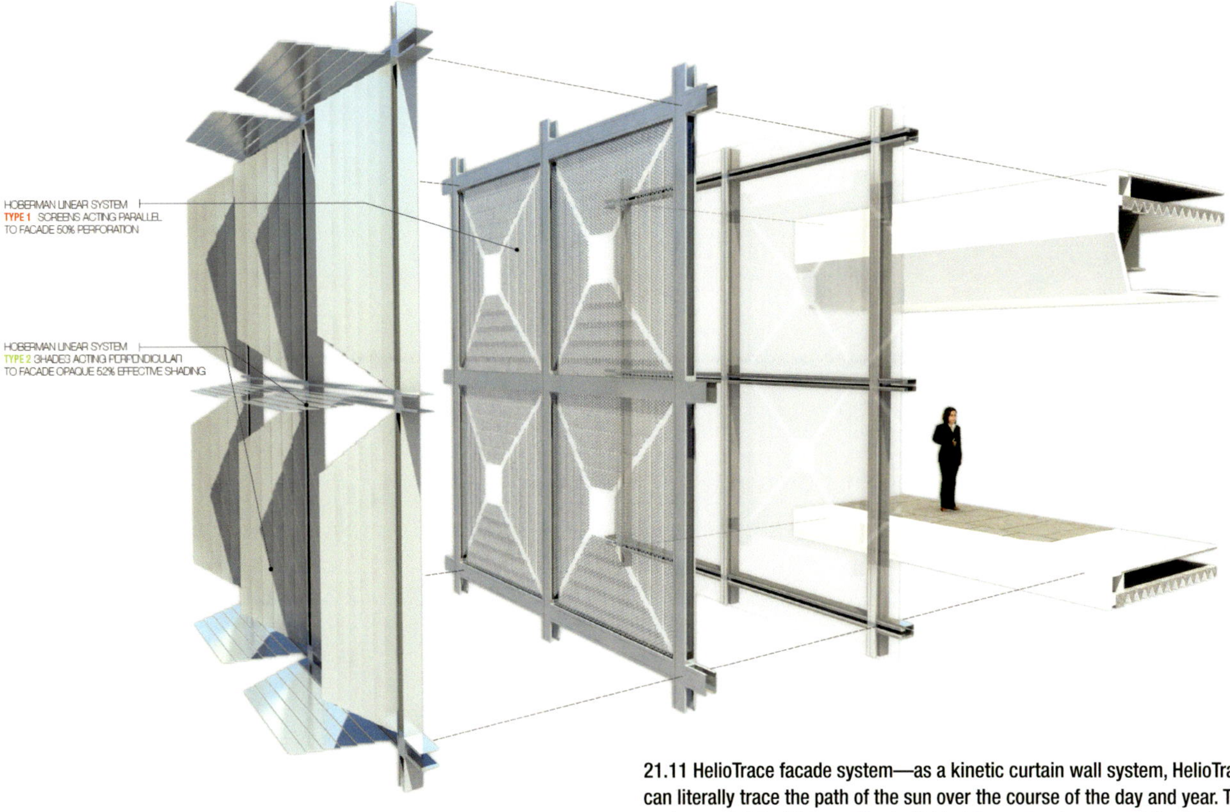

21.11 HelioTrace facade system—as a kinetic curtain wall system, HelioTrace can literally trace the path of the sun over the course of the day and year. The system maintains high-quality daylight at the perimeter zone while eliminating glare, reducing a building's peak solar gains
Images: © SOM

FACING PAGE

21.10 Technologies developed by CASE
(top) HeliOptix: Using ultra-high efficiency PVs, concentrating optics, and a fluid-based coolant system, HeliOptix is able to provide low-glare daylight, virtually eliminate solar heat gain, and generate power and usable hot water for a total system efficiency approaching 80 percent; this is in contrast to contemporary building-integrated PV technologies approaching 15 percent efficiency.
(middle) Solar Enclosure for Water Reuse (SEWR): By running filtered grey and black water through cast glass curtain wall components, SEWR can reduce solar heat gain, provide low-glare daylight, and pasteurize waste water via UV exposure; the water in turn is reusable on-site as grey water.
(bottom) Climate Camouflage: Using readily available ceramic materials in an innovative configuration and integrated with color, texture, and phase change materials, Climate Camouflage can tune a facade to either accept or reject heat at different times of the year. The phase change material can then be used in radiant floor heating or cooling systems to dramatically reduce HVAC loads in the building
Images: (top) © SHoP, (middle) and (bottom) © Center for Architecture Science and Ecology/Rensselaer Polytechnic Institute

Chapter 22
Towards a Double-Skin Solution

Werner Sobek, Thomas Winterstetter and Claus Peter Weller

Introduction

High-rise buildings are a challenge to the architect. They are equally – or perhaps even to a greater degree – a challenge to the engineer who designs the load-bearing structure and the facade. Wind forces, earthquake risks, the need to reduce the consumption of resources (which requires a minimization of building material used) and the maximization of transparency in the facade and in the structure of the building itself all have to be taken into consideration. So the planning of high-rise buildings is a process of extreme complexity that demands a very high degree of expert knowledge.

In this chapter the different specific influences on and requirements of the facades of high-rise buildings are identified and explained. These comprise the building movements due to earthquake and wind, the integration of sun-shading devices and aspects of erection and maintenance, as well as special requirements with regard to fire protection. Particular attention is paid to sustainability, an issue that has gained tremendously in importance in recent years.

It will be noted that the main focus is on double-skin facades, even though it is accepted that this is a solution not yet commonly used in all major markets. In the USA especially, single-skin facades are currently the standard solution for high-rise buildings. However, it is considered that double-skin facades in many cases offer advantages that compensate for higher initial costs, and are becoming widely accepted and utilized throughout the world.

The question as to whether a double-skin facade is an appropriate solution for a given situation can be discussed as follows.

Irrespective of the specific type of double skin chosen, the temperature in the interspace between the outer and the inner skin (facade) is generally higher than the outside (ambient) temperature. During the day, the higher temperature in the interspace is caused by absorption of solar radiation by the two skins and the air between them. At night, the temperature in the interspace is elevated (relative to ambient conditions) due to a retarded cooling down of the system. This effect is advantageous whenever a building needs to be heated and results in lower energy consumption.

When cooling is needed, however, this effect may lead to a higher need for cooling energy, a higher radiant heat load on the building's interior and uncomfortably warm air streaming in when any windows in the inner skin are opened. These negative effects can be countered, however, by a proper design of the facade and an intelligent use of aerodynamic effects. If the design of the double-skin facade takes into account local, regional and building-specific effects, the negative aspects described might immediately be turned into positive ones. In this context, the sunshades arranged within the interspace are a very important factor. It is important to point out that only double facades allow for a sunshade on the exterior of the (inner) building skin, with the outer skin protecting the sunshades against rain, dust, and wind.

Closed shades drastically reduce the solar heat gains of a building; they also reduce the heat radiated into the interior by the inner skin. The shades reflect part of the heat back to the exterior. Another part of the heat is absorbed by the shades themselves. The rest of the heat is absorbed by the outer skin and by the air in the interspace. These absorption effects may (despite the shades) lead to an undesirable rise of the temperature of and within the system, unless further counter measures are taken. The most important such counter measure is proper ventilation of the interspace.

Interspace ventilation is an important factor contributing to the thermal behaviour of a double-skin facade. There is a risk of the airflow within the interspace being slowed down by aerodynamic friction. This may be caused by the channel or the interspace being too small, by an insufficient size and inadequate aerodynamic behaviour of the air inlets and outlets, or by a closed sunshade blocking the airflow. In order to ensure an optimal working of the double-skin facade, unrestricted airflow between the closed sunshades and the inner skin has to be guaranteed. Special attention therefore has to be paid to these aspects early in the design phase.

On the one hand, a double-skin facade offers many advantages over the single-skin alternative with regard to user comfort, lifecycle costs, etc.; on the other, it also leads to higher installation (building) costs and increased cleaning costs (there are four instead of two surfaces to be cleaned). However, in general, the advantages of a properly designed and built double-skin system will in the long term more than counterbalance these drawbacks.

Natural Ventilation

Good ventilation is a major contributor to the usability of a building. In the long term, inadequate ventilation can verifiably affect the health of the building's users. The subjective quality of the ventilation does not depend solely on the amount of fresh air streaming into the interior – mechanical ventilation systems generally provide sufficient air supply – more important from a psychological point of view is the user's ability to actively make contact with the outside by individually operable windows or flaps. Despite this demand, many high-rise facades are still designed with completely fixed glazing, without any possibility of the user opening parts of the facade. It is, however, possible to provide natural ventilation even at great heights and without any particularly complex technology.

Opportunities to open windows individually increases user comfort. However, it may also affect building service systems, unless possible interactions are taken into consideration at a very early planning stage. Providing natural ventilation in high-rise buildings is an important task that requires careful planning and close cooperation between facade consultants, climate engineers and MEP. An equally close cooperation between architects and facade engineers is an essential prerequisite for minimizing the effect of draughts induced by open windows. Possible measures are the adaptation of floor plans and the insertion of internal partitions.

Increased wind speed at height is the most critical factor when considering natural ventilation. This can lead to very high pressure gradients around the facades, of up to 30 Pa. Such a high pressure makes it almost impossible to open a window unless further precautionary measures are taken – not only with respect to the facade, but also with respect to air movements throughout the building. The alignment of internal walls and doors therefore has to be modulated appropriately in order to avoid adverse airstreams. Particularly critical in this context are the elevator shafts, which act as "air accelerators" unless countermeasures are taken.

FACING PAGE

22.1 Interior view of a double-skin facade at the KfW Westarkade building in Frankfurt am Main, Germany. Architect: Sauerbruch Hutton, Berlin, Germany
Image: Brigida Gonzalez, Stuttgart

A common solution allowing natural ventilation in high-rise buildings is the use of double-skin facades. Users can thus open windows in the primary (inner) facade, while pressure differences are buffered by the secondary (outer) facade. This solution is well proven and generally leads to satisfactory results. However, particular attention has to be paid to the constant ventilation of the gap between the primary and the secondary skins, otherwise the air in the gap can overheat due to ambient radiation. Furthermore, the interstice of double-skin facades requires several additional precautionary measures, for example protection against undesirable sound transmission or the spread of smoke (the latter will be discussed in more detail in the section on fire protection).

One example of a recently developed double-skin facade is the ADAC Headquarters in Munich (2009, architect Sauerbruch Hutton, Berlin). Due to prevailing local conditions, wind speeds of 6 metres per second and more had to be expected at the top of the building during 70 per cent of the year. In order to prevent air suction, an airflow limiter was integrated between the outer and inner skins. An operable window aligned behind this airflow limiter allows natural ventilation independent of pressure differences between the interior and the exterior. Occupants are thus free to open windows at any time, no matter what the outside conditions might be.

Natural ventilation does not necessarily depend on the use of double-skin facades. The Highlight Towers in Munich (2004, architect Murphy/Jahn, Chicago) demonstrate that secure natural ventilation is also feasible in single-skin facades. Here, a sophisticated shutter system allows the intake of fresh air from outside while reducing flow speed to a well-defined maximum. This system could easily be integrated into other single-skin facades.

22.2 Natural ventilation in a single-skin facade: the Highlight Towers in Munich. Architect: Murphy/Jahn, Chicago, USA
Image: Rainer Viertlbock, Gauting, Germany

Sun Shading

The high proportion of glazed surfaces in modern facades allows for extensive use of natural light to illuminate the interior. At the same time it creates new exigencies with regard to sun shading and glare protection. Sunlight not only contains visible light but also significant levels of infrared and ultraviolet radiation, leading to solar heating of the interior. The larger the facade area in relation to the floor area, the more energy is gained through the facade (the exact amount, of course, depends on the heat transmission quotient of the glazing material used in the facade). Given the high transmissivity of glass for solar radiation, it is essential to provide sufficient sun-shading devices to reduce cooling loads to an absolute minimum.

Sun shading has to be considered at a very early stage of planning, since it has an important influence not only on the visual appearance of a building but also on the costs and on the thermal loads to be dealt with in the building itself. Once a facade is installed it can only be retrofitted at a relatively high cost. At a very early stage, therefore, sun shading has to be considered as an integral part of the facade and of the building services, in order to ensure a sustainable and functional solution.

High-rise facades are not usually (or are only insufficiently) shaded by surrounding buildings or trees. Moreover, internal thermal loads are usually higher than in comparable low-rises. This makes it all the more important to reduce external thermal loads by using outlying sun shading elements. However, the specific characteristics of a high-rise also limit the use of such elements. Typical venetian blinds can only be operated at wind speeds of up to 10 to 15 metres per second. This is much less than the speeds commonly occurring at the upper storeys of a high-rise building. Maintenance is another issue – it has to be possible to repair or replace individual devices with an acceptable effort, even at great heights. Last, but not least, sun-shading devices should neither affect the view out of the windows nor substantially reduce the desired daylight gain.

Basically, there are three methods to cope with this challenge:

1. Specially designed venetian blinds with strengthened lamellas
2. Stationary sun-shading elements (either metallic screens or lamellas, or glass lamellas with thin film photovoltaics)
3. Specially protected venetian blinds that are either integrated into a double-skin facade or into an IGU (Insulated Glass Unit).

The first method can be used at wind speeds of up to 20 to 25 metres per second. Reinforced venetian blinds withstand all but extreme weather conditions; however, they also require a higher initial investment and more maintenance than other systems. Using stationary screens or lamellas instead of Venetian blinds is a simpler and cheaper solution, but one that leads to other restrictions due to the immoveable nature of these devices: solar heating is reduced (even when desired), as is the amount of usable natural light. An alternative to stationary metallic devices is glass lamellas covered with a thin film of photovoltaics. This layer of photovoltaic cells is not completely opaque and lets some daylight into the interior. It also produces electric energy, making the glass lamellas a multifunctional element.

In Western and Central Europe, at least, specially protected venetian blinds are the most commonly used solution. Double-skin facades require a higher initial investment, but among the various advantages they offer is the protection of sun-shading systems by an additional glass pane. This method is well tried and has been used successfully in various projects, including the ADAC Headquarters in Munich and the Post Tower in Bonn (2002, architect Murphy/Jahn, Chicago).

22.3 Stationary screen made of titanium tubes in front of the Interbank building in Lima, Peru. Architect: Hans Hollein, Vienna, Austria
Image: Christian Richters, Münster, Germany

22.5 Sun shading in the Post Tower, Bonn, Germany: venetian blinds are integrated into the double-skin facade. Architect: Murphy/Jahn, Chicago, USA
Image: HG Esch, Hennef, Germany

FACING PAGE

22.4 Exterior view of the double-skin facade of the Post Tower, Bonn, Germany. Architect: Murphy/Jahn, Chicago, USA
Image: HG Esch, Hennef, Germany

Erection

Installation of high-rise facades is subject to various special conditions. One is the high wind loads that handicap the handling of facade elements. Another is accessibility: at great heights it is usually impossible to use scaffolding, so the facades have to be installed from inside the building or from a crane placed on a storey above. Moreover, requirements with regard to the accuracy of facade fixings are extremely exacting. Stack joints have to be able to accommodate building movements while simultaneously maintaining the tightness of the building envelope. Extreme precision of installation is essential.

In order to cope with the above-mentioned conditions, high-rise facades are usually designed as fully unitized systems, prefabricated at ground level and then delivered to central storage positions in the building (storage either directly on the storey for which they are to be used or on the storey above is usual). One of the most efficient methods for subsequent installation of the individual panels in the facade is to use a monorail crane mounted on the edge of the floor slab above which the facade is currently being installed. This is a convenient means of moving the units – which can weigh up to 1,000 kilograms, and even more – to their assembly location. Since the length of the ropes between the monorail crane and the facade panel can be kept quite short, the panels are less susceptible to swinging under wind loads. Panels can therefore be installed in a very precise way and the risk of accidents is significantly reduced.

Erection of high-rise facades is predominantly a matter of logistics. Panels should arrive as late as possible at their storage location in the building so as not to constrict other work going on at the site. Conversely, the panels have to be there in time so that the mounting will not suffer any delay. Moreover, the units have to be stored in the right order: even on one storey the panels may differ slightly from each other, depending on their individual position and function within the facade. Therefore detailed planning of the mounting is essential in order to make the building process fast and safe.

Besides the specific challenges linked with installation, high-rise facades also have special requirements concerning bracket design. Fixings for the facade units should be above the slabs and immediately behind the slab edge. This allows the units to be installed from inside the building, resulting in an important gain in convenience and safety. Fixing the facade units in front of the slab edge will retard erection and make it more dangerous for workers. All this illustrates that installation must be considered right from the start of design as an integral part of the facade construction.

22.6 Mounting of prefabricated facade panel with a monorail crane at the KfW Westarkade building in Frankfurt am Main, Germany. Architect: Sauerbruch Hutton, Berlin, Germany
Image: Carsten Costard, Budenheim, Germany

Fire Protection

High-rise buildings have to fulfil particularly strict requirements with regard to fire protection. This concerns not only the question of emergency exits and stairways, but also the need to keep the fire from spreading for as long as possible. The facade plays a leading role in this, since here the risk of floor-to-floor fire spread (flashover) is highest. As high-rise facades are usually designed as fully unitized systems, there is no parapet to prevent the fire from flashing over from one storey to the next above. Moreover, since the fire resistance of glazing is relatively low, additional precautions have to be taken.

There are four methods normally used to meet the special requirements of fire protection in high-rise facades. The first is closing the gap between the slab edge and the facade plane, since this is the weak point where fire and smoke can bypass the slab plane. Usually the gap is closed with silicate boards and mineral wool. The second measure (in the case of double-skin facades) is the integration of fire-resistant panels at the bottom of the facade elements. These panels reduce the duration of a flashover. Since the aluminium profiles of normal facade elements may collapse due to excessive heat, these fire resistant panels should not be connected to the profiles; they must instead be connected directly to the concrete slabs. This connection of the panels to the concrete slabs leads to additional construction work; moreover, the connection between the panels and the concrete slab must be protected against fire impact as well. Nonetheless, this is an important measure that can make an important contribution to the fire safety of the building.

A higher sprinkler density along the perimeter of the building is the third method of fire protection in high-rise facades. In case of fire, the sprinklers will help to keep the temperature at the facade as low as possible, thus extending the time until the facade panels collapse under the impact of heat. Glazing is typically the first part of the facade to be damaged by fire, and without preventive measures it normally fails 15 to 25 minutes after the fire has started. Flashover usually follows within 30 minutes. Practical tests, however, have shown that sprinklers at the facade will produce a low temperature zone near the glazing, so preventing failure due to thermal stress.

Glazing failure in a fire situation is particularly dangerous in double-skin facades. Unless adequate provisions are made, smoke may spread throughout the whole building through this gap in the facade. Hence, if double skins are used, the volume between the inner and the outer layer has to be partitioned horizontally and vertically into clearly defined fire compartments. This is the fourth method of fire protection in high-rise facades. The partitioning is usually done with silicate boards or fire protection glazing. Since the partitioning of the facade into fire compartments also influences the aeration within the facade interstice it has to be integrated into the design of the air ventilation concept.

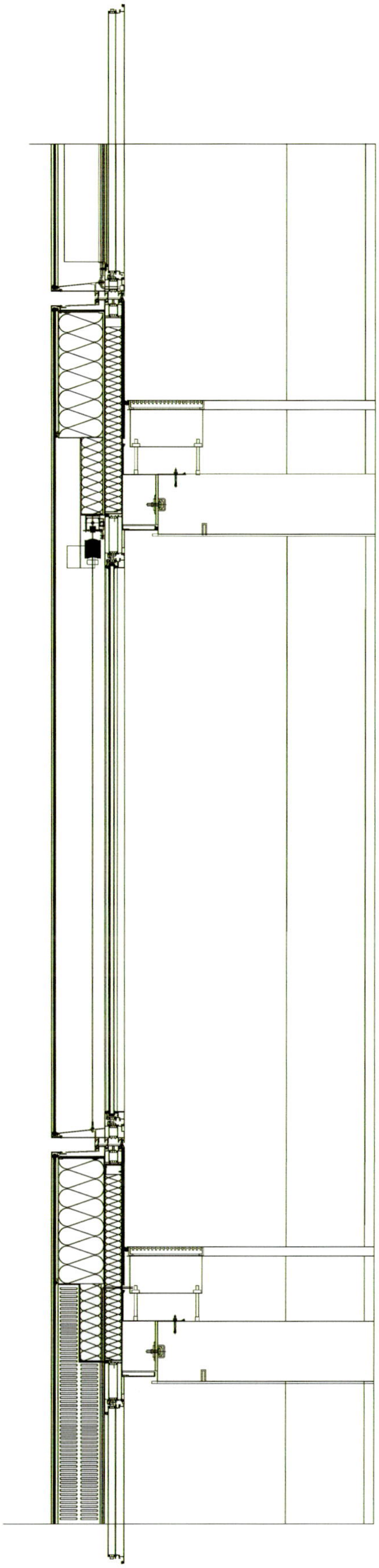

22.7 ADAC Headquarters, Munich, Germany: drawing of a fire-resistant parapet in the double-skin facade. Architect: Sauerbruch Hutton, Berlin, Germany
Graphic: Werner Sobek, Stuttgart, Germany

Cleaning and Maintenance

The question of how the exterior of a building is to be cleaned and maintained is all too often neglected in the design of high-rise facades. Usually this is not a major issue for low-rises, since most work can be done from the ground. In high-rises, however, facade cleaning and maintenance may become extremely difficult (and expensive) if no appropriate provisions are made. Proper planning can definitely help to extend the life of the whole facade construction, however cleaning and maintenance costs – including the cost of installation and servicing for the building maintenance unit (BMU) – can still add up to significant sums. They should therefore be clearly stated during cost estimation so as to ensure cost security for the client over the whole lifecycle of the building.

Maintenance and cleaning proposals must ensure maximum accessibility. This also includes the ability to transport (heavy) glass panes and install them in exchange for damaged panels. The most common system used for cleaning and maintaining high-rise facades is the BMU. A BMU's load-bearing structure is typically placed on the roof and/or on a service storey, and the BMU itself is lowered to the lower storeys by means of ropes. Such systems are an efficient and economical solution. It is important, however, to provide fixings for the bearing ropes, as otherwise the BMU will start to swing with high amplitudes. Any such fixings have to be taken into account at an early stage of planning, since they cannot be attached to the facade once it has been installed on the building.

Due to the increasing geometric complexity of many modern high-rises, it is often no longer possible to use classic BMUs. In this case there are two alternatives for cleaning the facade: it can be cleaned either manually (by climbers) or by robots. Both methods are relatively expensive, but nonetheless work quite well. However, they offer no way of replacing glass units from the outside. In cases like these the only option is to design the facades in such a way that it is possible to replace glazing from inside the building. This requires special facade profiles – another issue that has to be considered at an early planning stage.

22.8 Design study of a facade cleaning robot
Image: Werner Sobek, Stuttgart, Germany

Complex Geometries

In recent years geometry has become an ever more important issue for high-rise buildings. No longer is height the sole criterion that distinguishes a building from its peers. The increasing use of complex geometries and organic forms has significant repercussions for the facade construction: from cubic blocks, high-rise buildings turn into sculptures. This leads to steeply increasing exigencies with regard to functionality, mounting, tightness and visual appearance. Design of these buildings therefore starts from the outside and then works its way inwards. This in turn requires ever closer cooperation between the designers and the facade engineers, right from the very first sketches of the building.

3D design tools based on CAD are an absolute prerequisite to the technical planning of complex geometries. The geometries generated with architectural design software such as *Cinema4D* and *3ds Max* are not usually precise enough to deliver the information needed by the facade engineers. Hence the creation of a detailed building geometry is normally the result of the facade engineer's work. Such detailed data is needed not only to ensure technical feasibility, but also to minimize construction cost. Without optimization by the facade engineer, complex geometries generate high numbers of irregular facade panels. The more complex the geometry of the building envelope, the more complicated the facade construction; complex geometries may lead to hundreds of different panel types in one facade. Such

22.9 Baku Flame Towers: use of a standardised panel despite complex geometries and varying storey heights. Architect: HOK, London, UK
Image: HOK, London, UK

an extremely low level of standardization would make it very difficult to design the facade, and production and mounting of the individual panels would become a nearly insurmountable obstacle, quite apart from the exploding costs this would lead to.

The major challenge in designing complex facades, therefore, is to find methods of reducing the numbers of different facade elements without changing the overall appearance of the building. This optimization also concerns the glazing units, since irregular glass panes are more expensive than rectangular ones. The Baku Flame Towers (2012, architect HOK, London) are a representative example of how such an optimization works. All three towers have a similar shape. However, their facade panels would theoretically show great variation, not only because of the complex geometries of the towers but also because of their varying storey heights. During the design process the facade engineers therefore executed several examination rounds in order to find regularities and simplifications, always taking the local conditions – such as high earthquake loads and limited local production capacity – into consideration. Despite the difficult preconditions it was thus possible to develop a common panel design that meets all architectural and technical requirements, and which can also be produced locally.

Building Movements Due to Earthquake and Wind

Due to the greater building height, high-rise facades are exposed to magnified movements. These movements are induced either by wind or by seismic activity. The result of the movements is a floor-to-floor deflection. Consequent requirements regarding facade design are almost the same for both causes of building movement.

There are two main aspects of building movement that may result in severe damage if not considered appropriately in facade design:

1. Impact on the horizontal stack joints between the facade panels of two adjacent storeys – this impact may be generated by, among other factors, vertical movements resulting from slab deflection.
2. Impact on the vertical stack joints at the corner of one storey – this impact may be generated by, among other factors, horizontal movements resulting from thermal elongation.

A common solution for dealing with the first of these is the provision of slidable stack joints. Thus the vertical connection between two panels that are aligned with each other is omitted. This simple solution leads to completely different panel behaviour in terms of slab deflection, as the panel joints will spread at the lower edge of the panels. The potential extent of this spread has to be controlled very accurately: movements that are too large would result in a loss of air and water tightness. The spread may also turn into a 'negative' one, leading to a collision of two adjacent panels. This can be allowed to a certain extent, as the movement may be transferred to the next vertical stack joint by two panels leaning against each other. Such a mechanism is an appropriate solution for unitized facades. It has to be simulated very precisely, though, taking into consideration representative load cases and the particular movements resulting from these load cases. If necessary the panel profiles have to be strengthened in the areas concerned by the use of steel inlays. These steel inlays also help to maintain a minimum distance between two panels. This is especially important when structural glazing is used: direct contact between the glass panes of two adjacent panels may cause severe damage as the glazing is likely to break in case of high local mechanical stress. It is therefore important to protect the glass edges appropriately.

It must be kept in mind that the omission of vertical connections between facade panels does not offer a solution for the building corners. Here the sliding facade will collide with the aligned perpendicular facade, unless an appropriate solution is found. One possible solution is the design of special corner panels which are fixed at the bottom and allowed to rotate slightly around their lower joint line. When using this mechanism it is important to design the vertical joint at the edge of the corner panels so as to accommodate the whole horizontal movement resulting from the sliding of the facade. An alternative way of avoiding corner collision is to provide a large vertical corner stack joint with special EPDM gaskets. This solution is relatively simple, as far as the technical complexity is concerned, but like the first solution it also requires a special corner profile to be taken into consideration at an early planning stage.

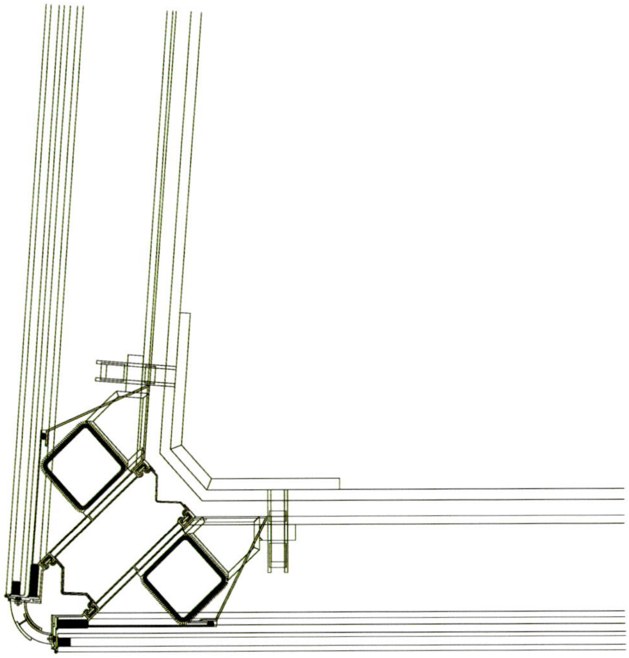

22.10 Baku Flame Towers: special corner profile with EPDM gaskets. Architect: HOK, London, UK
Graphic: HOK, London, UK

Chapter 23
Fire Safety and Evacuation

Simon Lay

Fundamentals of Fire Safety in Tall Buildings

Key fire and life safety objectives

Designers, approval authorities, developers, insurers and the general public all express both concern and interest in the fire and life safety aspects of tall building design. Perceptions of safety can often be skewed by isolated disasters or fictional accounts in films and books. However, tall buildings remain, on the whole, one of the safest forms of building, not least because the iconic nature of many tall projects ensures that they attract due care and attention in design.

The primary objective of fire and life safety design in tall buildings is to ensure the safety of building occupants, emergency responders[1] and persons in adjacent buildings. In addition to this primary objective, there is a secondary objective that relates to the preservation of commercial value, which can take the form of capital loss, consequential losses in revenue and negative public relations affecting the building tenants, the building owner and also the city or state where the building is located.

Design approaches for fire and life safety in tall buildings

Table 23.1 (overleaf) details a set of functional objectives that represent the basic objectives of fire and life safety design in tall buildings. A number of alternative design approaches can be taken in order to meet these functional objectives.

Prescriptive guidance on fire and life safety design is available in building codes and the associated handbooks. These would include, for example, the NFPA 101, NFPA 5000, the IBC, the Russian SNIPs and British Standards (NFPA 2011, 2012; ICC 2012; SNIP 1998; BSI 2008). Often individual cities or economic areas will set additional recommendations for the fire and life safety design of tall buildings. These would include, for example, Section 20 of the London Building Act, and the requirements of the Mumbai High Rise Committee (the Maharashtra Regional and Town Planning Act 1966, as amended) and the New York City Building Code (NYBC 2010).

Such prescriptive building codes can be mandatory in nature but more often they recognise that some flexibility in design requirements may be necessary, and allow for alternative or 'performance-based' designs to

Life safety objectives	• Ensure occupants at risk can escape • Ensure that occupants not at risk can evacuate before they are placed at risk • Ensure emergency responders can attempt appropriate firefighting and rescue activities • Ensure emergency responders can evacuate if required • Ensure that people outside the building or in buildings close by are not placed at risk
Commercial objectives	• Limit the loss of property • Limit consequential losses such as downtime • Prevent an adverse perception of the building, its owners or its tenants. • Prevent an adverse perception of the city or state where the building is located

Table 23.1 Basic objectives of fire and life safety design in tall buildings

be adopted, subject to the approval of the authorities having jurisdiction.

Often building codes are assumed to ensure safety in design, but more often the guidance comes with a caveat noting that a prescriptive building code is for a typical model building and hence needs to be validated for its applicability on individual projects. As a result of this, building codes are typically advisory and cannot be relied upon to ensure that safety is achieved.

Tall building advice within model codes typically starts at a height or number of storeys that may not intuitively feel tall to architects and engineers. For example, within the NFPA guidance, tall buildings are typically those greater than 7 storeys, and under UK guidance buildings more than 18 metres in height are typically considered to be tall. In many codes, the design guidance for tall buildings does not vary with height and there is very little if any guidance available for very tall buildings. It is also a feature of model building codes that buildings are typically considered to be homogenous in terms of occupancy type, or that the design guidance for the most onerous occupancy type is applied across the whole building.

With the tendency for modern building codes to be advisory in nature, a lack of guidance specific to very tall or mixed-use buildings and a recognition that advanced design methods can lead to innovative solutions which can enhance or maintain safety at reduced costs, there has been a move towards performance-based approaches to the fire and life safety design of tall buildings. It can be demonstrated that a performance-based approach can achieve the Table 23.1 objectives more reliably than the prescriptive code approach, which would tend to assume that by following the code advice the functional objectives will be satisfied.

While performance-based design is inherently non-prescriptive in nature, there are guidance documents on the methods to be applied and key performance targets to be adopted when assessing whether or not individual functional objectives have been met or not. Typical performance-based design guidance includes the relevant sections of NFPA 5000 (NFPA 2012), the BS 9999 (BSI 2008) or the Australian Fire Engineering Guidelines (ICC 2005).

For either a performance-based and prescriptive building code approach, it is essential that the fire and life safety design of tall buildings is holistic in nature. It must work with the architectural and operational aspirations of the project and it must recognise the need for buildings to change and adapt over their lifetime.

Evacuation of Tall Buildings

Objectives

Protection of life is the primary objective for fire safety design in tall buildings, initially the life safety of occupants, but also the life safety of emergency responders and people in the surrounding area. For the life safety of occupants, it is necessary to detect fires when they occur, initiate evacuation and provide adequate escape routes that are protected from fire and smoke.

Protection of escape routes is covered in later sections. This section addresses the sizing, location and number of escape routes required.

Evacuation design approaches

A number of different evacuation design approaches are possible for tall buildings. These are summarised in Table 23.2.

Simultaneous evacuation	• All occupants are instructed to leave the building at the same time. As a result, it is often thought that this will lead to the shortest overall evacuation times. • This solution is not a common design basis for tall building design, as it requires very large escape routes to avoid congestion. • Occupants who are at immediate risk in the vicinity of the fire are not prioritised over those who may be significantly more remote from the fire and not initially at risk. • Large numbers of occupants may be evacuated even if the hazard from the fire is contained or is low. • Emergency responder actions have to wait for evacuation to be substantially completed.
Phased or staged evacuation	• Occupants most at risk (typically those on the same floor as the fire and adjacent floors) are instructed to evacuate first. • Occupants remote from the fire incident are either notified of the potential hazard so that they can prepare to evacuate, or they are not made aware of the incident at all. • Once occupants initially at risk are away from the fire hazard, the next closest occupants begin evacuation and the process cascades through the entire building. • A phased or staged approach ensures that priority of evacuation is given to those most at risk. • In situations where the fire hazard is resolved, evacuation may be stopped, ensuring that unnecessary commercial loss is limited. • Escape route sizes can be based on a smaller number of occupants at any one time, ensuring greater efficiency. • Emergency responder actions can be coordinated with the needs of evacuating occupants. • Detailed management and training is required to ensure the phased or staged evacuation is smooth and effective. • The overall evacuation time can be very long and protracted, requiring special consideration of building collapse.
Defend in place	• Only the occupants initially at risk are evacuated. This is typically only those in the fire compartment where the fire is detected, and occupants on the same floor may be left in place. • No other occupants in the building are made aware of the fire or evacuated. • Extensive use of small compartment sizes is necessary to limit the potential for fire or smoke to affect other occupants. • The escape route provisions can be minimised. • Emergency responders can carry out their activities without compromising the evacuation of occupants.

Table 23.2 Evacuation design approaches

The terms 'staged evacuation' and 'phased evacuation' can often be interchangeable, although 'phased evacuation' is typically taken to mean an evacuation that takes place under control of a central fire marshal, whilst 'staged evacuation' is typically taken to mean a fully automated process based on a fixed delay between each part of the evacuation.

Not evacuating occupants while also keeping them unaware of a fire somewhere in the building can be considered controversial, but phased or staged and defend-in-place strategies are an essential feature of most tall building designs. Without adopting these approaches, tall buildings cannot be constructed economically and unnecessary evacuations can make occupants unresponsive when an incident occurs that does place them at risk.

Defend in place may be seen as the most controversial approach, but is commonly adopted in residential tall building design. It is used in these situations because the fire compartment can typically be kept small. Also, false alarms are common in residential buildings and it is very important that occupants do not become unresponsive to alarms. Another important factor in such buildings is that there can be a significant delay from the

fire starting to people becoming aware of the fire and evacuation being initiated. This is true of occupants in the fire compartment, but even more so for occupants in other remote parts of the building. As a result, occupants may attempt an evacuation while firefighting activities are taking place, which can expose them to considerable risk. The conclusion is that it is credible and arguably safer to have occupants who are not in the fire compartment remain where they are protected by compartmentation, rather than leave that place of safety.

Variations of the basic approaches set out in Table 23.2 are possible. For example, refuge floors – required by code in places such as Hong Kong – are a variation on the defend-in-place concept, with occupants being moved to a common, protected location. The use of refuge floors implies that occupants in the refuge await rescue by the emergency responders. However, this approach is considered flawed as there should be no reason why protected escape solutions cannot be put in place that ensure that all occupants can evacuate from a tall building without reliance upon outside help.

Designing for evacuation

The sizing and placement of escape cores can have a significant impact on the design of a tall building and must be considered at an early stage. Structural design plays a key role in the form of tall buildings and can in some cases determine the location and even the sizing of escape cores. Often, performance-based design methods are necessary to achieve a holistic design that ensures adequate escape provisions are available within the architectural form of the building.

Evacuation from a tall building takes place in five phases:

1. Notification of the fire event and initiation of the evacuation
2. Horizontal movement to reach exits
3. Entry to vertical escape routes
4. Vertical travel to ground level
5. Horizontal movement to reach the outside of the building.

FACING PAGE

23.1 The Beetham Tower in Manchester, UK incorporates a defend-in-place evacuation policy for residential occupants and phased evacuation for hotel residents. The project also incorporates an alternative to pressurisation for protecting escape and firefighting cores, as well as the use of alternative suppression technologies
Image: Simon Lay, WSP

In order to raise the alarm, tall buildings should be provided with fully addressable automatic fire detection and alarm systems. Fire location should be indicated at a central control panel or control room and transmitted automatically to the emergency authorities. However, care must be taken to ensure that false alarms do not result in unnecessary evacuations or nuisance calls to the emergency authorities. It should also be noted that many tall buildings, particularly residential towers, may not have 24-hour management and can suffer many false alarms. In such cases, fire detection systems should focus on the initial evacuation of those at risk and the activation of safety systems (such as smoke control), although communication of the fire floor location will be valuable to emergency responders. The use of appropriate fire detection and detection protocols such as voting or 'double knock', where multiple fire detection devices need to activate in conjunction with management investigation periods before an alarm sequence is initiated, are also important tools to be considered as part of an appropriate fire detection and alarm philosophy.

Fire alarms in tall buildings should be appropriate to their location within the building. Public address alarms which play recorded or live messages are an essential component of phased or staged evacuation and have particular benefits in terms of initiating evacuation or a readiness stance. In other situations, and once occupants are within evacuation routes, simpler alarms based on sounders or horns may be adequate.

Once occupants begin to move horizontally towards the vertical escape routes, they may need to cross open-plan areas or move along enclosed corridors. Entry points to vertical escape routes need to be well separated to ensure that a fire does not compromise access to more than one escape route at the same time, but designers should take into account the possibility that occupants making their way to an exit may find that route blocked by fire and have to use an alternative route.

Where escape is across open-plan areas, signage of escape routes is important, and distances to an exit should be limited so that locating an escape route is simple and quick. Where escape is through corridors, these corridors may need to be protected (typically to a fire resistance standard of 30 minutes), particularly if the corridor leads to a dead end.

Recommendations on maximum acceptable distances to reach a vertical escape route are available in building codes and can vary with occupancy type. For example, the NFPA guidance allows for the maximum travel distance to an exit in sprinklered offices to be 91 metres, with a limit of 30 metres for single direction or common path travel at the start of an evacuation route.

For hotel occupancies the equivalent figures are 30 metres and 15 metres, respectively.

Where a performance-based approach is to be adopted, designers would need to demonstrate that the time to reach an exit is less than the time taken before untenable conditions occur in the fire compartment.

Travel distance limitations to an exit are not fundamentally different in tall buildings compared to other building types and need not change for very tall or supertall buildings. However, in tall buildings it may be more critical to ensure that vertical escape routes are separated from each other. An approach recommended in the NFPA 101 (NFPA 2011) is to adopt the '$^1/_3$ rule', which requires vertical escape routes to be separated by a minimum of one third of the maximum building diameter. This can have a significant bearing on the location of building cores, but is intended to provide resilience in the case of multiple cores being damaged simultaneously in an incident. This separation is particularly relevant to major events other than fires, such as bomb blasts. The use of resilient core designs such as concrete cores may enable some flexibility in these separation requirements.

The width of entry points into vertical escape routes needs to be great enough to ensure that occupants do not have to queue for an excessive period of time to leave the floor plate from which they are escaping. It is commonly assumed that evacuation of the floor plate should be completed in less than two and a half minutes from the time that occupants begin their evacuation. By applying an assumed typical flow rate, an exit width per person can be calculated. A commonly used value is to allow 5 millimetres of exit width for each occupant.

An approach still adopted in some building codes (for example the National Building Code of India; BIS 2005) is a unit exit width approach, where exit routes are sized in terms of 500 millimetre units of exit (or half units thereof) and each unit exit width is assumed to be adequate for a specified number of persons. However, this approach is typically being phased out in newer versions of building codes, such as the NFPA 101 and IBC codes. In recognition of this, it is recommended that, in preference to a unit exit width approach, either an exit width per person approach be adopted or a performance-based approach applied by designers; for example, adopting the calculation methods set out in the SFPE handbook, the BS 9999 or the *International Fire Engineering Guidelines* (SFPE 2008; BSI 2005; ICC 2005).

Once occupants are inside a vertical escape route, they should be in a safe place. It is therefore essential that vertical escape routes be properly protected, using robust methods to prevent fire or smoke entering the route. (For smoke control and protection from fire, see the section on fire and structural design.)

Vertical escape routes need to be sized to maintain a flow of occupants and to enable emergency responders to move against the flow of occupants if required. Building codes typically recognise that occupants move more slowly on stairs than when walking to reach an exit, and hence stairs need to be bigger than the routes leading to them to maintain the same flow rate and to prevent overcrowding. For example, NFPA 101 (NFPA 2011) recommends that the 5 millimetres per person for exit from a floor plate be increased to 7.6 millimetres per person when sizing stairs (for typical high-rise occupancies). This simple approach can also be modified under NFPA guidance by adopting calculation methods from the SFPE handbook (SFPE 2008).

The number of occupants used as the basis for sizing vertical escape routes under NFPA guidance is the maximum number of occupants on any floor served by the escape route. This approach is considered to be appropriate for high-rise buildings where phased or staged evacuation is applied.

In addition to the sizing of escape routes based on a width per person, it is also typically appropriate to apply a minimum width. Recommended minimum widths are 1000 millimetres for entry to escape routes and 1100 millimetres for stairs. This minimum entry door width ensures that persons in wheelchairs can reach the safety of a stair enclosure, and the minimum stair width provides space for emergency responders to move upwards, against the downwards flow of occupants.

Simultaneous evacuation

While simultaneous evacuation is not commonly the basis of design for tall buildings, it is recommended that all tall buildings should have their evacuation times evaluated in the event that simultaneous evacuation takes place. This is to ensure that events other than fires that may require total building evacuation (including acts of terror and natural disaster) are properly considered. A simple approach to this is also recommended in the NFPA guidance, which requires a minimum stair width of 1420 millimetres to be adopted if the aggregate building occupancy exceeds 2000 people. However, in very tall buildings or mixed-use schemes, it is recommended that a comprehensive computational evacuation modelling exercise be conducted to check the escape times under simultaneous evacuation conditions.

When checking the total evacuation time under simultaneous evacuation conditions for a building that

	Recommended maximum egress time for all persons to reach a final exit (outside)		
	Up to 50 storeys	50–100 storeys	Over 100 storeys
Buildings designed for *simultaneous* evacuation	30 minutes	30 minutes	30 minutes
Buildings designed for *phased* evacuation	60 minutes	90 minutes	90 minutes

Table 23.3 Maximum recommended total evacuation times
Source: Lay (2007)

has been designed on the basis of phased evacuation, the evacuation time should meet the guidance set out in Table 23.3. In deriving the limits in the table, the importance of being able to achieve reasonable egress times under security alert conditions has been taken into account, whilst the basis for limiting the maximum recommended egress period to 90 minutes is the limitations of structural fire protection.

The provision of adequate egress for persons who are not fully ambulant needs to be specifically addressed in any tall building egress design. It is important that adequate provisions are made for self-rescue using evacuation elevators for occupants who cannot walk down stairs, and space needs to be allowed within building cores for occupants who are not fully ambulant to rest without blocking the egress of others.

Emerging technologies for vertical movement

Stairs are still the traditional, tried-and-tested evacuation solution, but much of the focus on the evacuation of tall buildings since the 9/11 collapse of the World Trade Center towers in 2001 has shifted to the use of elevators. Advice on when the use of elevators becomes useful or essential is provided in Table 23.4.

The recommendations in Table 23.4 take into account the fatigue of walking down stairs, the likely space saving from using elevators for evacuation, the complexities of management required to control elevator evacuation, the need for training when elevators are used for high numbers of occupants and the need for groups of people to stay together in some situations.

It should be noted that Table 23.4 is not intended to reflect the benefits of elevator evacuation for mobility impaired persons. It is considered that in any scheme over 3 storeys in height, elevators could have a real benefit in supporting the evacuation of mobility impaired persons, and in schemes over 6 storeys in height the use of elevators may be essential.

Building height	Building use	Use of elevators in evacuations
Up to 50 storeys	Offices	○
	Hotel	—
	Residential	—
	Public space	○
50–70 storeys	Offices	○
	Hotel	○
	Residential	○
	Public space	●
70–100 storeys	Offices	○
	Hotel	○
	Residential	●
	Public space	●
Over 100 storeys	Offices	●
	Hotel	●
	Residential	●
	Public space	●

— Elevators considered to be of limited benefit
○ Elevators considered useful to support evacuation
● Elevators considered essential to evacuation

Table 23.4 Guidance on the use of elevators in evacuation
Source: Lay (2007)

To protect the elevators in an evacuation, it is considered that the elevator cores should be of concrete construction (or concrete filled, permanent steel shuttering). Elevator system design should follow the guidance for firefighter elevators (such as the British Standard BS EN 81-72; BSI 2003). This approach will introduce requirements for standby power provisions, waterproofing of systems and additional control system requirements.

Although elevators can be used to evacuate occupants quickly from high level in a tall building, the time for occupants to reach a place of relative safety from an occupied floor plate can be longer, due to the cyclic, 'batch' nature of the elevator process, as against the continuous nature of a stair evacuation. This requires a refuge space to be formed where occupants can wait in safety for the elevator to arrive.

Refuge spaces need to be sufficiently large to make occupants feel comfortable and to enable them to move within the space (for example if they decide to use the stairs rather than wait for an elevator). The following key points should be taken into account:

- The minimum floor space per occupant should be no less than 0.5 square metres per person.
- The refuge should be separated by 120 minutes' fire resistance from the fire floor (doors to the refuge can be 60 minute doors with smoke seals).
- The refuge should be provided with smoke ventilation (either pressurisation or an air exchange 'flushing' system).
- The refuge should include a communication system for occupants to talk to the building fire control centre.
- The refuge should connect to both evacuation elevators and a stair core.
- The refuge should be well lit, to a typical day-to-day standard.

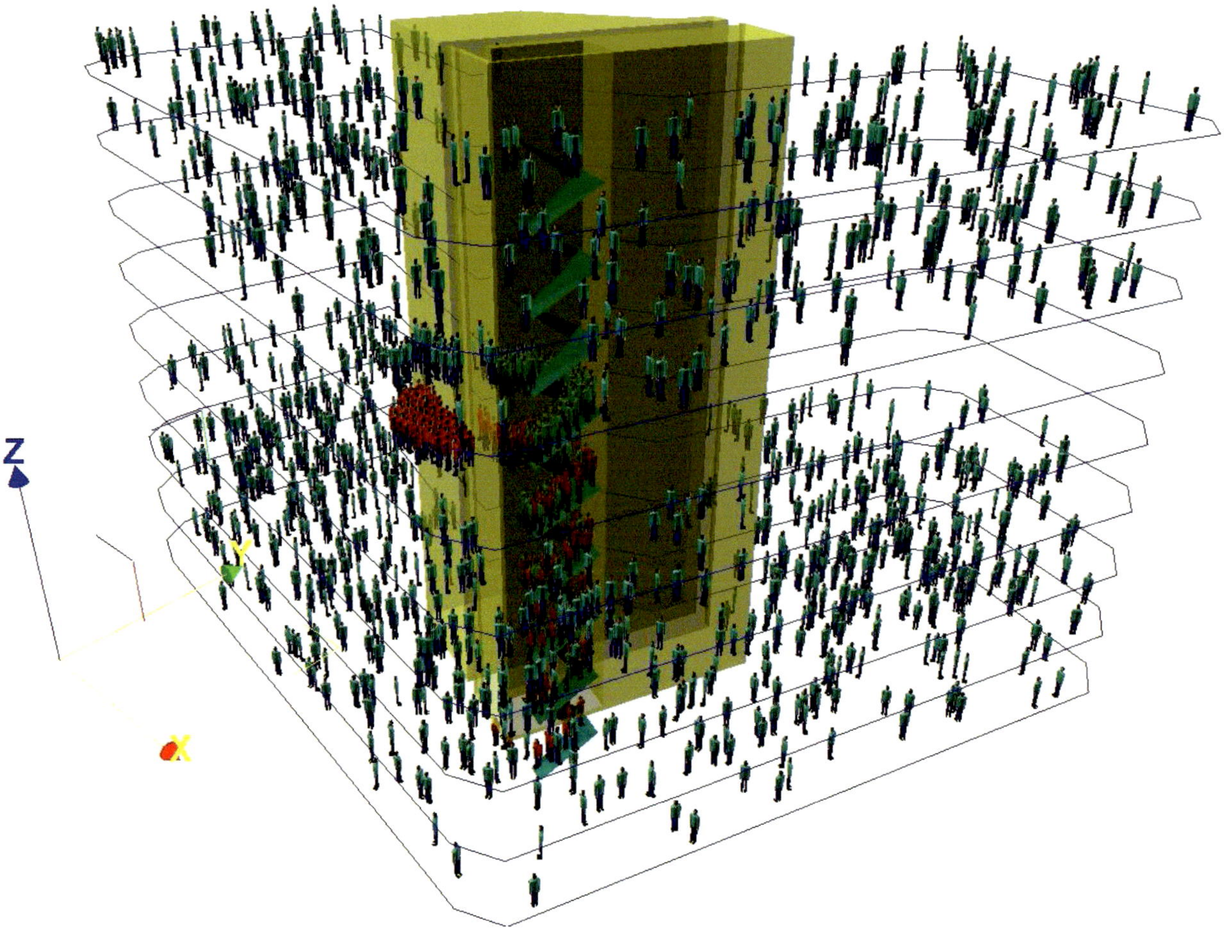

23.2 Computational Evacuation Modelling is a critical tool for the analysis of modern high-rise schemes and allows the simulation of a wide range of evacuation scenarios including elevator evacuation
Image: James Bertwistle, WSP

In many situations it can be difficult to engineer a sufficiently large enough day-to-day elevator lobby to accommodate occupants during an evacuation. This may therefore require additional leasable area to be set aside for what remains an unlikely event. Some use can be made of sanitary accommodation attached to elevator lobbies to extend refuge spaces, but a more common solution is to use stairs to move occupants to a floor below the fire floor and then, once in a place of relative safety, occupants can use the general office space or circulation areas to queue for elevator escape.

A natural extension of moving occupants initially by stair to a safe place to await elevator escape is to consider the use of sky lobbies in very tall buildings. Schemes in excess of about 70 storeys often include a sky lobby as part of the elevator network design (with MEP plant positioned above or below). This may require occupants to walk down up to 35 storeys, but this is commensurate with the guidance in Table 23.4 and is an effective strategy for many tall buildings.

It is also noted that, in designing for elevator evacuation, care should be taken to recognise that the general public have been trained not to use elevators in the event of a fire. This is the normal situation in most existing buildings. In an office building, staff training can overcome this challenge. However, in buildings where the public have access, it will be necessary to have trained staff to direct evacuation.

Alternative evacuation methods

In addition to elevators, there are numerous other novel evacuation solutions that have been proposed. These include solutions from helicopter rescue and personal rocket packs to externally mounted slide and rail solutions, as well as zip wires to neighbouring buildings and large rescue towers to be brought alongside the building. These systems are considered unlikely to play a significant role in the evacuation of occupants in real buildings; in many cases they may be unsafe, and they are considered unnecessary.

The elevators and stairs provided in buildings today are commercially tolerable and can provide adequate means of egress in fires and other emergency conditions, without resorting to design solutions which are not fully tested and would be potentially confusing to occupants. There are also significant concerns relating to the cost and reliability of novel solutions, and such designs may take up additional space in a building.

Firefighting Activities in Tall Buildings

Objectives and challenges of firefighter operations at height

Firefighting activities in high-rise buildings have been shown to be hazardous to emergency responders. Even if the extreme examples such as the World Trade Center 9/11 event are set aside, more typical fire scenarios in high-rise buildings still claim many lives in the emergency services.

The challenges of high-rise emergency activities can be summarised as follows:

- Moving equipment, personnel and firefighting water large vertical distances
- Remote evaluation of the fire hazard before committing personnel
- Command and control of multiple teams within a complex, hidden environment
- Maintaining an escape route for emergency responders
- Interaction between firefighting activities and evacuation procedures.

Most tall buildings are located in cities with established firefighting infrastructure. However, this can lead to challenges in reaching buildings through crowded streets.

A core principle for tall buildings is that effective compartmentation and automatic fire suppression can significantly reduce the hazard to emergency responders (as well as supporting evacuation), and all tall buildings should make use of such compartmentation and automatic suppression.

Rescue activities

It is commonly assumed that emergency responders should put their own lives at risk to rescue occupants. However, the focus first and foremost should be the safety of emergency responders so that they can enter potentially dangerous environments without undue risk. Self-rescue should be the basis of design for all occupants of all tall buildings, with rescue by emergency responders being the action of last resort. High-rise building design should not place an undue dependency on rescue of occupants by emergency responders.

Emergency responders can assist in encouraging egress and provide a critical communication link to an area commander who has a view of the wider tactical situation and occupants. However, their primary role should be to seek to contain or extinguish the fire so that

self-evacuation can continue unaffected. Where emergency responders are supporting the evacuation of individuals it is likely to be those of limited mobility and it is essential that firefighters be provided with emergency elevators so that they can reach occupants and remove them from the building without blocking stairs.

Water supplies

Water remains the primary medium for tackling fires. It is relatively simple to pump long distances and there are well-established solutions for providing firefighters with water. In very tall buildings (typically in excess of 150 metres) it is likely that multiple staging tanks will be required to move water to the firefighting locations. All firefighting water supplies for high-rise buildings should be designed to provide at least two high-pressure firefighting hoses from each riser. This is to provide coverage for a primary crew who will attack the fire and their backup or support crew.

It is important that the failure of automatic suppression be taken into account in the design of water supplies for firefighters. One of the consequences of smaller fire compartments is that the firefighting conditions can rapidly become very severe. Extensive research in the UK and other locations into firefighting tactics has suggested that higher pressure 'pulse' tactics can be very effective and reduce the risk to firefighters. As a result, high-rise firefighting water supplies should be designed to achieve a minimum 60 bar.

Transportation of personnel and equipment

Until relatively recently, many of the internationally recognised building codes did not fully recognise the need for a dedicated 'firefighting core' to be available to emergency responders. One of the longest-standing design standards for such provisions is BS 9999 (which replaced the previous BS 55588, Part 5; BSI 2008, 2004). A typical firefighting core arrangement as defined under BS 9999 is shown in Figure 23.4 (overleaf). It is recommended that all high-rise buildings incorporate cores to this standard or higher. Recent versions of building codes such as NFPA 101 (NFPA 2011) have recognised such a provision, as do local codes in some countries, including Russia, France and Australia.

A firefighting core comprises a stair, elevator and interconnecting lobby, which provide a safe base for firefighters to launch an attack on the fire, and a safe place to withdraw to and escape from. It should be noted that firefighting cores do not need to be dedicated to the use of emergency responders. Under normal circumstances they can be used for egress and day-to-day activities. However, the need for emergency responders to take command of these cores should be recognised.

It is particularly noted that emergency elevators should not be dedicated to the sole use of emergency responders and are much more reliable if they are part of the general, day-to-day vertical transportation provisions of a building.

A key feature of a firefighting core is that it places elevators and a stair in close proximity, linking them together and connecting to the building floor plate via a lobby. This is very different from many traditional core designs; in particular it differs from the cores typically derived under NFPA or IBC guidance, for example, where lifts and stairs may open directly onto an open floor plate.

FACING PAGE

23.3 The iconic AON Center in Chicago makes use of phased evacuation to reduce stair core requirements. Local city ordinances also require the provision of firefighter elevators
Image: Jan Klerks

Fire and Structural Design in Tall Buildings

Structural fire protection

While some building codes may suggest up to 240 minutes' fire resistance is required for tall buildings, the most common maximum fire resistance requested in the majority of codes is 120 minutes. This is commensurate with typical fire severity analyses (BSI 2002), which rarely generate a structural fire resistance period in excess of 120 minutes (based on likely compartment fires). Fixing the maximum egress time at a value below 120 minutes (90 minutes is recommended in Table 23.3) ensures that, even if the building design or operation inadvertently invalidates the assumptions of the fire severity analyses, there will still be time to complete the evacuation process. To achieve a total evacuation period of 90 minutes or less under simultaneous conditions, it is likely that measures other than simple stair evacuation will be necessary in a very tall building.

It is important that all tall buildings be designed with sufficient structural fire resilience to ensure that a localised fire cannot lead to a disproportionate structural effect such as collapse. The impact of a fire should remain local to the fire, not cascade through the

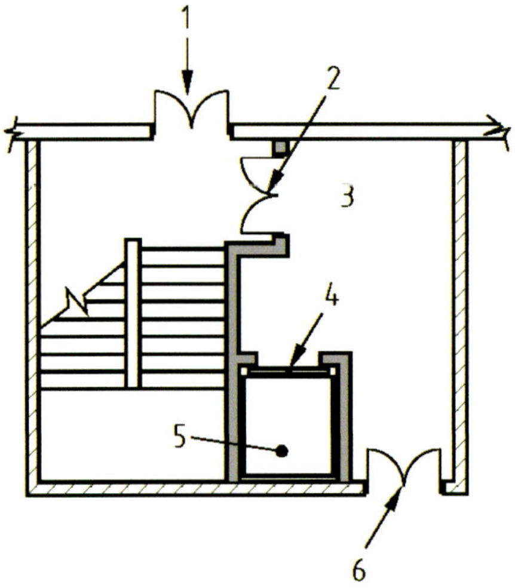

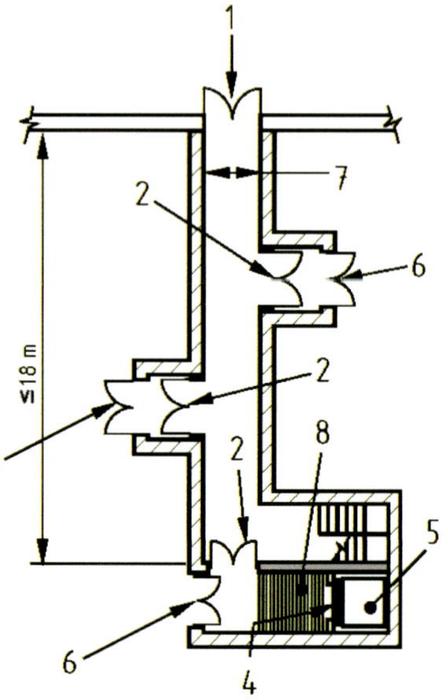

i) Access directly from open air

ii) Access via a corridor

a) Fire and rescue service access at lowest storey

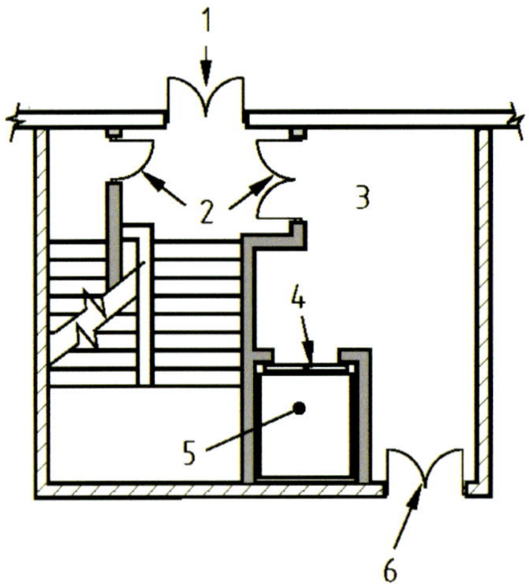

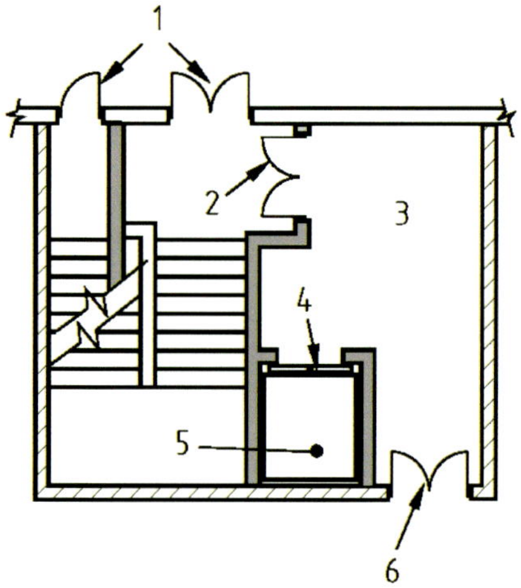

NOTE The corridor should not be used as a circulation space.

i) Access to basement via stair to upper storeys

ii) Access direct to basement

b) Fire and rescue service access directly from open air ground level in a building with basements

Key

1	Fire and rescue service access at fire and rescue service access level
2	Fire door with 30 min fire resistance with smoke seal
3	Fire-fighting lobby
4	Fire door with 30 min fire resistance without smoke seal
5	Fire-fighting lift
6	Fire door with 60 min fire resistance with smoke seal
7	Width of corridor 500 mm wider than needed for means of escape
8	Fire and rescue service mustering point (minimum 5 m^2)
(grey fill)	Minimum fire resistance of 1 h from both sides
(hatched)	Minimum fire resistance of 2 h from outside the fire-fighting shaft and 1 h from inside the fire-fighting shaft

23.4 Fire fighting core arrangements, from BS 9999:2008

structure. A widely accepted approach to ensuring that fire effects do not become disproportionate is to design structural fire resistance to be appropriate to the individual fire compartments. The fire severity that the structure is exposed to can be calculated using methods such as those set out in BS EN 1991-1-2:2002 (BSI 2002), and the structural fire resistance applied to match this 'burn-out' period. This approach can also yield benefits by reducing fire protection requirements in some structural members. In some cases, this approach can be taken further by considering how fire-imposed structural loads can be spread effectively from affected to unaffected members. Such analyses require detailed, advanced computational analyses but are becoming commonplace in tall buildings.

Any structural fire protection proposals should take into account the lifetime implications for the building. This is a particular concern in steel-framed buildings, as the movement of the structure can impose additional stress on applied fire protection products, which must be taken into account. Care should also be taken that maintenance and building alterations do not damage fire protection. In some cases (such as protection to cellular beams) small areas of damage can invalidate the protection to an entire member.

Compartmentation

Compartmentation is a critical element in reducing the fire hazard levels in tall buildings. Good compartmentation can effectively reduce even the tallest building to the same risk levels as a low-rise property. Effective compartmentation is the single most critical element of fire safety design in tall buildings.

To achieve a sufficiently robust level of protection, construction systems such as dry-walling need to be installed and maintained to the highest possible levels, and it is considered preferable to adopt concrete, blockwork or other inherently robust construction methods rather than lightweight partition systems to protect tall buildings.

Every floor in a high-rise building should be a compartment floor. Where features such as atria are introduced into a building, these can represent a specific hazard to be addressed. Openings between floors should either be protected by introducing compartmentation between hazard spaces and the void or through a combination of suppression and smoke control methods. However, the trade-off between active fire systems and passive compartmentation requires a highly robust design. For example, smoke control should take into account the partial or complete failure of suppression systems, and ideally a probabilistic risk study should be used to check the robustness of the active system approach. Care should also be taken in the use of empirical calculation methods for smoke control in very tall atria; it is typically more appropriate to use advanced computational fluid dynamics methods instead to achieve a valid design solution.

External fire spread

It is difficult to provide comprehensive guidance on solutions to resist vertical fire spread via the outside of buildings. The facade of a building can represent up to 35 per cent of the building cost and is critical to the building form and function. Tall buildings are often differentiated by their facades and each tends to be a unique design. There are so many variables to consider that prescriptive guidance on preventing external fire spread is not a practical proposition.

In many codes (for example NFPA 5000, the National Building Code of India and many building codes in Asia), fire-resisting spandrel panels are required to provide protection at the interface between floors, particularly in unsprinklered buildings. This is a simple, passive form of protection and can be effective. However, this does not adequately address the requirements of many modern facades, including more complex designs that incorporate deep ventilated spaces to act as climate barriers.

It is considered that a performance-based approach is the most credible method for assessing external fire spread on a building-by-building basis. Using performance-based methods, relatively simple changes in building design that could reduce the potential for external vertical fire spread can be assessed. For example, introducing close centred sprinklers at the facade, fed by a separate part of the sprinkler system, could provide a very robust active solution to protect the facade even if sprinklers in the occupied spaces fail.

Smoke control

Protecting cores for firefighting and evacuation

If escape cores and firefighting cores are to be available for use in the event of fire, it is essential that they are not compromised by smoke. Often fire fighting activities can introduce significant quantities of smoke into parts of escape routes, and it is essential that this does not compromise other parts of the building and that firefighters retain a safe escape route. Most building codes

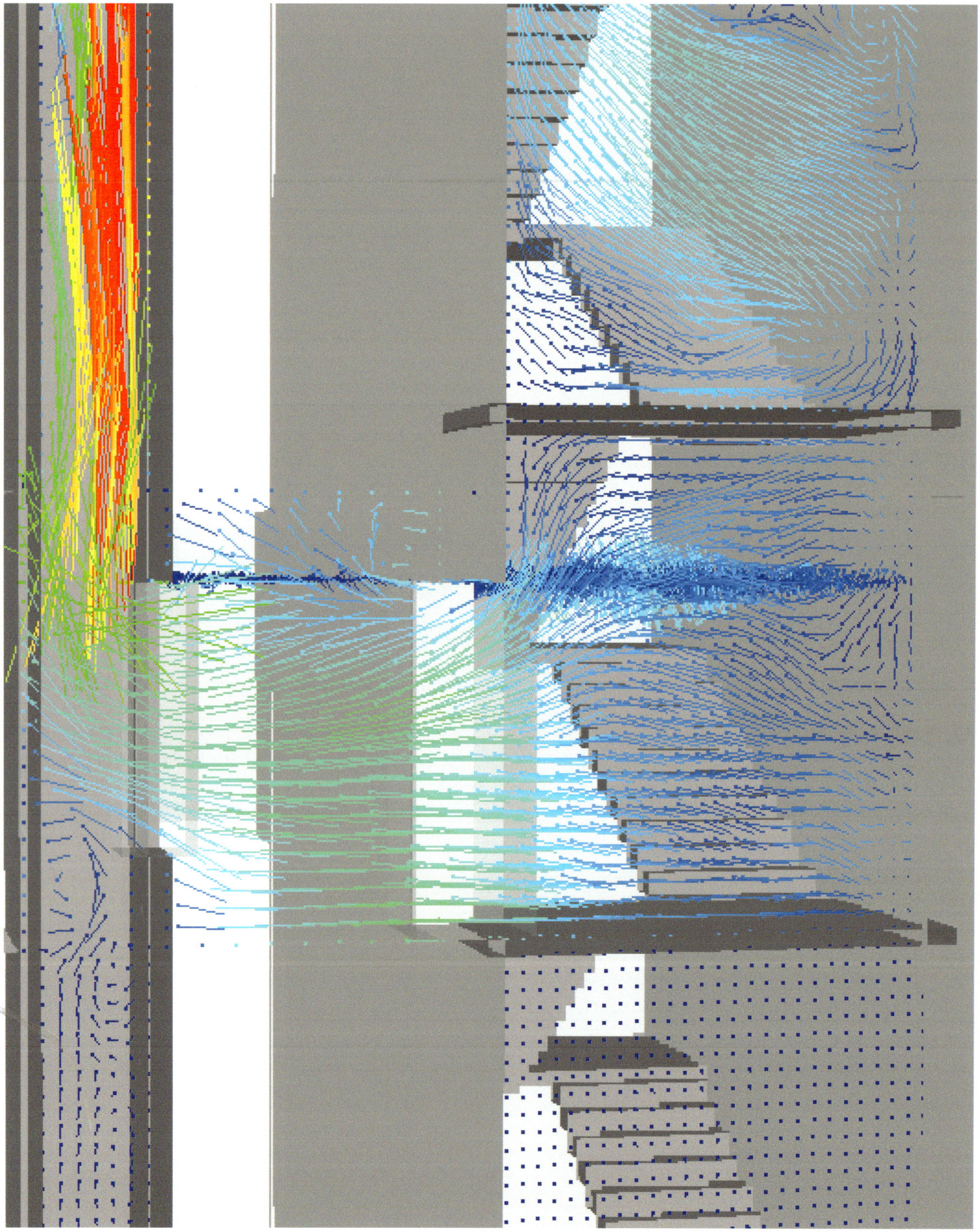

23.5 Computational Smoke Modelling is critical for studying smoke movement in high-rise buildings and for developing robust smoke management solutions
Image: James Bertwistle, WSP

recognise a difference between the smoke protection required for evacuation and that required for emergency responders.

Building codes recognise the need to protect cores from smoke ingress and this is typically provided for either through a lobby (vestibule) to the stair or by the use of pressurisation. Some codes (for example UK guidance) recommend a combination of pressurisation or ventilation and lobby. The NFPA guidance allows for lobbies to be omitted if pressurisation is provided.

Pressurisation and alternatives

Pressurisation is a solution frequently recommended to protect cores in high-rise buildings. It relies on maintaining the core at a pressure sufficiently greater than the adjacent space where the fire is located, such that smoke cannot enter the core. While the design principles of pressurisation systems are simple, the practicalities of design, installation, testing and maintenance are complex, particularly in a very tall building, where there can be competing pressure differentials created by wind and stack effects.

If a pressurisation system is to be relied upon, it is essential that a high quality of guidance, such as NFPA 92A or BS 9999 (NFPA 2009; BSI 2008), be applied.

There is also a concern that pressurisation systems are highly dependent on the quality of construction and fit-out. Even if the design provides redundancy in the pressurisation system, if it fails through poor workmanship, many floors can be left vulnerable simultaneously.

In recognition of the practical challenges associated with pressurisation systems, it has become common to consider alternative design solutions. Two such approaches are to extract smoke before it can reach the core (by direct venting from the fire compartment) or to provide an interception space in front of the core (for example a lobby or corridor) which is ventilated such that smoke is extracted or diluted before it can reach the core. Alternatives to pressurisation systems are not covered by standard design codes and have to be designed using a performance-based approach. This design process requires careful setting of fire parameters and performance requirements to reflect both escape and firefighting activities. Advice on typical performance-based requirements for such systems can be found in BS 7974 or the SFPE handbook (BSI 2001; SFPE 2008).

When properly designed, alternative solutions can offer flexible and robust protection to cores in a more efficient manner than traditional pressurisation solutions. The performance-based approach also ensures that the specific risks of the building in question and challenges such as firefighting on multiple floors can be taken into consideration. This enables a safe design to be demonstrably achieved, rather than assumed on the basis that code compliance will result in a safe design.

Fire suppression

Role of automatic suppression

Automatic fire suppression is a critical component to complement compartmentation in high-rise building design. When effective, it can reduce the fire hazard significantly. There are well-documented cases in which the intervention of automatic fire suppression has stopped even well-developed fires, for example the fire at One Meridian Plaza in Philadelphia in 1991, which was stopped when it reached a sprinklered floor.

There are very few cases in which the exclusion of sprinklers in high-rise buildings can credibly be justified.

Emerging technologies and system robustness

The best established means of automatic suppression is the fire water sprinkler. Contrary to popular mythology, sprinkler systems do not trigger throughout a whole building; standard building sprinklers activate only locally to the fire when the heat from the fire impacts directly on the sprinkler heads.

There are challenges to designing traditional sprinklers in high-rise buildings, most significantly the storage of large quantities of water. In response to this problem, alternative solutions such as water mist systems have been developed. These require less water or can use the potable water storage to support firefighting. Solutions like these can offer levels of protection equivalent to traditional sprinklers and should be regarded as credible design options.

Care should be taken in very tall buildings to achieve a good level of resilience in the design of the automatic fire suppression systems. This is a complex problem which is not well addressed by standard sprinkler design codes. While not typically required purely for fire safety scenarios, greater resilience may be necessary to address concerns relating to other extreme events.

Emergency Planning and Fire Safety Management

Fire safety management

Good fire safety design is only the starting point for fire safety in tall buildings. High-quality fire safety management and ongoing vigilance is equally important if fire safety is to be achieved. It is important for landlords, building operators and tenants to work together under the umbrella of a common fire safety management plan.

There is good evidence that regular fire safety drills and an awareness of fire safety matters can significantly improve the response of occupants in an extreme event. Perhaps the most compelling example of this is the well-organised evacuation of the World Trade Centre in 2001 during the 9/11 incident, which is often credited to the improvements in fire safety training and management that arose from previous terror attacks on the site.

It should also be recognised that fires will happen in tall buildings, but that they remain rare events compared to day-to-day operational activities. As a consequence, if fire safety design or management requirements obstruct operational activities, there is a tendency for fire safety requirements to become subservient to operational needs. It is therefore vital that the operational demands of buildings are taken into account in the fire safety design.

Emergency planning

Tall buildings are large, complex buildings and responses to an incident will involve emergency responders from a wide range of agencies both within and outside the building. As well as planning the management of safety within the building it is critical that a response protocol is established for dealing with incidents and that this is coordinated across all relevant parties. The response protocol will need to consider matters such as:

- Command hierarchy
- Communication strategy and methods
- Water supplies
- Street closures and access for emergency vehicles
- Dissemination of details of the building
- Interface with the media and the public
- First aid, triage and hospitalisation provisions
- Post-event investigations
- Building and business recovery.

Interaction with Extreme Events: Considerations

This chapter has focussed specifically on the fire safety requirements for tall buildings, but reference has been made to the need to plan for evacuation and for system robustness to be considered for scenarios other than a building fire. The high profile and high occupancy levels in tall buildings make the consideration of other extreme events such as terror attacks, earthquakes and tsunamis an important part of the design.

Many of the provisions for fire safety can have benefits in the case of other extreme events, but there can be additional requirements for extreme events that may clash with fire safety design. It should also be noted that, when designing purely for fire safety, the fire event is typically assumed to be accidental. Fire can also be a secondary consequence of an extreme event. When this is the case, fire safety systems and design elements may be severely compromised prior to the fire event. The risk profile for the tall building should be evaluated to determine whether or not the scenarios that include the failure of key fire safety systems need to be evaluated.

References

BIS (2005) *National Building Code of India 2005*. New Delhi: Bureau of Indian Standards.

BSI (2001) *BS 7974:2001, Application of Fire Safety Engineering Principles to the Design of Buildings: Code of Practice*. London: BSI.

BSI (2002) *BS EN 1991-1-2:2002, Eurocode 1 – Actions on Structures – General Actions – Actions on Structures Exposed to Fire*. London: BSI.

BSI (2003) *BSI BS EN 81-72, Safety Rules for the Construction and Installation of Lifts: Particular Applications for Passenger and Goods Passenger Lifts – Part 72: Firefighters Lifts*. London: BSI.

BSI (2004) *BSI BS 5588-5, Fire Precautions in the Design Construction and use of Buildings – Part 5: Access and Facilities for Fire-Fighting*. London: BSI.

BSI (2008) *BS 9999:2008, Code of Practice for Fire Safety in the Design, Management and Use of Buildings*. London: BSI.

ICC (2005) *International Fire Engineering Guidelines*. Washington, DC: ICC.

ICC (2012) *International Building Code*. Washington, DC: ICC.

Lay, S. (2007) 'Alternative evacuation design solutions for high-rise buildings', *The Structural Design of Tall and Special Buildings*, 16(4): 487–500. Online. Available at: http://onlinelibrary.wiley.com/doi/10.1002/tal.412/pdf (accessed 3 September 2012).

NFPA (2009) *NFPA 92A, Standard for Smoke-Control Systems Utilizing Barriers and Pressure Differences*. Quincy, MA: NFPA.

NFPA (2011) *NFPA 101, Life Safety Code*. Quincy, MA: NFPA.

NFPA (2012) *NFPA 5000, Building Construction and Safety Code*. Quincy, MA: NFPA.

NYBC (2010) *Uniform Fire Prevention and Building Code*. Washington, DC: ICC.

SFPE (2008) *The SFPE Handbook of Fire Protection Engineering*, 4th edition. Quincy, MA: NFPA.

SNIP (1998) *SNIP 21-01-97, Fire Safety of Buildings and Structures*. SNIP.

Note

1 Emergency responders can include city, state and military firefighters, ambulance crews, police, dedicated rescue teams and in-house response teams.

Chapter 24
Servicing the Tall Building

Alistair Guthrie and Robert Henderson

HVAC Systems: Introduction

Mechanical systems for ventilation and comfort have always been necessary for tall buildings. In some climates, low-rise buildings can rely on opening facades for ventilation and natural cooling. Historically this has not been the case for high-rise. More recently, designers have sought ways to provide high-rise buildings with breathable facades to lessen the impact of mechanical systems for ventilation. These buildings are still few in number because in most parts of the world climatic extremes of both wind and temperature make this alternative approach very difficult to achieve.

Early high-rise heating, ventilating and air conditioning (HVAC) systems were not very different to those installed today. There are examples from the early twentieth century of mechanical ventilation systems fed from centralised and decentralised plant using both constant volume and variable volume fans. Central heating and cooling were provided in the early skyscrapers from citywide steam mains or electrically driven chillers. Over the last 100 years these systems have been significantly refined, driven by mass production, cost and, most significantly in the last decade, by the need to improve energy efficiency. Many of these changes have been made possible by the use of better controls, both of individual components and of the system as a whole.

In addition, some recent designs have looked at the potential of the tall building to be a generator of energy from the wind or the sun. Other possibilities include using the building itself or the surrounding ground as a mechanism to store energy for either heating or cooling. This is explored further in Chapter 13; this chapter looks at ventilation and comfort control systems that apply specifically to tall buildings.

Climate

Standard climate design parameters for diverse individual locations can be identified using a variety of sources, such as ASHRAE and CIBSE guides. However, these do not take into account the temperature and wind speed variations that will occur at the heights achieved by tall buildings.

Wind

Environmental wind effects need to be evaluated at the very start of the project, when steps can be taken to eliminate or reduce the need for mitigation measures. Addressing issues of environmental wind may be an essential part of any planning application and wind tunnel testing may be appropriate to determine the exact nature of the wind effects on the building systems. The location of air intakes and exhausts, smoke clearance systems and natural ventilation systems will require careful consideration. Tests will be needed on the dispersal of plumes of moist air or smoke from open loop cooling towers, boilers, generators and suchlike, to ensure that they do not blow back onto the building. On the upper levels of tall buildings wind effects can be severe, creating ranges of fluctuating pressures that may affect HVAC systems operation.

Wind's random and sporadic nature makes designing for all conditions difficult, and designers should consider the principles of each system with regard to local wind effects. The one statement that will be consistent when dealing with wind is this: *wind velocity increases with height.*

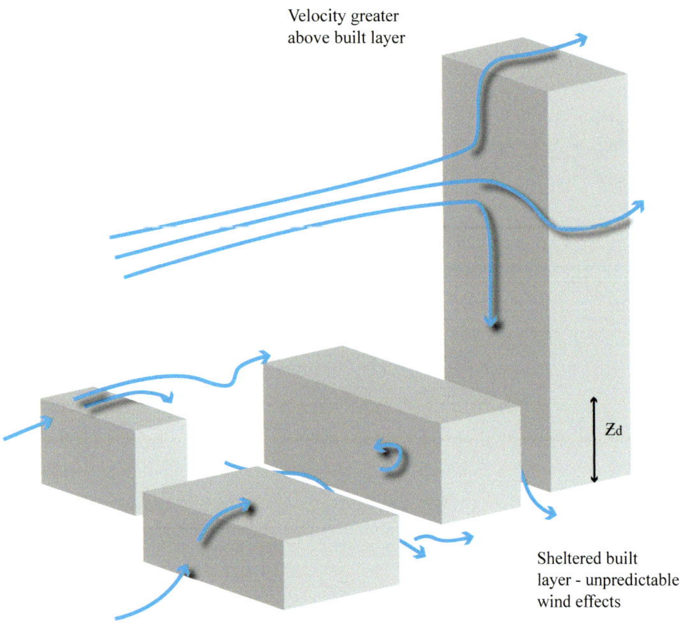

24.1 Wind effects on tall buildings within the built environment
Graphic: Arup

Consider a tall building with regard to its surrounding built environment, as illustrated in Figure 24.1. Buildings within the built layer *Zd* will be sheltered from the wind and velocity increase will begin from above this built layer. Wind effects within this sheltered layer will be unpredictable. Exposed single buildings without a built-up layer surrounding them will see this increase in velocity from a height close to ground level.

Generally there will be three types of building location to deal with:

- Where a building is the equivalent or average height of other buildings in its surroundings
- Where a building is the tallest building within its surrounding environment
- Where a building is situated next to another tall building.

Air pressure is proportional to velocity squared. Thus if wind speeds at 300 metres are twice those found below 30 metres, the air pressure acting on the ventilation louvres at 300 metres would be four times that at 30 metres. In ventilation design, the adverse effects of this high pressure should be considered for the location of intake or exhaust louvres, to reduce the impact of year-round strong winds. Appropriate pressures must also be included in the design of air handling units and fans.

Devices that shield louvres from the wind may provide only limited protection, given the randomness of wind-induced flow directions. Instead, consideration should be given to providing ducted connections to the same air handling plant from all building facades in order to cancel out the pressure regimes, as shown in Figure 24.2. A suitable additional pressure for wind should be calculated and included within the air-side systems. Where necessary, a means of measuring this pressure and varying the pressure development of fan equipment should be included.

Pressure effects will fluctuate across the face of the building with extremes at sharp edges and corners, as shown in Figure 24.3. Louvres should be placed away from these red zones, towards the centre of the building face.

Intake and exhaust distances

Intake and exhaust air paths should be kept far enough apart to avoid any short-circuit problems. A minimum separation distance is 5 metres (ASHRAE 2010), but standards and local regulations differ.

These recommendations form an acceptable starting point but a wind tunnel test may be required to check and modify these assumptions. Wind effects can cause exhaust air to attach itself to the building face and this air may then be sucked into intake louvres (Figure 24.4 overleaf). This may be a particular problem in the case of smoke and kitchen exhausts containing contaminants. High-velocity exhausts may reduce recirculation risks but designers should beware of introducing noise issues.

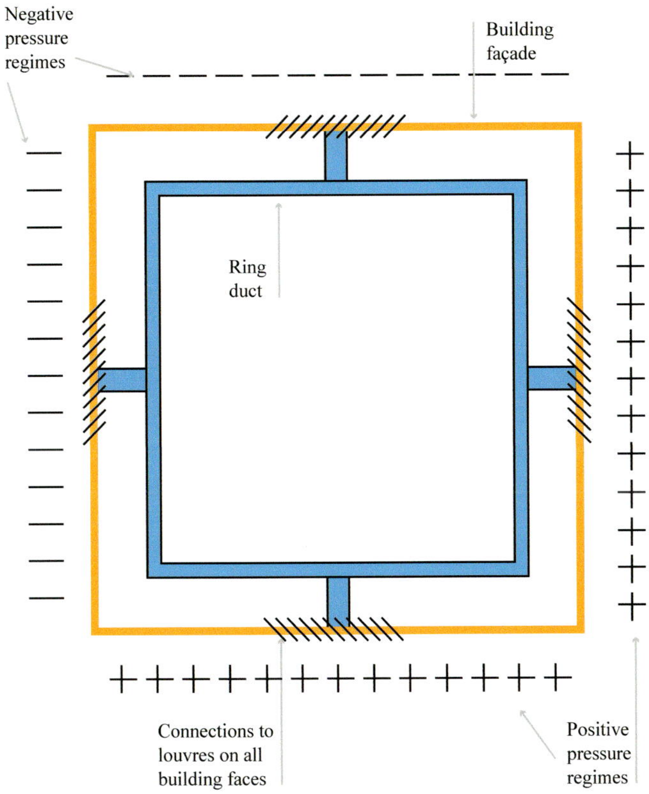

24.2 Ventilation connections (plan) to neutralise wind effects
Graphic: Arup

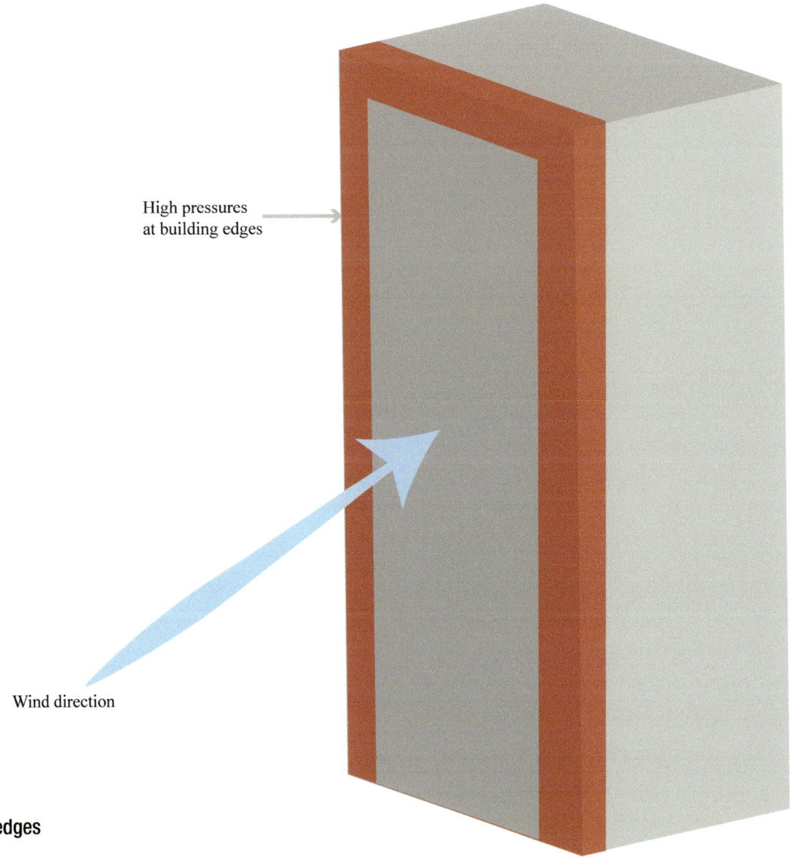

24.3 High pressure locations at building edges
Graphic: Arup

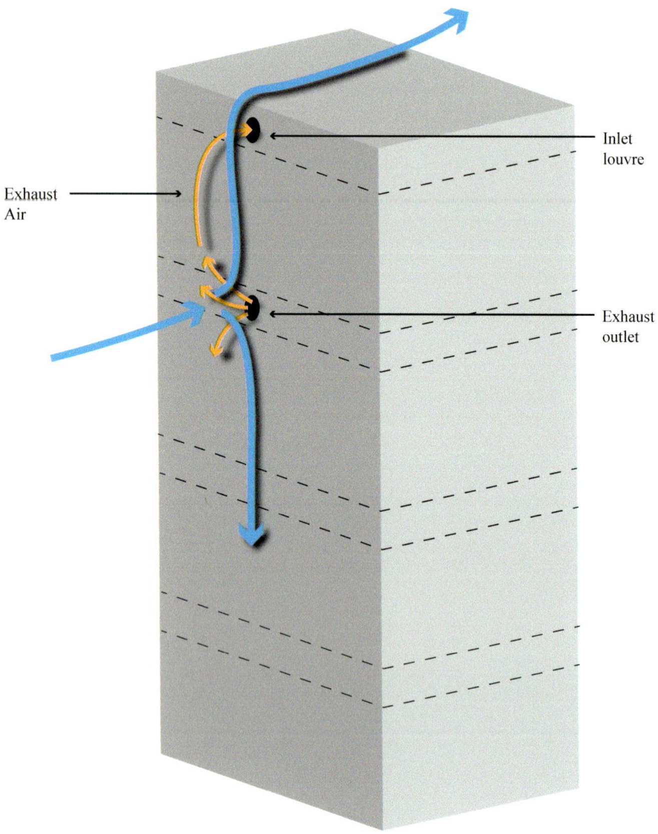

24.4 Exhaust outlets
Graphic: Arup

Intakes must be placed well away from external sources of contamination, such as chimneys of adjacent buildings or kitchen exhausts.

Temperature

External temperature is a key design consideration for HVAC systems, and decreases at an average rate of 0.6 to 0.7 degrees Celsius for every 100 metres of height. Based on this variation, the cooling load for tall buildings will decrease with height while the heating load may increase.

The performance of heat rejection equipment increases with lower ambient air temperature. In a particular project it was found that air-cooled chiller efficiency increased by 7 to 9 per cent due to the lower temperature at 450 metres. Further benefits are found with higher, constant winds assisting plume dispersion and reducing the likelihood of re-entrainment and interaction with other building systems. Any location of plant at height should also consider plant replacement strategies.

Stack Effect

Stack effect is the movement of a column of warm air upwards in tall buildings and can be significant. It is primarily a problem in cold climate conditions when the temperature differential between inside and outside is greatest. It can also occur in hot weather conditions, where the effects are reversed. For example, if the interior of a building is heated to 20 degrees Celsius, the air inside has a lower density than the colder air outside, and therefore tends to rise. The warm air tries to leave through a leaky facade or other openings at high levels, and cold external air enters through openings at lower levels.

This pressure difference can cause a large movement of air from outside into the building at ground level and consequences can include:

- Noise
- Lift doors not closing, preventing lift cars from moving
- Air flows up elevator shafts
- Difficulty in opening lobby entrance doors
- Large amounts of unfiltered air entering the building
- Difficulty in heating the lower levels of the building and in extreme cases a risk of freezing
- Energy loss
- Non-operation of staircase pressurisation (i.e. the fire safety system) due to incorrect differential pressures, and fire doors which are impossible to open
- Reverse flows in ventilation systems, such as toilet extract systems
- Very cold entrances
- Noxious fumes from below-ground car parks being drawn into the building.

Management of the Stack Effect

Passive solutions to the problem of stack effect should be used as far as possible. Two solutions are shown simplified in Figure 24.5. Practical solutions are a combination of both methods.

The first solution philosophy is to turn the building into a sealed box. In order to provide this sealed box solution as far as is practicable, two main design features will be required:

- A high-performance facade (2 to 5 $m^3/m^2/hr$ permeability at 50 Pa)
- Draught resistant entrances (such as revolving doors).

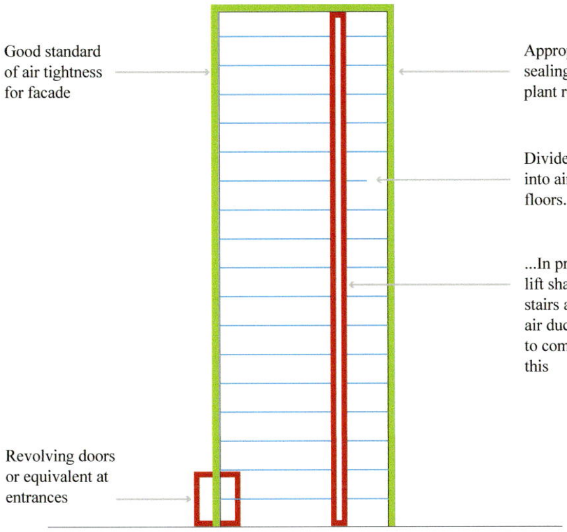

24.5 Stack effect strategies
Graphic: Arup

The second solution divides the building into sectional horizontal slices such that the stack height is reduced to levels that will not induce extreme pressure differences. Such sections can be achieved by floor-on-floor air conditioning and by floor-on-floor inlet and exhaust.

Cores and Risers

HVAC services are one of the major factors determining core size in tall buildings. With fewer HVAC services occupying the core area, more rentable area per floor can be achieved. Hence comprehensive design consideration on the core size becomes a significant task for HVAC engineers. The dominant HVAC items requiring space will be air distribution ductwork, although large-scale pipework will also require significant riser space, which can often be increased due to the need to accommodate sway and expansion fittings.

Cores may be situated centrally or at the building perimeter. Regardless of core position, risers within the core should be located at the core perimeter, preferably outside of the structural core, to minimise openings, maintain structural integrity and ease distribution.

Cores and risers located centrally within the floor plate maximise building perimeter and allow greater daylighting opportunities. Their central location can also help limit distribution runs and the impact on structural penetrations and structural and building services (SMEP) zone depth.

Perimeter or external cores and risers can provide benefits through placement on southern elevations (in the northern hemisphere) to provide thermal massing and shading effects. However, such locations may result in longer distribution runs and reduced daylighting opportunities.

In high-rise buildings, major plant will need to be located on several 'plant floors' distributed at intervals up the building. The location and frequency of plant floors is primarily dictated by pressure limitations within pipework systems and the size of ductwork. Optimisation will consider:

- Larger risers serving a greater number of occupied floors from fewer plant floors, but requiring more space within the core
- Smaller risers with decreased core space take, but requiring increased frequency and number of plant floors.

Designers should consider dividing risers to allow them to drop off with distance from the central plant floor.

A further alternative may be to distribute from different levels to keep overall riser space constant. However, such an arrangement makes heat recovery from return air difficult.

Figure 24.6 illustrates the riser space taken for a selection of projects. As shown, the percentage area for all surface risers tends to be around 3.5 to 4.5 per cent of building net area, with fluctuations within this range

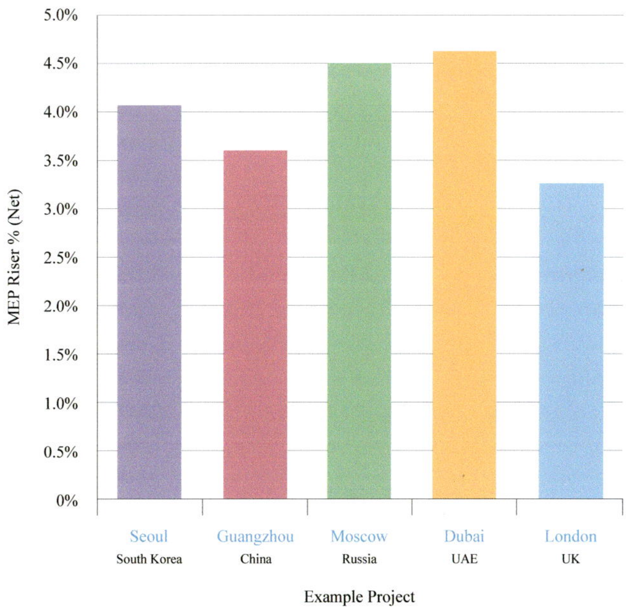

24.6 Riser percentage area (of building net area) for example projects
Graphic: Arup

being dominated by building height effects and air provision solution (for example, the Seoul project included an all-air VAV solution while the London project is serviced with minimum fresh air). The areas shown are for all risers, including fire, plumbing, electrical, IT and communications.

Services Voids and Horizontal Distribution

Floor-to-floor heights are a critical factor in the design of tall buildings. These will vary between the

- Occupied zone, where floor-to-floor height will be determined by marketing strategy, daylight penetration, floor plate depth, intended use and internal arrangements.
- SMEP zone, where floor-to-floor height will be dictated by building span, structural scheme and accommodation of power, data and plumbing services.

The SMEP zone build-up typically includes:

- Ceiling construction
- Plumbing
- Sprinklers
- Lighting
- Cable containment
- Ventilation ducts
- Structural beams
- Structural slab above
- Floor build-up above.

As the occupied zone is usually within a fixed range, SMEP zone optimisation will have a significant effect on the determination of the final floor-to-floor height.

Design coordination between the HVAC and structural solutions can be particularly significant in determining the final SMEP zone height and hence the floor-to-floor height. Minimising this will often affect the number of floors that can be accommodated within the building height limitations.

HVAC system selection plays a key role in this integration, as the air conditioning system and its associated distribution require the largest amount of space within the ceiling. Water-based minimum fresh air systems such as Fan Coil Units (FCU) or chilled beams will be more space efficient than a Variable Air Volume (VAV) system regardless of the structural solution. The differences in spatial efficiency between these two system types can be reduced through effective SMEP coordination.

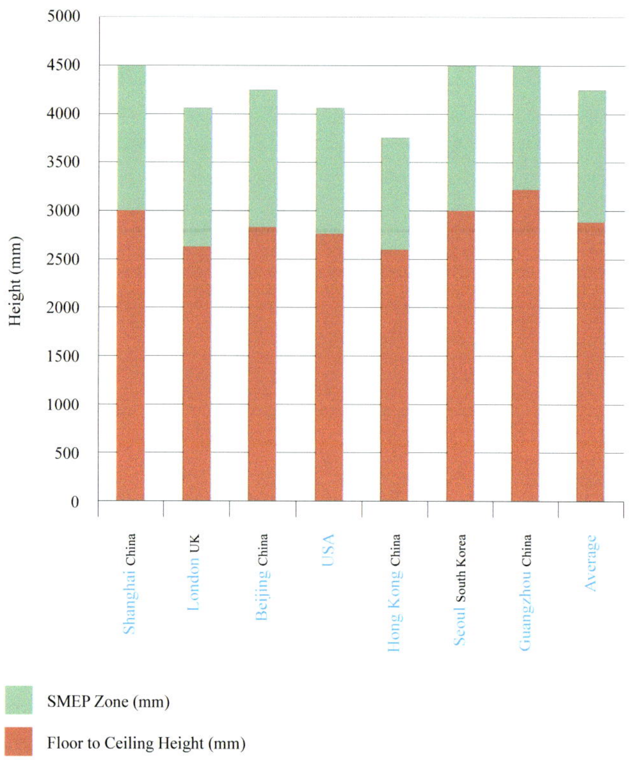

24.7 Typical floor-to-floor build-ups for example buildings
Graphic: Arup

Typical floor-to-floor build-ups for example buildings are shown in Figure 24.7.

Heating and Cooling Systems

The provision of heating and cooling within tall buildings is complicated by the large distances over which chilled and heating water have to be pumped from the central plant provision. This leads to a need for increased pumping power and water temperature gain or loss within the systems.

Consideration also needs to be given to the question of where heating and cooling plant is placed within the building. Tall buildings have reduced floor plates in comparison with low-rise, and as a result the scope for locating large amounts of plant at rooftop level is likely to be limited. Plant will therefore have to be placed at intermediate heights up the building. A primary factor dictating the location of these plant floors is the static pressure of pipework systems, and this is one of the main design differentiators of HVAC in tall buildings.

Pipework pressures

Pipework and associated joints and fittings normally used within tall buildings are pressure rated to PN16 (16 bar, approximately 160 metres static head). This provides a good starting point for determining the location of plant floors. Primary plant items such as chillers and boilers may be rated upwards of PN20 and can allow the first plant floor to be higher up the building, should this equipment be located within the basement. Pressures in tertiary end-use circuits may be dictated by final equipment such as FCU valves (typically PN6 or PN10) and careful consideration should be given to the cost implications of higher pressure equipment.

Limiting pressures on pipework allows for standardised equipment, reduces risk of failure, reduces health and safety risks, provides economic component selection and reduces weight and imposed forces. Increased pressure ratings also lead to higher maintenance costs, as high pressure operation will shorten equipment lifetime.

If lower pressure fittings are used, plate heat exchangers can be installed at intervals to break the pressure between circuits. Note that across each heat exchanger there will be an associated temperature loss between the exchanging fluids which is typically stated at around 1 degree Celsius. This reduces the efficiency of the system. Hence using higher pressure rated pipework and fittings reduces temperature drop through the inclusion of fewer pressure breaks and circuits in series, which can save on plant space, reduce the need for additional pumping stations, and can reduce operating costs. Compromise may be found in using central distribution with high pressure ratings, with standard pressure pipework beyond centralised plant floors using heat exchangers.

Pipework weights and support strategies

Pipework weight in tall building distribution will be significant and will increase with pressure rating, as will the weight of associated valves. Pipe cost will also rise sharply as the pressure rating increases.

Pipe support

Long runs of pipework in risers will expand and contract as temperature changes. These forces will require pipe anchors and a means of allowing for expansion. Typically all pipes will be supported on sliding (low friction) supports, to allow for pipe expansion without inducing stresses or failure. On vertical pipe risers, the pipe must always be supported and designers should be careful to coordinate with the structure. Concrete-framed buildings are all affected to a greater or lesser degree by post-construction creep and shrinkage, and even where the building is nominally steel-framed the core is likely to be structural concrete.

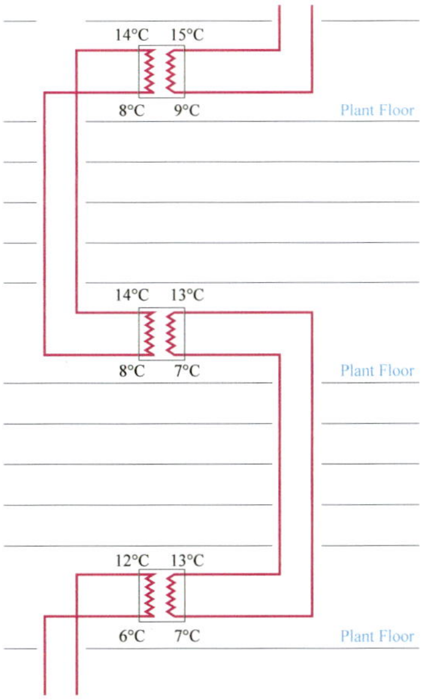

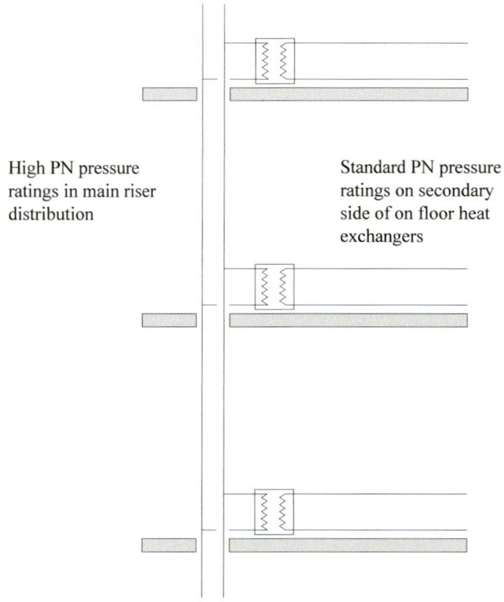

24.8 (left) Example HX temperature rise; (right) High pressure pipework
Graphics: Arup

Pipework dynamics, analysis and mitigation

Tall buildings sway, and this sway, together with the expansion of pipes within risers, can lead to issues with pipework fittings and fixtures. Both the building and the pipework system will operate dynamically and it is important to anticipate and mitigate any problems that might occur.

Pipework risers in tall buildings must be analysed with respect to thermal expansion, movements and resultant structural forces. In particular, designers should consider the worst-case warming up of pipes from cold when the system first begins operation, as the expansion of pipework due to temperature fluctuations will be most severe.

The degree of movement that is acceptable is different for each system. If it is found that the movement in the main riser is unacceptable, either the natural flexibility of the pipes or mechanical stress reduction bellows must be utilised to cope with the expansion. If the movement is restricted, stresses will be present, so it is best to leave the pipe as unrestrained as possible. Flexibility can be designed in by the use of bends and loops. Pipework is flexible but this flexibility is dependent on pipe size, material and orientation.

Mechanical flexibility

Providing mechanical flexibility is generally more suited to tall buildings, as such solutions provide for the absorption of large amounts of expansion in a confined space. Axial expansion joints (bellows or sliding joints) may be used in long straight lengths, since these are the most efficient methods of stress reduction; hence these may be most applicable to the risers. They operate by anchoring one side of the bellows and allowing the expansion in the other side to compress the bellows towards the anchor. However, since the movement in the pipe is restricted through friction, forces will be exerted by the expansion in the pipe. Articulated bellows may be used for the takeoffs since they allow the pipes to move independently and in so doing stop stresses building up at takeoffs or connections. Riser systems could operate with a combination of these methods: one (or more) expansion bellows in the vertical, with an articulated bellows on every takeoff.

The disadvantage of a mechanical system is that every expansion joint creates a weak point in the pipe work and the joints themselves must be well maintained. This means that, where possible, the number of expansions should be kept to a minimum. These joints can also generate noise during operation.

Bellows for these sizes of pipes are large and require a significant space for maintenance or removal. Table 24.1 gives approximate dimensions of the bellows for the different pipe sizes – maintenance space should be sufficient to allow removal or replacement of the whole expansion joint.

Plant replacement strategies

In tall buildings, much of the plant may be high up in the building and will be of sizes, weights and constructions that are not easily replaced. With this in mind, it is important that a suitable plant replacement and maintenance strategy be developed during the early phase design planning to ensure adequate delivery routes and infrastructure.

Adequate space to move the plant vertically and horizontally should be included and the logistics of replacement routes considered. Modular plant is an effective way of reducing problems inherent in the removal of large plant items from tall buildings. Such items can be delivered broken down and assembled and disassembled within the plant space at height. Sufficiently large space must be included within the plant room for the assembling and disassembling process.

Pipe size (mm)	Expansion joint		
	Length (mm)	Height, including base (mm)	Outer diameter (mm)
150	750	450	201
200	800	450	256
300	900	600	364
500	950	800	580

Table 24.1 Approximate expansion joint dimensions for different pipe sizes

For vertical plant movement the following replacement options should be considered:

- Lifts and lift shafts
- Temporary hoists
- Helicopters.

Heat Rejection

When considering heat rejection options, it is pertinent to consider both suitable chiller technology and where in the building the plant can be placed.

Chiller technology

A vapour-compression chiller is a device that produces chilled water by the phase change of a chemical refrigerant. As well as producing chilled water it also produces heat, which can either be used (heat recovery) or dispersed to atmosphere via a condenser circuit and heat rejection equipment (a cooling tower, for example). In some chillers all these components are housed in a single unit.

Centrifugal chillers will have better efficiency or Coefficient of Performance (COP) than reciprocating air cooled, although at low loads they may converge, and account should be taken of total energy use by fans and condenser water pumping.

Absorption and adsorption chillers are normally powered by hot water and use either water as the refrigerant and lithium bromide as an absorbent (absorption chillers) or ammonia as the refrigerant (adsorption chillers). Both types have very low COPs and should only be used in suitable situations, where waste heat is economically available. The lower COPs also lead to much higher condenser water flow requirements due to increased heat rejection; hence, for the same capacity, absorption chillers require a much larger cooling tower than conventional vapour compression chillers.

Total capital cost is likely to be comparable between air- and water-cooled solutions, with absorption significantly higher.

Space take for air- and water-cooled chillers is likely to be similar, overall (air-cooled may be slightly larger). Water-cooled plant offers the possibility of splitting the area between the cooling towers and chillers. However, single packaged units offer the advantage that there is no condenser water distribution outwith the unit. Space required for absorption chillers may be in the region of twice that required for air- and water-cooled solutions.

In tall buildings, which traditionally struggle for space, incorporating absorption chillers may be difficult to justify and, if these are selected as a solution, suitable space should be identified early in the planning of the scheme.

Chiller plant placement

Placement at the top of a tower may result in difficulties in maintenance and replacement, but will allow the plumes of warm, moist air that can form clouds in the winter to be dispersed. There may also be a further advantage in the lower temperatures experienced at height, enabling the cooling towers to operate at higher efficiency.

Placement at intermediate height may be difficult due to the need to find appropriate space for intake and exhaust air. Plumes forming against the facade of the building at certain times of the year can cause facade staining and deterioration. Cooling towers can be fitted with devices to deflect the warm, moist air away from the building, but these hoods will affect the efficiency. Downdrafts of wind across the face of the building will also interfere with the discharge, preventing proper dispersion, and may affect the microclimate at the bottom of the building.

Free cooling availability and impacts

Depending on the climate at the project location, there may be opportunities for free cooling within the heat rejection system. This means running the cooling towers to provide chilled water without use of the chillers. Once the external ambient temperature falls, the chiller can be turned off and the building cooled using the heat rejection circuit alone.

Free cooling systems should be designed to work with higher temperature chilled water, which may result in larger terminal units and additional capital cost. Higher chilled water temperatures which result from the free cooling may influence the choice of pressure break strategy, as heat exchanges will cause a further rise in the chilled water temperature.

Air Distribution

Consideration should be given at early design stages to the extent to which centralised or decentralised systems will be used for the distribution of air within the building.

Centralised systems

With a centralised plant system, all air handling unit (AHU) equipment is located within plant floors in the building, plus basement and roof as appropriate. Centralising AHU within these plant spaces frees plant from the typical floors and improves each floor's efficiency. The segregation of these areas also moves maintenance, noise and other issues away from the lettable floors and occupied zones.

Plant areas will generate noise and vibration so designers should be wary of placing plant adjacent to acoustically sensitive spaces. Appropriate acoustical levels must be achieved in the occupied floors adjacent to the plant levels, as these floors will represent the worst case for HVAC noise.

When laying out the plant space, the effect that floating floors, anti-vibration mountings, acoustical linings and other acoustical mitigation measures will have on the space available for plant must be considered.

Centralised AHU, along with other centralised plant spaces, provides these benefits:

- Simplified fire zones and means of escape plans
- Segregation of building maintenance from occupiers
- Elimination of extra acoustic and anti-vibration measures
- Mitigation of electromagnetic interference
- Overall net-to-gross efficiency improvement (depending on the plant room locations).

Decentralised systems

With a decentralised system, AHUs are located within separated plant spaces on each typical floor. Decentralising AHU plant reduces duct distribution run distances and thus distribution dimensions and riser size. Duct-borne noise may also be lower.

Decentralisation can also provide more flexibility for tenants, along with improved individual control. Such a solution also allows for spaces to be sold as complete units and reduces metering requirements.

On the negative side, decentralising can lead to higher costs due to greater numbers of AHUs and poorer economies of scale. A higher maintenance workload could be another drawback.

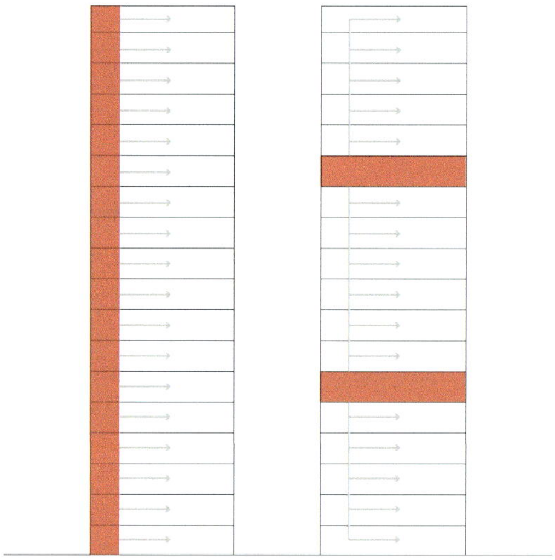

24.9 Decentralised (left) and centralised (right) plant spaces
Graphic: Arup

Two further considerations are the need to provide local acoustic isolation of the AHU plant and to find acceptable locations for supply and exhaust air louvres that are not unduly affected by wind pressures (see the section on wind).

AHU and fan selections and heat recovery systems

Fans should be selected on pressure losses calculated to the best available data, with a suitable commissioning tolerance included. As a guide, increasing the fan volume by 10 per cent will result in an increase in system pressure of around 21 per cent, which should provide enough tolerance to accommodate commissioning losses. Further scope for these fans to cope with change should also be considered. Tall buildings have long construction times, and ever changing modern tenant requirements can require a fan upgrade to meet the end market during construction itself. Selecting fans such that they are not at the limits of their fan curves can provide this additional flexibility, as can the provision of variable speed drives (VSDs) on the fan motors. These should be considered against the optimal fan efficiency operating point on the fan curve.

Limiting the size of distribution ductwork in tall buildings can seem an attractive proposal on first inspection: this can reduce riser and distribution space, leading to an increase in the net lettable area and a potential reduction in floor-to-floor heights. There are, however, significant penalties to be paid in fan power and noise issues. Mitigation is possible through the use of sound traps and attenuators, but these will require increased space within risers and ceiling voids, which can outweigh any benefits gained through the increased air speeds. Sound traps and attenuators also increase the pressure requirement of the fans, reducing the running efficiency.

Moving air at high velocity through smaller ducts also has a large impact on fan power requirements and hence energy consumption. Limits to fan powers are required to meet building code regulation in many countries, and can greatly affect the building sustainability rating for systems such as LEED, BREEAM and Greenstar.

Smoke

Life safety system operation requirements will vary depending on which country the building is located in and the fire code applied. Buildings require a means of keeping floors, escape staircases and lobbies free from smoke in the event of a fire. In high-rise buildings, opening the facades is not usually an option, hence a mechanical smoke exhaust and air supply system is required.

Mechanical system supply and extract are normally used to achieve the required air flow rates to the space, which usually exceed six air changes per hour and can exceed ten air changes per hour. These firefighting requirements may dictate the ductwork distribution sizes of the on-floor distribution, depending on flow rate and velocity/pressure requirements.

The smoke venting of lift shafts will have an effect on the building stack effect.

Requirements for the air pressurisation of stairs, lift shafts and lobbies to keep them smoke free will also depend on local code requirements. Generally, lift shafts and stairwells will be pressurised positively relative to lobbies. The lobbies will be then pressurised positively relative to the floor plates.

HVAC System Selection

HVAC system selection involves determining how and under what control mechanism the heating, cooling and ventilation is delivered to the floor plate. A myriad of parameters must be considered during HVAC system selection. These include tenant end-use, environmental considerations, building complexity, maintenance issues, architectural constraints, thermal quality, acoustics, cost, space take, fresh air requirements, ease of control, flexibility, lease agreements, plant space, equipment lifespan and many more. Table 24.2 and Figure 24.10 provide a summary (see overleaf).

Tenant HVAC Requirements

Tenant requirements

A feature of high-rise buildings is that they are likely to contain prestigious offices with valuable and highly demanding tenants. These tenants impose great demands on the HVAC services of the building, such as:

- High occupancy densities
- Higher ventilation rates
- Kitchen extract
- Increased cooling capacity, particularly for computer rooms
- Resilience for cooling, in the form of spare or standby risers, particularly for financial services operations.

High-rise buildings are more dependent on the services provided in the base building design. There is little scope for bringing in additional air through the facade, and the possibility of adding additional cooling plant on the small roof area available is low. Any additional risers will have to be accommodated within the existing landlord's risers.

Designers should anticipate that tenants are likely to want to move floor partitions, for example, and redistribute their layout between adjacent floors, as such redistribution may require alterations to HVAC systems. Tenants may also desire to increase ventilation provision either locally or across the whole tenancy. Space for tenant risers or connections should also be considered.

When considering tenant end-use, designers should also be aware of predicted changes in:

- Best practice
- Marketing trends
- Energy supply
- Climate
- Building loads
- Occupancy levels.

Designers should be aware of upward trends in occupancy levels within tall buildings. Research carried out by the British Council for Offices (BCO) in 2009 showed an increase in average workplace density of 40 per cent compared to 1997 levels, with an average

System type	Control	Noise level	Air distribution	Energy efficiency	Maintenance costs	Plant area	Ductwork size
Constant volume[1]	Good but limited	Low	Very good	Good to average	Low to average	High	Large
Variable air volume[2]	Good but complex	Low	Very good	Very good	Average to high	High	Large
Fan coil units[3]	Good	Can be high	Fair to good	Average	High	Low	Moderate
Chilled beams[4]	Good	None	See note 6	Very good	Low to average	Low	See note 7
Displacement ventilation[5]	Good	Very low or none	Good	Very good	Average	High	Large

1. Constant volume systems provide a constant supply of air, with a variable temperature to meet the loads within the space.
2. Variable air volume is an all-air system where air is supplied at a constant temperature with modulation of supply volumes to meet the loads within the space.
3. Fan coil units provide local control of spaces, usually heating and cooling, with a central fresh air supply system.
4. Chilled beams consist of water-cooled finned tube convectors which make use of natural and/or induced convection to provide sensible cooling. Again, fresh air is supplied from a central system.
5. Displacement ventilation systems use underfloor supply at higher temperatures to remove gains within the space and can provide better ventilation effectiveness than high-level systems.
6. Quality of air distribution is difficult to categorise since it will be influenced by the type of ventilation system installed. The effectiveness of chilled beams will also vary depending on whether they are active or passive, that is, whether they incorporate a fan to enhance air flow (active), or whether they are reliant on natural convection (passive).
7. No ductwork is required, although there is likely to be a separate ducted ventilation system.

Table 24.2 Characteristics of different HVAC system types

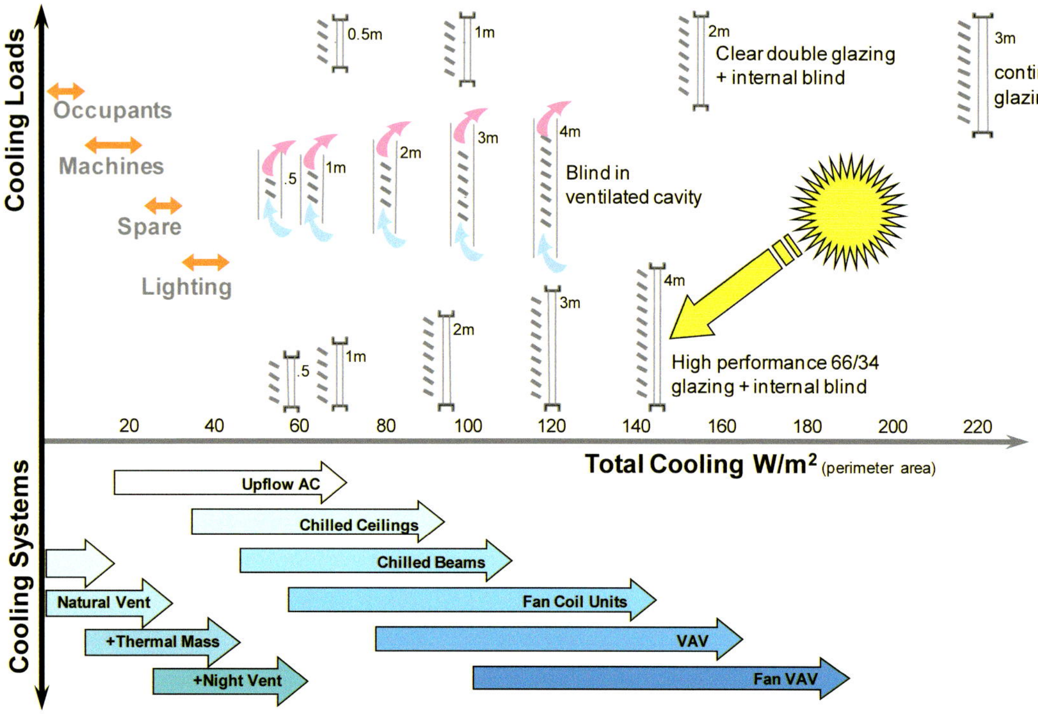

24.10 Matching facade solutions to system design
Graphic: Arup

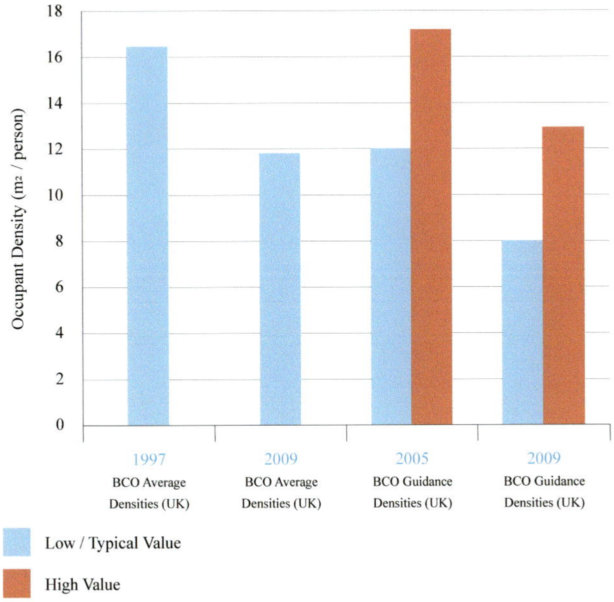

24.11 Historical office occupant densities
Graphic: Arup Data: BCO (2009)

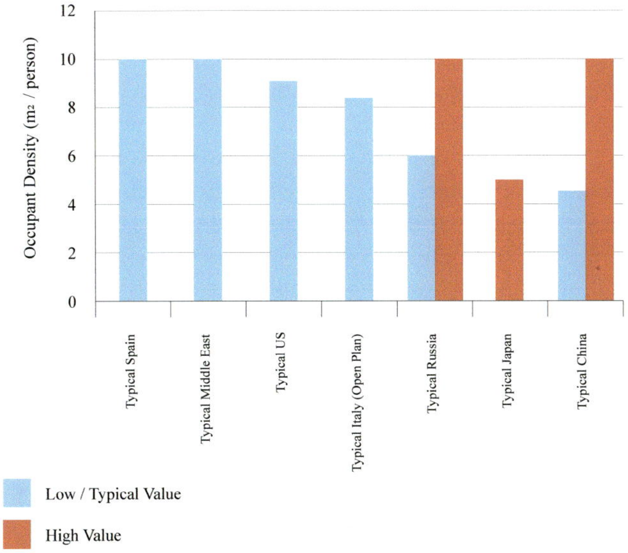

24.12 Typical office occupant densities
Graphic: Arup Data: CTE-SI3 (Spain); ANSI/ASHRAE Standard 62.1-2010, *Ventilation for Acceptable Indoor Air Quality* (USA); SNiP 2.08.02-89, *Construction Standards & Regulations (SNiP) Public Buildings and Structures* (Russia); UNI 10339 – *Air-Conditioning Systems for Thermal Comfort in Buildings* (Italy); *Building Design Standard* (Japan), *National Technical Measures for Design of Civil Construction* (China)

of 16.6 square metres per person in 1997 compared to 11.8 square metres per person in 2009. In countries such as Japan and China this density can be as low as 5 square metres per person.

Fresh air provision is fairly consistent at around 8 to 10 litres per second per person, but this will be affected greatly by increasing occupant densities.

Designers should be aware of issues with cellularisation. Office spaces with cellular offices or conference rooms require substantially more air than open plan. Guidance typically advises an additional 10 per cent fresh air volume to account for meeting spaces and cellular offices.

Increased cooling capacities and extra resilience

Increased cooling capacities over and above the base building provision are often required. In some cases tenants require dealer floors to be accommodated. In other cases a complete MER (Main Equipment Room) may be installed, including computer racks, UPS (Uninterrupted Power Supplies), and fire suppression systems.

A selection of tenants' requirements for financial services offices is set out in Table 24.3. The table shows the high cooling loads which result. These will have a major impact on the services provided and will require early consideration by the design team to be accommodated.

Building	Space type	Desk load
Building 1	General offices	120 W/person
Building 2	Financial offices	360 W/person
	Dealer floor	714 W/person
Building 3	Financial offices	300 W/person
Building 4	Financial offices	200 W/person
	Dealer floor	800 W/person

Table 24.3 Example desk loads for financial services offices

Resilience of HVAC systems

Consideration should be given to the operational needs of the building and the extent to which it can accept the shutdown of certain systems and plant items for maintenance and replacement.

Lack of inclusion of redundancy on condenser and chilled water risers can limit lettability. If a single chilled water riser is included, then should it suffer failure there is no way of shutting down the system to deal with the failure without taking the whole building offline. This would be an unacceptable scenario to most clients, in particular financial services clients.

Redundancy should also be considered on central plant. N+1 (single unit failure acceptance) would be considered a likely minimal provision. Less redundancy could be considered depending on building occupancy and usage patterns, but generally the design team will already have considered these in optimising the load capacity and, should a chiller fail during the summer, this may lead to a reduced cooling provision.

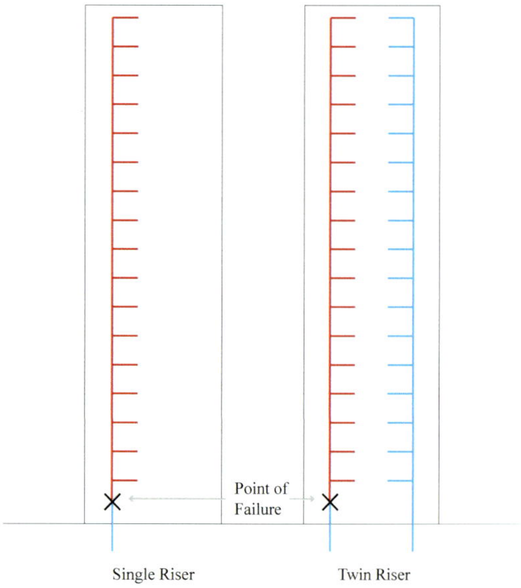

24.13 Riser redundancy
Graphic: Arup

Food and beverage impact within tall buildings

Including food and beverage (F&B) in tall buildings can take up large quantities of riser space and give rise to concerns regarding discharge of air. The ductwork can also be a fire risk.

Where high-rise buildings accommodate hotels or residential accommodation, kitchen extract must be provided. In the case of hotels this ductwork may be very large, and it is advantageous to locate hotel kitchens on floors adjacent to machine room levels.

If F&B ductwork is included then it should be laid out in a manner that allows for cleaning of the ductwork and risers. Designers should be aware of the risk of grease clogging up fire dampers and should include non-return shutters, shunt ducts, or, much preferably, a single fire-lined duct to the exterior from the F&B zone.

Discharge from F&B systems can result in noticeable plumes at the building exterior and should be located at roof level where possible. If the discharge occurs at lower levels the plumes can lead to staining of the building facade.

Low-energy HVAC strategies

As has been stated elsewhere, the secret to designing low-energy HVAC systems lies in minimising the load through efficient and intelligent facades and low-energy fit-out and occupation. Heating and cooling loads through the facade will determine the perimeter system, which in many high-rise buildings is a large proportion of the total area. The internal system will be determined by the efficiency of the lighting system, the density of occupation and the allowance for small power demands, mainly computers (see Figure 24.10 for how the facade and internal gains are reflected in the HVAC system design). If a low-energy air or water system is installed, it will work efficiently at these low loads. A high capacity system, however, is much less efficient at low loads.

In a high-rise, a considerable amount of energy is used in transporting the heating and cooling to each floor through pipework and ductwork. It is approximately four times more efficient to transport this energy via a water-based system than via an air system. So the main contributors to low-energy HVAC systems are short low-pressure air duct systems and/or systems such as fan coils or chilled beams which rely on water as the heating and cooling medium. Systems that vary the flow, such as VAV air systems or variable flow pumping systems, will reduce this 'transport' energy. These systems have been described previously.

If all these systems have reduced loads, then strategies for the efficient and low-energy generation of heating and cooling will have a much greater effect on the overall efficiency of the building.

Ground source heating and cooling

Ground source heating and cooling refers to a family of strategies which either extract or store thermal energy in the ground around or below the building. In some cases a well is dug and the ground water, which is at a relatively constant temperature, is pumped up and passed through a heat exchanger linked to a heat pump. In the heating season the heat is extracted from the water, which is then pumped back into the ground at a lower temperature. In the cooling season the reverse happens, and the system is used to provide cooling to the building.

A similar strategy is sometimes used when there is not much water below the building. In this case pipes are placed into deep boreholes where they are able to exchange the heat with the ground. These ground exchange piles work best when the heating and cooling load is similar across the seasons. The efficiency of these systems is maximised if the heat pumps do not have to raise the temperature of the heating water too high or reduce the temperature of the cooling water too low. This means that the systems are designed to work with low-temperature heating of water and higher-than-normal chilled water temperatures. In a high-rise building this will increase pumping energy, so a balance between all these factors needs to be found.

The value of these systems is quite limited in high-rise buildings, as the ground area is usually small compared to the building area that needs heating and cooling.

Wind catchers

There have been some attempts to design tall buildings to catch the wind and to use this wind for ventilation and/or cooling of the building. A strategy might be a funnel on the top of the building, which turns so that its mouth is always in negative pressure wind shadow and can pull air out of the building through a central shaft or atrium. This is often linked with openings in the facade to allow the air to be pulled through. It is very difficult and often impractical to make this work, as wind pressures are variable and can be too high or too low. The air that is pulled into the building is at outside temperature and this will inevitably be too cold or too hot for much of the year. Strategies that use this principle in limited areas such as winter gardens and atriums have been successful in some cases.

Solar chimneys

The idea of a solar chimney is to use the sun to heat the air in a tall chimney structure so that the hot air rises and pulls the air from bottom to top. This can in theory be used to ventilate the building. This concept seems an attractive one when we have tall chimney-like structures in hot climates. However, the pressures generated over the buildings are low compared to the pressures required to move air through filtration, heating and cooling systems. This means that unless the building is designed entirely around this principle, it is not applicable.

Electrical services

Unlike the other MEP services to be provided within a tall building, electricity does not have to overcome the problem of gravity, as electrons are unaffected by it. However, one of the major challenges is to achieve the most efficient and thus cost-effective way of distributing electricity through the building. Normally the best way of doing this is by using a higher voltage than is used at the user interface. This is typically in the tens of thousands of volts, as opposed to hundreds of volts, as this reduces losses in the system. Unfortunately, distributing at these higher voltages does represent a safety concern, as they can be dangerous under a fault condition. It is for this reason that, when deciding which voltage to distribute at, it is essential to ensure that the local authority will allow a high-voltage solution to be adopted. Although this is acceptable in the UK, there are many countries and regions around the world where such high voltages are not common practice and special agreement may be needed.

Another challenge in servicing a tall building is the requirement for an alternative electrical source for firefighting and life safety equipment during an event. This alternative source is normally provided by a standby diesel-driven generator or generators. Unfortunately such units are heavy, vibrate when running and need to discharge exhaust gases. One common problem is therefore where best to locate these units. Placing them on the roof ensures that the gases are safely discharged to atmosphere; however this means there will be a heavy and occasionally vibrating object very high in the building, with all that that entails in terms of structure and replacement at a later date. The opposite problem – ensuring the safe discharge of exhaust gases – occurs when locating the units at or below the ground floor.

Other electrical issues to consider are:

- Tall buildings require additional protection against lightning strike, as there is always the risk of a strike on the sides of the building as well as on the top.
- If the building is tall enough it may need aircraft warning lights.
- Due to the length of vertical risers, consideration needs to be given to the weight of cables, and it is common practice to introduce a twist into cables so as not to have a direct vertical run, thus reducing the effect of gravity.

Lighting

Good lighting must employ both daylighting and electric lighting in a fully integrated way. Only by following this integrated and holistic approach to lighting design can high quality and low energy consumption lighting installations be achieved.

Daylighting opportunities should be considered at the outset of a project, often before electric lighting design is considered. Daylight, or at least a good view, seems to be most people's preference, when given a choice. This, however, is not easy to achieve in city centres and financial districts where high-rise buildings and relatively narrow streets are the norm.

In the modern office, indeed in any environment where display screens are used, windows present the largest potential for glare and visual discomfort, but also the main source of daylight. Here is the first lighting design issue and, often, compromise. To improve access to daylight the tallest and widest windows should be used. But this can lead to glare, heat gain and excessive use of blinds and other shading elements. These shading elements reduce the amount of daylight available to the interior as well as the view out. There is therefore a need to carefully balance these often conflicting factors to ensure that energy consumption targets can be achieved while providing a comfortable environment.

When considering electric lighting it is essential that the terms 'lowest energy' or 'minimum energy' are avoided. Producing lighting design using minimum energy usually results in poor lighting and not necessarily low energy consumption. There are potentially large benefits for any organisation that invests relatively small amounts in good lighting: it has been shown that good lighting can improve staff wellbeing, reduce absenteeism and increase productivity. 'Low-energy' lighting is often incorrectly confused with 'sustainable' lighting. 'Sustainable' does not mean low-energy lighting; sustainability is also about people, and therefore about providing a visually interesting and comfortable lit environment. All lighting design should consider this, rather than rushing to produce the lowest installed lighting load possible.

The three basic objectives that need to be satisfied to achieve good lighting design are to make spaces safe, to make it possible to carry out tasks and to create the right mood. The last of these is often forgotten, which can result in spaces that lack visual interest. Applying these to electric lighting design, while ensuring that all electric lighting is controlled relative to daylight availability, can produce truly sustainable lighting installations that are low energy, and visually comfortable and interesting for occupants.

Water supply

Water supplies in tall buildings need to be boosted to relatively high pressures in order to overcome static pressure. Therefore distribution pipework must be suitably rated, as discussed previously.

A combination of boosted and gravity supply systems can be utilised to ensure a more energy-efficient system, through removing high pump heads and demand surges. By incorporating intermediate plant rooms at various floor levels throughout the high-rise building the overall energy usage is more uniform and pumping energy can be reduced.

Each group of floors can be supplied with water from gravity tanks above, at intermediate plant levels. Hot water systems at each plant level can be fed from the intermediate cold water tanks, ensuring equally balanced water pressures at outlets.

Sub-metering water supplies in tall buildings is an effective way of monitoring water usage and detecting leakage. Water meters serving separate tenancies or flats can be linked to the Building Management System and monitored centrally.

Water use reduction

As there is a relatively high energy use for supplying water in high-rise, as well as a high potable water demand, it is important to reduce water use by the selection of low-flow sanitary fittings.

Low water-use taps and dual flush water closets can be selected to minimise water consumption, utilising infrared switches, concussive taps or aerated fittings to avoid water wastage. Selecting low-pressure fittings will also minimise the pumping and energy requirements for the water supply system.

Rainwater harvesting

Rainwater falling on the building roof can be captured, filtered and used as a non-potable water supply for toilet flushing, irrigation or facade wash-down.

Typically rainwater is routed down through the building and collected in a central storage tank at ground or basement level. Treated rainwater is then pumped to a smaller day tank and distributed to the toilet fixtures from here. With intermediate plant rooms there can be a number of rainwater storage tanks at each level, which can serve the toilets by gravity and thereby reduce the pumping requirements.

There is obviously limited potential for rainwater harvesting on most tall buildings, compared to low-rise buildings, due to the relatively small available roof area. Nevertheless rainwater harvesting can usually help to minimise the potable water demand and reduce the overall building water consumption.

Rainwater can also be utilised for cooling tower make-up, but will require physical filtration as well as chemical treatment. In mixed-use or residential developments, where there is a low toilet-flushing demand, cooling tower make-up water demand can be supplemented by rainwater. However, cooling tower demands tend to be much greater than the available harvested rainwater supply.

Grey water reuse

'Grey water' refers to waste water collected from wash-hand basins and showers. This water can be treated, stored and reused within the building as a non-potable water supply for toilet flushing, irrigation or wash-down points.

Grey water from kitchen sinks cannot be reused due to its high grease content, which can damage the membrane filters that treat the grey water.

The advantage of this system in tall buildings is that large volumes of grey water can be collected, particularly in residential high-rise buildings, which have a relatively high proportion of showers and baths.

Condensate water from fan coil and air handling units can also be collected and reused within the building for wash-down and to supplement irrigation water, when mixed with potable water.

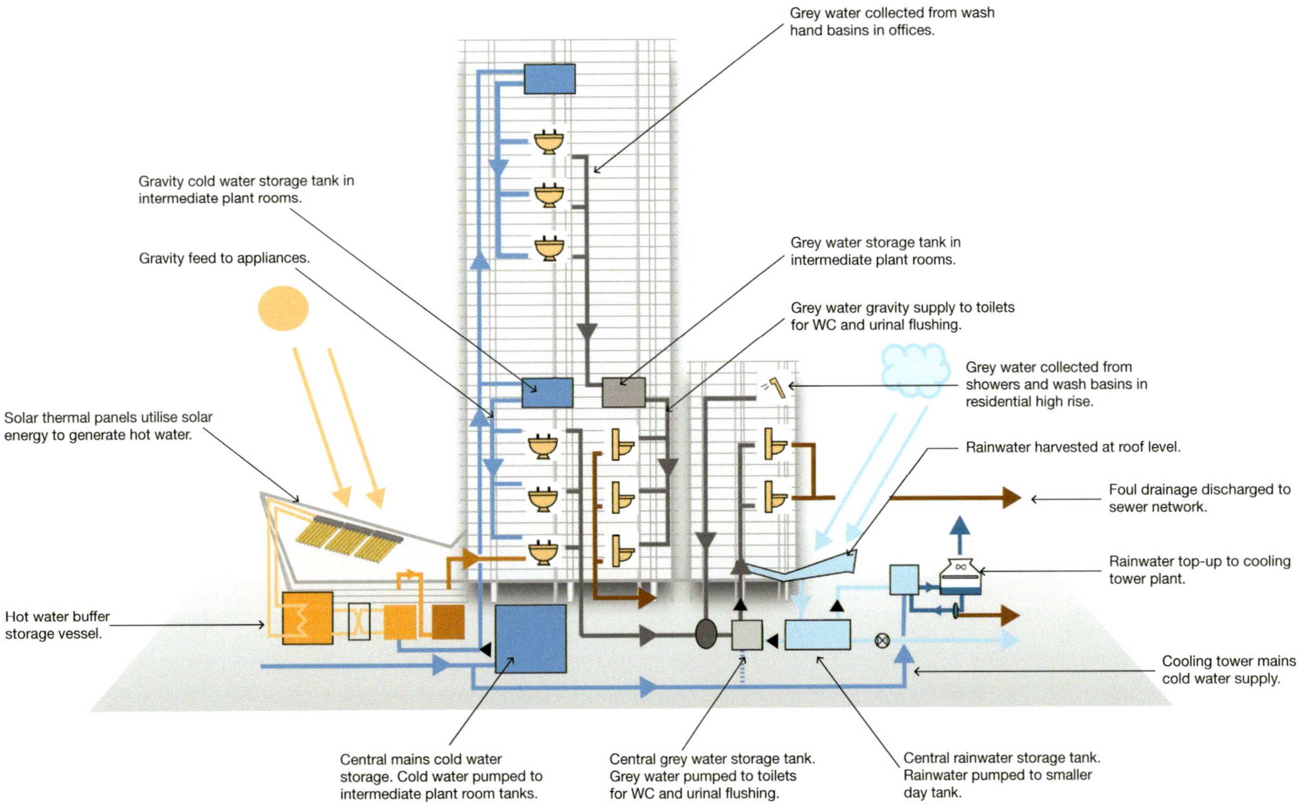

24.14 Water collection and re-use
Graphic: Arup

Green roofs

Green roofs can be incorporated effectively into tall building design, bringing benefits to urban areas in terms of increased biodiversity as well as attenuation of surface water flows. Tall buildings often have podiums, where green roofs can effectively be used to disguise roof plant as well as manage the rainwater flows.

Rainwater can be captured within the soil and drainage layer under a green roof and reabsorbed. The attenuation of surface water at roof level has a positive effect in terms of reducing flows into the surface water sewer network and relieving the pressure on the drainage infrastructure.

There are a wide variety of green roof types, ranging from intensive green roofs, requiring high levels of irrigation, to extensive green roofs, requiring minimal or no irrigation. Water-resilient sedums adapt to reduced rainfall levels throughout the year and do not require irrigation. This type of green roof planting is more sustainable since it does not increase the building potable water demand.

References

ASHRAE (2009) *2009 ASHRAE Handbook: Fundamentals*. Atlanta, GA: ASHRAE.

ASHRAE (2010) ANSI/ASHRAE Standard 62.1-2010, *Ventilation for Acceptable Indoor Air Quality*. Atlanta, GA: ASHRAE.

BCO (2009) *BCO Guide to Specification 2009*. London: BCO.

CIBSE (2006) *Guide A: Environmental Design*. London: CIBSE.

Chapter 25

Vertical Circulation: Past, Present and Future

Steve Edgett

The fundamental reason for the erection of a building is to enclose space to house people and the activities they undertake. The circulation systems of any modern building are thus a vital component of the building, unless the structure is erected solely as a monument. In all buildings, the circulation systems allow for entry, exit and use of the building.

While stairs and walkways are an important component in any building, the vertical circulation aspect of a tall building is obviously vital. The building occupants rely on a mechanized circulation system as the primary means to enter and depart their destination within the building. How is such a system planned? What are the criteria used in its planning? How are alternatives evaluated? What other objectives affect the planning of circulation systems? What are the specific types of equipment used to satisfy the circulation needs of the building? These are the questions that this chapter will answer.

Elevators and Escalators: An Introduction

Before the nineteenth century, large buildings were accessed almost exclusively via stair, ramp or walkway. Although elevators had appeared in Roman times, it was the development of steam power that spurred their use in factories, mines and warehouses. Early designs had a poor safety record, and thus did not see more generalized usage. In 1853, Elisha Otis invented the safety device that prevented the elevator from falling in the event the suspension cable failed. To many historians, this single event presaged the development of the modern city, as it made tall buildings far more usable than they had been in the past. From an economic standpoint, the combination of improved views and noise reduction offered in the upper floors of residential and commercial buildings alike changed the pricing of urban real estate.

Practical escalators followed soon afterwards. Although an escalator-like device was invented in 1859 and first demonstrated in 1891, Otis bought the rights to manufacture the escalator in 1899 and the foundations for the modern circulation system were in place.

Circulation Planning: A Primer

Circulation systems are planned solely around how many people are expected to utilize these systems, regardless of the type of building in which they are used. Factors determining whether elevators or escalators are used include the number of floors served and the function of the building. The quantity of floors to be served determines how much work is to be done in moving people to or from their desired destinations, and the function of the building (for example commercial, residential, hotel, retail, sports facility) determines the kind of traffic for which the system is to be tailored.

In office buildings, the world's most common tall building type, "peak traffic" typically occurs during the noon traffic period when, largely focused around the lunch break, people enter, leave and continue to travel among the floors of the building in a relatively short period of time. For residential and hotel occupancies, peak traffic more typically occurs in the evening, but it is again mixed, two-way traffic in which people are both entering and leaving the building. Retail facilities are arranged to ensure that circulation affords the greatest opportunity to experience the variety of options available to the shopper, whether these options are the multiple stores located in a large multi-level mall, or the variety of goods in a single department store. In sports and entertainment facilities, peak traffic typically occurs at the end of an event, when a large percentage of the population wishes to exit the facility.

Yet in each facility, the single most important factor is the quantity of persons for which the system is to be planned. Unlike code occupancy maximums, the estimated quantity represents the usual occupancy of the building. This estimate will have lasting effects on the building's usefulness over its life.

As an example, commercial buildings are occupied at average densities, based on usable area, of one person to ten square meters (1:10) to one person to twenty square meters (1:20), depending on location and specific occupancy. An office building planned for a population density of 1:15 may allow great leasing flexibility in New York or Chicago, where average densities are between 1:15 and 1:18, but the same building will not meet the circulation needs of the population if located in Shanghai or Singapore, where average densities range from 1:8 to 1:12. In practice, the building with a circulation system which does not meet the needs of the local market will exhibit queuing during peak traffic periods, and the time lost in queuing translates into cost to any business occupying that building.

Residential buildings can present similar challenges in the design of the circulation system; luxury rentals may be very lightly occupied, mostly by empty-nesters, while many residential towers must be planned for some occupancy by families. In the end, it is still the estimate of the number of persons expected which determines the nature of the circulation system.

Hotels similarly may have differing occupancy rates, from a heavily populated resort hotel to a five-star business hotel in the urban core. Hotel circulation systems can also be significantly influenced by the adjacency of restaurants or ballrooms to the primary vertical circulation systems of the building.

Important Criteria

Once the building's population and mission have been agreed upon, two central criteria are its projected cost and its investment value over time. These criteria will determine both the quantitative and qualitative criteria proposed for the building's circulation system.

Projected cost/investment value

In any market, there is a mean cost for a building which is dependent upon its placement in the marketplace, the nature of the marketplace and, perhaps most importantly, the timing of the owner's investment. Without an understanding of the owner's ideal investment in the building and his investment goals for the property, the fundamental design program for the building is incomplete. As an example, the owner's investment objectives may result in a building that is constructed to a very modest budget as a means for the owner to maximize short-term profit from his investment solely through rapidly appreciating values in the marketplace.

Such buildings are often disguised to look like true "Class A" properties, but are in fact only finished to complete an early sale or initial leasing. Similarly, in emerging markets, a building may be developed to meet all of the standards of a true Class A building, but then be unable to compete in a marketplace of lesser quality buildings, as tenant sophistication has not yet reached the point where the potential tenant recognizes the value of higher quality building systems.

The other side of this equation is the true long-term investment. Civic buildings, office buildings, owner-occupied buildings, even many residential and hotel properties may be constructed by an investment group whose investment objective is the long-term value of the property.

Without a thorough understanding of owner investment objectives, it is fruitless to design a circulation system that will often be the source of the greatest loss of building revenue. The single most important aspect of a building's circulation system is not the cost of the system itself, but the space lost from the building and the lost rent or use of that space over time.

Quantitative criteria

There are innumerable approaches to identifying the applicable quantitative criteria in the design of a tall building. For a commercial, residential or hotel facility, the criterion most frequently expressed is the "Peak Five Minute Handling Capacity". For a basement garage, the circulation system is most frequently designed to meet the maximum exiting for the garage. The quantitative criteria in retail facilities are an attempt by the designer to match traffic flows most often generated by a specialized retail consultant.

Qualitative criteria

Once the quantitative criteria have been agreed upon, the qualitative aspects of the circulation system are put to the test. For commercial, residential or hotel facilities, peak period waiting times are of significant concern. These are often based on a simplistic calculation of arriving or bi-direction traffic, and have traditionally focused only on the "average" waiting time. Today, sophisticated simulations can demonstrate that, in addition to average waiting times, the designer can also look at the distribution of long waiting times. This metric is far from universal, however, so the more common metric continues to rely on an obsolete and simplified model of passenger movement. It must be emphasized that the quality of service afforded to users is not premised on averages, but must take into account any significant percentage of long waits that may occur in a specific design. A by-product of simulation is the availability of "journey times," which identify the maximum and average travel times experienced by users. As journey times demonstrate the total system time used to transport passengers, they are indicative of the overall efficiency of a particular lift or circulation system in the building.

For any qualitative comparison of service, it is imperative that the analysis results are compared only during the identified traffic peak. Thus, even with flawed models, the primary issue is to achieve a qualitative service which matches other such facilities.

Evaluation of alternatives

After the population, investment model and criteria are identified, the designer begins test-fits against specific building massing strategies developed by the architect to meet zoning and client program requirements. In these test-fits, the circulation planner must consider adherence to the criteria, as a foundation for the evaluation of alternatives. From there, it is up to the designer of the circulation system to review alternatives which meet other, more subtle criteria. Which system results in a core generating the most usable space? Which system best fits the planning grid of the building? Which system meets the owner's leasing objectives? These criteria are often in conflict with one another and will be affected by regional differences in the real estate market.

As an example, in a mature market like New York or London, the building circulation system for a commercial building should be planned for multiple floor tenancies and provide leasing agents with the greatest flexibility in locating potential tenants. Such planning typically requires larger lift groups (to allow the maximum number of vertically adjacent floors). Larger lift groups result in a reduction in leasable space in the building, as the core is larger than one designed to maximize floor area due to the usual application of larger cabin sizes and a greater number of penetrations through the floor.

In a commercial market composed largely of multi-tenant floors, thus lacking the need for interfloor connectivity, lift configurations of smaller groups will often offer both superior efficiency and lower overall cost. This is the more typical market in urban China. In some commercial buildings with very high floor populations, escalators have been designed into multiple-floor applications, on the premise that "instant" transportation is provided. When the combination of lost space and, more importantly, lost time in passenger transport is considered, such arrangements are very expensive indeed. It is thus vital for the lift design engineer to consider such factors as local leasing and space loss in addition to the quantitative and qualitative criteria related to the performance of individual lift systems.

Residential towers often challenge the lift system designer. Where commercial building designs are premised on having a sufficient quantity of lifts to carry the desired population, residential buildings, with much lower populations and resultant lift traffic, are typically designed to maintain qualitative service. Thus, in a specific tower, two or three lifts may indeed move sufficient numbers of persons; four or more may be required to ensure waiting times are tolerable for occupants of the tower.

In any area of the building where high volumes of pedestrian traffic are anticipated, the design team must carefully consider the locations of circulation elements to avoid the potential for crushing traffic at stopping points (like security stations) or interference at crossing traffic points. Some of this is remedied by modern multi-level facilities, but the computer simulation of pedestrian traffic has increasingly become a required part of the design of such facilities.

Additional Objectives Affecting Circulation Systems

There are a number of additional factors that can vitally affect the design of circulation systems in modern buildings. Goods-carrying capacity, life safety, security, energy efficiency and the importance of a high quality passenger interface can all influence the basic building around which the circulation system is designed.

Goods-carrying capacity

The circulation systems of all modern buildings must take into account the nature of the goods that have to be moved both into and out of the building during its lifetime. Before a core is finalized, the design team must understand both the size and weight of the largest or most massive objects that will need internal transportation. In many modern buildings, the designed-in goods lifts must be capable of carrying items from large pieces of glass (for buildings glazed from the interior), to bulky and massive transformers and HVAC equipment, to furniture, to normal goods deliveries.

In most buildings, the service lift will be designed to carry a 610 by 2134 millimeter stretcher, as defined by the International Building Code,[1] a size too large to fit in the typical passenger lift. While it is unusual for tall buildings to be designed to allow goods lifts to carry a forklift, many institutional, retail and transportation buildings are designed for this feature. Not only does this make a difference in the assigned lift, it affects plan dimensions, as the structure required for a forklift-capable freight lift is significant and is not normally contained within a typical wall. For all of these reasons, it is imperative that the design team defines all of the parameters of the design of the service lift.

Life safety

Life safety plays an increasing role in the design of tall buildings as there is general recognition that the traditional practice of turning off the lifts in a fire will strand the disabled and require firefighters and other emergency personnel to climb the stairs in an alarm condition. Most building codes now require lifts to remain operable in the event of a fire. Firefighter's lifts are typically combined with the goods lifts in buildings, and include pressurized and fire-resistive vestibules which are pressurized in fire conditions and which offer protected lobbies (pressurization and advanced enclosure fire ratings are applicable to the surrounding walls). Such lifts will normally include measures to ensure that water from a plumbing leak or sprinkler activation does not drain into the lift shaft, thus disabling the lift.

Seismic response of lifts has become increasingly important, in an attempt both to minimize the number of persons trapped in lifts following an earthquake and to allow for rescue of injured persons on upper floors. Standby power is a necessity for any high-rise building, both to ensure that the occupants are not trapped in lifts during a power outage and to provide backup power for lifts during an emergency condition.

Security

Security in large buildings is becoming an increasingly important component of their design. Physical security is premised on the building's capability to withstand known threat levels, whether these are external threats like fire, earthquake or weather, or internal threats such as the violation of a tenant's space enabling threat of stolen goods or information. The properly designed circulation system is capable of meeting these threats in a manner consistent with the market for which the building is designed. Internal security can be supplemented by a variety of control measures on lifts that limit access to upper floors in the event of unauthorized entry. Additionally, the physical layout of lifts can form a vital means of limiting ground floor access to authorized users.

Energy efficiency

Great progress has been made in the design of more energy-efficient conveying systems in modern buildings. Escalators are now available that automatically switch to low speed when passenger traffic is not present, and motor controls are offered which, on heavily traveled down-traveling escalators, provide regenerated power from the escalator motor. With the advent of highly efficient variable voltage, variable frequency AC drives, lifts are perhaps 40 percent more energy efficient than lifts of just a generation ago. The incorporation of weight savings into commodity market lifts has also served to improve their energy efficiency. Lift system designers can, with proper specification, make the lift system more energy efficient than it might once have been by ensuring all lifts use more efficient gearless hoist machines and fully regenerative drives.

Controls

The lift control system is the heart of the lift system, and work has been ongoing since the introduction of microprocessor controls to improve the dispatching performance of multiple-lift groups in virtually every application. It is the advent of destination-based dispatching systems that have, however, signaled the greatest change to the circulation design in modern buildings. Employing a keypad or touchscreen interface at each floor, the destination-based dispatcher loads the cabins according to destination, instead of loading according to arrival time, during peak traffic periods.

This has the effect of decreasing the amount of lift-use time per passenger delivery, and can thus significantly aid dispatching efficiency during peak usage periods. Additionally, where individual passengers are recognized by proximity, RFID card or similar devices, the lift control system can then identify the passenger and their typical destination, or, in the case of some systems, it can actually "learn" the most frequent destinations of specific passengers. Because these systems are up to 30 percent more efficient in single-direction traffic, such as is found in the exiting of parking garages, they can even reduce the number of elevators required in these facilities.

Unfortunately, due to the proprietary nature of software algorithms in these systems, it remains unclear whether or not they can be viewed as equally efficient in real-world applications. Their widespread use is a relatively new phenomenon and there is to date little published data available that demonstrates any comparable performance.

25.1 Destination-based dispatcher systems have revolutionized circulation design in modern high-rise buildings
Image: Otis

Equipment Types

Any discussion on circulation systems is not complete without some description of the types of devices applied to create the circulation system of the building.

Lifts/elevators

Lifts come in a wide variety of types aimed at specific markets and uses, from the low-speed hydraulic style to the high-speed, high-rise gearless traction type. The traction-type lift is generally preferable in nearly any application where functionality and long-term life are considered important design criteria, but the hydraulic type still offers a cost advantage for low-use lifts in the North American market. In virtually all other world markets, traction type lift applications are predominant.

Hydraulic

This lift type is very simple, using one or more hydraulic pistons to raise and lower the lift cabin, either directly or in some cases using a side-roping arrangement which permits the cabin to be moved by steel cables looped over pulleys mounted to the top of one or more hydraulic pistons. In a hydraulic lift, a positive displacement pump driven by an AC motor pushes the weight of the lift cabin and its rated capacity in the up direction, and the lift travels in the down direction by relying on gravity; downward speed is controlled by a valve which controls the flow of oil. The most common hydraulic lift employs a direct-acting piston mounted on the bottom of the cabin. "Holeless"-type hydraulic lifts employ one or more pistons at the side of the lift cabin in a manner that eliminates the need to drill a hole beneath the lift.

"Roped hydraulic" types are also used for applications that may require travel distances outside those of the normal hydraulic lift. As noted, this type of lift drive system is used primarily in North America, although it has also seen a modest degree of success in Europe.

Traction

Traction lifts are moved up and down by means of a motor-driven traction sheave, which imparts motion to the lift via steel hoisting ropes. In recent years, composite materials of steel and polyurethane have been used for mid-rise lifts (in belts used by Otis Gen2-style lifts) and there is increasing evidence that full composite hoisting ropes using advanced aramid fibers will provide a lighter-weight alternative to traditional steel roping. Aramid hoisting ropes using DuPont™ Kevlar® were first applied to ThyssenKrupp's Isis product line.

The basic principle of the traction lift is unchanged from lifts produced 100 years ago. A lift cabin is attached to a counterweight by hoisting ropes. The counterbalancing of the total load is such that the counterweight is given a mass equal to the weight of the lift cabin plus approximately half of the rated load. Thus the motor for the traction lift is sized only to move a maximum load equal to half the capacity of the lift, producing a machine that is highly efficient at moving passengers or goods. If the drive system is appropriately designed, the lift under overhauling load conditions (an unloaded cabin traveling up or a well-laden cabin traveling down) will actually regenerate electric current. Modern variable frequency drives and permanent magnet AC motors have significantly improved the overall energy efficiency of lift operation, reducing energy usage by up to 50 percent when compared to older technology lifts employing DC motors and AC/DC generators for power conversion.

While the most common traction lift arrangement locates the hoisting machine at the top of the lift, the transmission of motion through roping allows the lift machine to be located at the side or bottom of the lift to conform to the specific design needs of a building.

Double-deck

Double-deck lifts have been successfully used to increase the quantity of passengers carried per unit shaft area. In certain conditions, such as shuttle duty from the base of the building to an upper floor sky lobby, double-deck lifts can provide very effective means of minimizing the

25.2 Composite steel and polyurethane belts have replaced traditional steel roping on Otis' Gen2-style lifts
Images: Otis

25.3 Double-deck lifts can be a very effective means of minimizing the size of the core in very tall buildings, but can struggle to cope with the noon peaks in modern offices
Image: Otis

25.4 Although not yet in widespread use, ThyssenKrupp's TWIN™ system offers designers a unique alternative to more traditional solutions
Image: ThyssenKrupp

size of the building core. Such systems require a change in lifts to reach the uppermost floors served by the sky lobby local lifts, but are really the only practical means to realize a viable core size in buildings of 60 or more floors. In the absence of good simulation data, double-deck lifts have been used in the past for local service, as they can efficiently satisfy single-direction traffic conditions. In more complex traffic, such as the noon peak in a modern office building, the qualitative service afforded by double-deck local lifts is inferior to that of alternatives employing single-deck lifts. In large part, the poor service afforded by double-deck local lifts is the result of longer journey times and "trap" incidents, in which the occupants of one deck must wait while the attached deck loads or unloads its passengers. As double-deck lifts are significantly more costly than single-deck alternatives, their usage must be carefully gauged, balancing space savings and inconvenience to users.

In response to more elaborate design requirements, variable floor height double-deck lifts have been developed. While the two cabins remain ganged together, they include a mechanism that allows for the two cabins to be moved closer together or further apart to match varying floor-to-floor heights. Although such systems are practical for shuttle lifts, the time necessary to move the cabins individually makes them impractical for use in local double-deck lifting schemes. In any event, the variable floor height mechanism adds significantly to the cost and overall complexity of the double-deck lift system.

Double lift, same shaft

ThyssenKrupp has pioneered a system employing two independently controlled lifts traveling in a single shaft. This system, designated TWIN, employs two lift motors and a unique roping arrangement that allows independent motion of two lift cabins within one shaft. The use of this concept requires highly complex controls to ensure that lifts do not "interfere" with one another, thus avoiding a lower cabin traveling up into space occupied by an upper cabin. There has been no widespread use of this system, thus its efficacy in real-world conditions remains largely unknown. As with double-deck lifts, the cost of the TWIN system needs to be considered in the light of the potential savings in floor space and reduction of inconvenience to users.

Machine-room-less

The advent of the so-called machine-room-less lift system is at last a recognition on the part of lift manufacturers that the space used by the lift is an important component of its total cost. These lifts, which should more appropriately be labeled the "Machine-in-shaft" type, use a variety of means to place the machine and, in most locations, the controls within the outline of the shaft. The machine-in-shaft system was pioneered by Finnish manufacturer KONE, who located a permanent magnet synchronous motor at the side of the lift and used an underslung roping arrangement to minimize the overrun space necessary above the lift. Variants on the KONE idea include an innovative arrangement developed by Otis Elevator, in which small-diameter steel cables are embedded in polyurethane belts. The use of smaller diameter cables allows for smaller diameter sheaves throughout the system, thus decreasing the overrun space required for the Otis machine-in-shaft system.

Circulation in Buildings over 60 Floors

As buildings reach over 50 to 60 floors, the travel time required per passenger delivered increases significantly, reducing the effectiveness of those lifts that serve the upper building floors. Only partly offset by increased speed of individual lifts, the number of lifts required for very tall buildings results in highly inefficient floor plates in the lower floors. The solution employed in most buildings is the introduction of intermediate sky lobbies where users must change from express lifts to local lift systems. The most efficient of these literally stack the local lifts originating at the sky lobby above the local lifts originating from grade. In mixed-use towers, such as those combining offices with a residential or hotel component, the change in usage often forms a natural point for a sky lobby and the opportunity to provide a differentiated vertical address for the secondary usage. Because the number of lifts is proportional to the population, mixed-use towers are always most efficiently served when the lowest population levels are located at the upper floors of the tower; this typically results in a stack with the office component at the base of the tower and the residential or hotel component in the upper portion.

In the case of a prototypical office tower of 36 floors with an average floor population of 120 persons (representing a net floor area which may be over 1900 square meters in North America or around 1300 square meters in China), a conventional single-deck lift system might comprise 18 passenger lifts arranged in three groups of six lifts each. Such a system will meet all of the criteria for a Class A office building and, given the total population of 4200 persons, the lifts are apportioned at an efficient rate of one lift to 233 persons. Doubling the height to 72 storeys with similar net floor areas will require 24 additional passenger lifts in four groups of six lifts each. This increase is due to the fact that lift efficiency drops enormously when lifts must travel 40 or 50 floors before loading or unloading passengers. However, a sky lobby can be located at a level above floor 37, effectively creating one 36-floor tower atop another, and the local lifts can be stacked in the same shafts as those below. This sort of arrangement requires some careful vertical planning to ensure that sufficient space is allocated for machine rooms and pits of the local lifts, but it is quite conceivable to stack such a tower. The number of express lifts that serve the sky lobby will be significantly lower than the number of local lifts serving the floors above. In this rather simple case, ten single-deck lifts can handle all of the passengers for floors 37 to 72, or six double-deck lifts can further reduce the shaft area while making both the ground floor and sky lobby levels more complex and less space efficient.

Note

1 See 3002.4, "Elevator car to accommodate an ambulance stretcher," in the *2012 International Building Code* published by the ICC.

FACING PAGE

25.5 Panoramic lifts and sky lobbies are features of many recent tall buildings
Image: Otis

Kasse
Kasse

Chapter 26
Logistics and Project Management

Ian Eggers, Phil Solomon, Warren Alexander-Pye and Malcolm Hannon

Introduction

Almost inevitably the decision to develop a tall building and progress it to the construction phase is driven by the desire to make a bold public statement. As a result, the architecture of these buildings is rarely anything less than remarkable. The market knowledge and expertise required to construct a tall building is a global asset and the skyline to which these monuments are added is also globally recognised.

Team culture has to be developed early and then nurtured throughout to ensure that everyone's specialism is recognised and inclusive team working is second nature. Although the definition of the main goal is normally clear, the time taken to deliver these projects is usually several years and, as such, the intermediate goals and milestones become those which have to be defined and shared with all.

This chapter highlights those areas of the construction process that make a difference to tall buildings. The techniques and tools may seem familiar to those used to development and construction, however the focus here is on those things which become critical to success at up to a thousand feet (300 metres) above the ground.

Planning, Programming and Scheduling

When embarking upon creating the plans, programmes and schedules for tall buildings, there are a number of key factors that should be considered, as follows.

Establishing tenancy splits

This is a first step. The mix of building uses will have a substantial effect on the planning of a tall building.

Office floors may sometimes be finished to a developer's specification with only floors and ceilings fitted. For other floors there may be tenants with pre-lets who have an agreement with the client either to incorporate their fit-out into the main build or to fit their areas out themselves following handover from the base building team.

Hotel operators will inevitably be looking to get early opening dates, and may opt either to push for early access to undertake their own fit-out (and maybe early opening) or may agree the client will fit out their space on their behalf. Incorporating furnishings, fixtures and equipment (FF&E) and early access to kitchens and other back-of-house space in the lead-up to opening become far more of an issue for hotels within towers. The requirement for an operator run-up period should not be forgotten; typically a three-month period following completion of fit-out would be needed, placing even more pressure on the front end of the fit-out programme. Residential units within towers place similar pressures on the programme where early or phased occupation may be required – time, space and access need to be allowed for marketing purposes, which may also involve early release of a dedicated passenger lift to a viewing gallery.

The variables that these options present make it essential, when planning a tower, to understand the scope of the works, the means by which access for men and materials can be achieved and the effect that any early milestone dates for access or occupation may also have on the sequence of works and the overall construction logistics strategy. This is particularly relevant to making the work area ready and services available in preparation for the tenant possession dates.

Establish a coordinated logistics strategy

This is essential for most tall buildings as the building footprint is usually maximised in relation to the main site boundary, and space at ground floor level is therefore at a premium. Construction operations need to be carefully planned to keep logistics routes open, and to find the most efficient routes to lifts and hoist without impacting too heavily on any early access requirements for tenants.

Care must also be taken in the selection of cranes to ensure that crane ties and the timing of raising or jacking cranes has minimal impact on the primary construction sequences, while keeping pace with the needs of the site.

The vertical materials hoisting strategy for any project is vital, and for tall buildings it is even more so. Crane locations and their impact on the structure and cladding sequences must be an early consideration. Early decisions need to be taken with regard to hoist positions, the minimisation of external hoists, and making the maximum possible use of internal shafts. Of course, all this must be offset in relation to the critical path for the project, which may favour completion of risers over completion of other elements. Early release of permanent lifts for beneficial use is almost always a key advantage in solving this trade-off.

Establish the services commissioning strategy early

For tower construction, the constructor needs to get involved early to influence the design to suit the proposed sequence of works. From early on the mechanical, electrical and plumbing (MEP) services strategy must be clear in terms of basement, intermediate and roof plant rooms, plus critical risers. Early handovers will influence commissioning strategy, and the systems must be designed for and capable of being installed to meet the needs of zoned area release.

These considerations, plus any required temporary provisions in the systems to enable sectional occupation, should be identified from the outset of the planning process. Often the zoning of the building will need to be further subdivided for construction purposes, not to mention the fact that certain systems are best installed early and used for servicing the construction phase, rather than devising costly temporary systems. Fire alarms, wet/dry risers and sprinklers are good examples of such systems.

Identify working cycles

Most tall buildings have a high degree of repetition in their core, floor structure, cladding and riser services, lending themselves to the establishment of working cycles where several trades employ the same sequence and interface management regime on several consecutive occasions as the tower rises. Programming of these works relies on achieving the most expedient turnaround of trades and the optimum follow-up of following trades – within this cycle some trades will need to be working as early as possible and others may be working as late as possible. The duration of cycles needs to recognise this without extending the overall duration and thus the associated time-related costs.

Construction planning of towers needs to treat the buildings as production lines using 'line of balance' techniques, whereby cycle times are developed to respect the progress of the slowest trades. The aim is not to be carried away by the fast pace that a slip-formed core can achieve, but to look at the cycle times for the floors that surround the core, and the ability to close floors with cladding in the most efficient way. This gains importance when a high level of finishing or fit-out is required.

It is important to recognise when looking at floor cycles that time will be lost with increasing height due to the time taken in hoisting or winching materials, and with the increasing effects of wind and local climatic conditions. Often the cranes may not be able to lift 'to height' due to wind, but at the same time they may be able to work quite happily 'at height'. The programme should aim to maximise this potential, and introducing additional plant such as column-mounted 'Chicago booms' or mini-cranes could benefit greatly in offsetting this risk.

Obtain specialist advice

The planning of towers must be carried out with a reasonable degree of knowledge of the processes that need to be adopted to achieve the optimum programme. The use of specialist advice to inform the design process and the planning of the site works is essential. Special techniques such as top-down construction (where the basement is excavated after the construction of the above-ground structure has started) may be needed to advance the core from the earliest date, which in turn may place limitations on how far the core can be advanced before it can be fully stabilised by the following floor structure works.

Crane and hoist suppliers should be consulted to review the site's logistics strategy. Cladding contractors will give advice on how they would prefer to hoist and erect their systems. Lift contractors should be approached to consider the viability of 'jump lifts' – using the permanent elevator shafts for access even as they are under construction, by installing a 'self-climbing' elevator system (see below) – and progressive shaft handovers. MEP contractors should be consulted to look at installation techniques and how they would approach systems completion and commissioning, especially where partial or progressive handovers and occupation of the building are required.

Use clear planning techniques

As previously mentioned, tower construction programmes run over several years and it is important to identify a number of intermediate milestones that will provide the team with short- and medium-term goals, and to establish a means of measuring critical progress.

The use of Critical Path Analysis (CPA) techniques is the norm on projects, and the type of software used is of little relevance as long as the planning team are able to use it to communicate the plan in a coherent manner at strategic and detailed levels.

Tall building planning lends itself more easily to the use of graphical techniques to illustrate the rate of progress of floor cycles and the lead time between trades.

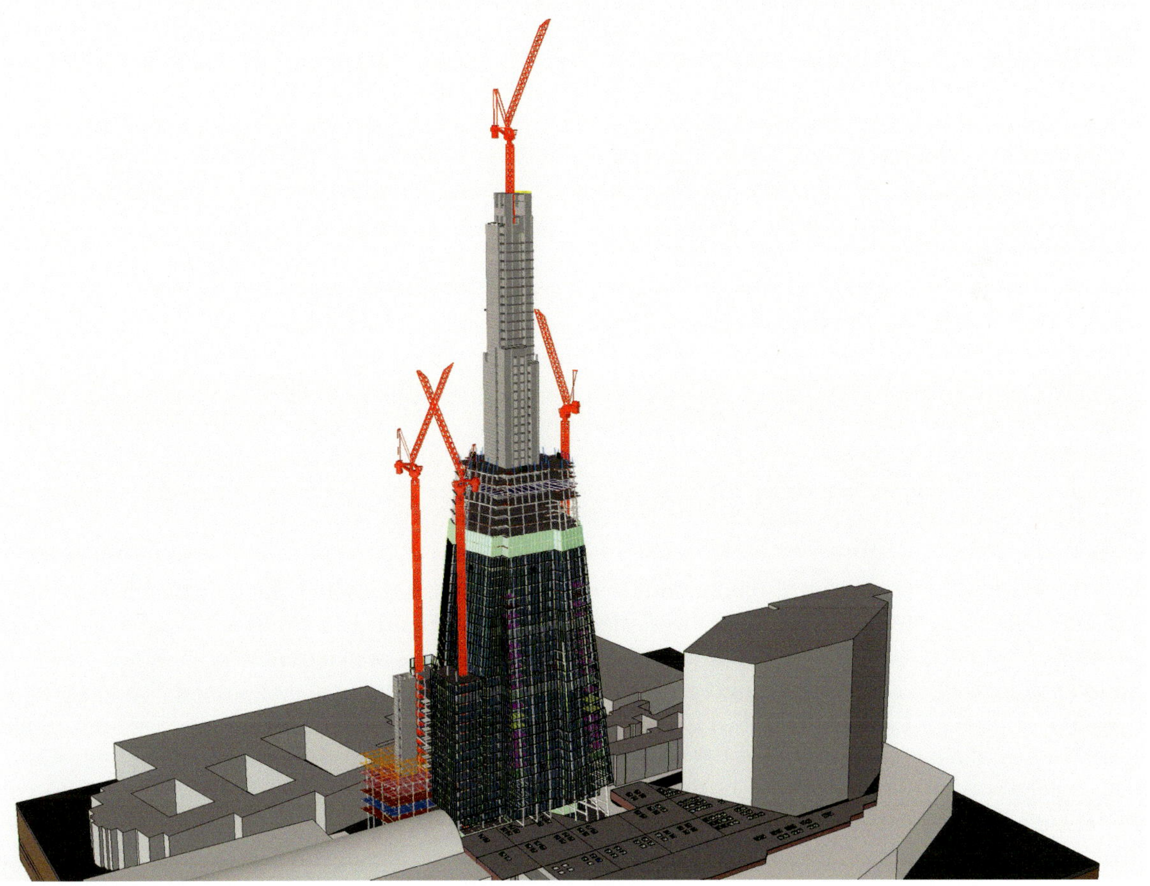

26.1 Example of a time slice from a 4D model showing the forecast progress of a build sequence at a given point in time
Graphic: Mace

The use of modern 4D and even 5D techniques (which combine the use of planning software with CAD-based images) further helps to communicate the plan; Figure 26.1 shows an example of this. Use of these techniques enables phasing drawings, real-time videos and snapshots to be produced. It is also of enormous benefit in de-risking the interfaces between trades.

These factors all contribute to the development of a coherent plan for a tall building. As construction techniques become more reliant on off-site production techniques and software develops that can improve design and construction integration, so planning methods will also improve.

Logistics

Traditionally, the main emphasis in construction has been given to scheduling and calculating production outputs, rather than ensuring the resources required to construct are complete. A large proportion of delay and additional cost in construction projects is attributable to resources not being deployed in the right place at the right time. By focusing on the creation of a logistics strategy in conjunction with the master project programme, the site-based execution can be more readily controlled, giving rise to greater efficiencies for those who construct the building and reducing site-based risk.

Construction logistics is more than just the movement of materials. Its scope includes the collection, distribution and storage of materials and the movement and welfare of all personnel and other resources from source to the workface. It can also involve backhaul – the removal of surplus, damaged or unwanted materials from the workface and their return to source or delivery elsewhere for reuse, recycling or safe disposal. The logistics strategy needs to encompass the entire supply chain to ensure timely deliveries of the right quantity and quality of labour, materials, security, safety information, etc. In the first instance, all of this requires smart planning.

The key tool for producing a smart plan is the Project Logistics Strategy. The output of this strategy will identify constraints, formulate demands and produce an integrated suite of procedures, policies and phasing diagrams. Figure 26.2 shows the critically important detail required in a ground floor logistics plan. Site logistics strategy and logistics scope are constantly evolving and need to be reviewed weekly and updated monthly. The primary elements that need to be included in the strategy are traffic management, people management, materials management, housekeeping and waste management, security, plant and equipment management and maintenance. Also critical is the interface with the Project Fire Strategy and Fire Safety Plan, to ensure that firefighting and emergency evacuation can be achieved successfully. Iconic buildings can become targets for terrorism and so the interfaces with the project Anti-Terrorism Plan also need recognition by the logistics team.

Selection of the correct plant and equipment through the life cycle of the construction project is a specialism in its own right. Procuring the most advantageous equipment for high-rise construction requires continual assessment and modifications to suit the fluidity of the project construction delivery and the impact of the external construction market. When a tall building requires highly specialised kit, the lead time may be increased by restricted global availability, or indeed the fact that an item of plant may have to come out of service from another project first.

Whether the structure is concrete, steel or a composite, a detailed analysis of concrete pump capacities and pipe routes is required so as not to hinder following trades. Planned maintenance and breakdown allowances need to be considered. During the fitting out period, beneficial use of lifts for people and materials management requires meticulous scheduling.

One of the major bottlenecks in high-rise construction is the movement of workers and materials to the workface through the traditional route of tower cranes and hoists. The risks on tower crane hook time provided by inclement weather, high wind speeds at upper levels, worker speed of component assembly and hook travel time put increased pressure on finding alternative methods of material delivery. The number of stops made by a hoist can have a major influence on the logistics plan – even to the point that express hoists covering several floors may actually work faster than the road network can deliver.

Detailed calculations are required for material volumes and speed of delivery. Freeing up crane hook time by introducing alternative equipment for cladding installation creates speed of construction. On the supertall Shard at London Bridge Tower project, external raking hoists delivered cladding panels to the floors, as shown in Figure 26.3 (overleaf). Final panel positioning was by means of manipulators, as shown in Figure 26.4 (overleaf). Monorail systems are also ideal for releasing crane hook time, although they do require additional testing and certification in situ, prior to going into service.

Delivering workers and materials to the right place at the right time is critical, but just as important is ensuring an effective strategy is in place for maximising productivity, producing quality work and providing a safe environment for workers. Successfully managing housekeeping and waste management is critical to

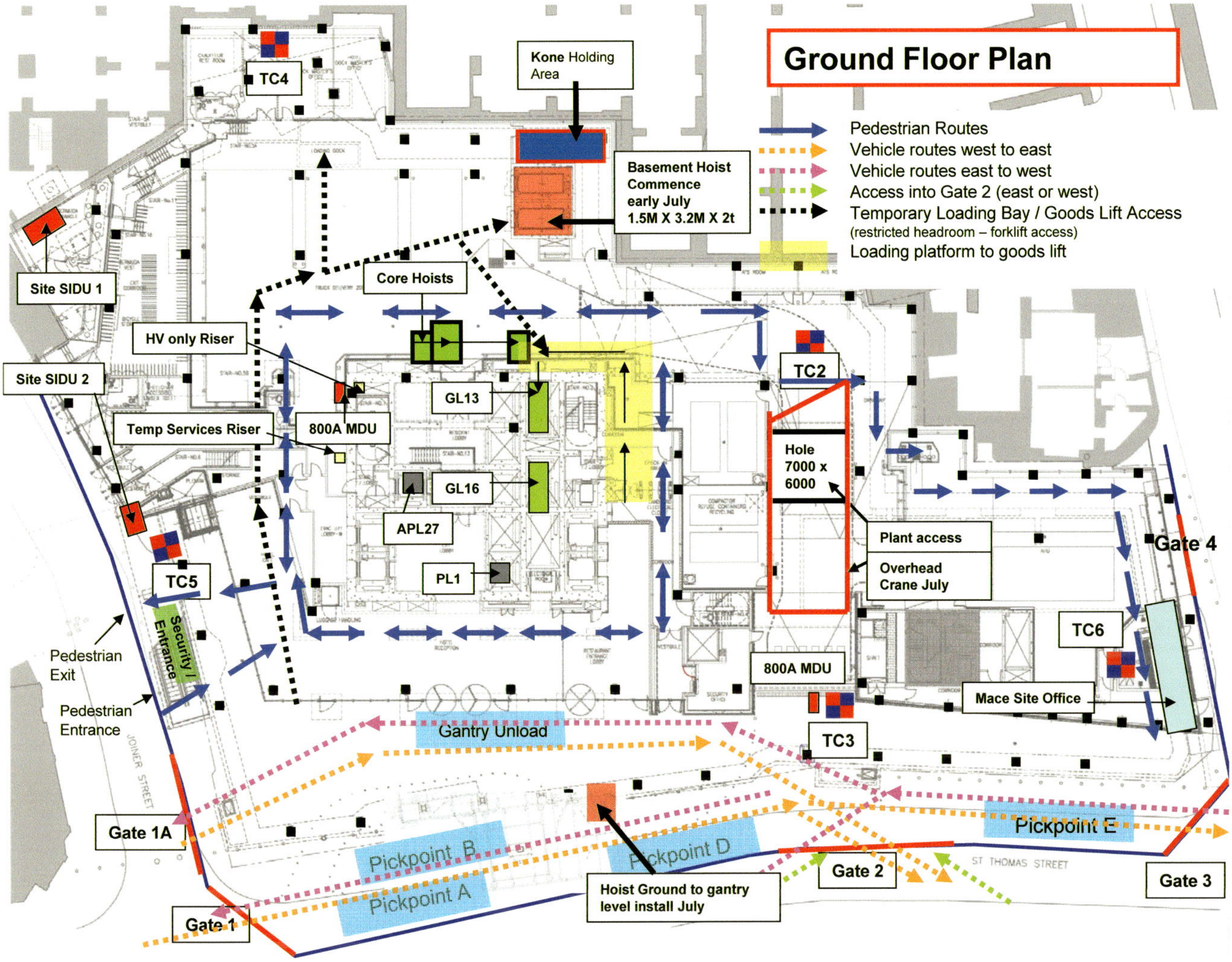

26.2 Example of a ground floor logistics layout showing critical operational features
Graphic: Mace

26.3 An inclined hoist, custom made to follow the glazing line to overcome changes in facade line at the upper levels of the building
Image: Mace

26.4 Typical curtain wall installation techniques used on tall buildings
Image: Mace

delivery. Vertical transportation equipment must be fully utilised by ensuring that downward movement is always with a full load. Consideration must be given to exactly how much packing material is permitted up the building. Often it is more expedient to strip the packing at ground floor, as the material would only have to come back down again, wasting effort. A properly run tower site will have clean and clear floor plates with minimal scope for materials damage on floor, such that protection materials become largely superfluous.

The traffic management element of the strategy investigates and defines the external routes of vehicles and workers to and from the project and the impact on the local neighbourhood, as well as segregation of vehicles and workers on-site. This ensures safe routes are provided from the security gates to the welfare areas and workface.

Externally (particularly for city-centre projects), construction routes need to be analysed to cause least disruption to the local neighbourhood and ensure timely arrivals at the project. It may be appropriate to liaise with road network officials from adjacent boroughs and districts where vehicles transfer from major trunk roads to local road networks, to ensure that they are aware of the additional movements and get the chance to offer advice, especially in relation to planned highway improvement or bridging works which may need avoiding.

From the traffic plan may also arise the need for materials consolidation centres. The simplest form is a lorry holding park – an area where vehicles can park without nuisance and loads can be checked and called forwards to site. The most advanced levels of consolidation can include warehouses and fleet distribution. Options for consolidation can vary and the steel fabricator's yard and steel painter's yard are both informal consolidation sites where material is laid up and called forward in sequence. The benefits in urban areas can be enormous, and the rescheduling of deliveries will not happen at the site gates, allowing all other parties to proceed unhindered and keeping the road network clear.

Delays on construction projects are costly. A well-planned project security strategy will allow for control of the use of the workforce, assist in ensuring the right materials are delivered and removed, and minimise loss from the site due to theft by the workforce or the public. The value of security needs to be compared with the potential losses to the project. Pressures on security can also come from terrorists, base jumpers, protest groups and immigration officials checking the site staff's credentials. The constructor has the obligation to manage these, and the security regime must reflect this.

When the factors described are combined with the more standard and routine aspects of logistics the smooth running of the project can be guaranteed. The ability of the logistics manager to keep these aspects under control will further allow the construction managers to keep their focus on the control of delivering good quality site work, as discussed later in this chapter.

Innovation

The definition of 'innovation' is hotly debated. It arises when there is a need to implement a change at some level. For tall buildings there must be a *desire* to change the way we do things long before the *need* arises. The uniqueness of tall buildings means that the delivery team must possess a pioneering spirit.

In its purest form, innovation not only delivers the desired change but also does it with less effort, in a shorter period and at a reduced cost. This is as rare as a lightning strike, so we need to be prepared to accept that all these improvements may not occur together. Innovation thrives when there is openness, honesty and respect between team members, each of whom is working inclusively with the others. The most likely occasions when innovation may arise are in design or programme workshops, key stage design reviews and value engineering or value management workshops. A vigilant and energised team will spot the opportunities, whereas those who merely follow due process are unlikely to harness innovative thought.

For construction and project managers the most likely area for innovation is technology. Prefabrication has long been established as a means of innovating in construction. Bathroom pods are prefabricated finished products that embrace the benefits of factory conditions and manufacturing techniques. Taking this a stage further, prefabrication of riser sections and also primary on-floor mechanical and electrical distribution are making innovative changes to the way site assembly, testing and commissioning is carried out. Off-site fabrication changes the balance between factory and site works, alters the course of the critical path through projects and guarantees the quality of the delivered product, thus commercially de-risking the construction process.

The ability to vertically integrate between trades is also an area where innovation can thrive. The smartest thinking decorator can innovate in the way in which they build relationships with the preceding trades. A new product has recently emerged which provides steel floor plates with a visco-elastic fill material. Bolting in

26.5 Innovative use of a tower crane mounted on top of a prefabricated steel core, used on an extremely confined inner city site
Image: Mace

a single plate of this material reduces the need for the chain of trades associated with concrete composite floor slabs on metal deck formwork. As a result, it is possible to reduce the frame weight and foundation loads when the engineering conditions permit. Figure 26.5 shows an innovative way of mounting a tower crane on top of a prefabricated core to overcome the difficulty of a lack of space on-site.

One of the key areas of innovative thinking in relation to tall buildings has been the development of lift technology. Not only has there been development in respect of the ride quality in high-rise lifts, but installation techniques have become simpler and safer too. One of the major innovations has arisen from the need to deliver better and more efficient service, which has resulted in an exponential development, in recent years,

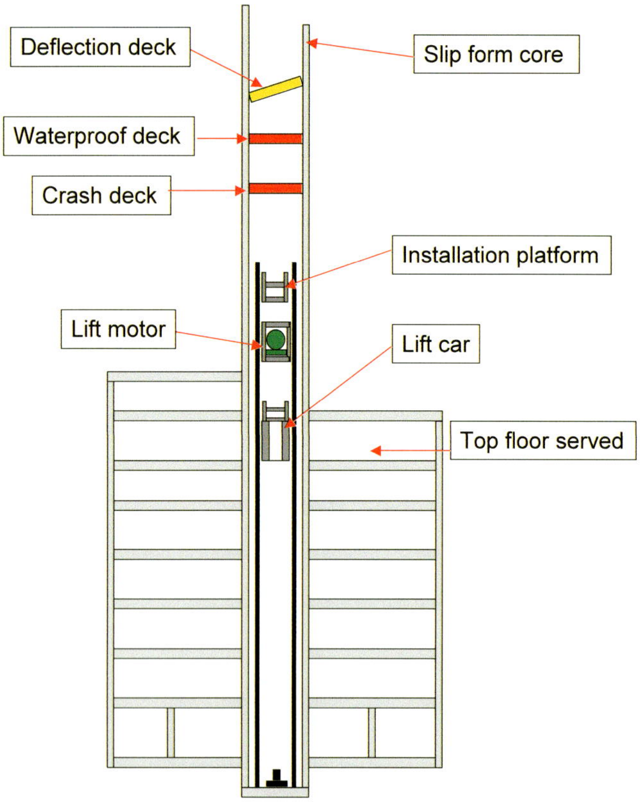

26.6 Diagram showing jump lift technology. This can operate in a lift shaft which is still being formed at upper levels
Graphic: Mace

of jump lift technology, double storey height lift cars and destination control systems. The builder's need to have high-rise lifts available early for workers and materials has also driven the systems to become very robust, and easily converted or refurbished under beneficial use. Figure 26.6 illustrates the concept behind jump lift technology.

Innovation comes with a note of caution. Where new ideas and creative thinking exist it may take time to convince others of the benefits. Clients, tenants, insurers and regulatory bodies may need persuading that the new ideas are the right ones – the programme needs to cater for this and where possible these people should be invited to follow the developmental journey with the team.

The harsh atmospheric conditions at a thousand feet (300 metres) up demand tougher and more innovative materials and specification. Innovative logistical techniques are driven by the need to get more workers and materials up in the air. The need to reach the top and bottom of these buildings in a timely fashion for the fastest release of critical plant areas demands the most innovative construction sequences. High-rise construction drives innovative thinking.

Design Management and Procurement

Design management and procurement are intrinsically linked. The common bond between them, which sets tall buildings apart, is that the scope for error by repetition is huge. A missed issue on one floor may appear four times on that floor, and over 50 floors becomes an issue in 200 places. If the solution is not procured correctly, those 200 issues, each costing $100, become a $20,000 cost problem. If the issue takes a week per floor to solve then an additional 50 weeks of labour time must be built into the programme. The multiplier effect cannot be underestimated. Anything that requires a trade to make supplemental passes across a floor needs to be managed out.

Design management needs a clear strategy throughout the design stages, as the drawings develop from concept to scheme to detail. For the constructor it needs to include a thorough validation for buildability. Each stage should have a report produced to which all parties contribute, with the client being asked to sanction the contents. This is the basis of gateway management, which will carry forward into the procurement phase. Each gateway is a confidence-building review and check which means that the client is reassured that the brief and budget are being adhered to, and the professional team know that the client is aware of their progress and is actively either accepting or rejecting their recommendations at each design hold point.

Typical areas where the design manager can make a difference are varied. Intumescent and fire protective treatments always need careful evaluation. Acoustic linings can often add huge amounts of time and cost. Cladding panellisation and bracketry also need close review. Staircases to cores can contain many details that can prove tricky, notably finishing down the string plates (indeed, why finish them at all?). The design manager can add value by suggesting options in all of these instances.

The next most valuable activity that the design manager can undertake is in conjunction with the procurement manager. Value engineering can be done at a minimum of two stages: first, design against cost plan; second, tender information against trade specialist advice. The multiplier effect can work to the team's advantage here, as contractors may be able to add significant value by proposing savings on simple, repetitive details.

The procurement manager must recognise that the tall building market requires experience and specialism that may not exist in their own company's supply chain and, as such, time should be allowed for more extensive project-specific prequalification interviews within

Tall Building - Anyplace

Tender Event Schedule

ReportDate 7-Oct-10

Works Package		Initial Design Release	Design Review/Pre-Tender Est	Design Release	Collate Bid Docs	Out to Bid	Bid Return	Bid Period	Bid Analysis	Recommend	Place Order	Lead In	Start on Site
3210 External Cladding	Planned	29-Sep-10	29-Oct-10	10-Jan-11	10-Jan-11	24-Jan-11	7-Mar-11	24-Jan-11	7-Mar-11	21-Mar-11	25-Mar-11	28-Mar-11	2-Dec-11
	Forecast	29-Sep-10	29-Oct-10	10-Jan-11	10-Jan-11	24-Jan-11	7-Mar-11	24-Jan-11	7-Mar-11	21-Mar-11	25-Mar-11	28-Mar-11	2-Dec-11
	Actual	6-Oct-10	-	-	-	-	-	-	-	-	-	-	-
2800 Structural Steelwork	Planned	3-Oct-10	6-Dec-10	14-Jan-11	17-Jan-11	28-Jan-11	11-Mar-11	31-Jan-11	14-Mar-11	28-Mar-11	1-Apr-11	4-Apr-11	12-Sep-11
	Forecast	3-Oct-10	6-Dec-10	14-Jan-11	17-Jan-11	28-Jan-11	11-Mar-11	31-Jan-11	14-Mar-11	28-Mar-11	1-Apr-11	4-Apr-11	12-Sep-11
	Actual	3-Oct-10		-	-	-	-	-	-	-	-	-	-
2400 Superstructure Concrete	Planned	10-Dec-10	13-Dec-10	21-Jan-11	24-Jan-11	4-Feb-11	18-Mar-11	7-Feb-11	21-Mar-11	4-Apr-11	8-Apr-11	11-Apr-11	18-Jul-11
	Forecast	10-Dec-10	13-Dec-10	21-Jan-11	24-Jan-11	4-Feb-11	18-Mar-11	7-Feb-11	21-Mar-11	4-Apr-11	8-Apr-11	11-Apr-11	18-Jul-11
	Actual	-	-	-	-	-	-	-	-	-	-	-	-
7400 Lift Installations	Planned	17-Dec-10	20-Dec-10	28-Jan-11	31-Jan-11	11-Feb-11	11-Mar-11	14-Feb-11	14-Mar-11	28-Mar-11	1-Apr-11	4-Apr-11	16-Apr-12
	Forecast	17-Dec-10	20-Dec-10	28-Jan-11	31-Jan-11	11-Feb-11	11-Mar-11	14-Feb-11	14-Mar-11	28-Mar-11	1-Apr-11	4-Apr-11	16-Apr-12
	Actual	-	-	-	-	-	-	-	-	-	-	-	-
5500 Window Cleaning & Gantries	Planned	10-Jan-11	10-Jan-11	4-Feb-11	7-Feb-11	18-Feb-11	18-Mar-11	21-Feb-11	21-Mar-11	4-Apr-11	8-Apr-11	11-Apr-11	-
	Forecast	10-Jan-11	10-Jan-11	4-Feb-11	7-Feb-11	18-Feb-11	18-Mar-11	21-Feb-11	21-Mar-11	4-Apr-11	8-Apr-11	11-Apr-11	-
	Actual	-	-	-	-	-	-	-	-	-	-	-	-
3300 Atrium Glazing	Planned	17-Jan-11	17-Jan-11	11-Feb-11	14-Feb-11	25-Feb-11	8-Apr-11	28-Feb-11	11-Apr-11	2-May-11	6-May-11	9-May-11	26-Mar-12
	Forecast	17-Jan-11	17-Jan-11	11-Feb-11	14-Feb-11	25-Feb-11	8-Apr-11	28-Feb-11	11-Apr-11	2-May-11	6-May-11	9-May-11	26-Mar-12
	Actual	-	-	-	-	-	-	-	-	-	-	-	-
6300 Mechanical Systems Infrastructure	Planned	21-Jan-11	24-Jan-11	18-Feb-11	21-Feb-11	4-Mar-11	15-Apr-11	7-Mar-11	18-Apr-11	9-May-11	13-May-11	16-May-11	5-Dec-11
	Forecast	21-Jan-11	24-Jan-11	18-Feb-11	21-Feb-11	4-Mar-11	15-Apr-11	7-Mar-11	18-Apr-11	9-May-11	13-May-11	16-May-11	5-Dec-11
	Actual	-	-	-	-	-	-	-	-	-	-	-	-
7000 Electrical Services	Planned	28-Jan-11	31-Jan-11	25-Feb-11	28-Feb-11	11-Mar-11	22-Apr-11	14-Mar-11	2-May-11	16-May-11	20-May-11	23-May-11	27-Jan-12
	Forecast	28-Jan-11	31-Jan-11	25-Feb-11	28-Feb-11	11-Mar-11	22-Apr-11	14-Mar-11	2-May-11	16-May-11	20-May-11	23-May-11	27-Jan-12
	Actual	-	-	-	-	-	-	-	-	-	-	-	-
7100 Commissioning Services	Planned	31-Jan-11	31-Jan-11	25-Feb-11	28-Feb-11	11-Mar-11	22-Apr-11	14-Mar-11	2-May-11	16-May-11	20-May-11	23-May-11	27-Aug-12
	Forecast	31-Jan-11	31-Jan-11	25-Feb-11	28-Feb-11	11-Mar-11	22-Apr-11	14-Mar-11	2-May-11	16-May-11	20-May-11	23-May-11	27-Aug-12
	Actual	-	-	-	-	-	-	-	-	-	-	-	-
6200 Sprinkler Installations	Planned	4-Feb-11	7-Feb-11	4-Mar-11	7-Mar-11	18-Mar-11	15-Apr-11	21-Mar-11	18-Apr-11	9-May-11	13-May-11	16-May-11	9-Dec-11
	Forecast	4-Feb-11	7-Feb-11	4-Mar-11	7-Mar-11	18-Mar-11	15-Apr-11	21-Mar-11	18-Apr-11	9-May-11	13-May-11	16-May-11	9-Dec-11
	Actual	-	-	-	-	-	-	-	-	-	-	-	-
6500 Ductwork	Planned	14-Feb-11	14-Feb-11	11-Mar-11	14-Mar-11	25-Mar-11	22-Apr-11	28-Mar-11	2-May-11	16-May-11	20-May-11	23-May-11	13-Jan-12
	Forecast	14-Feb-11	14-Feb-11	11-Mar-11	14-Mar-11	25-Mar-11	22-Apr-11	28-Mar-11	2-May-11	16-May-11	20-May-11	23-May-11	13-Jan-12
	Actual	-	-	-	-	-	-	-	-	-	-	-	-

Legend

Date Type	Compared Against	Colour Codes			
Planned Dates	Report Date	Complete	Due Soon	Over Due	Not Due
Forecast Dates	Report Date	Complete	Due Soon	Over Due	Not Due
Actual Dates	Report Date	Complete	Due Soon	Over Due	Not Due

Programme Details

Created From:	Strategic Programme
Revision Number:	B
Programme Number:	RPTB/Str/01
Programme Date:	7-Oct-10

26.7 A basic example of a tender event schedule showing red/amber/green (RAG) status
Graphic: Mace

a global market. The need to make overseas visits must be recognised, and the management resource for these visits must reflect the fact that they can become wearisome and reduce the availability of the procurement manager to work regular hours.

Another priority for the procurement manager is to continue the gateway management regime and set a series of gateways based on tender events identified in the master programme. A simple Tender Event Schedule (TES) is one of the best ways to control procurement activity, using a red/amber/green (RAG) system to provide an 'at-a-glance' summary of procurement status. An example of such a document is shown in Figure 26.7. Typical team and client sign-offs would then exist at tender document issue, tender recommendation and contract placement, to ensure that the inclusive working culture is continued.

Once the contracts are placed, the importance of a robust change control mechanism cannot be understated. The real purpose of change control is to empower the professional team to offer suggestions for changes with impunity, to allow the entire team to evaluate the validity of a change and to ensure that only the viable tier of changes are presented to the client. The client is then empowered to either accept or reject the change. The underlying principles are that no member of the team can spend against the budget in isolation and that the whole team should choose the route forward for the project as a single entity.

Managing Construction

It must be realised that the site-based construction manager is the last person in the chain to have the opportunity to set the scene before tools are put to work. He or she is also the person who, having prepared correctly, will ensure that the quality standards are met, the costs do not overrun and the milestones are achieved as planned. The construction manager is also pivotal in ensuring safe working.

A common feature of tall building construction is the inability of any one person to manage the site and the ability of the site to absorb a large number of section managers in the execution of the work. If a single manager is given a 10-floor slice of a tall building, the chances are that they will only be able to visit a limited number of target areas in any given working day. Tall buildings require discipline in boiling down the role of the construction manager to its bare essentials and working efficiently.

This process starts with the early involvement of the construction manager in the procurement process. Checking the scope of works, commenting on the contract provisions for site attendances and understanding the quality aspirations of the project are essential activities. Attending the kick-off meetings with the package managers and then holding the package safety and quality inductions are all key tasks in ensuring the construction manager is fully connected with the work that is about to be delivered. Figure 26.8 summarises the pre-commencement thought process for any given task.

The construction manager should consider their role with regard to the following areas:

1. Establishing what conditions are required to start the work. Scheduling and completing the pre-start needs will focus the mind and ensure that the stumbling blocks are removed before they become a problem.
2. Ensuring the work area is ready. Double checking the completion requirements of the previous trade and making sure they have been met. If not, then making corrections – fast!
3. Ensuring the personnel are the right ones for the job. Is the trade contract manager the one whose CV was reviewed at tender stage? If not, why not? Are the operatives all qualified and experienced?
4. Forecasting the materials-in and waste-out regime for the entire package and making sure the logistics team are briefed. Likewise ensuring that critical plant and materials deliveries are booked and that the lay-down areas on site are cleared and marked in advance.
5. Deciding the conditions required for completion. Ensuring that the quality standards are understood by all, and accepting nothing less than what the specification requires. Working out what the start conditions for the next trade are and ensuring these are delivered.

These duties apply to any building, but for a tall building the construction managers must be focused and must work smart. Successful tall building construction cannot support 'floor walkers' – it needs dynamic and efficient managers. There will be many distractions en route every day and only the most flexible, adaptive and proactive managers will do.

Summary and Conclusion

The essence of effective logistics and project management for tall buildings is thus focus and control. The control aspect can be engendered through planning, strategy and process; the focus aspect, however, is down to the quality of the leadership and management staff allocated to the project. Defining roles and responsibilities is essential and the effective project manager will need effective logistics, planning, design, procurement and construction managers to focus on their respective areas, coordinate with others and grind out the detail.

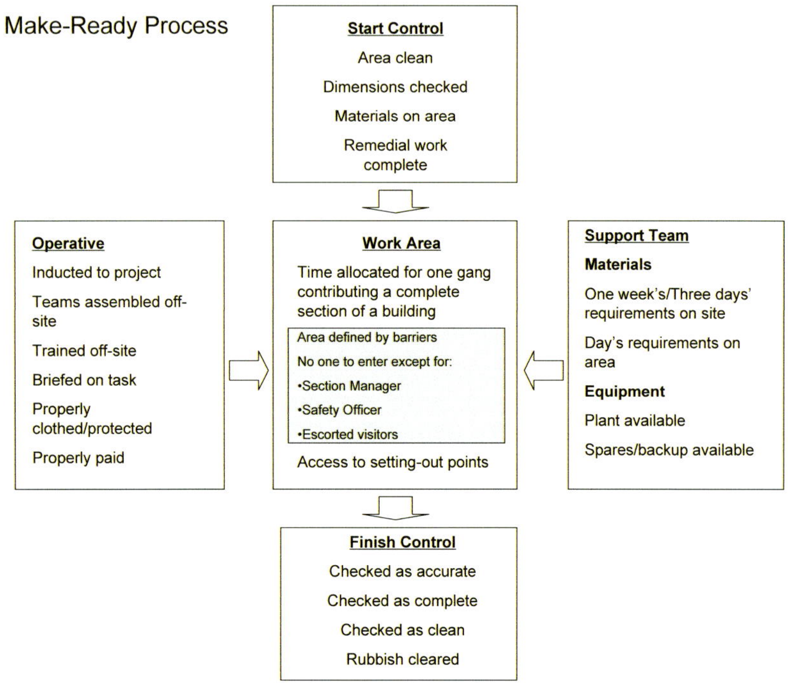

26.8 Schematic diagram showing the process for preparing an area and delivering finished work on-site. This is the core of the construction manager's focus on-site
Graphic: Mace

Tall buildings are visionary, and as such the people involved need to be visionary too. The project manager must be able to build a cohesive team. He or she must also be able to draw the other members of the professional team in close and make sure that they too share the best of their work and coordinate thoroughly with others. The topics discussed within this chapter will allow the project manager to set focused intermediate milestones and goals which will unify the team and build confidence while also delivering a truly unique building.

As a final note, it is important to remember the words of Richard Halpern, the project director for the Sears Tower. When asked in conversation what the secret of constructing a tall building was, he replied, "It's the accumulation of small things very well done, no one big ticket item!"

Afterword

The Future of Tall?

Antony Wood

This book clearly demonstrates the advances that have been made in the typology of tall buildings over the past two decades or so. However, despite 130 years of development, many believe that tall buildings have not yet advanced to a satisfactory state, especially on environmental and sustainability grounds. Most tall buildings historically seem to have been designed as either vertical extrusions of an efficient floor plan (the "commercial" approach), or as stand-alone pieces of high-rise urban "sculpture" (the "sculptural-iconic" approach). In both cases the main relationship with the urban setting is either a commercial or a purely visual one, with the tall building usually dominating.

This has led to the syndrome of tall buildings as "isolationist" architecture—stand-alone, non-site-specific models that are readily transportable around the cities of the world. This, in turn, has served to create an alarming homogeneity across global urban centers—a creation of a "one-size-fits-all" skyscraper "mush" which in some places rejects thousands of years of local vernacular tradition. The skyline may quickly become synonymous with a place, but that does not necessarily mean it is inspired by it, or relates to it. This is especially true of cities in developing nations, where the import of all things "western" is often seen as progressive and modern. Thus the vast majority of tall buildings internationally follow the standard template of the rectilinear, air-conditioned, western "box."

In addition, tall buildings have become synonymous with the greatest excesses of energy expenditure, in both embodied construction and operation. Though there are definite energy advantages tall buildings can offer, both in creating more sustainable patterns of life through higher density and through the potential for greater renewable energy generation at height, there is no doubt that, as single buildings in their current form, most tall buildings are energy profligate.

It needs to be acknowledged that tall buildings are by no means accepted by all as a sustainable building type, and the reasons for this investigated. Some consider the typology to be anti-sustainable—and most if not all of the reasons cited have significant substance (these reasons are summarized in Table 27.1, together with the case "for" tall buildings). At the same time, many of these individual points fail to recognize the bigger picture beyond the single building: as mentioned throughout this book, denser cities with concentrated land use and infrastructure offer more sustainable patterns of life.

"Against" tall buildings	"For" tall buildings
Higher embodied energy in constructing at height—structure, materials, etc.	Denser cities = reduced transportation (and consequential impact on environment).
High energy consumption in operation—elevators (up to 15 percent of building energy use), services, etc.	Efficient land use and population concentration = reduced suburban spread and loss of countryside.
Higher energy consumption for both maintenance and cleaning (e.g. replacement of facade silicon joints).	Concentrated cities = reduced size of infrastructure networks (urban/suburban, power, services, waste, etc).
Impact on urban scale—wind downdrafts, overshadowing (solar rights), wind rights, right to light, etc.	Proximity of residence and workplace = less travel time (less wasted time?).
Overpopulation in certain localities; greater demand on existing urban services and infrastructure.	Greater potential for mixed-use = less travel time, less duplication of building form and resources.
Antisocial internal environment—lack of open, recreational, communal space (especially in residential buildings).	Standardization of floor plates and use of materials = material (prefabrication?) efficiencies.
Greater wind loading at height—impact on size of primary structure, facade design, etc.	Higher wind velocities at height = greater potential for harnessing wind energy.
"Sealed" environments at height; requirement for air-conditioning, artificial lighting, etc.	Higher atria/volume of space = potential for natural ventilation through increased "stack effect," etc.
Less net usable area to gross area and restrictions on internal planning—vertical circulation core, etc.	High "thermal mass" = potential use in natural ventilation, heating and cooling strategies.
Safety and security fears (especially post-9/11), including safety during construction.	Long, narrow floor plates = potential for good internal day lighting (and thus reduced energy).
Low ratio of external building surface area per unit floor area—impact on potential for solar arrays, etc.	Space in the sky = potential for "secure" communal or recreational spaces, away from traffic, pollution, etc.
Implications of power failure—impact on vertical circulation, safety, etc.	Potential for more efficient energy production and distribution systems.
Increased travel time (wasted time?).	Urban densification adds value and vitality to cities.
People suffering from vertigo—building occupation/human rights legislation?	Urban signposting and way-finding.
	Increased access to view, light and air at height.
Recycling potential, urban impact of demolition, disposal of materials after demolition.	

Table 27.1 Summarized cases "for" and "against" tall buildings as an appropriate typology in urban centers

The main challenge for the typology for the future is to create tall buildings that are relevant to the specifics of place—physically, environmentally and culturally. To do this, we need tall buildings that maximize their connection to the city, climate and people. Questions over what has inspired the recent diversification of approaches to building form and whether they can be justified in energy and carbon terms are valid, and need to become a more essential part of the industry's dialogue. The sustainability discussion in recent years has been focused almost exclusively on operating energy, which, while vitally important, has resulted in the neglect of discussion of embodied energy in building construction. Even the very definition of "net zero energy" seems to omit the materials and construction process entirely. Numerous exemplar tall buildings have recently made great strides in the reduction of operating energy. However, the energy expended to create building forms in the first place is by no means constant across buildings, with iconic-sculptural forms clearly requiring more material gymnastics (and hence more carbon) to deliver the same quantity of floor area as a more regular form.

But there is another side to this equation: that of a building's contribution to society, beyond delivering maximum floor area with the minimum energy and carbon expenditure. What do iconic-sculptural forms bring to our cities in terms of beauty, or impact on urbanity and the human senses? Do we want to live in a world full of ultra-energy- and cost-efficient but rather dull boxes? What about the impact on social sustainability and urban diversity, and a whole range of other, less quantifiable aspects of "sustainability"? As with all things, there will be an optimal balance point in this equation, but the debate thus far, for obvious reasons, has been focused on quantifiable metrics rather than subjective judgments.

What is clear is that there is still much to be done before tall buildings deliver their full potential—in their contribution to dense cities and urban form, reduced energy consumption (embodied as well as operating), and social diversity and inclusivity. By way of an afterword, eight principles are offered for consideration in the design of future tall buildings, together with illustrations from the "Design Research" projects undertaken by the author in conjunction with architectural students, in his role as a professor.

1. Variation in Form with Height

Tall buildings should not be monolithic vertical extrusions of an efficient floor plan, but should vary in form with height. This variance in form should be inspired by both the building program internally and the city externally, both physically and environmentally. A tall building could be considered as a number of stacked communities or horizons, with each horizon having a different potential to relate to aspects of the site or city. The climate is not the same externally throughout those horizons (the external air temperature outside the Burj Khalifa is several degrees cooler at the top than at the bottom!); neither should the form and skin be monolithic, as a response to this stratified climate. Similarly, a tall building potentially has a visual relationship with many places in the city far and wide, at differing horizons within its form; a visual dialogue with these distinct places (and other buildings) can help inform a variance in form to further connect the building to its locale.

27.1 Skybox, London. An example of how a tall building could be considered as a number of stacked communities or horizons, with each horizon having a different potential to relate to aspects of the site or city. The Skybox tower, which challenges typical high-rise design, is comprised of individually stacked "boxes" inspired in function and orientation by the differing physical aspects of the city (e.g. horizon views). Each "box" responds to the individual needs of the occupants and the surrounding urban context, creating a unique, varied form
Image: CTBUH/IIT, Prairna Gupta

2. Variation in Texture and Scale

There should also be a variance in skin and texture throughout the building, depending on the responsibilities of each different horizon within the form. The concept of scale should be introduced throughout the building—a tall building could be thought of as a number of small buildings placed on top of each other within an overarching framework of structure, systems, aesthetics, etc., rather than as one extruded, monolithic form inspired by a single plan, and designed accordingly.

27.2 Swadeshi Tower (textile tower), Mumbai. An example of how there should be a variance in skin and texture throughout a building, depending on the responsibilities of each different horizon within the form. In the Swadeshi Tower, inspired by the Dobi Ghat washing areas in Mumbai, textiles are integrated into the building aesthetics and materiality through the application of a weave-like cladding to the facades, providing residential spaces with shading from solar gain and privacy
Image: © CTBUH/IIT, Nishant Modi and Hiren Patel

3. New Functions

Traditional programs for tall buildings need to be challenged, to increase the usefulness of the typology in sustainable cities of the future. This challenging of program should occur on two levels: (1) the type of functions that are traditionally accommodated within tall buildings, and (2) the number of functions that are accommodated in a single tall building. Tall buildings have the versatility to accommodate uses other than the standard office, residential, and hotel functions that currently predominate. We could see the radical incorporation of functions such as sports (external solar control skin as rock-climbing wall?) or agriculture (hydroponic greenhouses, facade farms?). In addition, cross-programming or mixed-use within tall buildings should be encouraged, to give opportunities for more sustainable activity patterns (for example dualities of car parking, support functions, and servicing) as well as variance in tall building design and expression to diversify urban form. The challenges of climate change require us to intensify every expenditure of carbon—we need to have multiple uses for every element created.

27.3 Annapurna Tower (food tower), Mumbai. An example of challenging traditional programs for tall buildings, to increase the usefulness of the typology in sustainable cities of the future. The Annapurna 'Vertical Farm' Tower aims to create a new vertical residential community with high quality living space bound together by aspects of urban agriculture and food provision
Image: CTBUH/IIT, Cindy Duong and Shin Young Park

4. Communal Spaces

More open, communal, recreational spaces (internal or external, hard or landscaped, large and/or small) need to be introduced into tall buildings, rather than maintaining an insistence on the maximum financial return on every square meter of floor space. Such spaces have been proven to improve the quality of the internal environment, which has an impact on saleable/rental return, occupant satisfaction, productivity of workers, etc. In addition, the inclusion of these spaces will make tall buildings more suitable for socioeconomic groups often marginalized in tall buildings due to the lack of such vital spaces where a sense of community can develop—families, the young, the old, etc. Social sustainability on an urban scale is a major challenge for our future cities.

27.4 TATA Tower (urban development of the TATA Corporation), Mumbai. An example of the introduction of more open, communal, recreational spaces into tall buildings. Tata Tower is an urban parking development for India's largest car company. The building acts as a vertical parking resource for the neighborhood, and through renewable energy generation technologies is able to provide energy for itself, its vehicles, and neighboring towers as well
Image: CTBUH/IIT, Seth Ellsworth and JaYoung Kim

5. Envelope Opacity

Tall buildings should be designed with more envelope opacity, not as all-glass transparent boxes requiring significant external shading devices to control the excessive light, heat, and glare. Although the impact on internal daylighting and views out needs to be balanced, all-glass towers do not make sense, especially in intensely hot, solar environments. In addition, greater facade opacity gives an opportunity for greater thermal mass, allowing the envelope to be more insulated from external temperature and climate variations. More opacity also gives the opportunity for greater facade variance and expression.

6. Organic Matter

Vegetation should become an important part of the material palette for tall buildings, both internally and externally. The presence of vegetation will improve environmental quality on both the local scale (as part of the shading or air cooling system of the building itself) and the city scale (through quality of air, reduced heat-island effect, etc.).

27.5 Gyoyook Tower (education tower), Seoul. An example of a tall building designed with more envelope opacity, not as an all-glass transparent box. Gyoyook Tower is aimed at developing a high-rise habitat to foster the education of Seoul's children
Image: CTBUH/IIT, Kevin Ford and Stevie Brummer

27.6 Moksha Tower (vertical cemetery), Mumbai. An example of vegetation as an important part of the material palette for tall buildings, both internally and externally. The Moksha Tower project takes traditional burial methods from four major religions in Mumbai and translates them into a vertical context
Image: CTBUH/IIT, Yalin Fu and Ihsuan Lin

7. Skybridges

It seems completely nonsensical that cities are making a push for ever denser, ever taller urban form, but allowing the ground plane to be the sole physical plane of connection between towers. Skybridges have the potential to enrich both tall buildings and cities, improve evacuation options, and reduce energy consumption through allowing horizontal as well as vertical movement. Every tall building should be considered as a vital element in an overall, three-dimensional urban framework, rather than as a stand-alone icon superimposed on a two-dimensional urban plan.

27.7 Khel Tower (vertical athletic center), Mumbai. An example of towers connected through skybridges, Khel Tower is a vertical sporting facility in the heart of Mumbai. A series of hotel towers lean past each other to allow views of the city and ocean, with suspended sporting facilities and fields in between. These large recreational facilities are greatly lacking in Mumbai
Image: CTBUH/IIT, Kent Hoffman and Mark Swingler

8. We Need to Bring ALL Aspects of the City Up into the Sky

If cities are looking to concentrate perhaps ten or a hundred times more people on the same land through building tall, then they need to replicate the facilities that exist at the ground plane up in the sky, including the parks and the sidewalks, the schools and the doctors' surgeries, and the other public and civic functions. The ground plane needs to be considered as an essential, duplicable layer of the city which needs to be replicated—at least in part—at strategic horizons within and between buildings in the sky, not in place of the ground plane but in support of it.

27.8 Yatra Towers (procession towers), Mumbai. An example of the ground plane considered as an essential, duplicable layer of the city which needs to be replicated—at least in part—at strategic horizons within and between buildings in the sky. Yatra Towers is a series of linked towers in the heart of Mumbai that retain the sense of community and industry of the low-rise area, which is swept into the sky. Each tower aims to enhance the quality of life of its residents while addressing many of the urban issues of Mumbai
Image: CTBUH/IIT, Irene Matteini and Nathaniel Hollister

Many of the principles outlined above are starting to happen, as evidenced by the projects and concepts displayed in this reference guide. Time will tell if the tall building will ever reach a completely satisfactory state of evolution in both energy and cultural terms but, as the global recession puts paid to some of the more excessive tall ideas of the past decade, I genuinely believe it is heading in the right direction.

Antony Wood
Executive Director,
Council on Tall Buildings and Urban Habitat
Chicago, USA

Part Six

Case Studies

Case Study 1

Aqua Tower

Chicago, USA

HEIGHT:	262 M
NUMBER OF STORIES:	81 + 6 PARKING LEVELS
GROSS FLOOR AREA:	184,931 M^2
PRIMARY USE:	MIXED USE: RESIDENTIAL, HOTEL, OFFICES, AND RETAIL
START ON SITE:	2006
COMPLETION:	2010
PROJECT COST:	$475 MILLION; CONSTRUCTION COST $325 MILLION
OWNER/DEVELOPER:	MAGELLAN DEVELOPMENT GROUP
DESIGN ARCHITECT:	STUDIO GANG ARCHITECTS
ARCHITECT OF RECORD:	LOEWENBERG ARCHITECTS LLC
STRUCTURAL ENGINEER:	MAGNUSSON KLEMENCIC ASSOCIATES
CIVIL ENGINEER:	IE CONSULTANTS, INC.
MEP ENGINEER:	ADVANCED MECHANICAL SYSTEMS, INC.

Project Description

Totalling over 1.9 million square feet, Aqua Tower is an 81-story mixed-use high-rise in downtown Chicago that includes a hotel, apartments, condominiums, parking, retail and offices. As a "topographic" tower, the building is designed to make specific and unexpected views available from different parts of the tower. The large terraces make it possible to inhabit Aqua's facade and the city simultaneously. They also work to shade the glass and provide a foil to wind pressure, two significant concerns for high-rise towers.

Architecture

When the full 11.3-hectare (28-acre) site of Lakeshore East Development is complete, Aqua will be located within a forest of high-rises. Unlike a tower in an open field, new towers in urban environments must negotiate small view corridors between existing buildings. In response to this, the tower's sculpted facade makes unexpected views possible from different parts of the tower. "Hills" created on the building's vertical surfaces allow sightlines that connect to specific Chicago landmarks around corners of, and through gaps between, existing buildings. Its outdoor terraces, cantilevered up to 3.6 meters, differ in shape from floor to floor, and inflect based on criteria such as the view, solar shading, and size and type of dwelling.

When viewed together, these unique terraces make the building appear to undulate, presenting a highly sculptural appearance rooted in function. Aqua Tower includes 15 levels of hotel, 34 levels of rental apartments and 28 levels of condominiums. Mixed use continues at the podium building, which includes residential lobbies, retail stores, restaurants, hotel ballrooms, and commercial space with a large outdoor garden on its roof.

At 6,811 square meters this roof garden is one of Chicago's largest, and contains an outdoor pool, running track, gardens, fire pits and yoga terrace. From below, Aqua's plinth quietly navigates the site's complexity by spanning over pre-existing elements such as an electrical substation, and by aligning with existing infrastructure, including an adjacent three-level roadway. The plinth physically connects pedestrian areas with stairs and elevators linking street level, to park level, to lakefront, and beyond.

Early and close collaboration between architect and builder, as well as the use of contemporary digital tools, allowed the variation in the shape of the floor slabs to be achieved without increasing the building's construction

RANDOLPH
Aon Center

timetable. The result is a site-specific, 82-story tower with no two floors alike.

Sustainability is a key component of the design. An east–west orientation maximizes its winter solar performance. Its balconies extend further on the southern facade to provide shading, reducing solar exposure in summer and allowing passive warming in winter. In addition to Low-E coatings on all glass, the design team modeled seasonal sun patterns to identify remaining areas of glass that needed higher performing glazing to increase energy efficiency throughout the tower. Glass on the east and south facades is reflective in areas without a protective balcony, while glass facing west has a tinted coating that improves its shading coefficient. In total, Aqua employs six different types of glass—clear, tinted, reflective, spandrel, and translucent—the placement of which is determined by the orientation and function of interior space. Fritted glass is also used and, combined with the balcony handrail design, used to minimize bird strikes.

Chillers and cooling towers have been eliminated from the project thanks to the availability of a district cooling system in downtown Chicago. Thermo Chicago pipes in water at 4.5 degrees Celsius from its central chillers, which is warmed as necessary by heat exchangers in the tower. A largely concrete structural frame has significant thermal mass, which minimizes internal temperature variations and reduces heating and cooling peak demands.

The building is also constructed on a former brownfield site, and 50 percent of its site is dedicated green open space, exceeding Chicago's standard zoning by 25 percent. Its green roof features a drainage system that captures rainwater for plant absorption. Furthermore, Aqua exceeds the City of Chicago's minimal requirements for natural ventilation and sunlight by more than 50 percent in over 90 percent of its spaces.

To diminish reliance on cars, canopied walkways lead visitors to the building's main entrance, while two grand public stairs bring pedestrians from Upper Columbus Drive down to a park at grade level, providing access to Chicago's downtown area and lakefront. The tower also connects to Chicago's extensive underground pedway system, linking users and residents to restaurants, retail, cultural activities, and jobs in the Loop and on the Magnificent Mile. Additional consideration was given, when designing the tower's garage exits below grade, to minimize congestion at pedestrian levels. To further reduce traffic and confusion, the garage's six levels have different access points that correlate with the tower's specific uses and users.

Structure

Next to gravity, the most demanding effects considered in the design of tall buildings are those associated with wind forces. For most buildings, providing enough strength to resist these forces is relatively simple. Of greater concern is the sway of a tower and the resulting impact on occupant comfort.

Aqua combines traditional concrete shear walls with outrigger and belt walls strategically located throughout the tower's height to effectively manage the building's motion. Two shear walls and a central concrete core extend from the tower's foundation to the roof. Outrigger walls are provided at Levels 55 through 57 and 81 through 82, with supplementary belt walls between Levels 57 and 58. The outriggers and belt walls activate all of the columns in the tower in resisting sway, while additional and detailed wind tunnel studies were completed, and confirmed that the undulating slab edges disrupted or "confused" the flow of wind around the tower, effectively reducing the wind demands.

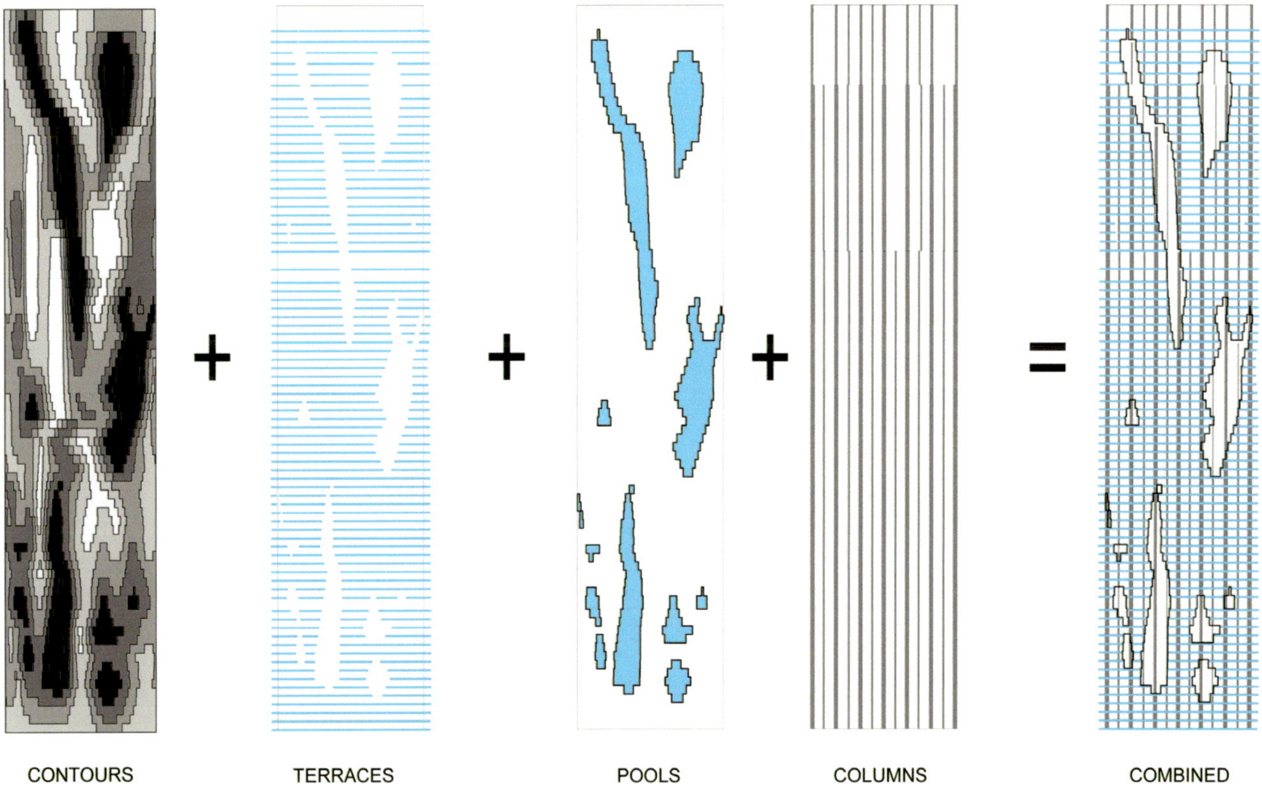
+
+
+
=
CONTOURS
TERRACES
POOLS
COLUMNS
COMBINED

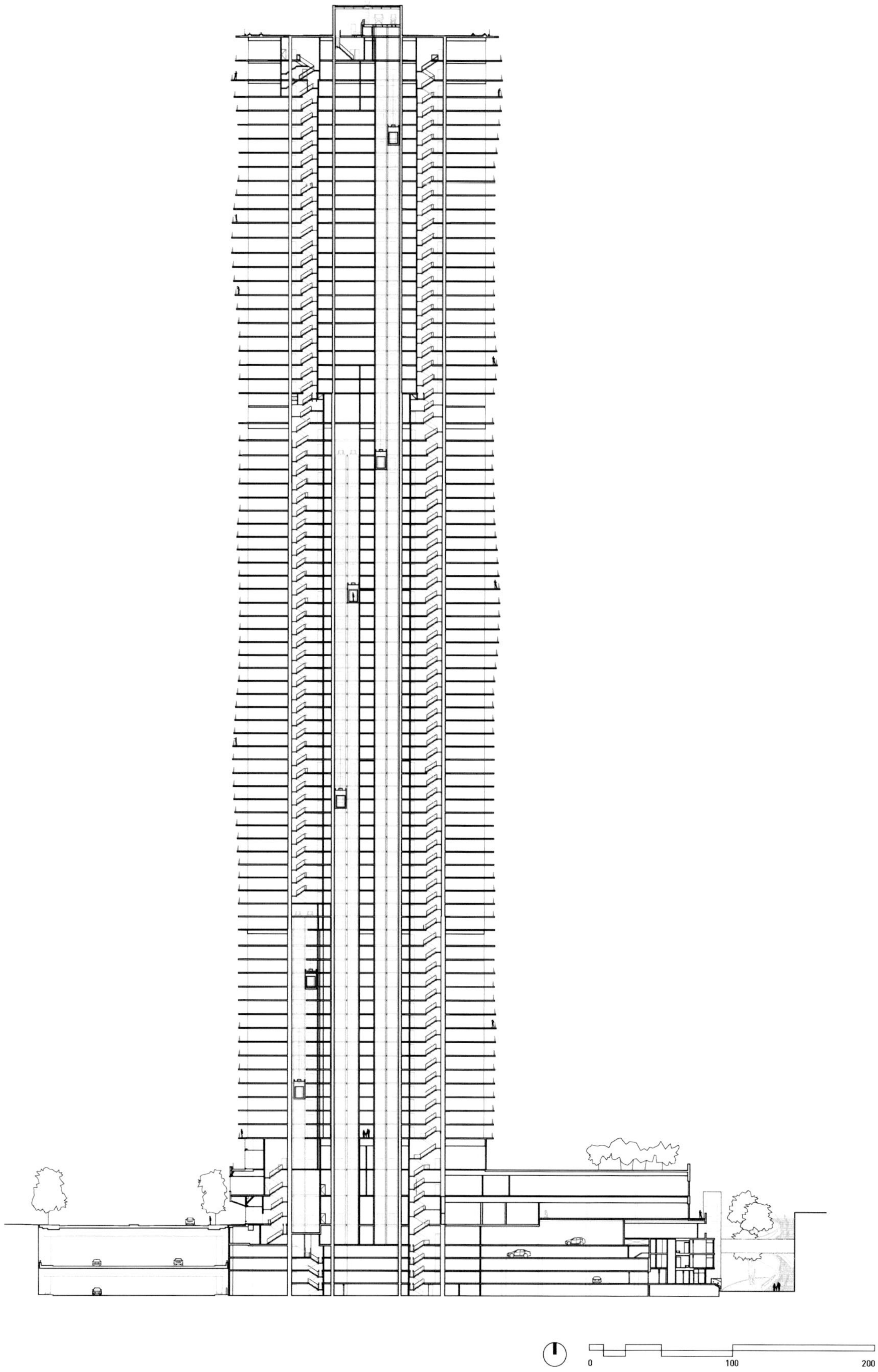

EAST-WEST SECTION

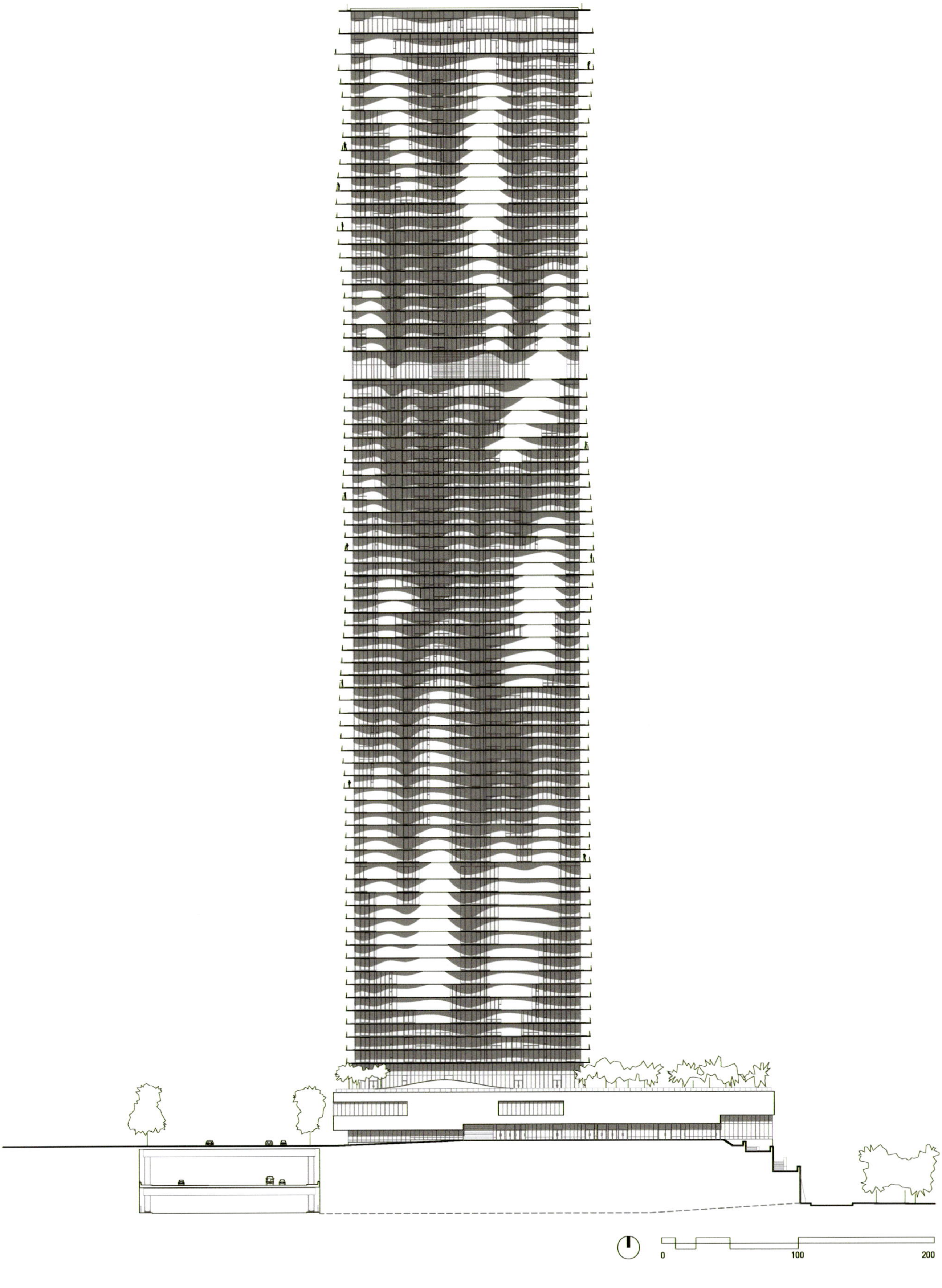
0
100
200

Case Study 2

Bank of America Tower

New York, USA

HEIGHT:	366 M
NUMBER OF STORIES:	55
GROSS FLOOR AREA:	195,098 M^2
PRIMARY USE:	COMMERCIAL
START ON SITE:	2004
COMPLETION:	2010
PROJECT COST:	$1.2 BILLION
OWNER/DEVELOPER:	BANK OF AMERICA AT ONE BRYANT PARK, LLC, A JOINT VENTURE BETWEEN THE DURST ORGANIZATION AND BANK OF AMERICA
LEAD ARCHITECT:	COOKFOX ARCHITECTS
ASSOCIATE ARCHITECT:	ADAMSON ASSOCIATES ARCHITECTS
STRUCTURAL ENGINEER:	SEVERUD ASSOCIATES
MEP ENGINEER:	JAROS BAUM & BOLLES
MAIN CONTRACTOR:	TISHMAN CONSTRUCTION CORPORATION
INTERIORS:	GENSLER

Project Description

Designed to set a new standard in high-performance buildings, both for the office workers who occupy the tower and for a nation awakening to the modern imperative of sustainability, the Bank of America Tower at One Bryant Park is a 55-story, 2.2 million-square-foot project and the first commercial high-rise to achieve LEED Platinum certification, the highest rating available from the US Green Building Council.

Drawing on concepts of biophilia, or people's innate need for connection to the natural environment, the vision was to create the highest quality modern workplace by emphasizing daylight, fresh air, and an intrinsic connection to the outdoors. At the urban scale, the tower addresses its natural setting as well as the context of midtown Manhattan, appearing to rise naturally from its iconic skyline.

Architecture

In response to its dense urban context, the building challenges the boundaries of public and private space with a highly transparent corner entry. A daylit and neutral space, the lobby creates a layered connection to the public realm of Bryant Park, whose restorative green spaces extend into the building through green roofs and the publicly accessible Urban Garden Room. Solid, natural lobby materials anchor the tower to the earth; small, tactile details such as white oak door handles, fossil-embedded Jerusalem stone, and leather paneling keep the massive tower intelligible to the human hand and eye.

As it rises from the uniform street grid, the massing of the tower shears into two offset halves, increasing the verticality of its proportions as well as the surface area exposed to daylight. From these two Cartesian volumes, mass is then sliced away, producing angular facets that open up oblique views around and through the urban forest of Midtown skyscrapers. The crystalline form—inspired by the legacy of the 1853 Crystal Palace which once stood adjacent in Bryant Park, and by a quartz crystal from the client's collection—suggests an appropriate natural analogue, at once organic and urban in nature. Through the building's exquisitely clear facade, the natural elements are experienced in an immediate and almost sensory way; from the outside, the facade changes with the sun and sky. On its southeast exposure, a deep double wall orients the building in its full height toward Bryant Park, its namesake and the most intensively used open space in the United States.

Structure

Concrete with 45 percent of its Portland cement content replaced by ground granulated blast-furnace slag is used for significant sections of the building's structure. At its heart is a lightweight structural steel core encased in reinforced concrete shear walls, with reinforced concrete slabs and beams framing the interior, including the stairs. Floors, by contrast, are concrete on metal deck permanent shuttering, framed by structural steel. Perimeter moment-resisting frames complete the lateral structural system.

Sustainability

One Bryant Park represents a shift in thinking about modern building design, achieving at scale many of the green building movement's most transformative ideas for water and energy conservation, material efficiency, and indoor environmental quality. It has fundamentally transformed the market for large-scale commercial developments, and practices implemented at One Bryant Park have changed the local landscape for contractors, suppliers and city agencies. Instead of depending solely on the grid, an onsite 4.6 megawatt natural gas-fired cogeneration (combined heat and power) plant provides a clean, efficient power source for nearly 70 percent of the building's annual energy requirements. Heat from the generators is used for space heating and domestic hot water; it also drives an absorption chiller as required for cooling. Recognizing its impact in the heart of a dense metropolis, 44 thermal ice storage tanks in the building's cellar produce ice at night, reducing the building's peak demand on the city's overtaxed electric grid. During the day, the melting ice supplements the building's cooling system.

This system was chosen after site monitoring indicated that wind conditions would be too gusty for the effective deployment of wind turbines, there would be too much shading for solar photovoltaic cells to function efficiently, and there was not enough space on the site for a geothermal (ground source energy) installation. Methane generation from an anaerobic digester utilizing tenants' waste paper was also dismissed, mainly on the grounds of document security.

Water-saving measures, including waterless urinals, gray water recycling, and rainwater harvesting systems, reduce the building's consumption by nearly 50 percent, saving millions of gallons of potable water per year.

With a state-of-the-art underfloor air system and 95 percent filtration, fresh air delivered to offices can be individually controlled and is actually cleaner when it is exhausted from the building. The building's floor-to-ceiling glass curtain wall minimizes solar heat gain through Low-E glass and heat-reflecting ceramic frit, while simultaneously providing exquisite views. The fitted pattern, which covers 60 percent of the glass where the curtain wall meets the floor and ceiling, gradually decreases in density toward the vision portion of each panel. Non-metallic spacers in the aluminum mullion system and extra mineral wool insulation at the floor slabs help achieve a U-value for the assembly of 0.38—a thermal resistance that is better than most glass towers built in New York City over the last decade. In the city defined by the modern skyscraper, the tower makes a highly visible statement on regeneration, urban stewardship, and global citizenship for the twenty-first century.

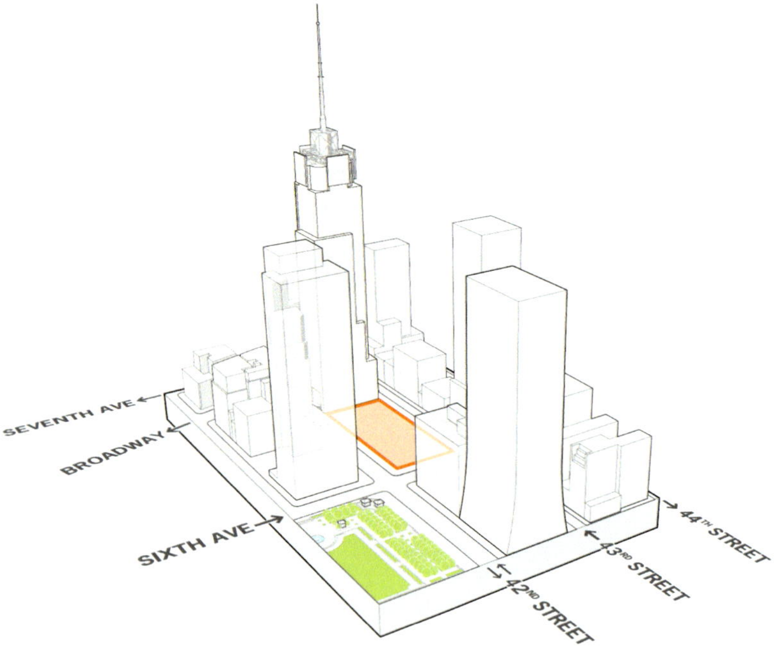
SEVENTH AVE
BROADWAY
SIXTH AVE
42ND STREET
43RD STREET
44TH STREET

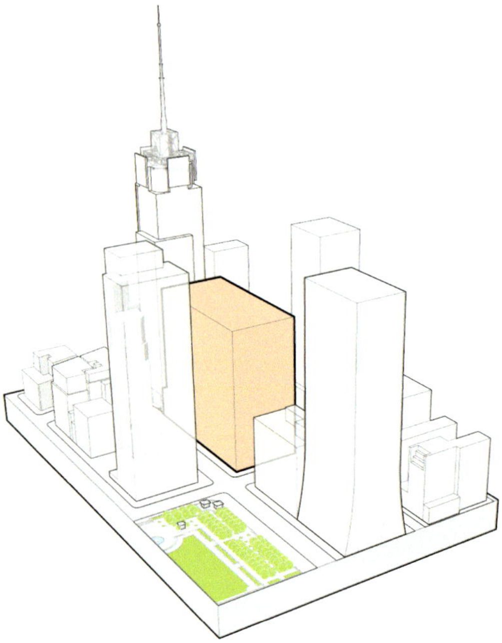

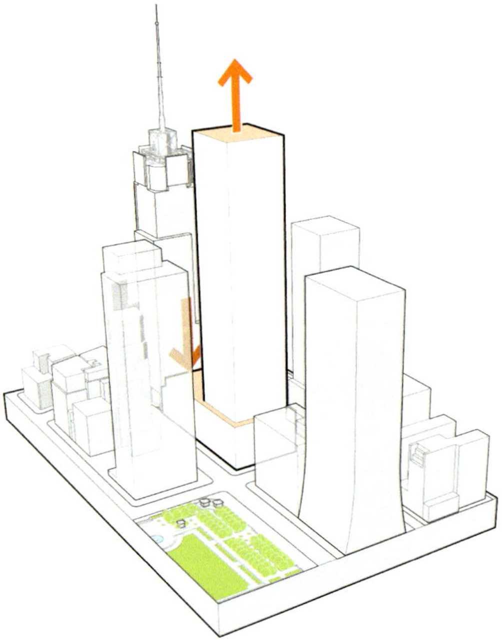

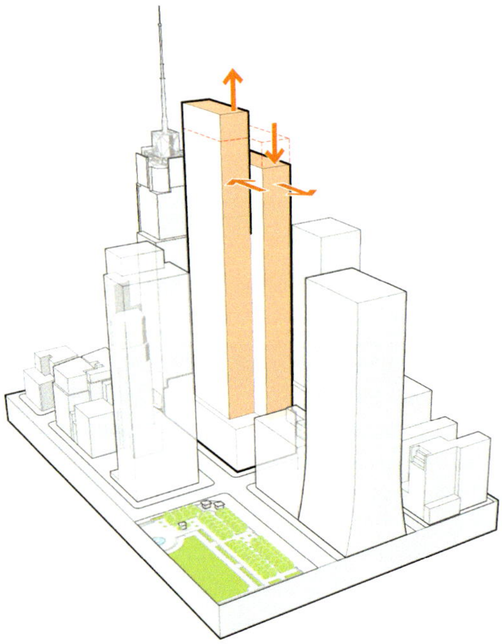

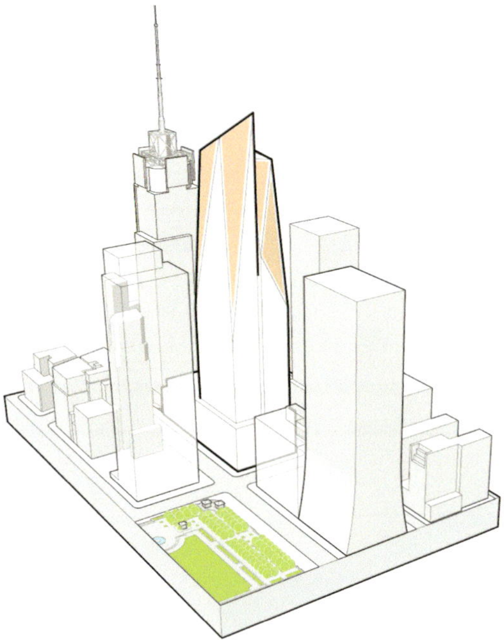

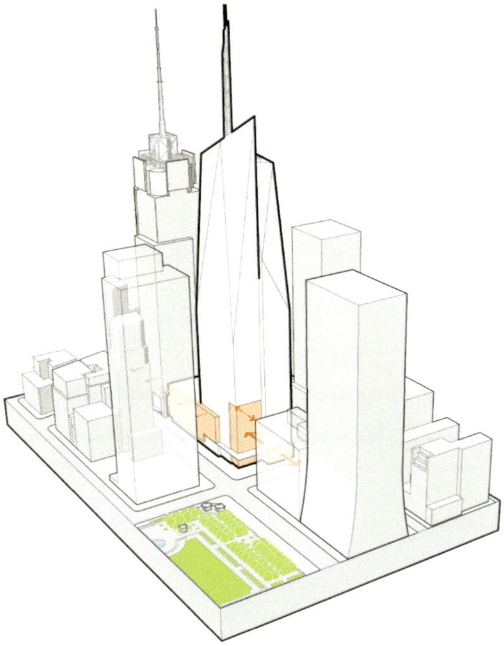

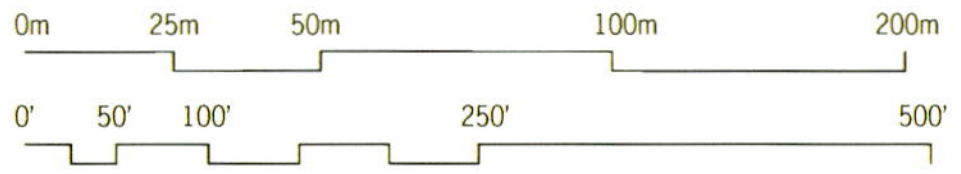

EAST-WEST SECTION
SCALE: 1/128" = 1'-0"

LEGEND

- OUTSIDE AIR
- FILTERED OUTSIDE AIR
- CONDITIONED AIR
- EXHAUST AIR
- CHILLED WATER LOOP
- CONDENSER WATER LOOP
- GLYCOL LOOP
- ELECTRICAL POWER
- NATURAL GAS
- HEAT

KEYED NOTES

1. 95% PARTICULATE AIR FILTER
2. AIR HANDLING UNIT ON EACH FLOOR
3. GAS TURBINE & GENERATOR
4. HEAT RECOVERY STEAM GENERATOR
5. ABSORPTION CHILLER
6. TRANSFORMER
7. ICE MACHINE
8. CHILLER
9. THERMAL STORAGE SYSTEM
10. HEAT EXCHANGER FOR WATER-SIDE FREE COOLING CYCLE
11. COOLING TOWERS

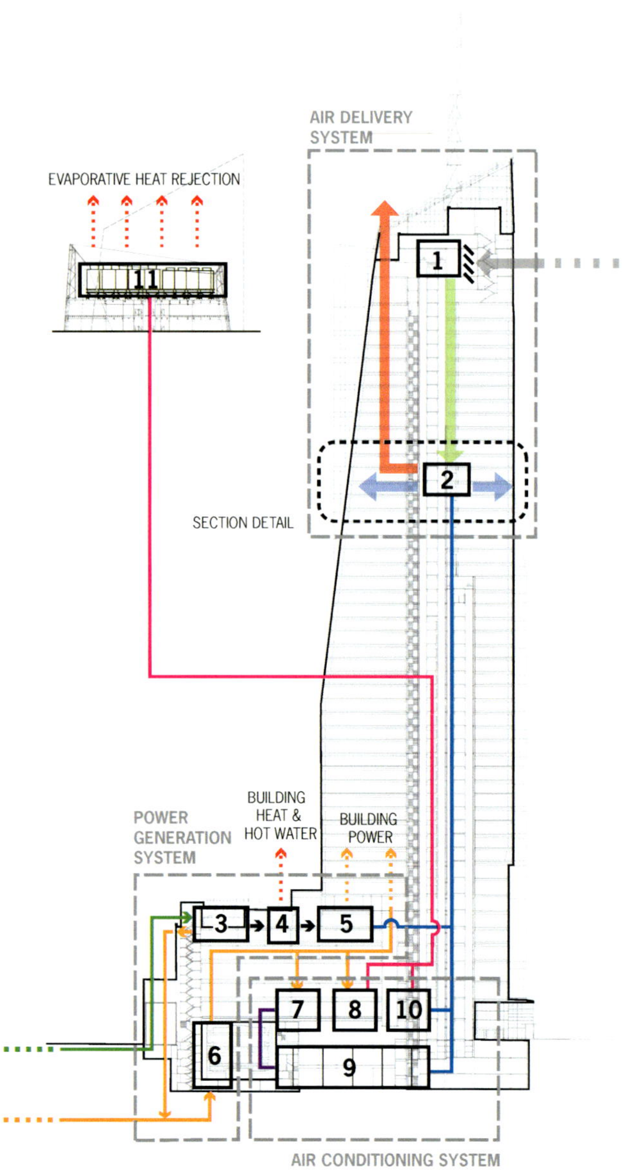

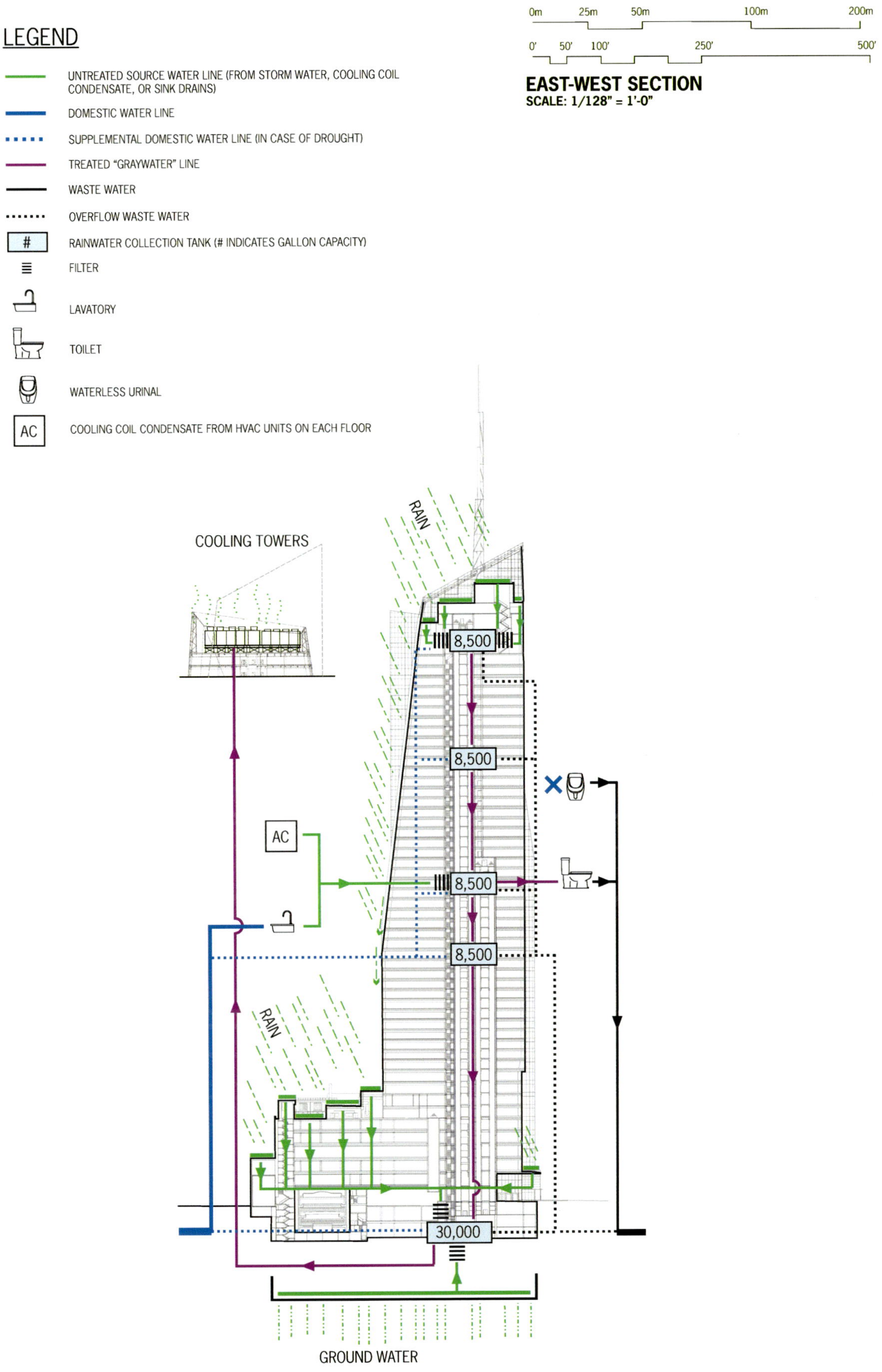
LEGEND
UNTREATED SOURCE WATER LINE (FROM STORM WATER, COOLING COIL CONDENSATE, OR SINK DRAINS)
DOMESTIC WATER LINE
SUPPLEMENTAL DOMESTIC WATER LINE (IN CASE OF DROUGHT)
TREATED "GRAYWATER" LINE
WASTE WATER
OVERFLOW WASTE WATER
#
RAINWATER COLLECTION TANK (# INDICATES GALLON CAPACITY)
FILTER
LAVATORY
TOILET
WATERLESS URINAL
AC
COOLING COIL CONDENSATE FROM HVAC UNITS ON EACH FLOOR
0m 25m 50m 100m 200m
0' 50' 100' 250' 500'
EAST-WEST SECTION
SCALE: 1/128" = 1'-0"
COOLING TOWERS
RAIN
8,500
8,500
AC
8,500
8,500
RAIN
30,000
GROUND WATER

Case Study 3

Hearst Tower

New York, USA

HEIGHT:	182 M
NUMBER OF STORIES:	46
GROSS FLOOR AREA:	79,500 M²
PRIMARY USE:	OFFICES
START ON SITE:	2003
COMPLETION:	2006
PROJECT COST:	NOT AVAILABLE
OWNER/DEVELOPER:	HEARST CORPORATION
LEAD ARCHITECT:	FOSTER + PARTNERS
ASSOCIATE ARCHITECT:	SHELL AND CORE: ADAMSON ASSOCIATES ARCHITECTS; FIT-OUT: GENSLER
STRUCTURAL ENGINEER:	CANTOR SEINUK GROUP
MEP ENGINEER:	FLACK+KURTZ
MAIN CONTRACTOR:	TURNER CONSTRUCTION
LIGHTING:	SHELL AND CORE: GEORGE SEXTON; FIT-OUT: KEUGLER ASSOCIATES

Project Description

Hearst Tower revives a dream from the 1920s, when publishing magnate William Randolph Hearst envisaged Columbus Circle as a new media quarter in Manhattan. Hearst commissioned a six-story Art Deco block on Eighth Avenue, anticipating that it would form the base for a tower, which was never realized. Echoing an approach developed in the Reichstag and the Great Court at the British Museum, the challenge in designing such a tower at seventy years remove was to establish a creative dialogue between old and new.

The original "U"-shaped structure was commissioned to house the twelve magazines William Randolph Hearst owned at the time. Construction began in 1927 and was completed in 1928 at a cost of $2 million. Originally the 3,700-square-meter (40,000-square-foot) building was named the International Magazine Building. Built in an unusual style, the building is classified as outside the Art Deco norm of the times and is a combination of multiple styles.

The facade is cast stone with a two-story base and four stories set back from the base. The design consists of columns and allegorical figures representing music, art, commerce, and industry. From the beginning, the building was structurally reinforced for an office tower and plans were filed for nine additional stories in 1946, but never executed. Considered an "important monument in the architectural heritage of New York," by the New York City Landmarks Preservation Commission, the building was designated a Landmark Site in 1988.

Architecture

The 42-story tower rises above the old building, linked on the outside by a skirt of glazing that encourages an impression of the tower floating weightlessly above the base. A lobby that occupies the entire floor plate of the old building and rises up through six floors is the main spatial event. Like a bustling town square, this dramatic space provides access to all parts of the building. It incorporates the main elevator lobby, the Hearst cafeteria and auditorium, and mezzanine levels for meetings and special functions.

The building is also significant in environmental terms. It was built using 90 percent recycled steel and is designed to consume 26 percent less energy than its conventional neighbors. As a result, it was the first office building in Manhattan to achieve a gold rating under the US Green Buildings Council's Leadership in Energy and Environmental Design (LEED) program. As a company, Hearst places a high value on the quality of the working environment—something it believes will become increasingly important to its staff in the future—and it is hoped that Hearst's experience may herald the construction of more environmentally sensitive buildings in the city.

Apart from the very high proportion of recycled steel, foreign-sourced materials and labor account for less than 10 percent of the total construction cost. The diagrid frame of the tower contains roughly 20 percent less steel than would a conventional perimeter frame, saving approximately 2,000 tons of steel. High-performance, low-emission glass set in diagrid form allows for internal spaces to be flooded with natural light, while keeping out solar radiation causing heat gain. Light sensors control the amount of artificial light on each floor based on the amount of natural light available at any given time; motion sensors allow for lights and computers to be turned off when a room is vacant.

High-efficiency heating and air-conditioning equipment utilize outside air for cooling and ventilation for 75 percent of the year. These and other energy-saving features, such as CO_2 sensors for demand-controlled ventilation, are expected to increase energy efficiency by 26 percent compared to a standard office building.

The roof has been designed to collect rainwater, which will reduce the amount of water dumped into the city's sewer system during rainfall by 25 percent. Harvested rainwater is stored in a 64-cubic-meter (14,000-gallon) reclamation tank located in the basement of the building, and is used to replace water lost to evaporation in the office air-conditioning system. It is also fed into a special pumping system to irrigate plantings and trees inside and outside the building. It is expected that the captured rain will produce about half of the watering needs. This water is also utilized for "Icefall," the grand atrium's water feature, which has the environmental function of humidifying and chilling the atrium lobby as necessary. In addition, electrically actuated faucets are expected to reduce potable water use by 25 percent.

Structure

A major design requirement was to preserve the six-story landmark facade and incorporate it into the new tower design. The new design incorporates two underground levels and a seven-story-high interior atrium, formed by a horizontal skylight system spanning approximately 12 meters from the tower columns to the existing facade. The tower houses offices from the 10th to the 46th floors above the sky-lit lobby atrium, an auditorium, a cafeteria, and public spaces.

Apart from the facade on three exterior faces of the building, all existing construction was removed. New foundations for the tower were positioned behind the facade, which is supported by its original perimeter steel columns and spandrel beams, and a new additional grid of vertical and horizontal framing.

A steel core is used, with a perimeter diagonal structural system (diagrid) forming four-story triangular frames. Concrete-reinforced steel supercolumns reach up to the 10th floor. The diagrid system developed by the design team WSP wraps around all four faces of the tower, providing increased structural stability and giving the tower its distinctive faceted appearance. Each triangle in the diagrid is 16.5 meters (54 feet) tall. This highly efficient structural system consumed 20 percent less steel than conventional moment frame structures. Composite steel and concrete floors are used, with 12-meter (40-foot) interior column-free spans for open-plan offices.

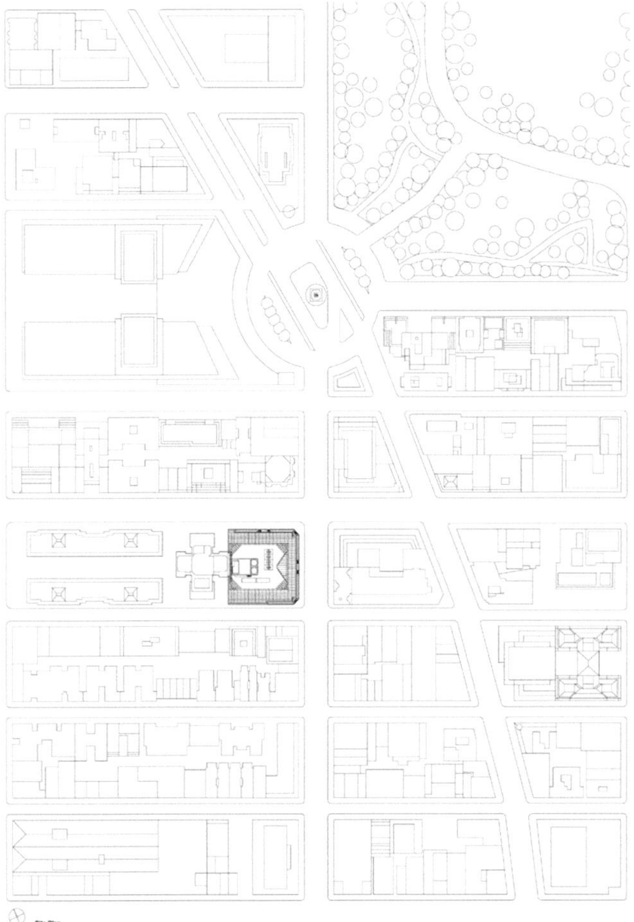

Site Plan

Elevation

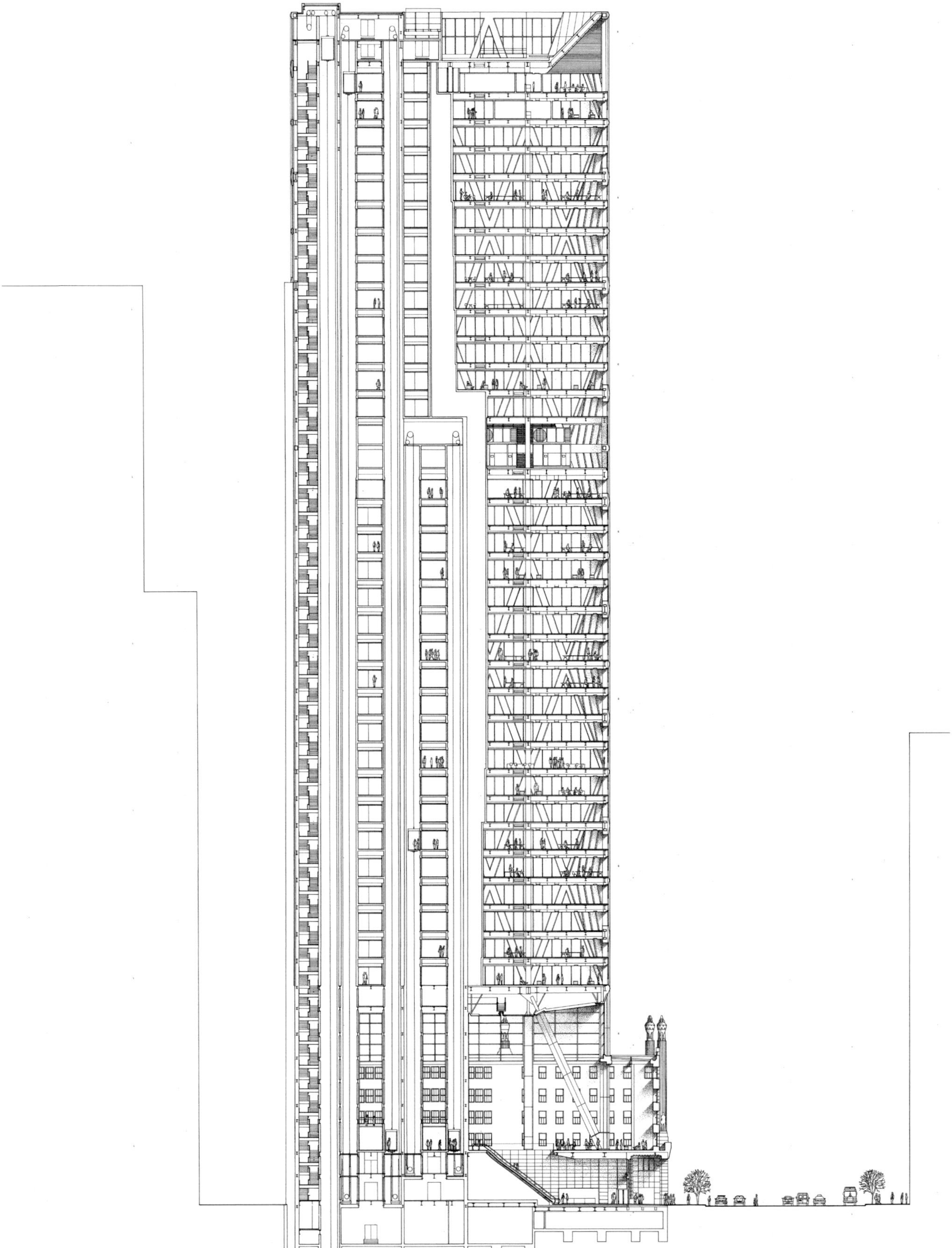

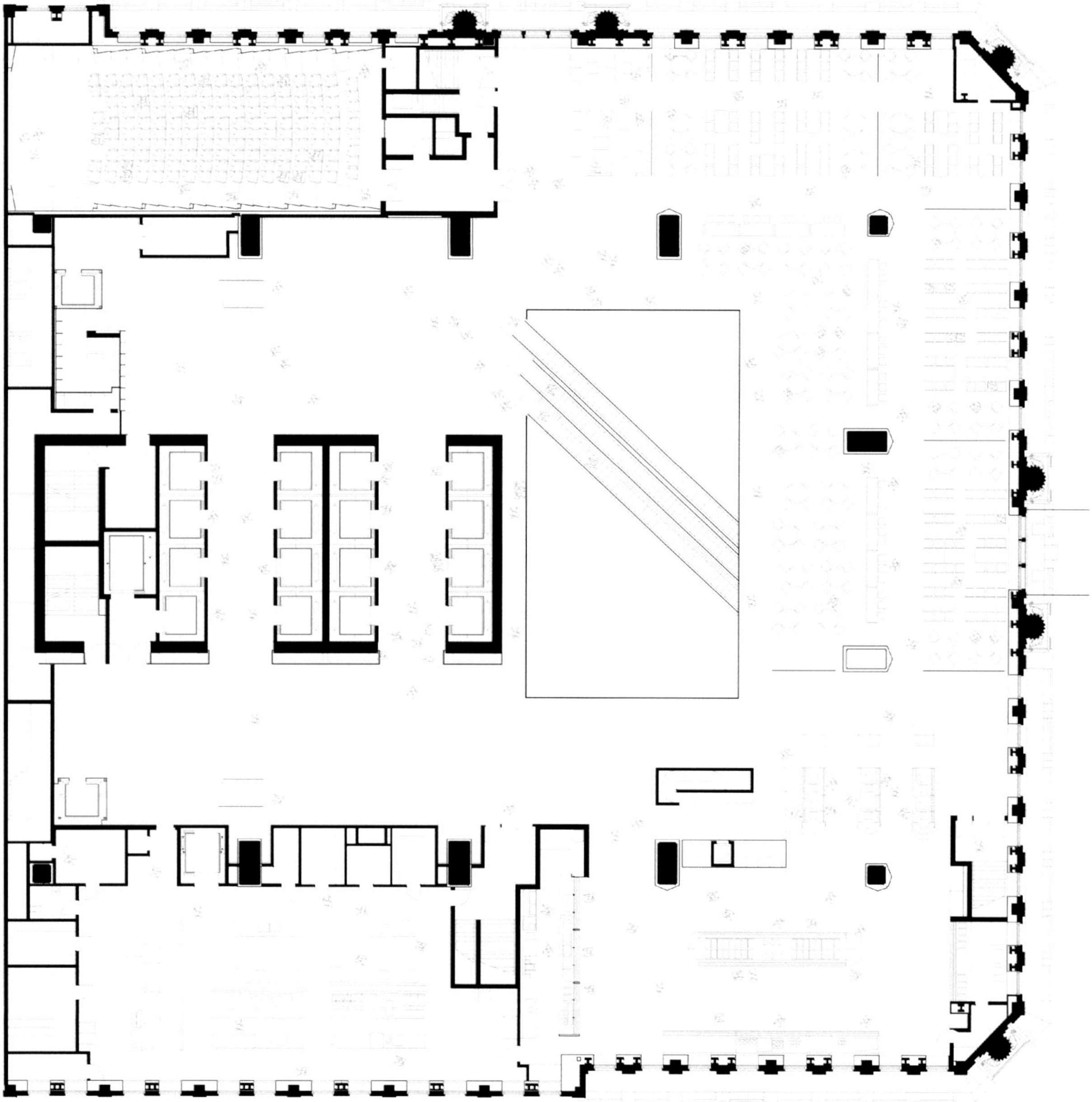

0 5 10

0 30

Lobby Level Plan

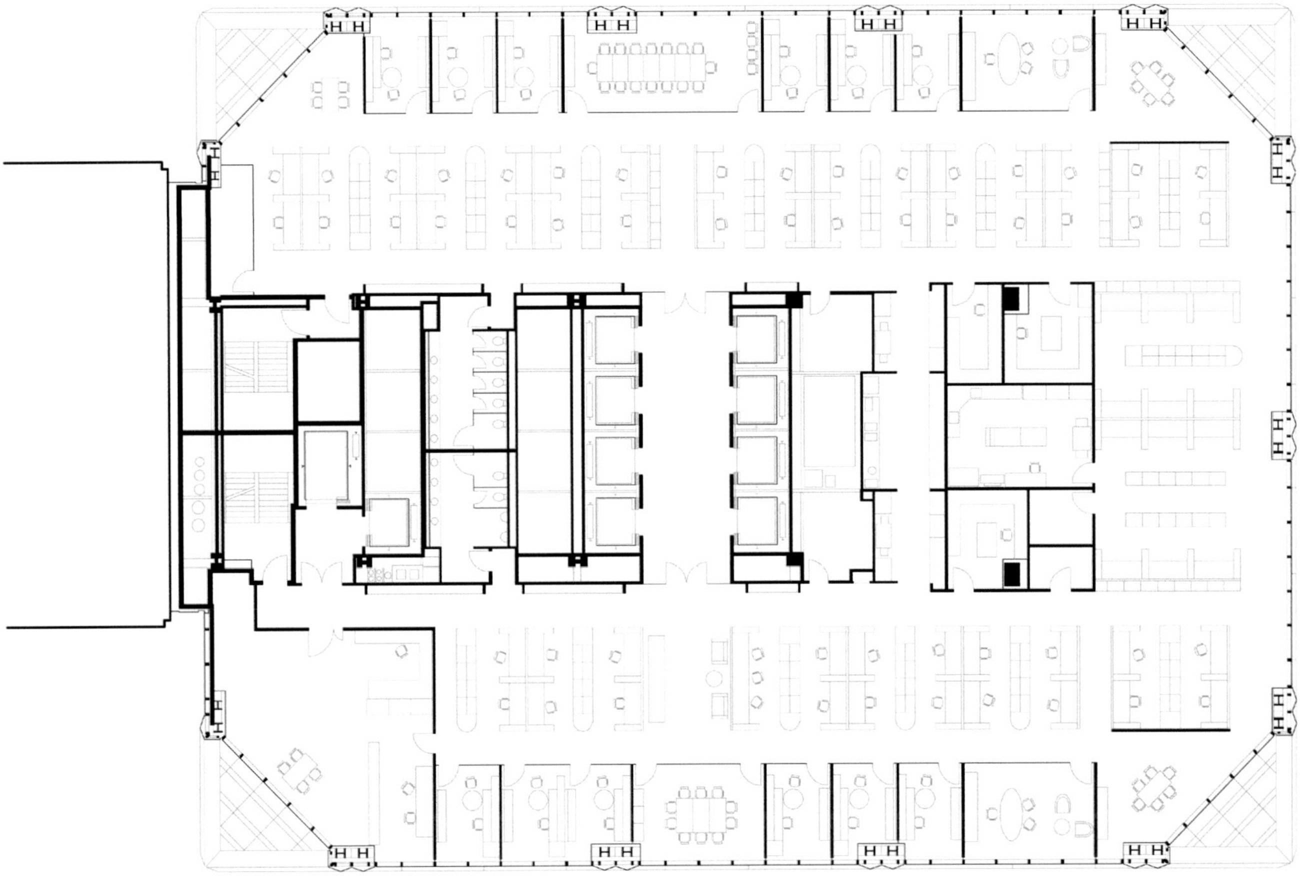

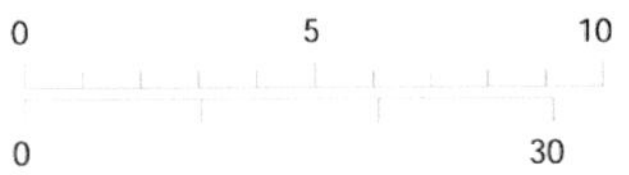

Typical Plan

Case Study 4

Manitoba Hydro Place

Winnipeg, Canada

HEIGHT:	88.6 M (WITH ROOF CANOPY 98.6 M AND 115.45 M TO TOP OF SOLAR CHIMNEY)
NUMBER OF STORIES:	22
GROSS FLOOR AREA:	64,590 M^2
PRIMARY USE:	OFFICES
START ON SITE:	2005
COMPLETION:	2009
PROJECT COST:	$283 MILLION
OWNER/DEVELOPER:	MANITOBA HYDRO
LEAD ARCHITECT:	KUWABARA PAYNE MCKENNA BLUMBERG ARCHITECTS (KPMB)
ASSOCIATE ARCHITECT:	SMITH CARTER ARCHITECTS AND ENGINEERS
STRUCTURAL ENGINEER:	CROSIER KILGOUR & PARTNERS LTD. / HALCROW YOLLES
CIVIL ENGINEER:	MTE CONSULTANTS INC
MEP ENGINEER:	AECOM (FORMERLY EARTH TECH)
MAIN CONTRACTOR:	J.D. STRACHAN CONSTRUCTION LTD
CONSTRUCTION MANAGER:	PCL CONSTRUCTORS INC
LANDSCAPE CONSULTANTS:	HILDERMAN THOMAS FRANK CRAM / PHILLIPS FAREVAAG SMALLENBERG
CLIMATE ENGINEERS:	TRANSSOLAR ENERGIETECHNIK GMBH
QUANTITY SURVEYORS:	HANSCOMB LTD

Project Description

In 2002, as part of the negotiated purchase of Winnipeg Hydro by Manitoba Hydro, the City of Winnipeg stipulated that Manitoba Hydro construct a new office tower to bring 2,000 employees into the downtown as part of its urban revitalization strategy. Manitoba Hydro Place now occupies an entire urban block with a front address on Portage Avenue, the widest thoroughfare in Canada, equivalent to the scale of Michigan Avenue in Chicago. The site was selected for its direct connection to the city's transit system and raised walkway system.

Winnipeg is a city characterized by an extreme cold and hot climate, where outdoor temperatures swing from −35 to +35 degrees Celsius. During winter months, the city also experiences an abundance of sunlight and strong southerly winds. At first glance the new building looks like a classic, modern, glass office tower. In actuality, it is one of North America's most energy-efficient, large-scale buildings, representing a new generation of bioclimatic architecture, with goals of 60 percent energy efficiency, urban revitalization, a healthy workplace, and architectural excellence. What distinguishes Manitoba Hydro Place is that it achieved these goals in an extreme climate (ASHRAE Zone 7). Submitted for LEED Platinum Certification, it is on track to outperform its original 60 percent energy reduction goal of 90 kWh/m^{2}a (28.5 kBtu/ft^{2}a).

Architecture

Form, orientation, and massing capitalize on Winnipeg's extreme climate for passive energy, while creating a new public icon for Winnipeg. A "Capital-A" form comprises two office towers set on a three-story podium. Transparent glazing systems help to mitigate the overall mass and scale of the building on the streetscape. The towers converge at the north and splay open to the south for maximum

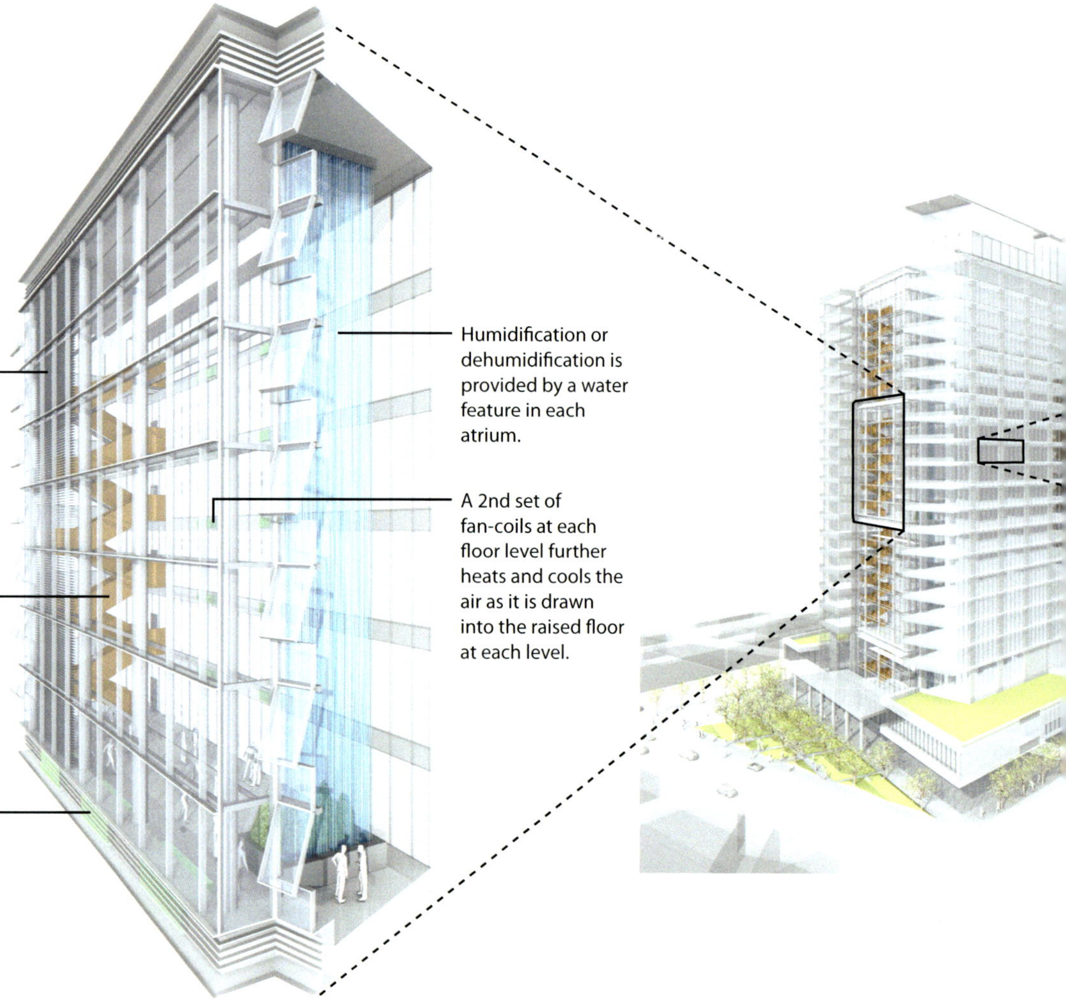

exposure to the abundant sunlight and consistently robust southerly winds unique to Winnipeg's climate. At the north end the 115-meter tall solar tower marks the identity of Manitoba Hydro on the skyline, and the main entrance to the building on Portage Avenue. The tower setbacks also mitigate shadow impact on Portage, the city's historic main shopping street. Siting the south end of the building on a 45-degree angle also created space for a new urban park on the Graham Street transit corridor.

A publicly accessible galleria that offers a sheltered pedestrian route through the full city block bisects the podium and represents one of the major urban design initiatives. Originally the program called for a "lobby," but in the course of the IDP (integrated design process) the architects proposed the generous three-story space as a way of both organizing the daily flow of 2,000 employees and addressing urban revitalization goals by providing the city with a new indoor public gathering space.

By organizing the office program into two separate towers and introducing multi-floor stacked atriums, the floor plate depth was kept shallow to guarantee natural daylight penetration deep into the building and access to views throughout. The energy-efficient design, which prioritizes the delivery of fresh air 24/7 and maximizes daylight to the work space, has made people appreciate the benefits of the extreme climate, particularly the abundant sunlight through the winter and the calming effects of strong southerly winds. The floor-to-ceiling windows and the south-facing views from the winter gardens allow previously unimagined views of Winnipeg's grand historic fabric, the promise of its future, with new buildings such as the Canadian Museum of Human Rights in the near distance, and Manitoba's vast blue skies and prairie horizon beyond.

Structure

Caissons socketed into rock at depths of between 40 and 100 feet (12 and 30.5 meters) support the main concrete frame, in which part of the Portland cement content was replaced by pulverised fuel ash. In situ concrete floors, typically 190 millimeters thick, had hydronic tubing cast in for radiant heating and cooling—these were utilized

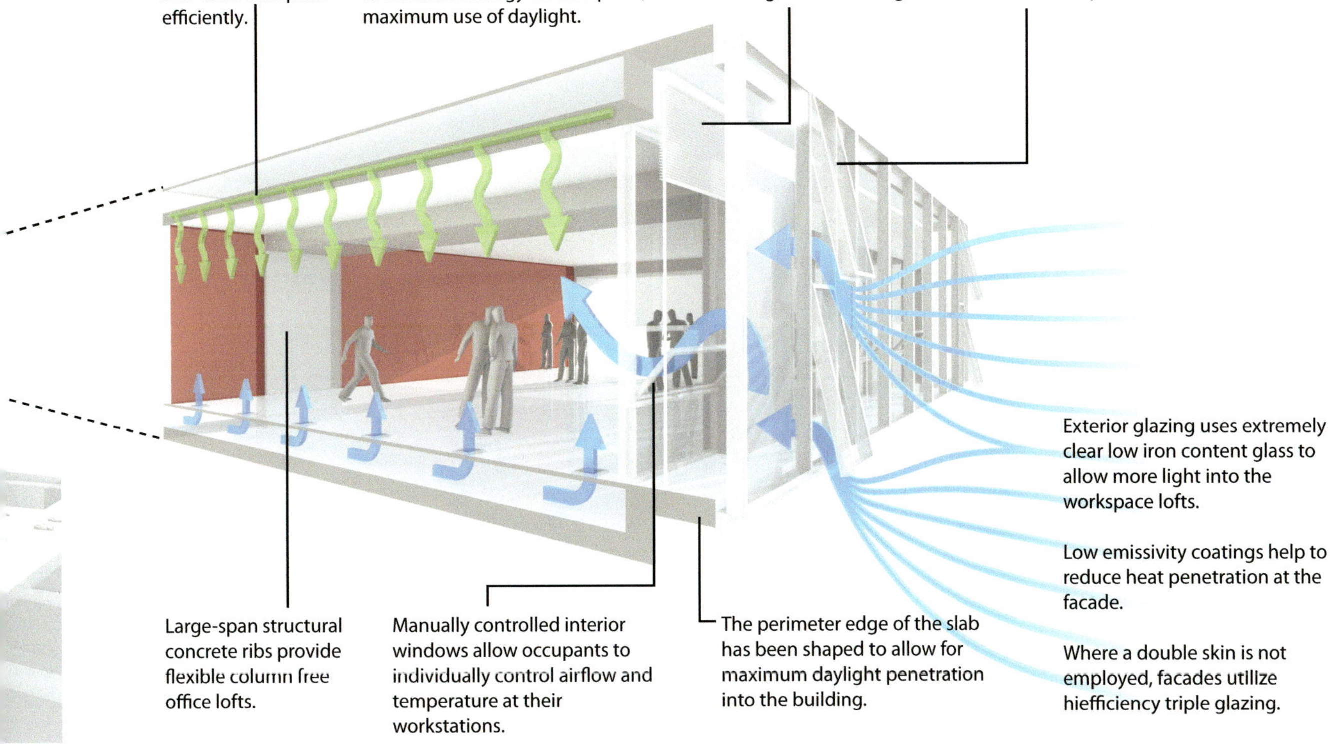

to circulate heated water through the freshly poured concrete during the worst of the winter weather. Steel was used internally for the atriums and the stairways.

Sustainability

Ironically, a glass tower in the extreme climate proved the most effective solution. When it is extremely cold it is also very sunny, ideal for solar gains. "Triple glazing" is used on all surfaces. The envelope is delaminated into single- and double-glazed walls, with a buffer zone between. Between the two walls, temperatures fluctuate naturally for most of the winter months, maintaining the performance of a triple-glazed facade. While buffer zones are configured in the winter for thermal insulation and fresh-air heating (in the case of the south atrium), their configuration changes with the seasons. One of the most recognisable architectural features of the building is the solar chimney at the northern end. At 115 meters high, the chimney is critical to the passive ventilation system, relying on the "stack effect" to draw air out of the building.

A biodynamic double facade, on the west and east faces of the building, creates a high-performance envelope which reduces heating and cooling loads by providing a tempered buffer to extreme outdoor temperatures. Operable windows on the inner and outer walls of the double facade permit natural ventilation at seasonally appropriate times.

Three stacked, six-story high winter gardens facing due south act as large, unconditioned spatial volumes. Unique in the context of the conventional hermetically sealed North American office building, they perform as the "lungs" of the building and work in combination with the solar chimney to provide 100 percent fresh air every day of the year. Each winter garden features a 23-meter tall water curtain composed of 280 Mylar ribbons, which condition the air before it enters the building.

A closed-loop geothermal (ground source energy) system consisting of 280 boreholes, 400 feet (125 meters) deep, provides approximately 60 percent of the heating, with highly energy efficient condensing boilers providing the balance during the coldest months. Primary energy sources are hydroelectric power and

natural gas. Thermal mass is high, the in situ concrete floor slabs provide radiant heating and cooling, and heat recovery is used to preheat incoming fresh air.

A Building Management System (BMS) with over 25,000 observation points enables employee connection and feedback to control lighting and solar shading, and thermal comfort. It also monitors local climate and uses the data to adjust set points automatically (floor slab temperatures, operable windows, solar shading, etc.).

Beyond energy conservation, the glazed envelope system is also critical to creating a highly supportive, comfortable, and healthy workplace. Everyone has access to the facade and receives natural lighting during 80 percent of normal office hours. Occupants control their personal environments using operable windows, task lighting, and shading devices.

Potable water demand is minimized by the use of low-flow fixtures and waterless urinals.

Conclusion

Since opening, the building has catalyzed significant cultural and civic shifts. In the previous suburban offices of Manitoba Hydro, 90 percent of employees drove alone to work; now 90 percent are taking public transit. Employees are enjoying the collaborative, open workspace, fresh air and views, and opportunities to meet and interact with colleagues across departments. The influx of 2,000 employees from the suburbs to downtown Winnipeg is already having an impact on the economy and, more importantly, on civic pride.

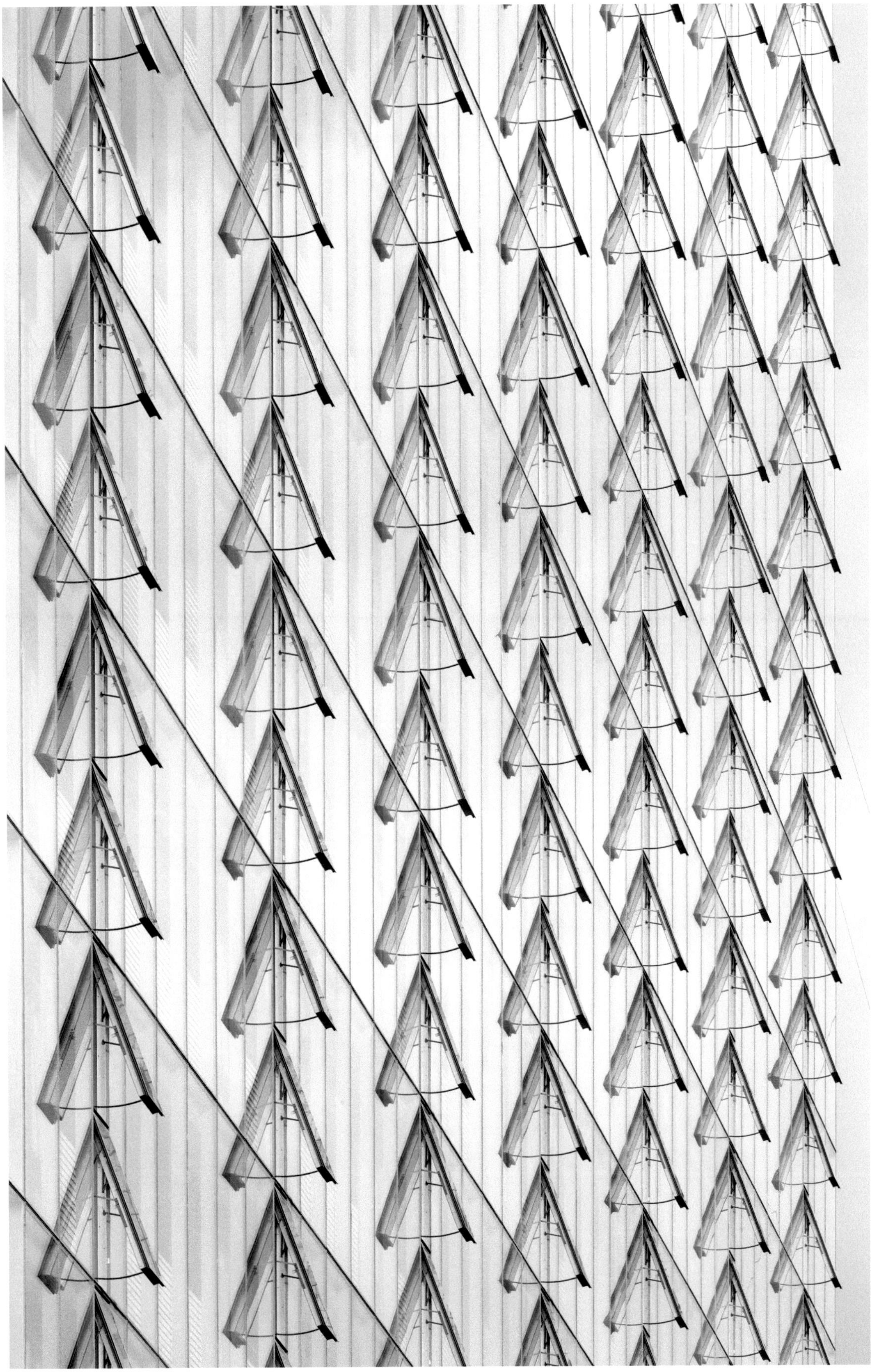

CONTAINED SPACES
8'-0" demountable solid & glass partitions

ENCLOSED SPACES
8'-0" demountable solid & glass partitions, with glass transoms to ceiling

ATRIA
north / south multiple floor atriums, with communication stairs

CORE
elevators, services, & storage

OPEN WORK STATIONS
50" tall panels

MANITOBA HYDRO
TYPICAL TOWER FLOOR
- east and west faces offer double skin facade with occupant controlled natural ventilation
- floor to ceiling height of 3.31m (10'-10")
- floor to ceiling glazing, typical
- overhead radiant heating and cooling
- air and services distributed through raised access floors

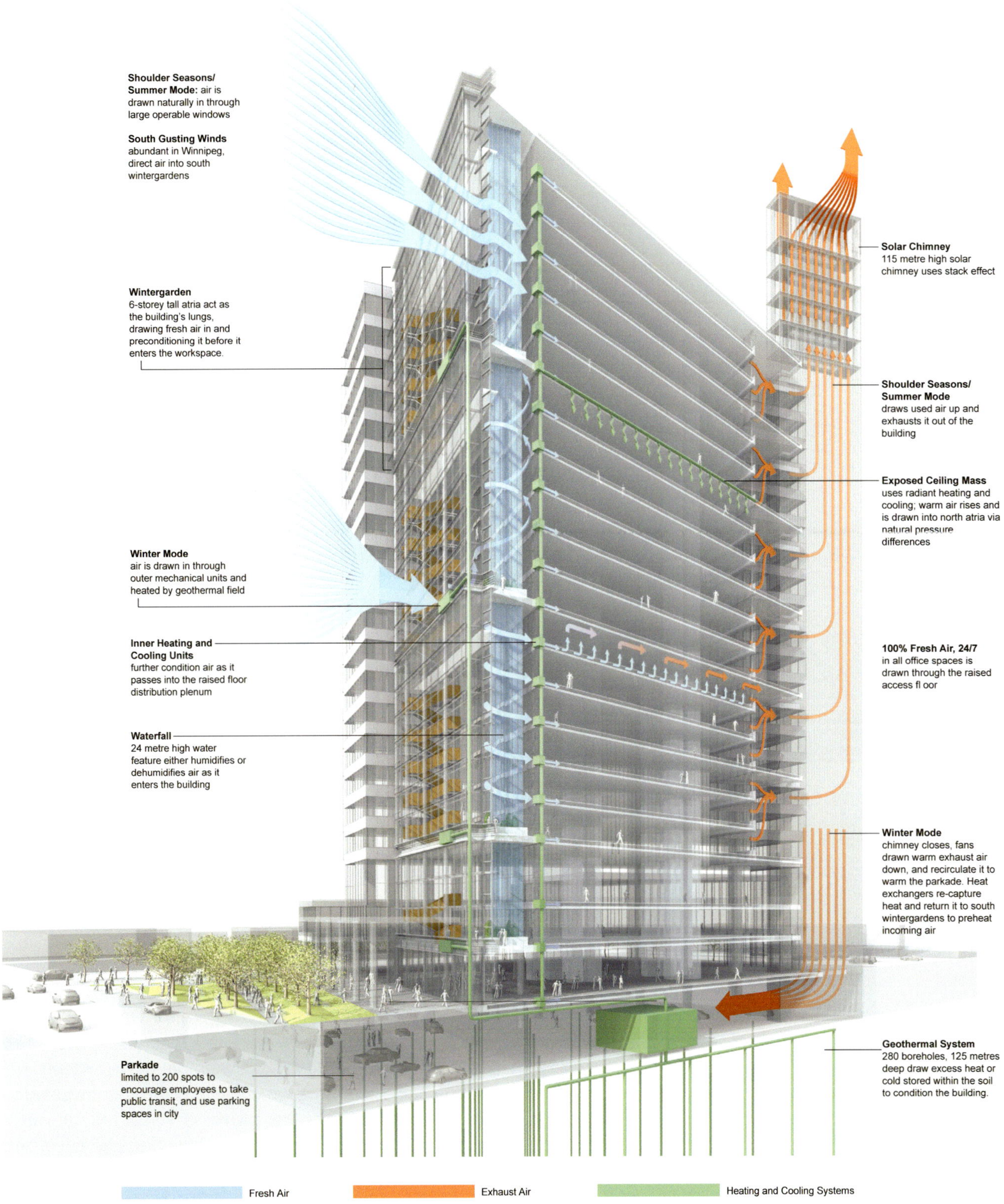
Shoulder Seasons/ Summer Mode: air is drawn naturally in through large operable windows
South Gusting Winds abundant in Winnipeg, direct air into south wintergardens
Wintergarden 6-storey tall atria act as the building's lungs, drawing fresh air in and preconditioning it before it enters the workspace.
Winter Mode air is drawn in through outer mechanical units and heated by geothermal field
Inner Heating and Cooling Units further condition air as it passes into the raised floor distribution plenum
Waterfall 24 metre high water feature either humidifies or dehumidifies air as it enters the building
Parkade limited to 200 spots to encourage employees to take public transit, and use parking spaces in city
Solar Chimney 115 metre high solar chimney uses stack effect
Shoulder Seasons/ Summer Mode draws used air up and exhausts it out of the building
Exposed Ceiling Mass uses radiant heating and cooling; warm air rises and is drawn into north atria via natural pressure differences
100% Fresh Air, 24/7 in all office spaces is drawn through the raised access fl oor
Winter Mode chimney closes, fans drawn warm exhaust air down, and recirculate it to warm the parkade. Heat exchangers re-capture heat and return it to south wintergardens to preheat incoming air
Geothermal System 280 boreholes, 125 metres deep draw excess heat or cold stored within the soil to condition the building.
Fresh Air
Exhaust Air
Heating and Cooling Systems

Case Study 5

One Madison Park

New York, USA

HEIGHT:	620 FT (189 M)
NUMBER OF STORIES:	50
GROSS FLOOR AREA:	163,500 FT^2 (15,190 M^2)
PRIMARY USE:	RESIDENTIAL
START ON SITE:	2006
COMPLETION:	2014
PROJECT COST:	NOT AVAILABLE
OWNER/DEVELOPER:	SLAZER ENTERPRISES/ THE RELATED COMPANIES
LEAD ARCHITECT:	CETRARUDDY (CETRA/ CRI ARCHITECTURE PLLC)
ASSOCIATE ARCHITECT:	N/A
STRUCTURAL ENGINEER:	WSP CANTOR SEINUK
MEP ENGINEER:	MG ENGINEERING
MAIN CONTRACTOR:	BOVIS LEND LEASE
LANDSCAPE:	HM WHITE

Project Description

Taking cues from the Metabolist movement of the late 60s and early 70s, One Madison Park employs a modular, plug-in design concept articulated through "pods" cantilevered to the north and east of the main tower shaft, providing residents with 360-degree views of New York City and beyond. Inspiration for the design team came from architectural influences such as Habitat (1967), NYU University Plaza (1967), and the Nakagin Capsule Tower (1972). Other influences include One Madison Park's historical neighbors, the Flatiron building (1902) to the west and the Met Life tower (1909) to the northeast. Designed to enhance its urban context, One Madison Park counterbalances the Met Life tower, creating a graceful, linear progression between the three towers.

Creating a building that transforms the Manhattan skyline is a rare opportunity for an architect and a developer. One Madison Park's developer purchased all the remaining air rights on lots to the south, east and west of the mid-block site, located at the base of Madison Avenue, enabling him to build a taller tower. At 620 feet (189 meters) high with a 5,175-square-foot (480 square-meter) footprint and 50-foot-wide (15 meter) base, One Madison Park is one of New York City's tallest and most slender skyscrapers, with a slenderness ratio of 12:1. Its 360-degree views are preserved through the purchased air rights, affording the 165,000-square-foot (15,329-square-meter) tower dominion over its neighbors.

Facing Madison Square Park, at the intersection of two major thoroughfares, 23rd Street and Madison Avenue, One Madison Park is uniquely situated on the Manhattan grid. Like Grand Central Station, the Time Warner Center, and the Arch at Washington Square Park, One Madison Park's axial position on Madison Avenue makes it visible from great distances at the street level, and bestows on the building an unusual prominence and stature.

The Brief

The developer set high expectations for One Madison Park, intending to create a premier luxury residence that would rival the top condominium developments in Manhattan. Adding to the challenge, the client wanted to begin construction on the tower before the full site was available, requiring a phased construction sequence. Site constraints limited the building's floor plate, necessitating a clever design that would maximize space and views.

The desired unit mix was intended to create a variety of apartments, ranging from 930-square-foot (86.4-square-meter), half-floor, one-bedroom units to

3,310-square-foot (307.5-square-meter), full-floor, three-bedroom units, and a 4,267-square-foot (396-square-meter) triplex penthouse. In addition, the property would feature luxurious amenities, including a spa, swimming pool, fitness center, media room, wine cellar, and outdoor terrace park. To create a serene and more intimate entry, residents would enter the building on the predominately residential 22nd street. This configuration would also allow for a ground floor commercial space to complement the existing retail activity of 23rd street.

Design Concept

Seven volumetric pods, constructed from clear white and green glass, delineate the tower's modular form. These cube-like forms deconstruct the building's mass and give it a sense of lightness. Unlike typical residential and commercial buildings, this modular concept creates varied units and a combination of effects, allowing for a unique and expressive building form. Cantilevered from the main shaft, the pods lend functionality beyond their aesthetic appeal, extending the tower's 2,700-square-foot (250-square-meter) floor plate to 3,300 square feet (307 square meters) and modulating the building form. This unique feature grew out of the need to expedite construction on the main tower, and from the desire to offer a mix of apartment types, some with terraces.

In designing the apartment layouts, lateral bracing was placed in the center instead of around the perimeter. The structural design forms a cruciform of shear walls buried between rooms and shafts, minimizing the impact on room layouts. This integration of efficient space planning and engineering gives each room an open, loft-like feel with an outward-radiating energy. Floor-to-ceiling glass walls draw the eye outwards to the city, creating a dramatic and picturesque backdrop that changes with the seasons. Designed to extend the visual connection of each home to the city outside, each apartment offers multiple places where southern and northern views can be seen simultaneously. Typical floors feature 360-degree views, with unique panoramas from each room. Looking out at the southern tip of Manhattan, or across to Long Island or New Jersey, residents will feel a special intimacy with the city, as if watching the world from their own sky-high castle.

Part of the challenge in developing the design, materials, and aspects of the structure was to create a modern building that was in concert with the fabric of Madison Square Park and the classic architecture of the neighborhood. The main tower shaft is clad in earth-toned bronze glass, enabling the tower to blend with the older surrounding limestone and masonry buildings. Similar in proportion, One Madison Park and the Met Life tower rise above the neighborhood, creating a dialogue between traditional classicism and contemporary architecture. Together they are visual and physical gateposts to Madison Park, anchoring the southern and eastern ends of the square. This physical dialogue becomes two dimensional with nightfall, as the Met Life building is reflected in the glass facade of the dark bronze tower.

Constructing One Madison Park would never have been possible without its innovative use of technology and structural engineering. Due to the building's height and slenderness ratio, dealing with wind and seismic forces was a major engineering challenge. Shear wall stiffness and strength had to be maximized while supporting the architectural design. The building's lateral dynamic movement is mitigated by the design of a Tuned Liquid Damping system incorporated into the building structure at roof level. When high-wind situations occur, the water sloshes back and forth and collects in the two chimneys to dampen the effect of the wind.

Creating One Madison Park was like fitting in the last piece of a puzzle; its contoured form, unique attributes, and placement are a culmination of the steps before, an endnote of harmonized perfection.

One Madison Park's progressive design and ingenious engineering has drawn appreciation from the architectural community and beyond. The poetic form, simple yet striking, conveys an understated elegance not often found in contemporary architecture. This distinctive, slender shape pushes the limits of today's skyscrapers, captivating viewers from miles away. It is this human connection that makes One Madison Park compelling and earns the tower its place in the iconography of New York City towers.

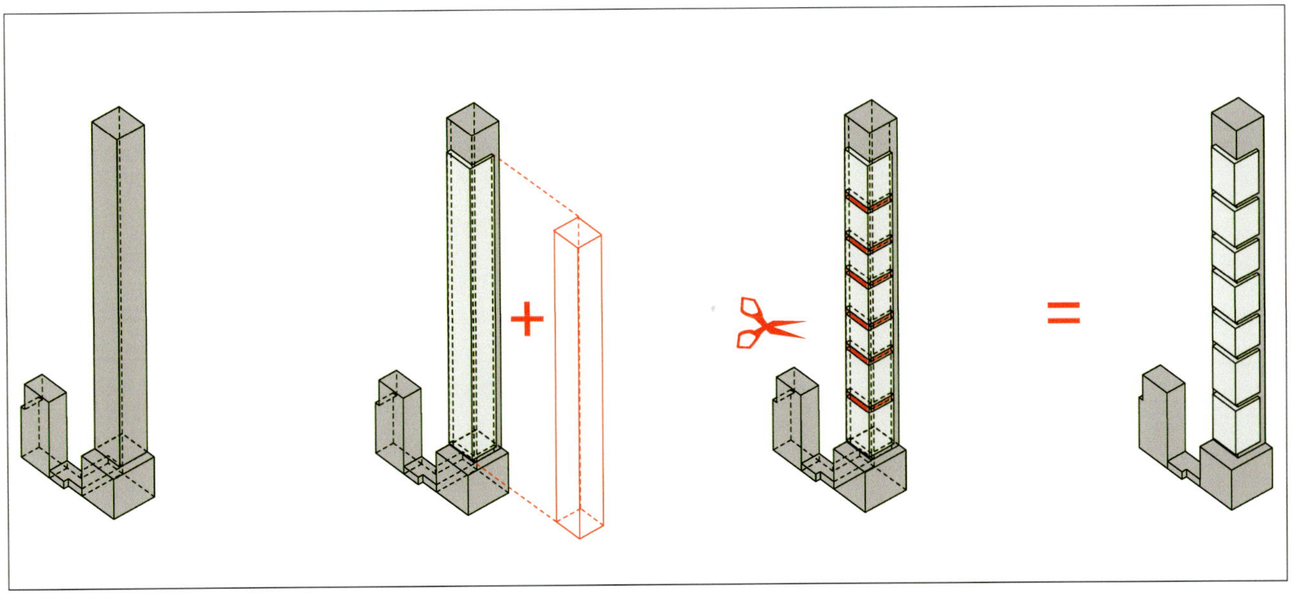

Massing Design

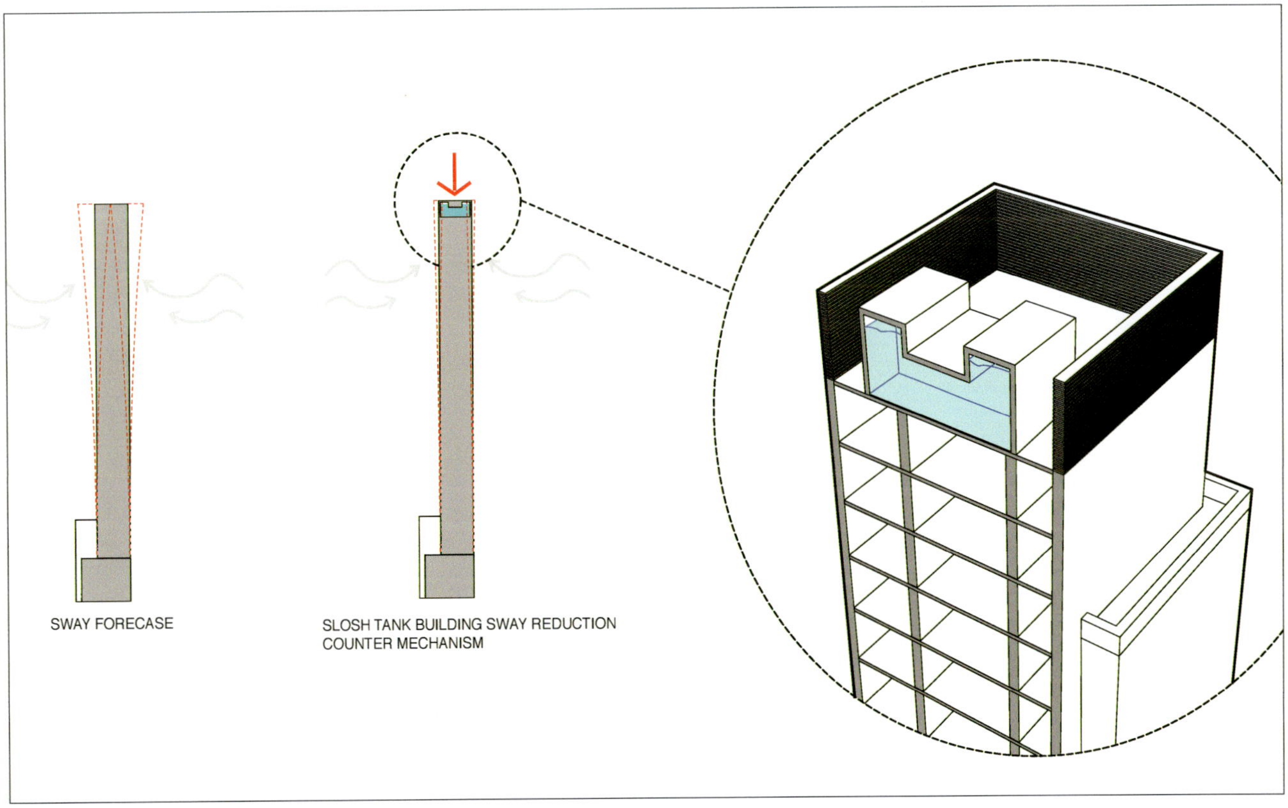

Slosh Tank Engineering

Case Study 6

Veer Towers

Las Vegas, USA

HEIGHT:	137 M
NUMBER OF STORIES:	37
GROSS FLOOR AREA:	841,844 FT^2 (78,210 M^2)
PRIMARY USE:	RESIDENTIAL, RETAIL
START ON SITE:	2006 (DESIGN)
COMPLETION:	2010
PROJECT COST:	NOT AVAILABLE
OWNER/DEVELOPER:	MGM RESORTS INTERNATIONAL
LEAD ARCHITECT:	MURPHY/JAHN
ASSOCIATE ARCHITECT:	AAI ARCHITECTS, INC.
STRUCTURAL ENGINEER:	HALCROW YOLLES
SPECIAL STRUCTURES:	WERNER SOBEK
MEP ENGINEER:	WSP FLACK + KURTZ
MAIN CONTRACTOR:	PERINI AND TISHMAN CONSTRUCTION

Project Description

Skylines can be described as the "fingerprints" of a city. Unlike Chicago or New York, the Las Vegas skyline is a collage rather than a profile. For its designers, the challenge of Veer Towers was to create an integral part of the massive CityCenter project and contribute to its unique and iconic character. Making CityCenter the centre of life in Las Vegas was the goal—the place to go, the place to shop, the place to live. The underlying strategy was to regenerate Las Vegas through a new symbol at its core, just like the Guggenheim in Bilbao, the Pompidou in Paris, or the Sony Center in Berlin.

Architecture

The Veer Towers lean at 5 degrees in opposite directions. Residential uses float above the retail and the 24-meter tall lobbies. Each building is at once robust and delicate. With no reflective glass in its cladding, Veer will be the first truly transparent building in Las Vegas. In itself this represents a great technological and even cultural challenge. Staggered panels of clear and fritted yellow glass animate the facades and give the complex a welcome shot of color, while horizontal louvers give shade from the desert sun.

Structure

The load-bearing structure is a simple and repetitive system with a "Z"-shaped central core. In both towers, the core is strategically positioned on the building's footprint to minimize gravity overturning effects, and it continues vertically up the entire building height. While all interior columns rise vertically, the tower columns on the north and south building elevations are inclined to follow the lean of the towers.

The south facades of the main building lobbies are expressed with slender 1,200-millimeter and 1,370-millimeter diameter concrete columns, free standing 24 meters plus, and inclined to articulate the lean of the towers. Due to space constraints and the requirement to maximize usable lobby space by minimizing column dimensions, composite column construction was introduced.

Architectural design of the main lobbies for the Veer Towers required a unique solution for the heating, cooling, and ventilation, due to the distinctive nature of these spaces. Each lobby is a multi-level space with a large expanse of glass on the south facade. This facade is almost 24 meters in height and provides large quantities of natural light to the lobby, resulting in large solar heat gains and losses. After studying the space loads and using computational fluid dynamics (CFD) analysis, it was determined that the best solution for conditioning the space efficiently was a radiant floor system using chilled and heating water, with displacement ventilation providing the required outside air ventilation and supplemental cooling or heating. A radiant cooling surface allows the space temperature to be higher than traditional all-air design solutions, reducing energy consumption while maintaining occupant comfort.

Heating and cooling of the apartments is provided by vertical fan coil units. The use of natural air and light is maximized throughout the building. Horizontal sunscreen blades provide shading on the east, south and west facades, and reduce energy consumption while minimizing the technical equipment requirements and maximizing occupant comfort. Additionally, approximately 50 percent of the glass lites on all four facades are color frit-coated vision glass, which further contribute to the energy efficiency of the facade while imparting a distinct character and aesthetic to the design.

Sustainability

The manifesto for the project is to exhibit urban responsibility, pay attention to the building's performance in terms of function and systems, use advanced and available technology, accept the aesthetic of construction and elevate it to the level of art, be sensible with regard to energy and ecology through the use of natural resources like daylight and fresh air combined with minimal technical equipment, and maximize user comfort.

Perhaps the most visible of the Veer's sustainable elements is the facade. Extensive use of high-performance low-E glazing maximizes daylighting and views to the outside, which, in conjunction with the use of exterior shades and a 57 percent ceramic frit in 50 percent of the building's envelope, provide all the shading to control and reduce solar loads. Although the fixed shading devices and high-performance glass control solar heat gain, they are not sufficient on their own to meet the project goal of exceeding the requirements of the American Society of Heating, Refrigerating and Air-Conditioning Engineers, ASHRAE 90.1-1999,1 by 20 percent. Other energy-efficient design strategies utilized within the Veer Towers and the wider CityCenter campus, such as high-efficiency central plant and cogeneration systems combined with a high-performance envelope, have achieved a building that exceeds ASHRAE 90.1-1999 by 37.6 percent.

Responsible use of appropriate technologies provided an expressive means of realizing this project in a sustainable way. The use of construction waste management techniques, including diverting 50 to 75 percent of construction waste from landfills, and the use of materials produced and manufactured locally or regionally, recycled materials and wood certified products, results in a significant reduction in environmental impact. Stormwater filtration systems controlling flow drainage, use of stormwater for irrigation, and gray-water systems all contribute to water conservation and the reduction in the use of potable municipal water, resulting in lower utility charges and reduced impact on natural resources.

In 2009, as a key component of CityCenter, Veer received US Green Building Council LEED Gold certification.

Conclusion

Veer Towers' design solution strives for simplicity and dynamism, reinforcing the iconic character of the whole complex. CityCenter is in fact generating "tissue" for the development of true city fabric. If anything, City Center is modern and visionary, and is constructing an open-ended micro-urban system. Veer is urbanistically significant, formally simple and elegant, technologically advanced, and environmentally responsible. The Veer Towers are full of spirit and optimism, with a strong iconic value; their leaning masses add an element of sculpture to the Las Vegas collage.

Note

1 ASHRAE Standard 90.1-1999: *Energy Standard for Buildings Except Low-Rise Residential Buildings.*

Case Study 7

Guangzhou International Finance Center (GZIFC)

Guangzhou, China

HEIGHT:	438.6 M
NUMBER OF STORIES:	103
GROSS FLOOR AREA:	448,371 M^2 (INCLUDING TOWER, PODIUM AND BASEMENT)
PRIMARY USE:	OFFICES
OTHER USE:	HOTEL, RETAIL
START ON SITE:	JANUARY 2006; TOPPING OUT: DECEMBER 2008
COMPLETION:	OCTOBER 2010
PROJECT COST:	£280 MILLION
OWNER/DEVELOPER:	YUE XIU ENTERPRISES (HOLDING) LTD
LEAD ARCHITECT:	WILKINSON EYRE ARCHITECTS
ASSOCIATE ARCHITECT:	SOUTH CHINA DESIGN INSTITUTE
STRUCTURAL ENGINEER:	ARUP
MEP ENGINEER:	ARUP
MAIN CONTRACTOR:	CHINA STATE CONSTRUCTION ENGINEERING CORPORATION / GUANGZHOU MUNICIPAL CONSTRUCTION GROUP
LIGHTING DESIGNER:	LICHTVISION DESIGN & ENGINEERING GMBH
LANDSCAPE DESIGNER:	ASPECT STUDIOS

Project Description

The city of Guangzhou, the powerhouse of Chinese manufacturing and a symbol of the success of Chinese industrialization, had outgrown its existing central business district (CBD) by the early part of the twenty-first century. It therefore set about planning a new CBD, the Zhujiang New Town.

Located on farmland to the east of the city, the Zhujiang New Town was developed to provide Guangzhou with a new central business district in time for the city's hosting of the Asian Games in 2010. The master plan created a central axis running north–south, extending southwards from the city's existing railway station down to the banks of the Pearl River. This axis manifests itself at the heart of the master plan as a central green space from which commercial and residential buildings radiate. Inherent in the master plan is a new cultural quarter with views over the Pearl River, where a new opera house, provincial museum, and library have now been constructed. Central to the master plan concept was the creation of landmark twin high-rise towers. The towers define the threshold and form a gateway between this cultural district and the commercial district to the north.

In 2005 the City of Guangzhou announced an international design competition for the design of these super-high-rise towers. Wilkinson Eyre, teamed with engineers Arup, was shortlisted for the project with 10 other practices. The Wilkinson Eyre team went on to win the competition and were appointed to the Yue Xiu group to complete the design for construction of one of the towers—the west tower, now renamed the Guangzhou International Finance Center (GZIFC). At this time the tower height was increased to 438.6 meters and 103 stories, while the concept for the form of the tower was retained. Within the tower, office floors occupy levels 2 to 66 and a Four Seasons hotel is on levels 67 to 103.

Design Concept

The towers were conceived as smooth, aerodynamic, and elegantly shaped, quite unlike other existing high-rise buildings. In elevation, the towers' identical profiles are elegantly curved, utilizing a series of 5.6-kilometer radii. From a narrow base their form flows to its widest point at about a third of the overall height, before tapering to its narrowest at the top. Their striking diagrid structure is clearly expressed through the buildings' transparent glass skin.

In plan, the tower form is a three-sided "triangular" shape, with three gently curving sides that meet at smooth, rounded corners. Each facade is formed from the sweep of an arc with a 71-meter radius, while 9-meter radii form the three corner "noses."

The aerodynamic shape, smooth cladding design and orientation of each tower has been carefully studied and is influenced in part by prevailing monsoon winds and the need to reduce wind loads. Each tower is orientated to maximize southerly views to the Pearl River for its occupants.

Architecture

A triple-height 12-meter high entrance lobby rings the base of the tower and allows secure access to the building's double-decker shuttles and standard lift groupings. This main lobby also connects via escalators to a secondary office lobby located at the lower basement level, which in turn allows access to below-ground retail and a mass transit railway station. A further dedicated lobby and set-down has been formed at ground level for the hotel.

The plan of the building has been designed to provide approximately 165,000 square meters of efficient and flexible office floor plates. Floor plates range in size from approximately 1,500 square meters to 2,500 square meters gross to allow for larger international tenants. However, the flexible nature of the design also allows for the creation of a series of smaller tenancies on each floor, which is more the typical market condition in China. The office plates have good usable depths from facade to core, and typically range from 11 to 15 meters, so that good daylight penetration is achieved.

While the typical floors are designed as single fire compartments, the larger floors, those exceeding 2,000 square meters, were designed to provide two compartments in order to meet fire codes. The 4.5-meter structural floor-to-floor height allows clear heights of 2.8 meters to be achieved. "Cellform" beams located above the ceiling soffit allow for services integration and a 245-millimeter raised floor allows for distribution of power and IT services.

Level 66 is the highest office floor and above this the five-star-plus Four Seasons hotel accommodation begins. Here the floor-to-floor height has been reduced to 3.375 meters for economy and in order to provide 2.8 meters clear height in the hotel guest rooms.

A health club, complete with infinity pool and spa, is located at level 67, above which back-of-house and kitchen areas have been zoned. The hotel's impressive main lobby is located at level 70 and is accessed via shuttle lifts directly from the ground. A vast atrium, taller than St Paul's Cathedral in London, rises up through the heart of the hotel from this lobby level. This spectacular 120-meter high atrium space is top lit and ringed by restaurants and bars at the lower level, then by the 344 hotel guest rooms and suites which occupy levels 74 to 99. Hotel plant rooms sit above and a helicopter landing pad at 437.5 meters provides fire brigade and VIP access to the tower.

In high-rise towers the design of the core assumes even greater importance than in a traditional building. The core is the building's lifeline and must deliver people and services up and down the building in a highly efficient manner, while allowing for escape in emergency and firefighting access. In addition, the core generally does the work of supporting the vast majority of the building's mass. The core design was therefore the subject of a great deal of refinement in order to maximize net lettable space.

All of the building's lifts are contained within the core, including firefighting/goods lifts and three escape stairs. Toilets serving the office floor plates are located between non-serving lift banks or where lifts have "dropped off." Two on-floor plant rooms have been incorporated within the core and link back to a series of dedicated plant floors. At level 70, in order to allow the hotel atrium to occupy the centre of the plan, the building's central core is transferred to three satellite cores that locate onto the outer walls of the main core below.

Plant floors have been carefully arranged in double-height zones up the height of the building, to coordinate with the building's structural node points while meeting M&E engineering spacing requirements, which seek to prevent excessive build-up of "pressure" on equipment. The plant floors also contain the two sky lobbies and fire refuge requirements for the building's occupants.

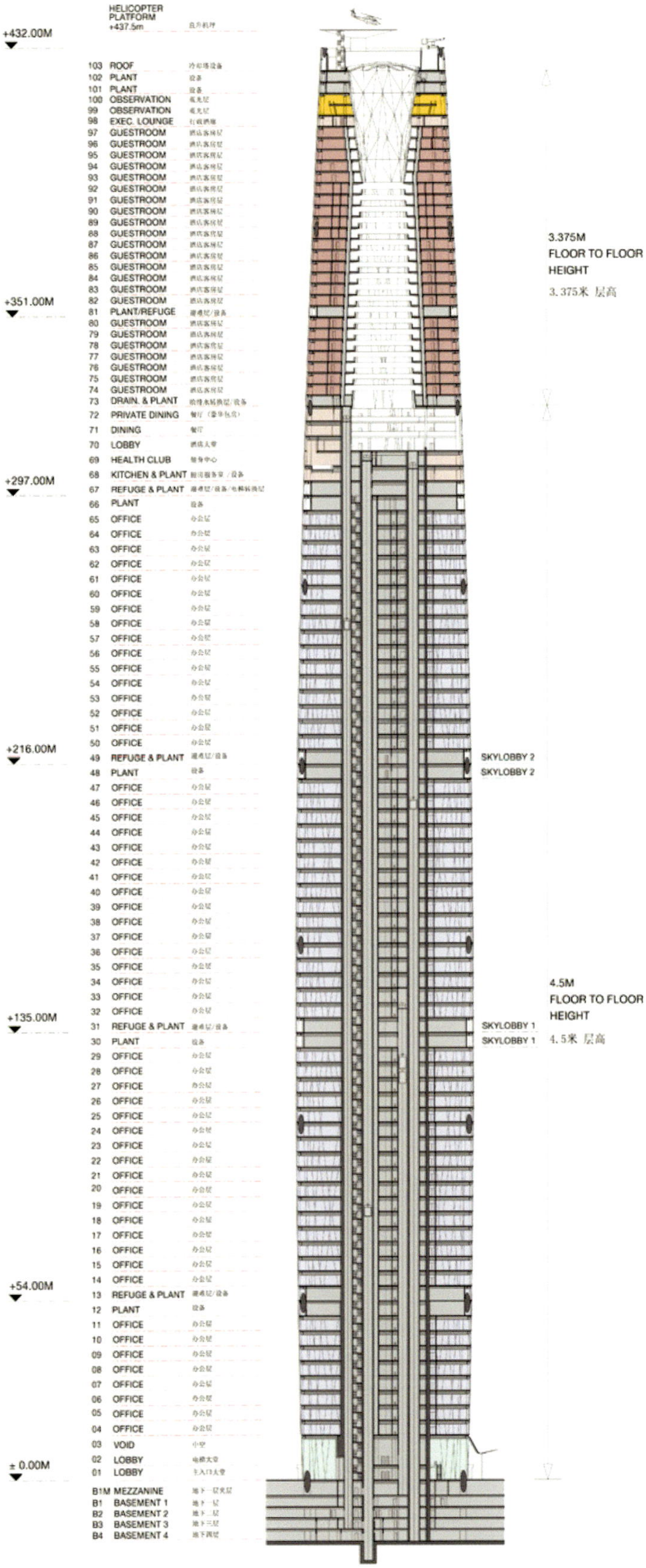
HELICOPTER PLATFORM +437.5m
+432.00M
103 ROOF
102 PLANT
101 PLANT
100 OBSERVATION
99 OBSERVATION
98 EXEC. LOUNGE
97 GUESTROOM
96 GUESTROOM
95 GUESTROOM
94 GUESTROOM
93 GUESTROOM
92 GUESTROOM
91 GUESTROOM
90 GUESTROOM
89 GUESTROOM
88 GUESTROOM
87 GUESTROOM
86 GUESTROOM
85 GUESTROOM
84 GUESTROOM
83 GUESTROOM
+351.00M
82 GUESTROOM
81 PLANT/REFUGE
80 GUESTROOM
79 GUESTROOM
78 GUESTROOM
77 GUESTROOM
76 GUESTROOM
75 GUESTROOM
74 GUESTROOM
73 DRAIN. & PLANT
72 PRIVATE DINING
71 DINING
70 LOBBY
69 HEALTH CLUB
+297.00M
68 KITCHEN & PLANT
67 REFUGE & PLANT
66 PLANT
65 OFFICE
64 OFFICE
63 OFFICE
62 OFFICE
61 OFFICE
60 OFFICE
59 OFFICE
58 OFFICE
57 OFFICE
56 OFFICE
55 OFFICE
54 OFFICE
53 OFFICE
52 OFFICE
51 OFFICE
+216.00M
50 OFFICE
49 REFUGE & PLANT
48 PLANT
47 OFFICE
46 OFFICE
45 OFFICE
44 OFFICE
43 OFFICE
42 OFFICE
41 OFFICE
40 OFFICE
39 OFFICE
38 OFFICE
37 OFFICE
36 OFFICE
35 OFFICE
34 OFFICE
33 OFFICE
+135.00M
32 OFFICE
31 REFUGE & PLANT
30 PLANT
29 OFFICE
28 OFFICE
27 OFFICE
26 OFFICE
25 OFFICE
24 OFFICE
23 OFFICE
22 OFFICE
21 OFFICE
20 OFFICE
19 OFFICE
18 OFFICE
17 OFFICE
16 OFFICE
15 OFFICE
+54.00M
14 OFFICE
13 REFUGE & PLANT
12 PLANT
11 OFFICE
10 OFFICE
09 OFFICE
08 OFFICE
07 OFFICE
06 OFFICE
05 OFFICE
04 OFFICE
03 VOID
±0.00M
02 LOBBY
01 LOBBY
B1M MEZZANINE
B1 BASEMENT 1
B2 BASEMENT 2
B3 BASEMENT 3
B4 BASEMENT 4
SKYLOBBY 2
SKYLOBBY 2
SKYLOBBY 1
SKYLOBBY 1
3.375M
FLOOR TO FLOOR
HEIGHT
3.375米 层高
4.5M
FLOOR TO FLOOR
HEIGHT
4.5米 层高

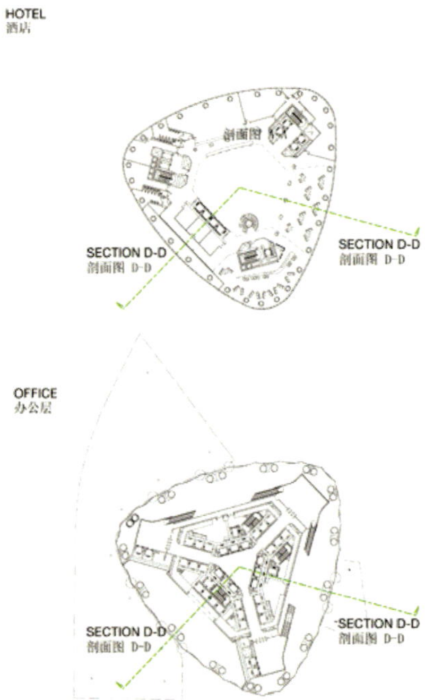
HOTEL
酒店
SECTION D-D
剖面图 D-D
SECTION D-D
剖面图 D-D
OFFICE
办公层
SECTION D-D
剖面图 D-D
SECTION D-D
剖面图 D-D

Structure

The building utilizes the world's tallest constructed diagrid structure, which is clearly expressed though the building's facade and gives it considerable "character." Concrete-filled steel tubes (CFTs) form the diagrid members, providing both good stiffness and fire protection to the structure. However, two-hour fire protection was still required in order to meet codes and this was trowel applied directly to the building's primary structure. Every 12 stories the tubular diagrid structure "nodes out" to form 54-meter high giant steel diamonds. At the base of the tower the structural members are 1,800 millimeters in diameter, and progressively reduce in size progressively to become 900 millimeters at the top of the building.

Much of the gravity load of the building's floors is taken by the structural core, which links back to the diagrid perimeter structure via floor beams to create a stiff "tube-within-tube" system. Inherent stiffness in the structure minimizes steel tonnage while providing resistance to acceleration and sway, thereby maintaining high comfort levels for the building's occupants. This stiffness and resistance to acceleration means that no additional damping of the structure is required.

The GZIFC project also features a large podium which connects at the base of the tower by an impressive four-story glazed entrance space, allowing direct access to the conference and banqueting facilities and in turn to the 45,000-square-meter high-end retail mall. A six-story retail podium wraps around the main tower and supports two smaller residential towers, each 100 meters high, containing 286 apartments. A further 18,000 square meters of below-ground retail space connects into the above-ground mall and tower lobby, while allowing direct access to the MTR system.

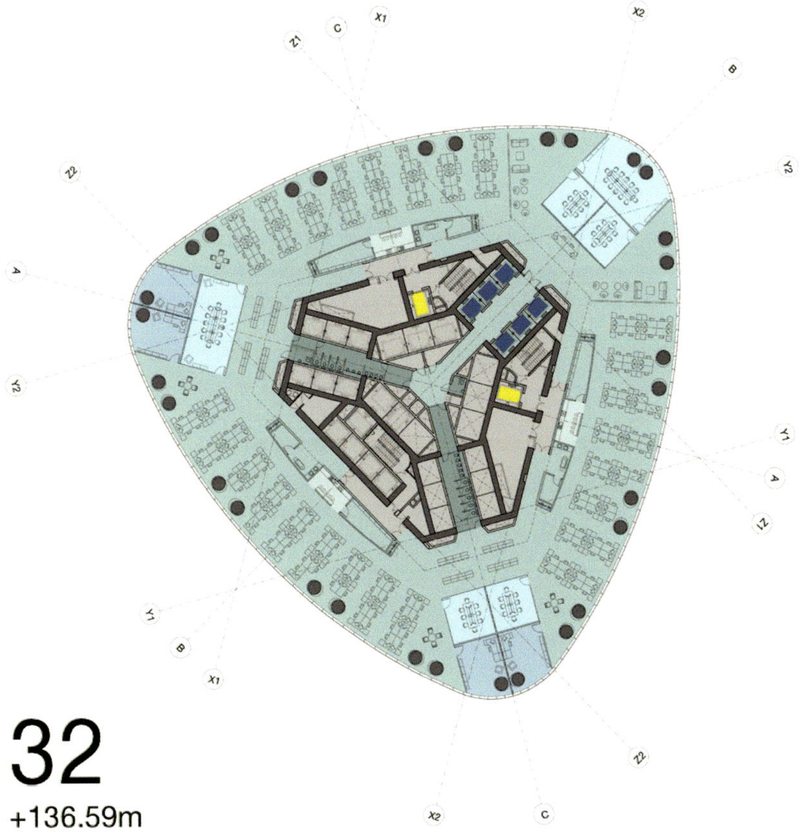

32
+136.59m

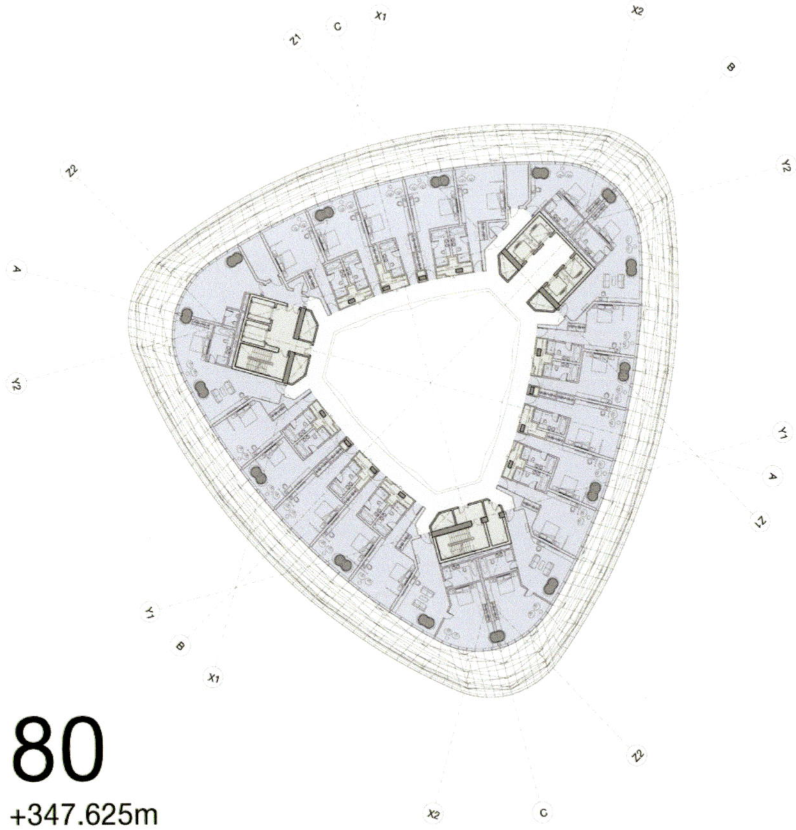

80
+347.625m

Case Study 8

International Commerce Centre (ICC)

West Kowloon, Hong Kong

HEIGHT:	484 M/1,588 FT
NUMBER OF STORIES:	108
GROSS FLOOR AREA:	262,176 M²/2,822,039 FT²
PRIMARY USE:	MIXED USE: OFFICE, RETAIL, HOTEL AND CONFERENCE FACILITIES
START ON SITE:	2003
COMPLETION:	2010
PROJECT COST:	NOT AVAILABLE
OWNER/DEVELOPER:	SUN HUNG KAI PROPERTIES AND MTR CORPORATION LIMITED
LEAD ARCHITECT:	KOHN PEDERSEN FOX
ASSOCIATE ARCHITECT:	WONG & OUYANG (HK) LTD.
STRUCTURAL ENGINEER:	ARUP
MEP ENGINEER:	J. ROGER PRESTON LIMITED
MAIN CONTRACTOR:	KAI SHING MANAGEMENT SERVICES LIMITED
LANDSCAPE:	BELT COLLINS & ASSOCIATES

Project Description

Soaring 490 meters above Victoria Harbor, the tower is the essence of Hong Kong in one destination: high-powered finance, global tourism, luxury shopping, and world-class hospitality, all gathered in a single tower built over a sophisticated transportation network spanning the Pearl River Delta. The client sought to catalyze development in this new urban center by accommodating a dynamic program mix in a supertall high-rise, and by connecting to adjacent development and multi-modal transport. ICC was also designed to complement the existing International Finance Tower, which is similar in scale, so the two could work as beacons or lighthouses framing the gateway to Victoria Harbour. The resultant ICC tower is the centerpiece of the Union Square reclamation project, which includes a new urban center with residential, office, retail, hotel, and recreation spaces, as well as a new transportation hub, Kowloon Station, connecting Hong Kong to the Hong Kong International Airport.

Architecture

The concept behind the design of ICC was to connect the ground with the sky and the water—the three main elements of Hong Kong. Soaring 490 meters above Victoria Harbour, ICC is the essence of Hong Kong in one destination: high-powered finance, global tourism, luxury shopping, and world-class hospitality, all gathered in a single tower built over a sophisticated transportation network spanning the Pearl River Delta.

A major priority in the design was to create a tower form that would combine the best possible structure with the best possible floor plate. For instance, a tower geometry based on a circular floor plate would perform well in the wind, but would be undesirable to Hong Kong's financial tenants, who prefer the efficient layout of square floors. Conversely, a perfectly square floor plate would perform poorly in the wind and lead to an increase in steel and concrete—an unsustainable proposition. An analysis of preliminary wind tunnel studies indicated that a square with notched, or "re-entrant," corners would exhibit nearly the same wind response as that of a circle.

From ICC's initial form, the massing was refined by gradually widening the re-entrant corners towards the top and inclining the upper third of the main facades by one degree to create the tower's elegant silhouette and improve its wind response. The tower's eight mega-columns splay out three degrees to widen the tower's dimension at its base, significantly reducing its overturning moment while providing longer clear spans for hotel and exhibition facilities.

At its base, the tower's chiseled facades give way to gently sloped curves. These curves lift off from the structure as a cascade of overlapping shingles to create sheltering canopies for the office and hotel entrances on the three sides overlooking the harbor. At the north, the facade sweeps down in a dramatic gesture towards the center of the Union Square development, enclosing the "dragon tail" atrium. This atrium serves as the public face of the tower and the primary connection to the rail station.

Main facades are articulated as four planar elements extending partially beyond the re-entrant corners and rising above the tower roof as sheets of glass to form the tower crown. Initially designed as cantilevered curtain wall panels, the facade extensions later incorporated a triangular return to create enclosed bay windows at the corner offices, giving direct views of the harbor. At the tower base, the triangular returns split from the main facade to form distinctive markers framing the lobby entrances.

The aesthetic appeal of the tower's external envelope is matched by its environmental performance. Sheathed in silver low-emissivity insulating glass, the tower's single-layer skin provides the maximum protection from solar heat gain while deploying a minimum amount of facade material. The silver coating has the unique quality of reflecting the heat-generating spectrum of sunlight (infrared, ultraviolet), while allowing the desirable visible light spectrum to transmit through the facade. Optical properties of the glass, supplied by Shanghai Yaohua Pilkington, include an emissivity rating of 0.15, a visible light transmission of 40 percent, and a shading coefficient of 0.27—more than three times the protection of uncoated glass. Moreover, the shingled panels provide self-shading of the main facades, with horizontal baffles in the re-entrant corners providing additional shading.

Structure

Winds of up to typhoon strength are a major design constraint in Hong Kong. Given the location of the tower above a cliff-like bedrock profile varying from 60 to 130 meters in depth, there were significant uncertainties attached to the use of traditional pile types. Instead, shaft-grouted barrettes were adopted as the foundations

of the building. There are 241 rectangular barrettes with an average depth of 70 meters. Grout was injected into the barrette/soil interface to increase the friction capacity of the barrette. ICC is the first tall building in Hong Kong to use this type of foundation.

A realistic wind profile was studied and adopted for the first time in Hong Kong to design a tall building. Taking into consideration the site topographic effects and the wind directionality effects, a wind profile with a mean hourly gradient design wind speed of 59.5 meters per second and a gradient height of 500 meters was adopted to design the structural stability system of the building. The wind design profile has enabled a more cost-effective design than the traditional design wind speed of 64 meters per second.

At the heart of the main structural skeleton of the ICC is a robust high-modulus grade 90 concrete central core wall, coupled with eight perimeter concrete columns through four outrigger trusses at four strategic levels up the height of the building. All steelworks are protected against fire by applying insulating materials such as cementitious spray, to ensure that in a fire the temperature in the steel sections do not exceed the limiting value within the designed fire resistance period.

Perimeter columns are up to 3.5 by 3.5 meters in cross-section, perimeter edge beams span 30 meters, and steel composite floors are used outside the core walls. These composite floors act as diaphragms connecting the core and the perimeter columns, a solution which not only resists lateral wind and gravity loads but also offers maximum clear sightlines to external views.

The grade 90 concrete with selected aggregate and elasticity value at 39 GPa helped to reduce the core wall and column sizes, resulting in less self-weight, more net floor area and better cost saving. Its fluid-like and self-compacting characteristics are also useful for the reinforced concrete elements with congested steel reinforcement.

Prestressed concrete was specified for the lowest level of outriggers, which helped to save cost in the uncertain market in 2003, when Hong Kong was hard hit by the panic surrounding a major outbreak of severe acute respiratory syndrome (SARS). Outriggers at the three upper levels are structural steel, designed to anchor in the central core wall using the interaction between steel and concrete, an approach that facilitated rebar fixing in congested zones of the core wall and left core wall openings available for building services installation. Different computing tools for structural optimization studies were adopted to reduce material cost and increase usable floor area.

Services

The tower is divided into five lift zones for the office and dedicated lift zones for the observation deck/OB floor above the office floors, and the hotel on the top floors of the skyscraper. Four mechanical floors service each individual zone of the building in the most economical and energy-saving manner.

There are altogether 85 lifts, providing vertical transportation for a working population of over 20,000 people, and the capacities and speeds of the lifts range from 900 to 4,500 kilograms and 1.5 to 9 meters per second.

A good lift design plays an important role in achieving a flexible and expedient elevating service. The following outlines the key features adopted for ICC:

- *Sky-lobby approach* With a total office floor area of 150,000 square meters and 67 office floors, direct transportation between the main lift lobbies and all the local lift zones would require excessive lift-shaft spaces, reducing the efficiency of the building core drastically. A sky-lobby approach is the best way to overcome such inefficiency. In ICC, the office floors are divided into five lift zones with two sky lobbies, located at the 48th and 49th floors (between lift zone 2 and lift zone 3) and the 88th floor (between lift zone 4 and lift zone 5).

- *Double-deck lifts with destination control system* To further increase the efficiency of the building core, double-deck lifts are provided to serve most of the lift zones, in addition to the sky lobbies. There are altogether 40 double-deck lifts provided for ICC: thirty-two 1,600- to 1,800-kilogram double-deck lifts with speeds up to 9 meters per second are provided for the office tenants, while the Observation Deck and the Ritz Carlton Hotel are respectively served by four 1,600-kilogram, 9-meter-per-second and four 1,800-kilogram, 9-meter-per-second double-deck shuttle lifts. In order to maximize the efficiency of double-deck lifts, a destination control system is adopted, whereby passengers place their calls in the lift lobbies, thus enabling the lift control system to analyze the demand holistically in real time and assign the most appropriate lift deck to serve the passengers. It will boost the handling capacity by 15 percent, in general.

The plumbing system in ICC comprises potable and flush water supply systems, for which fresh water and seawater are used, respectively. Seawater is supplied by the local water authority via city main and fed into the building for flushing sanitary fitments, including water closets and urinals. Fresh-water cooling is applied in the HVAC system, and bleed-off water from an evaporative cooling tower is re-used via water treatment for flushing purposes. Potable water serves basin, shower, and kitchen needs only, resulting in a significant reduction in potable water consumption. Recycling of bleed-off water from the cooling tower plant for use as flushing water also reduces the consumption of flushing water. The estimated peak daily bleed-off volume is 180 cubic meters. Condensate water from the air-conditioning systems is collected and recycled as make-up water in the cooling towers. It is estimated that four 50- by 25-meter swimming pools of condensate could be collected and recycled for cooling tower use per year.

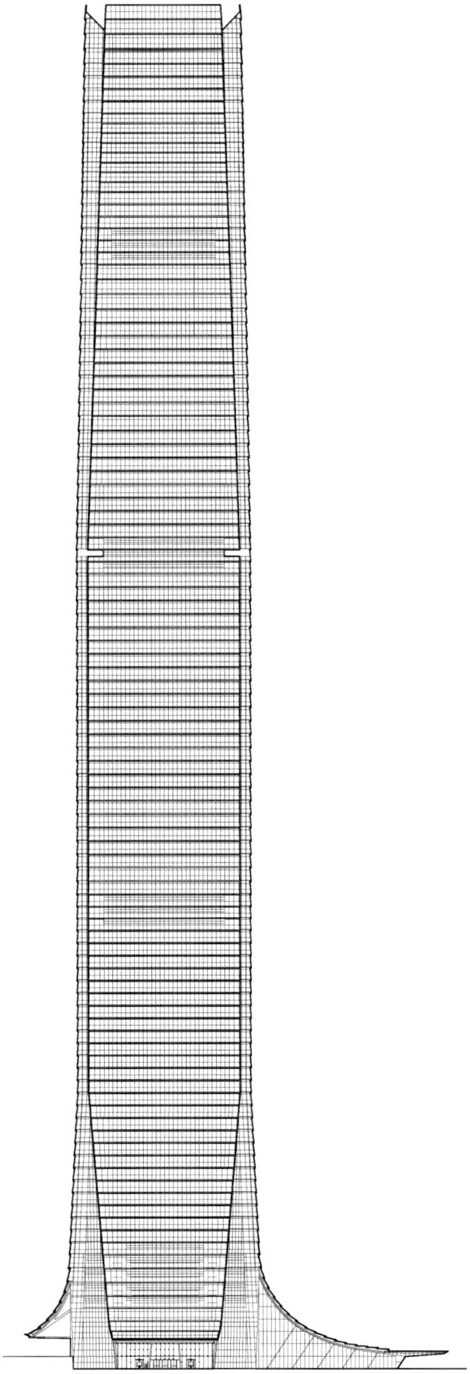

EAST ELEVATION

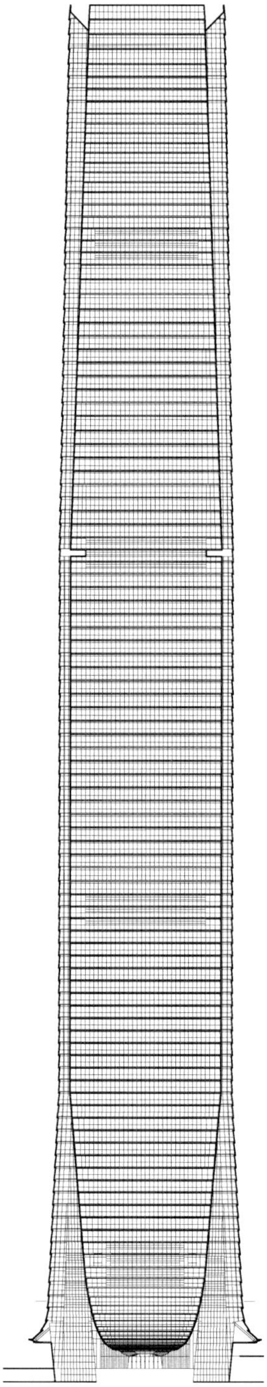

NORTH ELEVATION

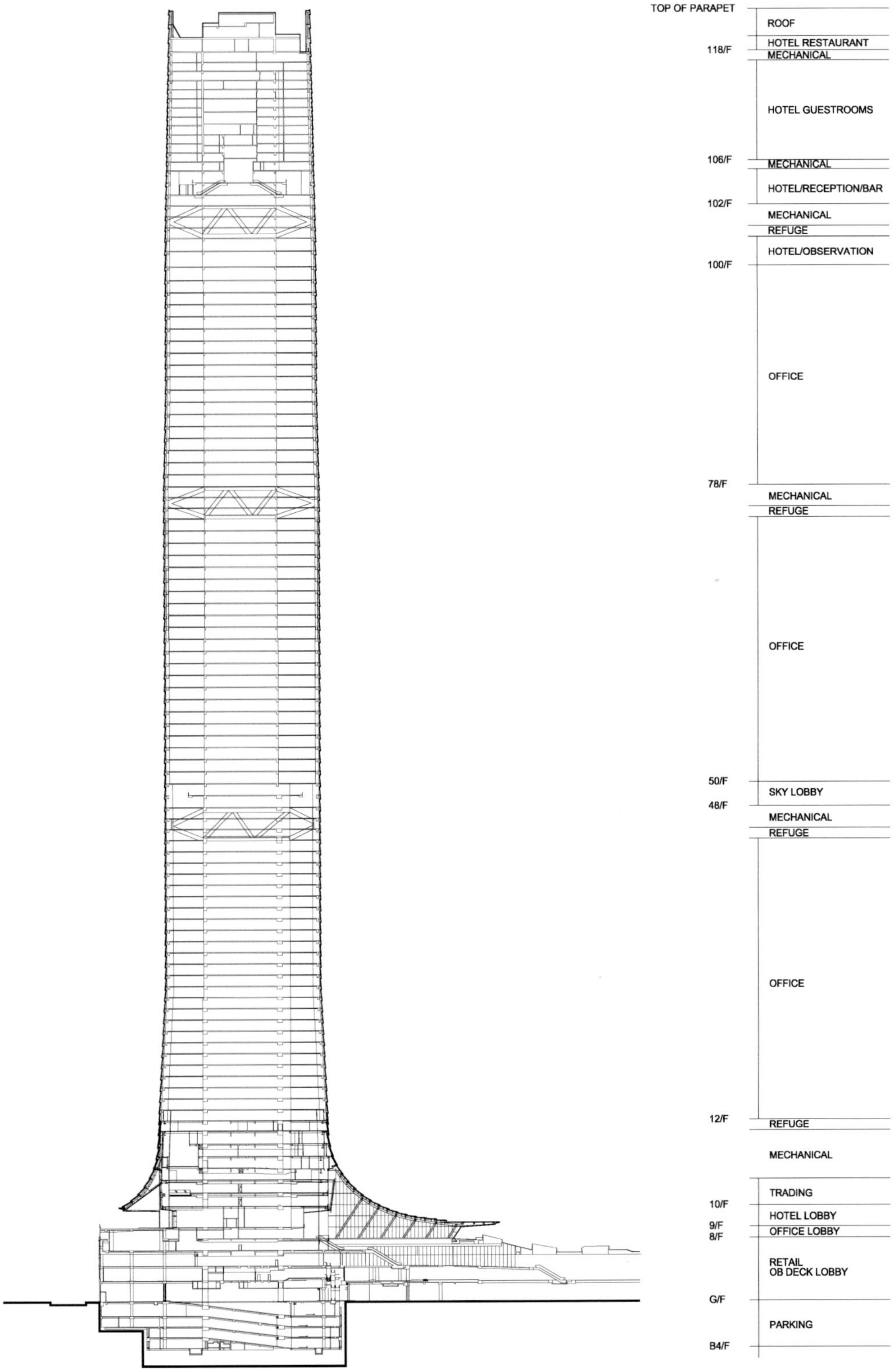
TOP OF PARAPET
ROOF
HOTEL RESTAURANT
118/F
MECHANICAL
HOTEL GUESTROOMS
106/F
MECHANICAL
HOTEL/RECEPTION/BAR
102/F
MECHANICAL
REFUGE
HOTEL/OBSERVATION
100/F
OFFICE
78/F
MECHANICAL
REFUGE
OFFICE
50/F
SKY LOBBY
48/F
MECHANICAL
REFUGE
OFFICE
12/F
REFUGE
MECHANICAL
TRADING
10/F
HOTEL LOBBY
9/F
8/F
OFFICE LOBBY
RETAIL
OB DECK LOBBY
G/F
PARKING
B4/F

Case Study 9

Regent Place: Lumière Residences and Fraser Suites

Sydney, Australia

HEIGHT:	LUMIÈRE RESIDENCES (TOWER A): 151 M; FRASER SUITES (TOWER B): 114 M
NUMBER OF STOREYS:	TOWER A: 48; TOWER B: 33
GROSS FLOOR AREA:	TOWER A: 54,050 M^2; TOWER B: 15,100 M^2; TOTAL: 116,500 M^2
PRIMARY USE:	TOWER A: RESIDENTIAL; TOWER B: HOTEL AND SERVICED ACCOMMODATION
START ON SITE:	2003
COMPLETION:	2007
PROJECT COST:	NOT AVAILABLE
OWNER/DEVELOPER:	GREENCLIFF (CPL) DEVELOPMENTS PTY LTD
LEAD ARCHITECT:	FOSTER + PARTNERS
ASSOCIATE ARCHITECT:	PTW ARCHITECTS
STRUCTURAL ENGINEER:	TAYLOR THOMPSON WHITTING/ROBERT BIRD GROUP
MEP ENGINEER:	CONNELL MOTT MACDONALD
MAIN CONTRACTOR:	MULTIPLEX CONSTRUCTIONS PTY LTD

Project Description

The Regent Street Theatre occupied a prominent position on George Street, an important axis through Sydney, lined with some of its most significant civic buildings, including the Town Hall and St Andrew's Cathedral. After the Theatre's demolition in 1989, the vacant site had become the 'missing tooth' in the city's public face. Balancing a sensitive response to this historic context with a bold departure from its low-rise, nineteenth-century neighbours, the towers of the Regent Place development established a new precedent in the area that has led to the construction of further tall buildings.

The strategic importance of this location presented one of the key challenges for the project – the City had very specific objectives regarding the massing and the floor plate for the towers, and the architect chose to develop a scheme that matched these requirements exactly. Bordering Chinatown and the Central Business District, the area had been in need of some regeneration. Regent Place has therefore had a significant impact. By introducing mixed-use and bringing in 1,000 residents, the development has contributed to the animation of this urban quarter.

Architecture

Regent Place comprises two towers: the 48-storey Lumière Residences, with apartments arranged in eight slender volumes around a central core; and the 33-storey Fraser Suites, which contains a hotel and serviced accommodation. In the Lumière Residences, vertical slots between the eight volumes rise the full height of the tower and bring daylight and natural ventilation into the central core, comprising the circulation corridors and lift lobbies. Daylight can also penetrate into otherwise windowless rooms towards the back of the apartments, and the slots also allow for natural cross-ventilation.

A five-storey sandstone podium at the base of the towers reflects the scale and materials of adjacent heritage buildings, aligning with the roofline of the street, and combining offices with a variety of retail and recreational facilities, including a swimming pool suspended dramatically over a triple-height entrance lobby. The two towers appear to float above the podium building by means of a two-storey shadow gap. Introducing a community of around 1,000 new residents to a predominantly commercial and governmental district, the design of the Lumière Residences harmonises with the different building types in the area, echoing the clean lines of neighbouring offices. The choice of internal loggia rather than balconies preserves a similar flush and seamless appearance, with folding screen doors and sliding windows in a dark glazing that appears opaque externally, yet completely transparent inside.

The project is located in a part of Sydney characterised by its dense urban grain, and the development spans three streets on a city block. Drawing daylight into the heart of the scheme and increasing permeability, the base is divided into a series of smaller blocks that open up onto internal glazed routes. These connections through the site reference Sydney's Victorian arcades, whose narrow top-lit covered streets were an earlier response to the high density of the historic city quarter.

Structure

No fewer than nine levels of basement were excavated into the underlying sandstone. Foundations for the towers were shallow concrete pad footings on sandstone with a capacity in excess of 4,000 kPa.

Bracing against wind and seismic effects is accomplished by large, central, in situ concrete cores, utilising concrete of up to 80 MPa. At the centre of each core 'box' is a lift lobby flanked by banks of lifts connected by 'coupling slabs' for extra stiffness.

Load-bearing precast concrete wall panels are used throughout, making the Lumière Residences the tallest load-bearing precast concrete building in Sydney. There are neither columns nor beams in the structural frame: flat slab floors up to 220 millimetres thick span up to 9 metres between precast wall supports ranging from 200 to 300 millimetres thick. This approach led to an estimated three-month reduction in construction schedule.

Mechanical Systems

Apartments are all air conditioned by reverse cycle water-cooled package units ducted to bedroom and living spaces. All the units are located in cupboards accessible from the corridor so it is not necessary to access the apartment to change filters, for example. Outside fresh air is supplied to each air conditioning unit, ducted from facade intakes on a floor-by-floor basis. All the air intakes are located in the deep vertical slots that divide the apartments into the eight separate Lumière tower volumes.

The two towers each have condenser water loops serving the package units. The temperature of the water is maintained within a range of 20 to 30 degrees Celsius by cooling towers and gas-fired water heaters located in the roof plant rooms in each tower.

Fire stairs are pressurised and the towers also feature stair pressurisation relief systems. Corridors are provided with conditioned air for peak summer cooling from roof plant room-located reverse-cycle water-cooled package units, operating on 100 per cent outside air with condenser water preconditioning coils.

Apartments are provided with a separate dedicated toilet exhaust and kitchen odour exhaust ducted to the roof. Each floor also includes a garbage room with chute, with garbage exhaust ducted to the roof. All apartments are fitted with low-water-use fittings and hardware.

Podium

Chilled water and heating hot water fan coil units serve retail and commercial areas in the podium. Chillers are located on level 8 and water heaters in the Fraser Suites roof plant room. Smoke exhaust is provided to the podium levels.

The condenser water loops also serve reverse cycle package units in the entrance lobbies, theatres, studio, bars, spa, gym, multifunction rooms and lounge. The two pool enclosures have dedicated air-conditioning units with heat recovery on the exhaust air and 100 per cent outside air supply with heating coil.

Car Park

The car park is mechanically ventilated, however an engineered solution allowed design airflows to be reduced below the standard AS 1668.2-1991[1] allowances. The system also includes CO monitoring to further reduce airflows during low demand.

Note

1 Standards Australia AS 1668.2-1991: *The Use of Ventilation and Airconditioning in Buildings: Ventilation Design for Indoor Air Contaminant Control.*

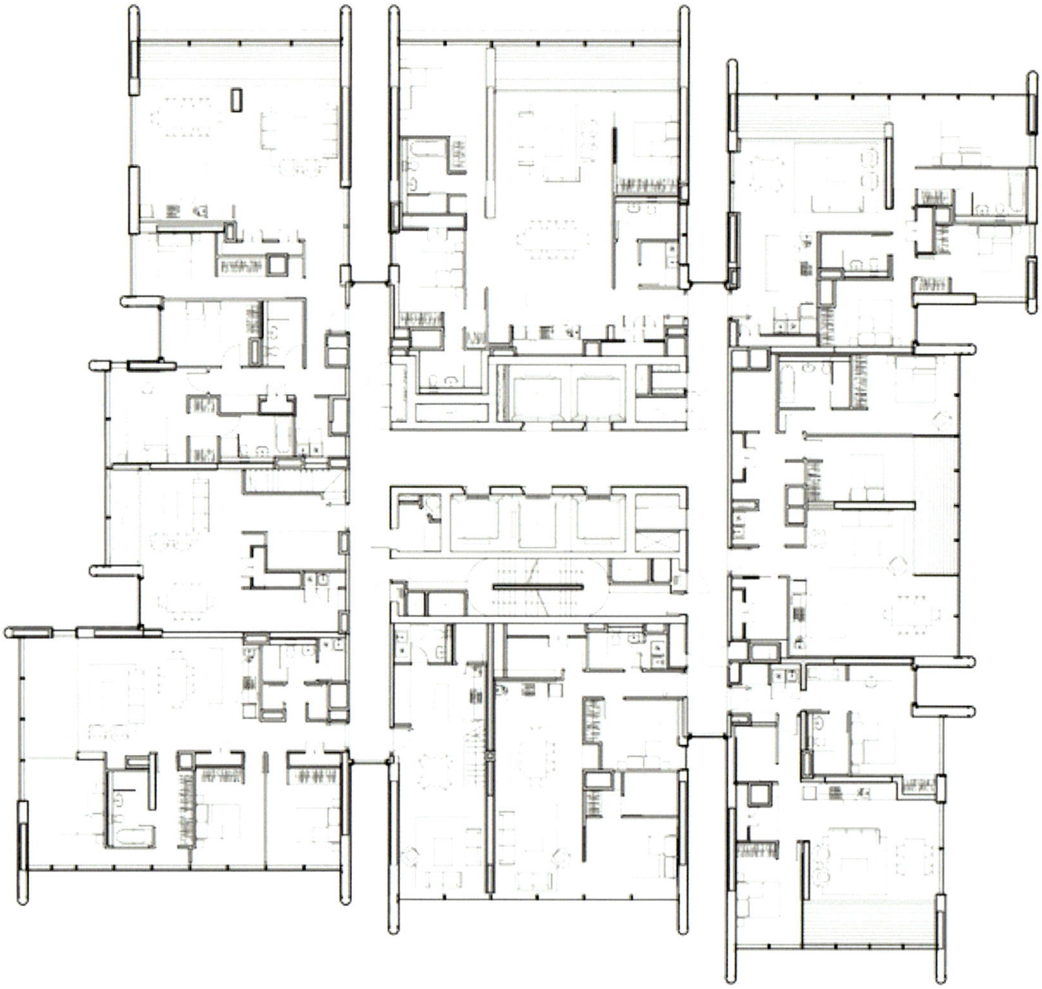

0 5 10m

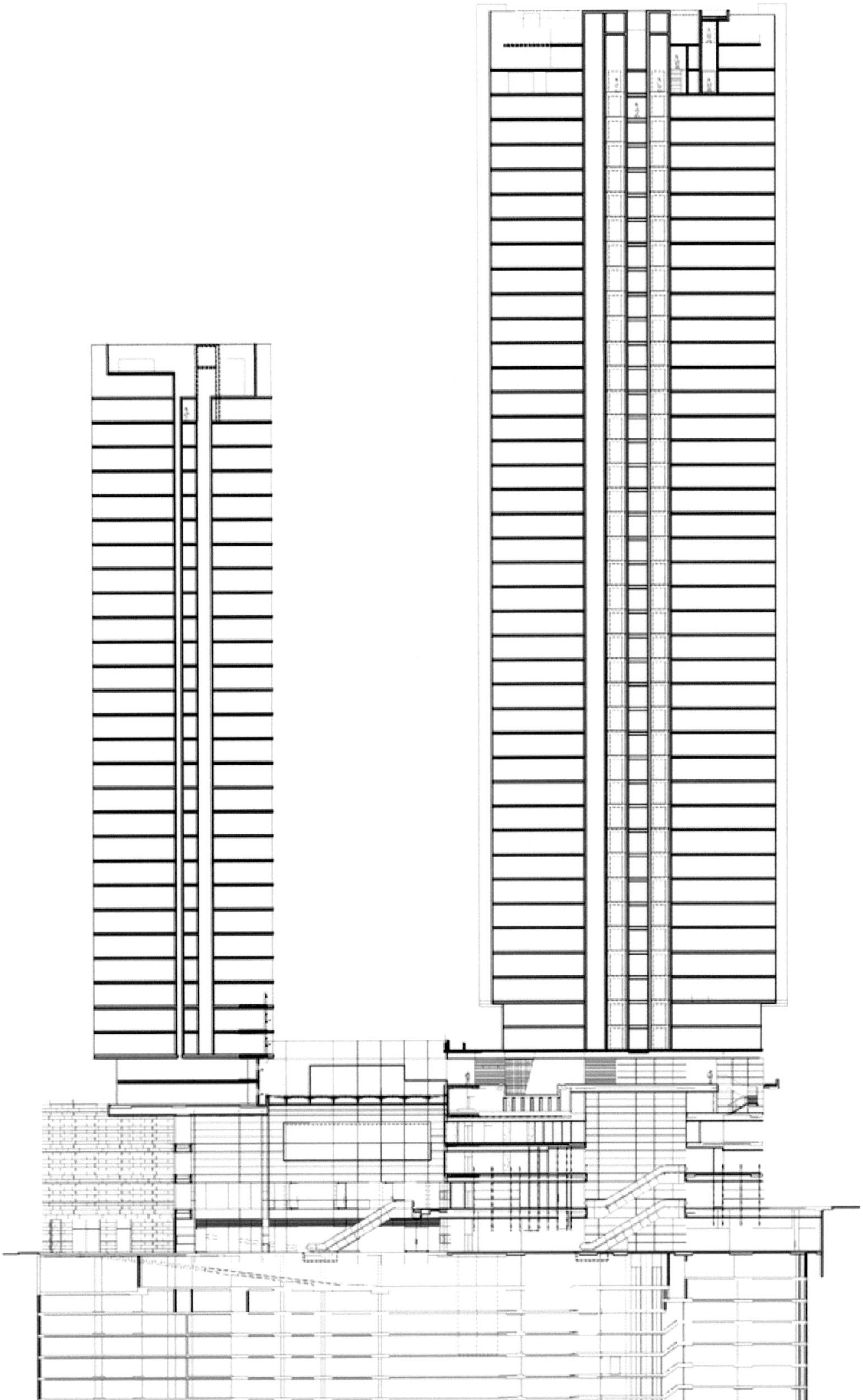

Case Study 10

Marina Bay Sands

Singapore

HEIGHT:	THE SANDS SKYPARK, AT ITS HIGHEST LEVEL, IS 195 METERS ABOVE GROUND LEVEL
NUMBER OF STORIES:	55 STORIES IN EACH OF THE THREE TOWERS
GROSS FLOOR AREA:	929,000 M^2
PRIMARY USE:	INTEGRATED RESORT: HOTEL, CASINO, CONVENTION CENTER, PERFORMING ARTS THEATER, ART/SCIENCE MUSEUM, RETAIL AND RESTAURANTS, WATERFRONT PROMENADE, AND SKYPARK
START ON SITE:	2005
COMPLETION:	2010
PROJECT COST:	$5.7 BILLION (INCLUDING LAND COST)
OWNER/DEVELOPER:	LAS VEGAS SANDS CORPORATION
LEAD ARCHITECT:	MOSHE SAFDIE
ASSOCIATE ARCHITECT:	AEDAS
STRUCTURAL ENGINEER:	ARUP
MEP ENGINEER:	R.G. VANDERWEIL, LLP (DESIGN); PARSONS BRINCKERHOFF (PRODUCTION)
MAIN CONTRACTOR:	(HOTEL, SKYPARK) SSANGYONG
LANDSCAPE ARCHITECT (DESIGN):	PETER WALKER & PARTNERS
LANDSCAPE ARCHITECT (PRODUCTION):	PERIDIAN ASIA PTE LTD
LIGHTING CONSULTANTS:	PROJECT LIGHTING DESIGN
WATER FEATURES:	HOWARD FIELDS & ASSOCIATES
HOTEL/CONVENTION INTERIOR DESIGN:	HIRSCH BEDNER ASSOCIATES
CASINO INTERIOR DESIGN:	ROCKWELL GROUP WITH SAFDIE ARCHITECTS
THEATER CONSULTANTS:	FISHER DACHS ASSOCIATES
GRAPHICS:	PENTAGRAM

Project Description

Located in Marina South, a peninsula of land reclaimed from the sea in the late 1970s across the bay from Singapore's Central Business District, Marina Bay Sands is a 929,000-square-meter (10-million-square-foot), high-density, mixed-use integrated resort complex that brings together a 2,560-room hotel, a 120,000-square-meter (1,292,000-square-foot) convention center, a shopping mall, an ArtScience museum, two Sands Theaters, numerous restaurants, and a casino. Conceived not just as a building project but as a city microcosm rooted in Singapore's culture, climate, and contemporary life, the project anchors Singapore's waterfront, creating a gateway to the city and providing a dynamic setting for vibrant public life.

Due to the project's prominent location, it was decided that three towers would be created instead of one. Each tower hotel is designed at a height of 55 stories. Spanning the top of the three towers, 200 meters (656 feet) above the sea, is a 1.2-hectare (3.0-acre) SkyPark, a new type of public space, framing large "urban windows" between the towers. Lavishly planted with trees, the SkyPark accommodates a public observatory, garden spaces, a 150-meter (492-foot) long infinity swimming pool, restaurants, and jogging paths, and offers sweeping panoramic views—a formidable resource in a dense city like Singapore.

Architecture

Two converging slabs of east- and west-facing rooms make up each tower. At the base the voids are connected by a continuous and conditioned glazed atrium, which fills the space between the towers with restaurants, retail spaces, and a public thoroughfare. Each tower slab form is also twisted slightly in relation to its pair, creating a dance-like relationship between the two parts and accentuating the slenderness of the buildings.

The tower slabs also give further character to the massing and relate to the site context: the glazed western facade looks towards the city center, while the east side, which faces the botanical gardens and ocean beyond, is planted with lush bougainvilleas. In plan, as the parcel varies in width, the cross section decreases from one tower to the next.

As the greatest heat gain occurs on the western facade, it was of paramount importance that an innovative solution be developed to maintain energy efficiency,

without limiting the view from the hotel rooms to Singapore's downtown. The design solution proposed and implemented was a custom double-glazed unitized curtain wall. Glass fins perpendicular to the facade were installed to provide shading. These use a 30 percent reflective glass and are responsible for shading the facade for up to 20 percent of all solar gain. The outer skin follows the natural curved shape of the buildings, and the use of reflective glass creates a taught mirrored facade.

Structure

Deep, soft marine clays dominate the ground conditions beneath the complex. A further complication was the need to construct a 35-meter (115-foot) deep "cut-and-cover" tunnel within the site, close to Singapore's longest bridge. Excavations under such conditions had the potential to be extremely risky, and to require extensive shoring and propping of the excavation walls.

A series of large cofferdams was the solution preferred by the structural engineers. Within the podium zones were two circular cofferdams 122 meters (400 feet) in diameter, while the hotel zone featured two different approaches— a 100-meter (330-foot) diameter donut and a twin-cell 76-meter (250-foot) diameter cofferdam without cross-walls. This solution minimized shoring and propping needs and speeded construction.

Superstructure design was dominated by the complex building dynamics arising from the chosen layout. How a single tower responds to loading from wind and seismic effects can be hard to analyse effectively. Linking three towers together at their highest points introduced a new set of challenges that required extensive wind tunnel testing, particularly as the towers are not dynamically identical. Each tower had to be analysed individually and in relation to its neighbor before strategies could be developed that would allow the SkyPark to function as planned and to ensure the comfort of the building users.

As a result of the towers' asymmetrical form, especially the spread at the base of each one, gravity dominates the primary lateral loads on the buildings, rather than wind loads or seismic effects. Reinforced concrete shear walls at 10-meter (33-foot) centres act as the primary vertical and transverse structure, restrained in the longitudinal direction by additional concrete core walls located within both legs of each building. These cores also act to restrain out-of-plane buckling of the slim shear walls, which taper from 710 millimeters thick at their base to 500 millimeters thick at the highest point.

At ground level the inclined tower legs generate significant horizontal forces, and post tensioning of the base slab is needed. Large shear forces are generated where the tower legs meet above the central atrium. These are taken by storey-height steel trusses located within the MEP floors at Level 23.

In the interests of structural simplicity and speed of construction, post-tensioned 200-millimeter (8-inch) in situ concrete floor slabs were preferred, spanning directly between the shear walls. This solution also gives flexibility in layout and services distribution, and accommodates a relatively tight 3-meter (9-foot-10-inch) floor-to-floor height.

Structural options for the SkyPark had to be evaluated early in the design process. Many were considered, including both steel and concrete variants, with weight and ease of construction the primary parameters. Eventually a system of post-tensioned segmental steel box girders 10 meters (33 feet) deep by 3.7 meters (12 feet) wide with 35-millimeter ($1\,^3/_8$-inch) sidewalls and 60-millimeter ($2\,^3/_8$-inch) flanges was adopted for the 66.5-meter cantilever, and a system of primary steel bridge trusses for those segments spanning between the hotel towers. Raking steel "V" struts mounted on the tower roofs directly above the shear walls support the bridges above each tower, with the segments between the towers simply supported by the ends of the above-tower segments. Well established bridge construction technology was also adopted: SkyPark sections were strand-jacked into position atop the towers.

Accommodating the differential movements between the towers caused by wind, thermal expansion, shrinkage, and creep required particular attention to movement joint design. Architectural detailing associated with the infinity pool helped conceal these joints and give the impression of a monolithic deck running across all three towers.

Another potential problem for a leisure-oriented structure at such a height was vibration, be it from wind or from such rhythmic human activities as running or dancing. To prevent any such movements from developing to uncomfortable or even dangerous levels, a 5-ton tuned mass damper was installed in the cantilever, and post-completion dynamic testing has shown that this should keep structural movement within comfort levels.

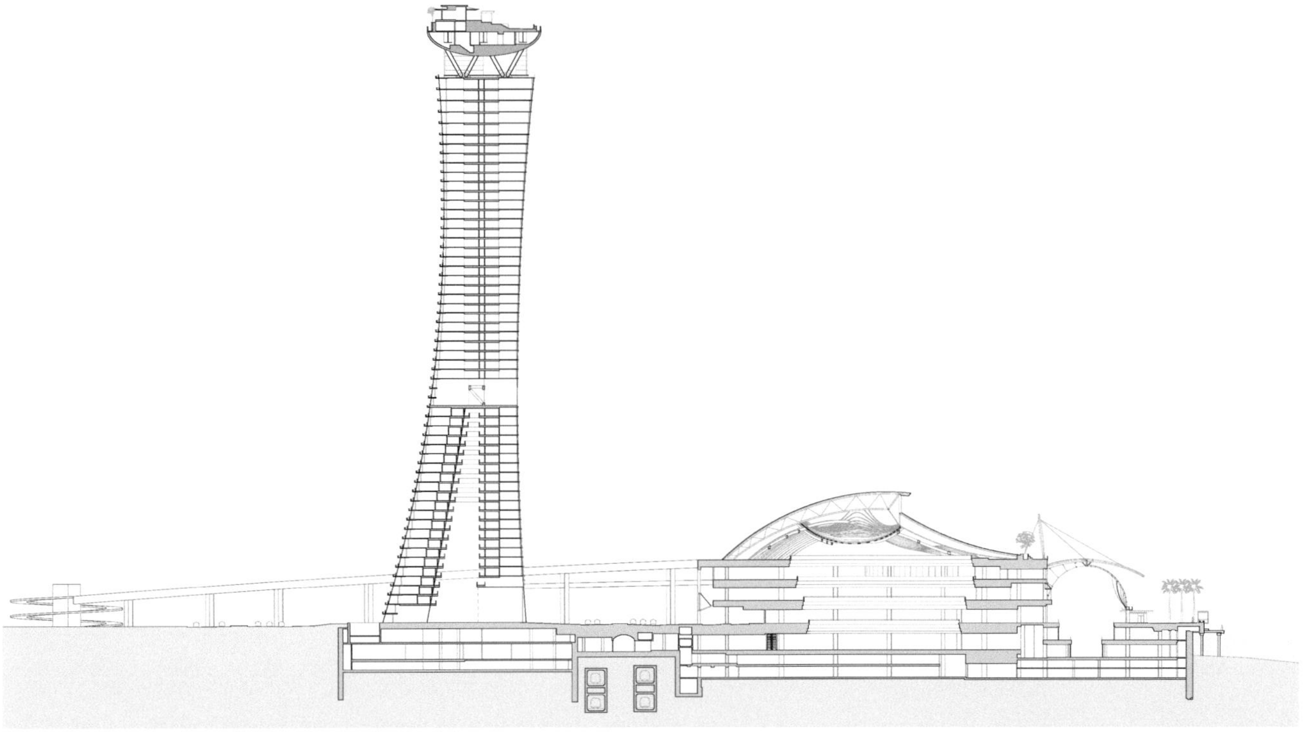

Date:
Scale: 1:1000 0 10 25 50 m
Moshe Safdie and Associates, Inc. ©

Section through Hotel, Casino, Retail and Promenade

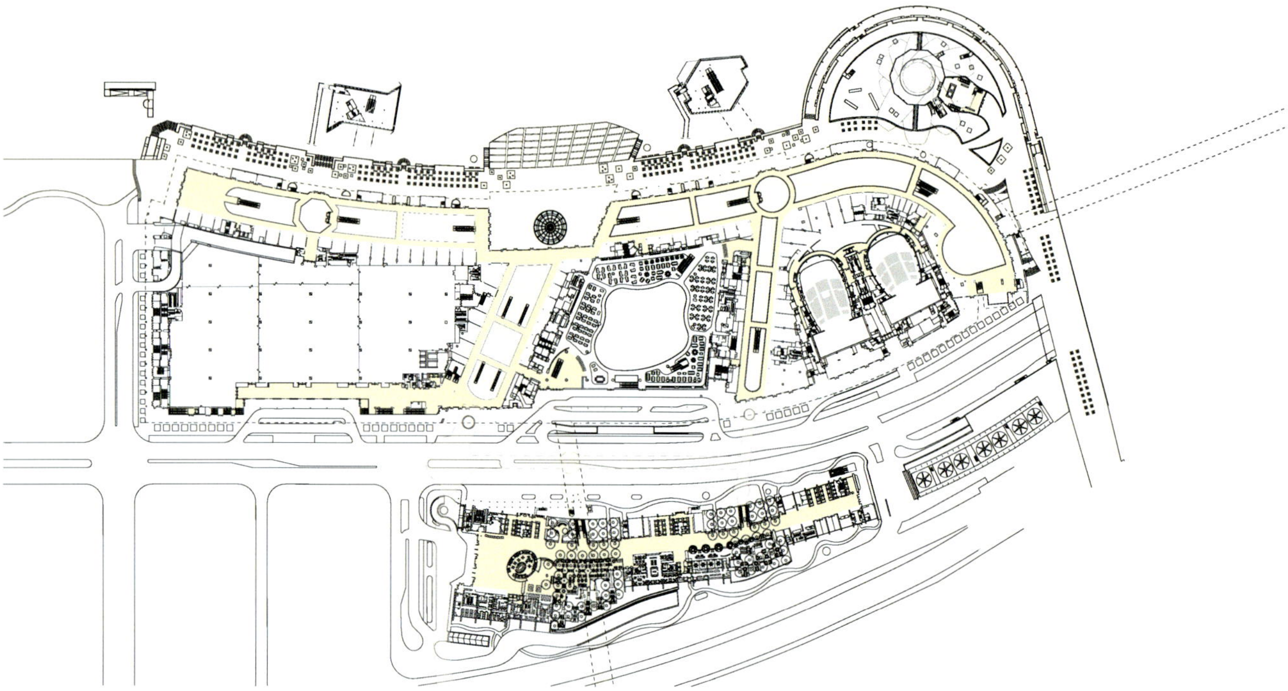

Date

Scale 1:1000 0 10 25 50 m

Level 1 Plan

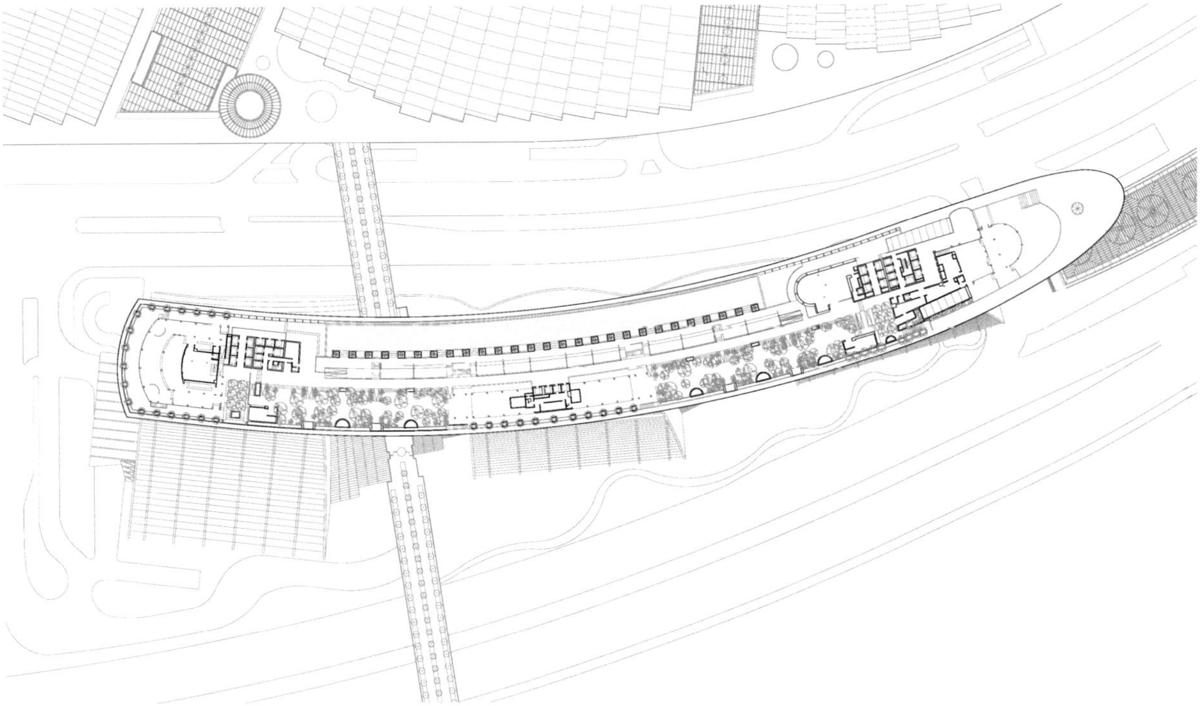

Date:

Scale: 1:1000 0 10 25 50 m

Moshe Safdie and Associates, Inc. ©

Level 57 Plan

Case Study 11

Missing Matrix Building: Boutique Monaco

Seoul, South Korea

HEIGHT:	100 M
NUMBER OF STORIES:	27
GROSS FLOOR AREA:	55,000 M²
PRIMARY USE:	RESIDENTIAL, OFFICES
START ON SITE:	2004
COMPLETION:	2008
PROJECT COST:	NOT AVAILABLE
OWNER/DEVELOPER:	BUMWOO CO. LTD, LEADWAY CO. LTD
LEAD ARCHITECT:	MASS STUDIES
ASSOCIATE ARCHITECT:	ZONGXOO U
STRUCTURAL ENGINEER:	TEO STRUCTURE
MEP ENGINEER:	HANA CONSULTING ENGINEERS CO., LTD
MAIN CONTRACTOR:	GS E&C
LANDSCAPING CONSULTANTS:	ENVIRONMENTAL DESIGN STUDIO

Project Description

In Korea, the ratio of the urban to non-urban population 50 years ago was 20 percent to 80 percent. Now, the ratio has inverted: 82 percent of the entire Korean population lives in urbanized areas, leaving 18 percent residing in non-urban areas. This makes Korea the most metropolitan populace in the world—and the number is expected to grow to over 90 percent by the year 2030.

These conditions exist as a result of systematically focusing on growing ever vertically. So far, the sole method for expanding a building vertically has been to stack any number of identical flat slabs, allowing for a maximization of production efficiency. Over the last 40 years this approach has, however, produced vast, homogeneous urban spaces, particularly in Seoul, which is now one of the densest cities in the world.

Countering this systematic force, a completely opposite force has begun operating in Korea in the last decade. When constructing vertical buildings—be it housing or offices—heterogeneity or "difference" achieved by design has emerged as a significant virtue in this icon-obsessed economy—as it has in other large metropolises throughout the world.

This project involves the construction of a tower with a floor area of 55,000 square meters, the lower levels of which are composed of commercial, cultural, and community spaces, while the upper floors, from the fifth to 27th, are "officetels," residences which could also be used as offices during the day. To ensure the maximum building footprint ratio (40 percent) as well as optimal natural light conditions (with southern exposure), a "C"-shaped plan is extruded into a 27-story tower reaching a height of 100 meters, the maximum height allowed by law. As such, when this plan with a maximum footprint ratio is simply repeated vertically, the floor space of the entire tower mass exceeds the legally allowed amount by approximately 10 percent. To reduce this mass systematically, missing matrices are introduced to meet maximum floor area ratio (97 percent) throughout the building mass, with a pattern of carved-out space. With these 15 missing spaces, the building gains more exterior surface and corners for enhanced lighting and viewing conditions.

Architecture

Spaces created by missing matrices are landscaped with trees visible from the inside and outside of the building. Inside the tower, a total of 49 different types of units, 172 units in total, are arranged heterogeneously to reflect and exploit rich spatial conditions. For example,

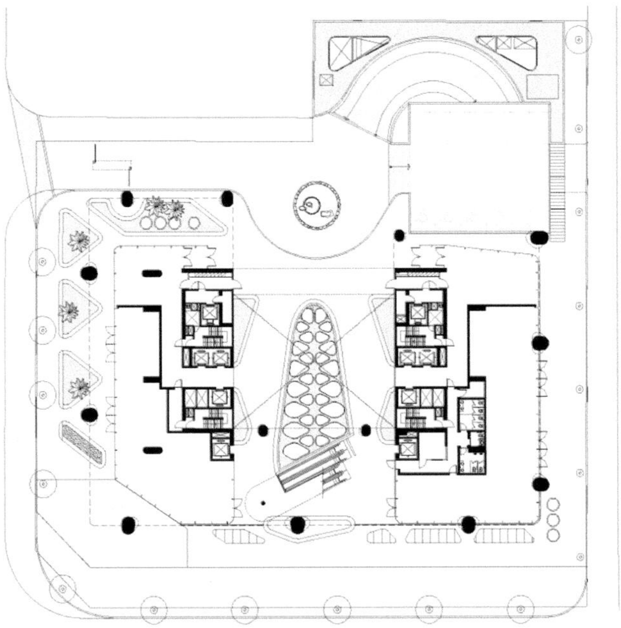

in the area created by the 15 missing matrices, there are 40 units with bridges that divide public (living/dining) and private (bedroom) spaces, along with 22 units with gardens. In addition, protruding spiral staircases are planned within the missing matrices, further adding to the heterogeneousness of the interior.

On the ground floor, the building opens up and connects to its surroundings through its urban park. Sidewalks are spacious, at 9 meters from its southern end, 10 meters from the west, and 6 meters from the east, allowing for small gardens and benches to provide a space of rest. The tall spaces of the first floor are occupied by retail stores and coffee shops, further encouraging the use and liveliness of this exterior space. At both extremes of this "U"-shaped floor plan are the lobbies for residents, which are also strategically placed to be the closest to the parking drop-off zone. The center of the building front opens up to elevators and escalators connecting to the cultural functions below and more retail stores above, as well as a small pocket park in the center of the site, to be appropriated as a cozy enclosure to rest in the midst of this metropolis.

Luxurious retail shops occupy the second and third floors. Along with the fourth floor, a glass bridge with a bamboo garden on both sides spans 20 meters to turn the "U"-shaped floor plan into a full loop, enhancing its spatial functionality by completing the floor's circulation.

The fourth floor has community facilities for the residents of the building. A bar and lounge space offering opportunities for meetings and gatherings is placed in the center, with a business center, conference room, and two guest rooms branching out to the east, and a fitness center, locker room, shower facilities, and a maintenance office to the west.

From the outside, Boutique Monaco seems to be a single "C"-shaped tower, but the interior organization of circulation around the two cores sometimes allows the building to act as two separate towers, depending on the spatial functionality. The ground floor separates the retail stores and the public spaces from the private spaces for the residents appropriately; 22-meter bridges are installed on the second to fourth floors to create a continuous circulation through the public spaces of retail stores and community spaces; the upper floors, from the fifth to 27th, have been split into east and west, each using their corresponding core, to enhance privacy for each of the five to 10 units per floor. This separates the building's 172 units into two groups of 81 and 91, ensuring that only half of the residents use each core, or only two to five residents per floor. Another, private garden can be found on the roof, for residents' use only. As on the second, third, and fourth floors, a trussed bridge connects the opposite ends to provide a spectacular view from more than 100 meters above ground.

Structure

Boutique Monaco is formed by stacking three key structural types. The upper floors (the fifth to 27th) have a reinforced concrete structure with post-tensioned slabs to support the long cantilevering spans, and the total weight of these floors is transferred by (maximum) 1.9-meter deep transfer girders on the fifth floor, to be distributed among the branch-type trusses along the building's exterior on the second to fourth floors, as well as the two cores and the remaining columns. This weight load is then taken underground by 13 large composite steel/concrete piers along the exterior of the building's ground floor, completing the building's overall composition.

This special structure and the transfer floor on the fifth level allows the relatively small apartment or "officetel" modules (8.8 by 9.5 meters in this case), to span twice as wide in the spaces below, opening up a spacious 17.6-meter to 22-meter span for the public on the ground floor.

In many ways Boutique Monaco is light-years ahead of the typical commercial high-rise tower. Its dynamic massing, with 15 voids introducing green spaces throughout the tower, the sheer variety of units with different spatial configurations, and the inclusion of appropriate retail and leisure facilities provide a rich living experience for the tower's residents.

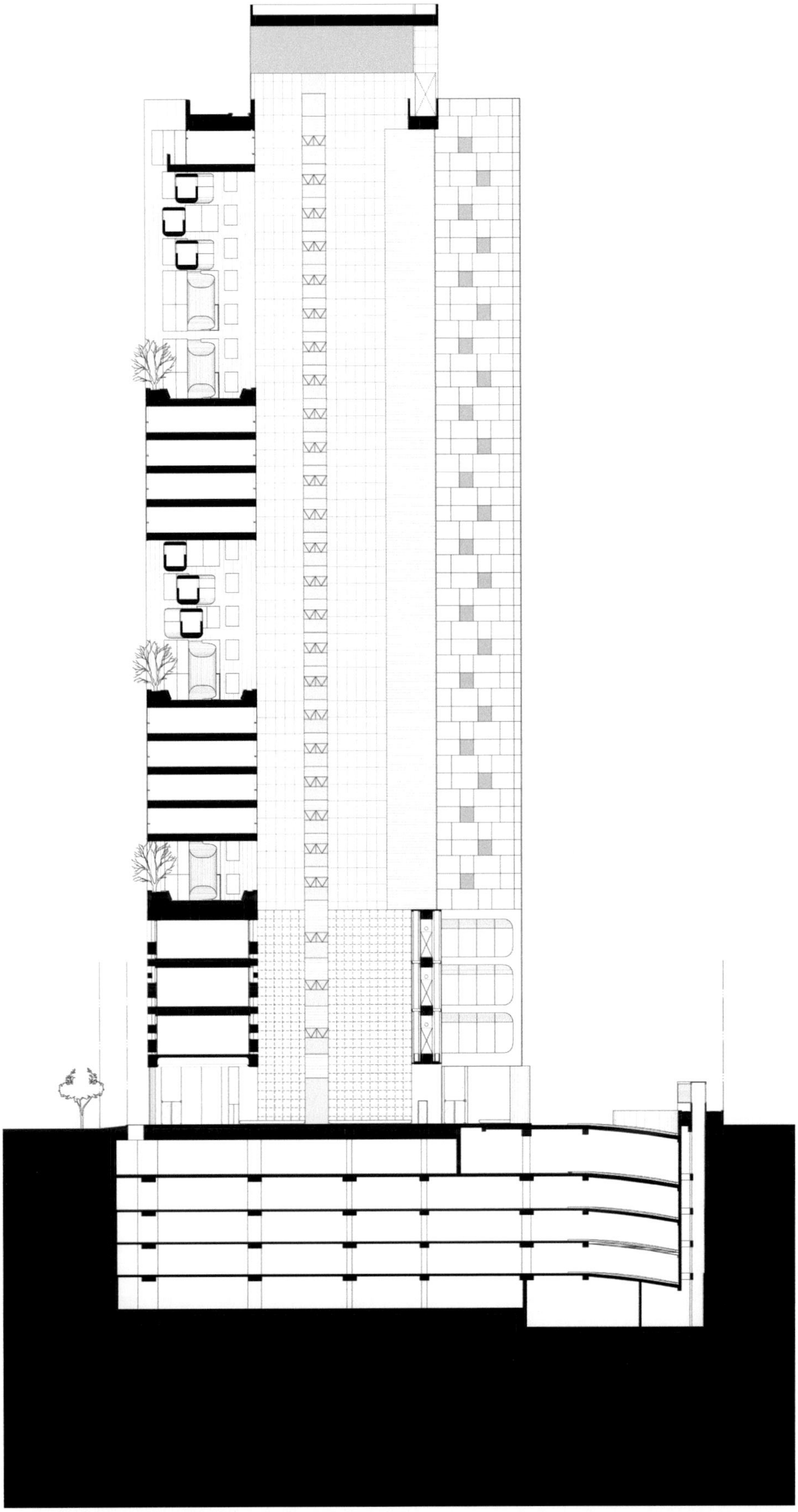

Case Study 12

Mode Gakuen Spiral Towers

Nagoya, Japan

HEIGHT:	GL PLUS 170 M
NUMBER OF STORIES:	38 PLUS 3 BASEMENTS
GROSS FLOOR AREA:	48,989 M^2
PRIMARY USE:	EDUCATION (DESIGN AND VOCATIONAL SCHOOLS), RETAIL
START ON SITE:	OCTOBER 2005
COMPLETION:	FEBRUARY 2008
PROJECT COST:	NOT AVAILABLE
OWNER/DEVELOPER:	MODE GAKUEN
LEAD ARCHITECT:	NIKKEN SEKKEI LTD.
ASSOCIATE ARCHITECT:	NIKKEN SEKKEI LTD.
STRUCTURAL ENGINEER:	NIKKEN SEKKEI LTD.
MEP ENGINEER:	NIKKEN SEKKEI LTD.
MAIN CONTRACTOR:	OBAYASHI CORPORATION

Project Description

Mode Gakuen, a design and vocational school with locations in Tokyo, Osaka, Nagoya, and Paris, held a design competition, with this design selected as the winner. Mode Gakuen Spiral Towers is a high-rise building located on a busy street in front of Nagoya Station. The building is named "Spiral Towers," plural, because there are three unique towers that intertwine into a spiral form. Three vocational schools are housed in the building: Mode Gakuen for fashion, HAL for computer, information technology, and design, and ISEN for medicine and welfare studies.

Following a development principle common to the whole Nagoya Station area, the design was planned to optimize the pedestrian network and spaces. As a result, the streets have a unified setback to the building walls, greenery was planted, and the sidewalks widened. In addition, cafés and retail stores are placed in front of the building to make the city come alive. An underground walkway is also featured, strategically planned with improvement of safety standards in mind.

The goal was to create a uniquely shaped building with an open and airy interior as outlined by the client's design concept, which called for a unique and attractive shape to promote the creativity of its students, and also to create a school building of which students could be proud. The three wings were chosen for the outside of the building to reflect the three different disciplines of the schools housed within it.

Architecture

While many high-rise buildings have been built around Nagoya Station in recent years, the Spiral Towers stands out as a truly unique building, a new landmark on the city skyline. The design appears to change shape when viewed from different angles and exudes an elegant yet dynamic appearance. A sunken garden created at the base of the building acts to connect and provide continuity between the floors above and below ground level. The building also contributes to the future prosperity of Nagoya by providing a highly creative study environment for more than 8,000 students.

At nighttime the beauty of the building is highlighted by lighting; white lighting in the windows is gradually replaced by blue lighting after classes finish. Also, three laser beams housed within the building provide passersby with a light show once every hour, bringing the city alive with light and color.

Choosing a spiral form results in a unique and interesting space for the interior. Most of the classrooms are fan-shaped, and extra space remains behind the platform and in the corners of the rooms which is used as storage space and allows ease of use. Moreover, a multi-purpose hall with 396 seats arranged in stadium seating takes full advantage of the unique space, illuminating flat sounds using the spiral shape incorporated in the interior.

The entrance hall, library, and job guidance rooms were designed as common areas in which students from the three different schools can interact and communicate with each other. Lounges are also created on each floor, providing spaces where students can meet and relax.

Students can also socialise in the high-ceilinged atriums that are located on every other floor.

Many ecological features can be found in the building, including double-glazed windows with airflow systems between the window panes to maintain transparency of the exterior and provide clear views from the interior. The inside sliding window frames are narrow, at 20 millimeters, to accommodate the design of outside triangle frames, and horizontal blinds were installed between the double glazed windows. Blinds are colored white on the outside to reflect the sun's rays, and black on the inside to create a good view and comfortable classroom environment. Furthermore, highly efficient fluorescent lights, outside air cooling systems, and natural air ventilation systems are also used to conserve energy.

Structure

The building is a planar structure with three fan-shaped wing sections connected radially to an oval-shaped center core. The spiral appearance of the building is a product of a floor-by-floor rotation (3 degrees per floor) of the wing sections, with a gradual reduction (1 percent per floor) in their size.

In designing a building with an organic spiral form in earthquake-prone Japan, it was decided not to adopt a structural frame that would depend only on its structural configuration, but to make a structural framing system that would offer a rational and strong yet delicate structural expression. A structural design in harmony with the structural plan was realized by adopting a strong, rigid truss-

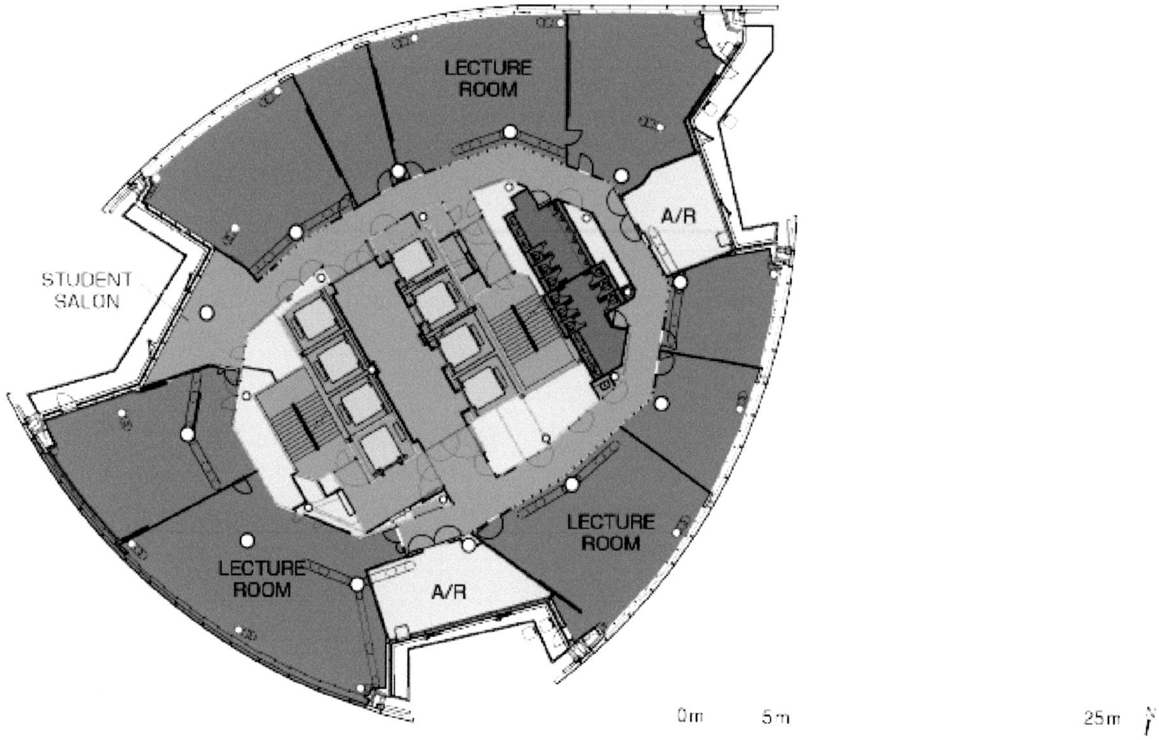

tube structure for the center core and framing structures with slender columns intertwined in a spiral manner.

Twelve straight concrete-filled steel tube (CFT) columns are arranged around this core, and steel tube braces are connected to these columns in a mesh network, forming the inner truss tube. This inner truss tube is highly strong and rigid with regard to the horizontal and twisting forces exerted on the building by earthquakes and high winds, providing the necessary structural performance. With no braces around the outside, a transparent appearance is achieved, and minimal, thin-diameter columns provide lower rigidity for a light frame that need not bear seismic forces.

Due to gravity, horizontal forces are constantly being generated in the inclined columns of the building's outer periphery. The forces thus generated exert torsion on the entire building structure and therefore demand the addition of members to prevent collapse of the columns. The generated horizontal forces are transferred to the inner truss tube via girders that form a horizontal truss, thereby counteracting the torsion that is exerted on the entire building structure.

Vibration-control columns fitted with viscous dampers on the column position are arranged at 26 locations, and thus it has been possible to achieve a large damping effect. Braces are installed on the uppermost section of the vibration-control columns to form a cantilever truss that extends from the inner truss tube and supports the constant load.

A rooftop seismic damping system was also adopted. In this system, which makes full use of the increases in deformation that occur in the tops of buildings during earthquakes, 1 percent of the total weight of the upper structure is loaded on the upper surface of rotation bearings on the rooftop, and seismic energy is absorbed using deformation of lead dampers. The cycle of the additional mass is synchronized with that of the building by using laminated rubber isolators. Also added as a fail-safe mechanism are fenders of the type normally used as shock absorbers between ships and quays. Dynamic analysis has confirmed that it is possible to obtain a maximum reduction of 22 percent in earthquake-induced deformation by using vibration-control columns and the adopted rooftop seismic damping system.

As regards the foundation work, it consists of cast-in-place concrete diaphragm walls and piles (enlarged base) supported by a layer of sand and gravel (with pile top at ground level – 41 meters). Steel-frame construction was adopted for the underground section of the building, while steel-reinforced concrete construction was selected for the third basement floor. The lowest layer of the inner truss tube is connected to the diaphragm wall at the building's outer periphery by rigid bearing walls that secure the building against rotational collapse.

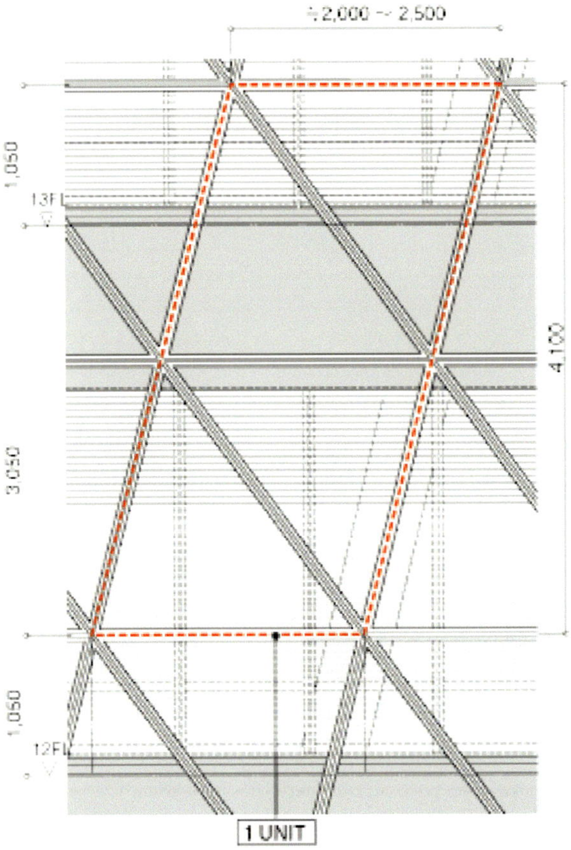

FACADE : UNIT CURTAIN-WALL

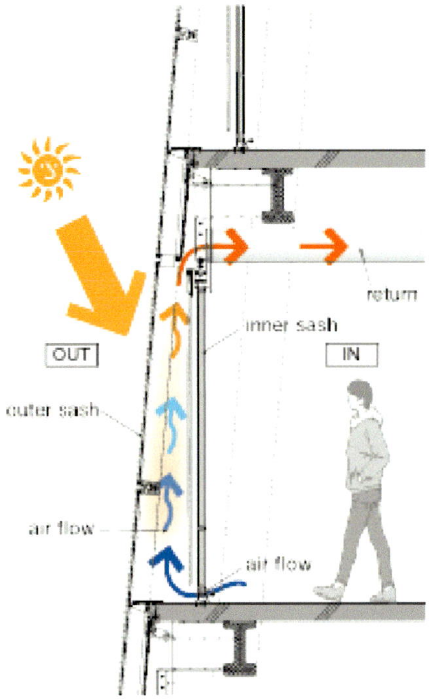

DOUBLE SKIN AIR FLOW WINDOW SYSYTEM

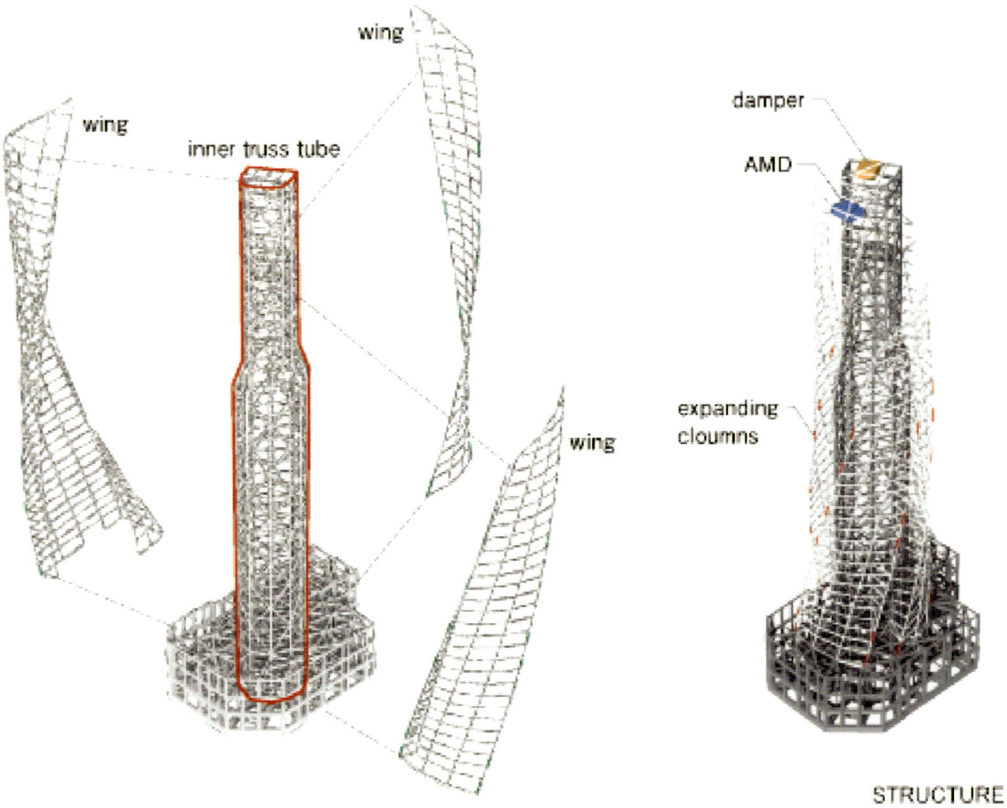

STRUCTURE DESIGN

英会話
&留学
英語・中国語・韓国語

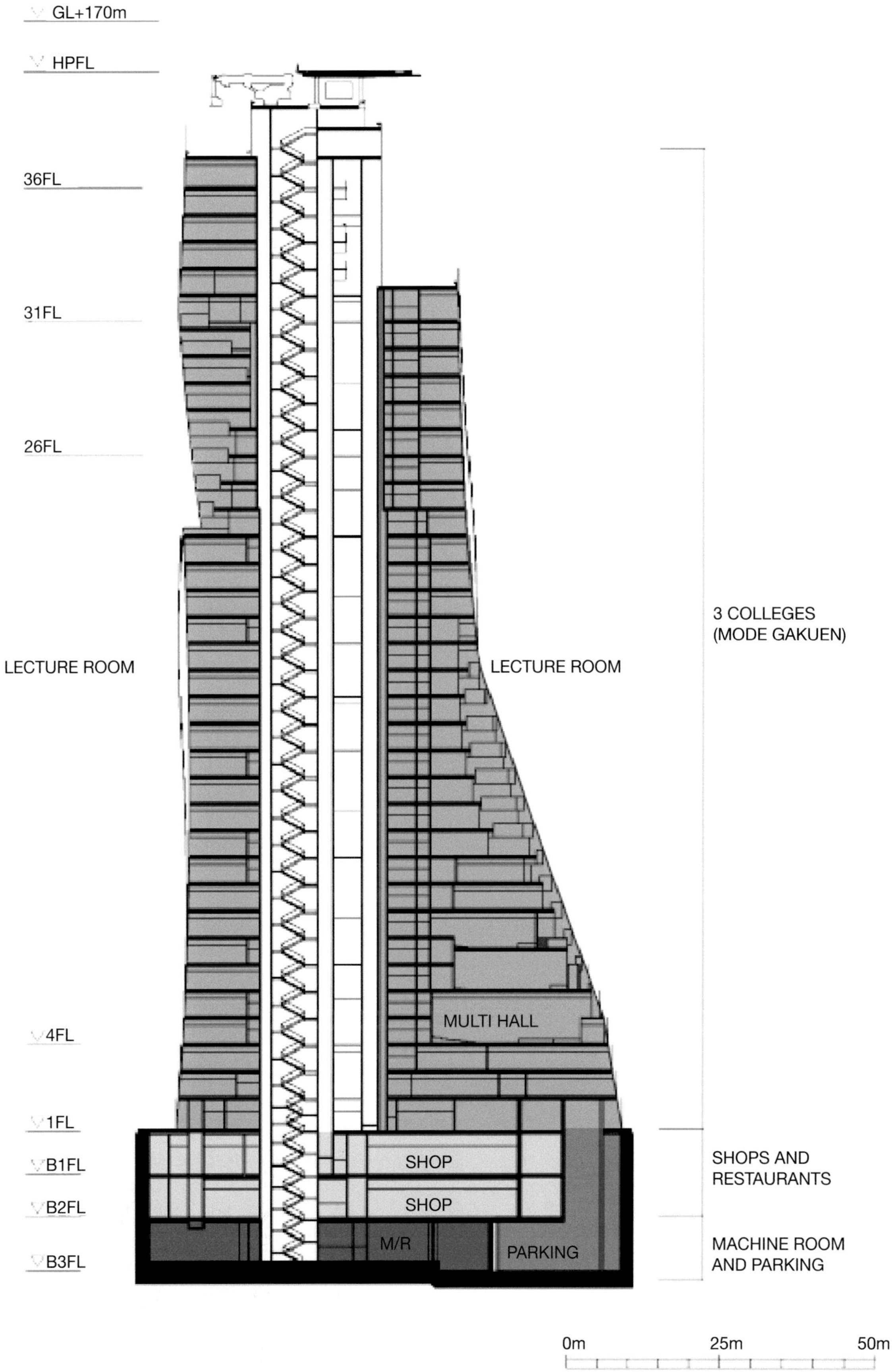
GL+170m
HPFL
36FL
31FL
26FL
LECTURE ROOM
LECTURE ROOM
MULTI HALL
4FL
1FL
B1FL
B2FL
B3FL
SHOP
SHOP
M/R
PARKING
3 COLLEGES
(MODE GAKUEN)
SHOPS AND
RESTAURANTS
MACHINE ROOM
AND PARKING
0m
25m
50m

Case Study 13

The Met

Bangkok, Thailand

HEIGHT:	231 M
NUMBER OF STORIES:	66 HABITABLE, 3 M&E ABOVE
GROSS FLOOR AREA:	124,885 M^2
PRIMARY USE:	RESIDENTIAL
START ON SITE:	AUGUST 2005
COMPLETION:	DECEMBER 2009
PROJECT COST:	$132 MILLION
OWNER/DEVELOPER:	PEBBLE BAY THAILAND CO. LTD.
LEAD ARCHITECT:	WOHA
ASSOCIATE ARCHITECT:	TANDEM ARCHITECTS LLC
STRUCTURAL ENGINEER:	WORLEY PTE. LTD.
MEP ENGINEER:	LINCOLNE SCOTT NG PTE. LTD
MAIN CONTRACTOR:	BOUYGUES THAI LTD.

Project Description

Currently (2012) Thailand's tallest condominium and Bangkok's fourth tallest building, The Met features alternatives to high-rise living models developed first in North America, where cold weather and strong winds are to be expected. Temperate climate residential apartments are typically compact, well insulated, and separated from the exterior. In tropical areas like Bangkok, however, the balmy climate encourages outdoor living, and height offers less noise and dust, lower humidity, stronger cooling breezes, and greater security. This project explores how aspects of low-rise tropical housing can be applied to create outdoor-indoor spaces in the sky. Residential density is extremely high—a plot ratio of 10:1—and the building is located between two train stations to encourage the use of public transport in a city notorious for its near grid-lock traffic jams.

Architecture

Inspiration for the design came from traditional Thai ceramic tiles, textures, and timber paneling. Temple tiles influenced the cladding, while the distinctive arrangement of the balconies recalls the staggered teak paneling on traditional Thai houses. Random inserts of faceted, polished stainless steel are incorporated into the walls, creating a delightful glittering effect at a scale appropriate to the vast city but harking back to the sparkling mirrors seen in Thai temples. Departures from the traditional western high-rise solid rectilinear box can also be seen in the tower's massing. Sky terraces link three staggered blocks every five stories, creating dramatic yet human-scaled external public and private spaces. This unique massing design creates tropical "houses in the sky," complete with breezeways, full exposure to light and spectacular views on all four sides, outdoor living areas, planters and high-rise gardens, open-air communal terraces, libraries, spas, and other facilities.

Every horizontal surface of the building is planted. Additionally, vertical faces are shaded by green creeper screens, rising up the full 66 residential stories. Residents are also provided with balconies containing private planters. All apartments are cross-ventilated, and all face north and south, making life without air-conditioning both possible and acceptable.

Common areas are spread throughout the towers, offering inhabitants a variety of experiences, from the intricately designed carpet of water, stone, and vegetation at ground level, to the extensive indoor-outdoor facilities at the pool level, to libraries, barbeques, and function areas at sky terraces.

A wide range of passive strategies are used to reduce energy consumption and minimize environmental impact. Shading from overhanging ledges and perforated metal screens minimize heat soak from sunlight, while living green screens at the east and west walls and the car parking areas shade the building and cool it through transpiration. Further cooling comes from the water gardens at ground level and on recreational floors, which also store rainwater, while photosynthesis from the extensive planting improves local air quality.

Together with the building's height and slenderness, lighting design transforms The Met into a dramatic and iconic statement on the city skyline. Highlighted sky gardens further emphasize the building's unique features, creating a desirable, lush appearance with distinctive Thai overtones, and bringing cool, dark, natural relief to the gray concrete vistas of central Bangkok.

Structure

Structural engineering is fully integrated with architectural design. The perimeter columns for the three towers are conventional reinforced concrete (RC) and set out on a 4.5-meter grid, making up 9.0-meter wide apartment modules. As columns increase in size as the loads accrue, the columns enlarge on the exterior of the building, creating protected indoor-outdoor spaces for balconies and terraces, and allowing apartment layouts to be standardized, even at lower levels. These exposed buttress columns are both structurally rigorous as well as an architectural expression. Using a precast concrete floor system for the towers, a construction cycle time of five days per typical floor was achieved.

The slender profile of the towers results in a height-to-width ratio on the order of 8:1 in the north–south direction. This requires that the full width of the building be utilized to resist lateral forces due to wind and seismic effects. Wind tunnel tests were carried out to ensure safety and comfort on the sky terraces.

Wind forces govern the design of the lateral system in the north–south direction, while seismic forces govern in the east–west direction. In the north–south (transverse) direction, the lateral load-resisting system comprises an assembly of RC core walls and other shear walls interconnected by RC coupling beams and RC outriggers. The stiffness of the perimeter moment frames is also mobilised to activate the intermediate tower columns above Level 10. This effectively enhances the lateral stiffness of the building and makes efficient use of the vertical structural elements. The three towers are also structurally tied together at sky deck and skygarden levels at every fifth floor, in areas used for private gardens, private pools, and common areas. Floor diaphragms are designed to transmit loads horizontally so that torsional loads generated due to dynamic wind and seismic effects are resisted effectively. In the east–west (longitudinal) direction, a dual lateral load-resisting system is utilized, comprising perimeter RC moment-resisting frames combined with coupled shear walls.

The Met's foundation comprises 400 one-meter diameter bored piles in conjunction with a continuous 3-meter deep reinforced concrete pile cap. Piles are founded at depths upwards of 50 meters.

THE MET
SECTION A-A
0 5 10 20 50 M
1 : 1000
WOHA DESIGNS PTE LTD / WOHA ARCHITECTS PTE LTD
OCTOBER 2008
© WOHA Pte Ltd 2008. All rights reserved.

Case Study 14

Shanghai Tower

Shanghai, China

HEIGHT:	632 M
NUMBER OF STORIES:	128
GROSS FLOOR AREA:	521,000 M^2
PRIMARY USE:	MIXED USE: OFFICES, HOTEL, RETAIL
START ON SITE:	NOVEMBER 2008
COMPLETION:	2015
PROJECT COST:	NOT AVAILABLE
OWNER/DEVELOPER:	THE SHANGHAI TOWER CONSTRUCTION & DEVELOPMENT CO. LTD
LEAD ARCHITECT:	GENSLER
ASSOCIATE ARCHITECT:	TONGJI UNIVERSITY (LOCAL DESIGN INSTITUTE)
STRUCTURAL ENGINEER:	THORNTON TOMASETTI
MEP ENGINEER:	COSENTINI
MAIN CONTRACTOR:	THE SHANGHAI TOWER CONSTRUCTION & DEVELOPMENT CO. LTD
LANDSCAPE:	SWA

Project Description

When completed in 2014, Shanghai Tower will anchor the city's Lujiazui district, which has emerged as one of East Asia's leading financial centers. Designed to embody Shanghai's rich culture, the 632-meter high mixed-use building will complete the city's super-high-rise precinct. The new tower takes inspiration from Shanghai's tradition of parks and neighborhoods. Its curved facade and spiraling form symbolize the dynamic emergence of modern China. By incorporating sustainable best practices, Shanghai Tower is at the forefront of a new generation of super-high-rise towers, achieving the highest level of performance and offering unprecedented community access.

The Lujiazui zone in Shanghai has been transformed from farmland to financial center in two decades, resulting in a skyline and architectural landscape that need a unifying landmark. With a rounded triangular footprint derived both from the bend in the nearby Huangpu River and from its relationship to the Jin Mao Tower and the Shanghai World Financial Center, Shanghai Tower will be a signature icon for the city of Shanghai. At the same time, it completes the precinct's harmonious trio of buildings and gives it its defining silhouette.

Architecture

Shanghai Tower is a city within a city, comprising nine vertical zones, each 12 to 15 stories high. With its emphasis on public space and its shops, restaurants, and other urban amenities strategically located at the floors with public atriums, Shanghai Tower envisions a new way of inhabiting supertall towers. Each of the building's neighborhoods rises from a "sky lobby" at its base—a light-filled garden atrium that creates a sense of community and supports daily life. The sky lobbies hearken back to the city's historic open courtyards, bringing people together throughout the day.

The upper floors will house hotels, cultural venues, and an observation deck with sweeping views of the Shanghai skyline. Central floors will house office space. A six-story retail podium concentrates shopping and dining near the base. And the ground floor will serve as an "urban market."

Both gateway and connector, the Shanghai Tower retail podium will be a world-class destination for shoppers, office workers, and hotel guests. Clad in luminous cast-glass tiles, the podium incorporates a mix of premium brands, one-of-a-kind specialty retailers, and high-concept dining. Strategically placed entrances funnel pedestrians into public spaces connected by a

network of concourses, escalators, stairs, and balconies. A five-story atrium with a glass facade opens toward the city, making an important connection between inside and out.

The design team anticipated that three important design strategies—the asymmetry of the tower's form, its tapering profile, and its rounded corners—would allow the building to withstand typhoon wind forces common to Shanghai. Using wind tunnel tests, the designers refined the tower's form, pinpointing a 120-degree rotation as the optimum for minimizing wind loads. Findings from the testing produced a structure and shape that reduce wind loads by 24 percent, ultimately yielding savings of $58 million in construction costs.

Structure

The tower's scale and complexity have created so many "firsts" for China's construction industry that more than 100 expert panels have been established to analyze every aspect of the design. Yet the structural system is relatively straightforward, designed in response to a windy climate, an active earthquake zone, and clay-based soils. At the heart of the structure is a concrete core. This core acts in concert with an outrigger and supercolumn system, with double-belt trusses that support the base of each vertical zone. The result is an easier and faster construction process and significant cost savings for the client.

To carry the load of the transparent glass skin, Gensler designed an innovative curtain wall suspended from the mechanical floors above and stabilized by a system of hoop rings and struts. And the strategic division of the tower into distinct vertical zones will supply the lifeblood of the building's heating, cooling, water, and power throughout, using less energy and at lower cost.

An innovative feature of the design is the incorporation of two independent curtain walls—the outer skin is cam-shaped in plan, the inner one is circular. The space between them forms atria that will house the landscaped sky gardens at regular intervals throughout the building. These bright atria will improve air quality, conserve energy by modulating the temperature within the void, create visual connections between the city and the tower's interiors, and provide places where building users can interact and mingle.

Sustainability

Shanghai Tower will be one of the most sustainably advanced tall buildings in the world—designed to achieve both LEED Gold certification and a China Green Building Three Star rating. To achieve the targeted ratings, many strategies that will generate a positive environmental impact have been incorporated. The foundation of this approach is state-of-the-art water resource management practices and high-efficiency building systems. A full 33 percent of the site is green space, with landscaping that breathes fresh air into the city and shades paved areas that radiate heat. Locally sourced materials with high recycled content are being used when available. And the building's heating and cooling systems will tap the power of geothermal (ground source energy) technology to deliver energy from fluids maintained at the earth's constant temperature.

Additional sustainable strategies embedded in the project include:

- *Daylighting* The continuous glass skin admits the maximum amount of daylight into the atria, reducing the need for artificial lighting. Floor-to-ceiling glass in the office and hotel floors yields similar benefits to those spaces.
- *Sun-shading* To reduce heating and cooling loads, both the inner and outer curtain walls will have a spectrally selective Low-E coating. Fritted glass on the outer wall provides additional sun-shading, aided by horizontal ledges at each floor level that will block high summer sun.
- *Building controls* Shanghai Tower incorporates intelligent building controls that lower energy costs by monitoring and adjusting systems such as lighting, heating, cooling, ventilation, and self-generated power. Lighting controls alone will save more than $556,000 each year in energy.
- *Cogeneration system* The 2,200-kilowatt natural gas–fired cogeneration system provides electricity and heat energy to the low zone areas. In addition to providing site-generated power, the system produces 640 tons of refrigeration during the cooling season, and heat during the winter months.
- *Wind turbines* In keeping with the client's desire to demonstrate cutting-edge technologies, wind turbines at the top of the building will power the exterior lighting for the building and some of the park areas. The turbines will produce an estimated 54,000 kilowatt-hours per year in renewable energy.
- *Regional materials* The team seeks out building materials that are harvested and manufactured within an 800-kilometer radius of the site. Local sourcing of products is sustainable because it reduces transportation-related environmental impacts and boosts local economies.
- *Building envelope* The building's two curtain walls

create atria that act like an insulating blanket, reducing energy costs. Used indoor air is circulated through each atrium to temper the space, keeping the sun's heat out during summer and the building's heat in during winter.

- *Landscaping* One-third of the site will be dedicated green space, with extensive planting to lessen the heat-island effect of paved areas. Efficient irrigation systems, combined with plant materials requiring low watering, reduce overall water consumption.

All told, Shanghai Tower's sustainable strategies will reduce the building's carbon footprint by an estimated 34,000 metric tons CO_2 equivalent per annum.

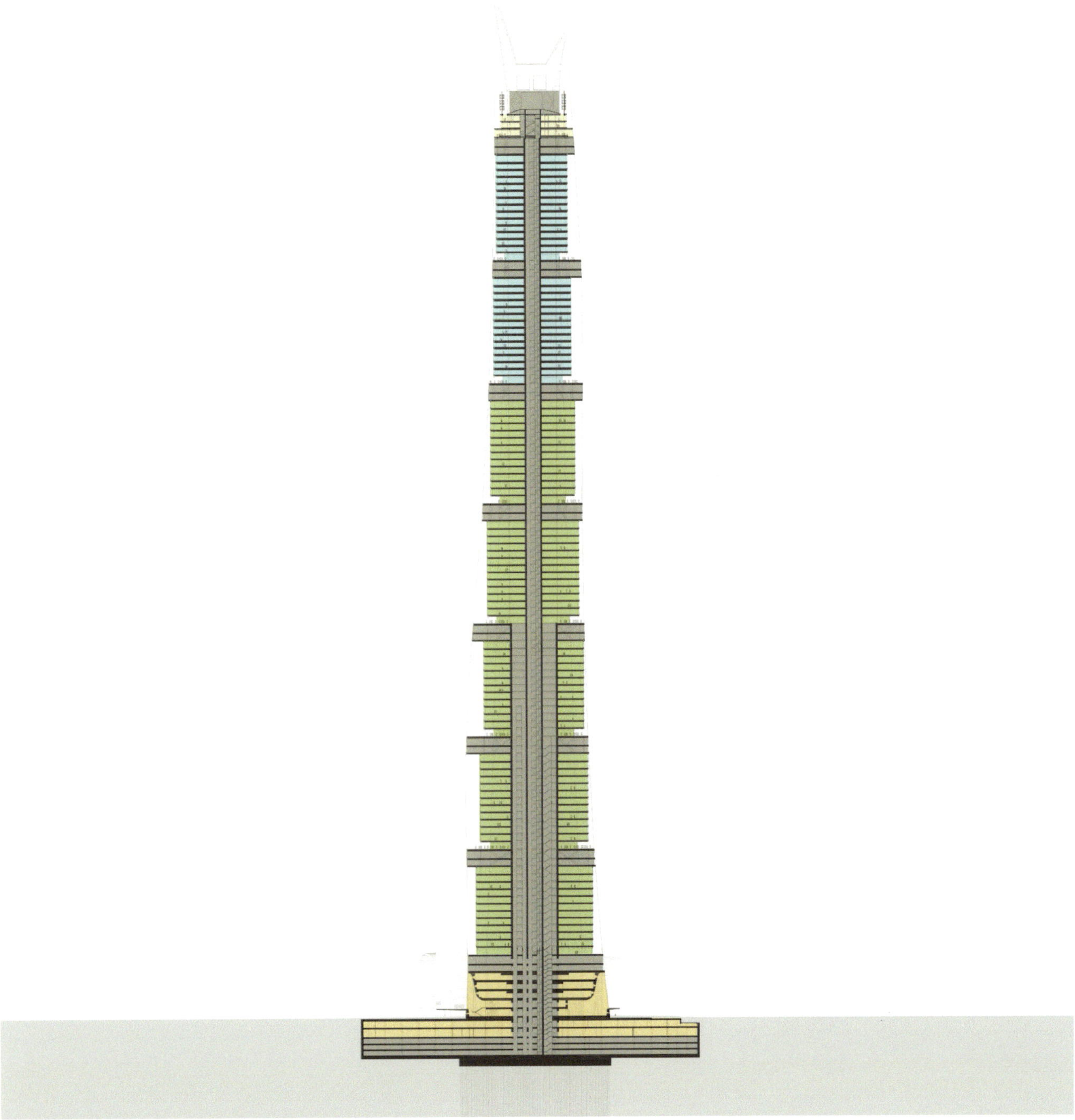

Case Study 15

The Troika

Kuala Lumpur, Malaysia

HEIGHT:	TOWERS A, B AND C: 204 M
NUMBER OF STOREYS:	TOWER A: 38; TOWER B: 44; TOWER C: 50
GROSS FLOOR AREA:	95,000 M^2
PRIMARY USE:	MIXED USE
START ON SITE:	2005
COMPLETION:	2011
PROJECT COST:	NOT AVAILABLE
OWNER/DEVELOPER:	BANDAR RAYA DEVELOPMENTS BERHAD
LEAD ARCHITECT:	FOSTER + PARTNERS
ASSOCIATE ARCHITECT:	GDP ARCHITECTS
STRUCTURAL ENGINEER:	WEB STRUCTURES SINGAPORE
MEP ENGINEER:	VALDUN (JURUTERA PERUNDING VALDUN SDN. BHD.)
MAIN CONTRACTOR:	IJM CONSTRUCTION SDN. BHD.
LANDSCAPE ARCHITECT:	SEKSAN DESIGN

Project Description

Located at the north-eastern corner of the Park, within the precincts of Kuala Lumpur City Centre – Kuala Lumpur's 'city within a city' – The Troika development combines apartments, offices, shops and restaurants within a single complex, with the aim of promoting a more densely planned approach to living and working in the twenty-first century city. The project comprises three apartment towers – of thirty-eight, forty-four, and fifty storeys respectively – which together form the tallest residential development in Malaysia.

Architecture

The geometry of the three towers evolved gradually through detailed modelling analysis, their forms being sculpted to respond to the surrounding buildings and to maximise the dramatic views of the Park, the Petronas Towers and the surrounding city. The unusual external structure consists of a number of slender concrete shear walls, which support a series of stacked blocks that rotate subtly to allow the primary living areas and balconies in each of the 230 apartments to focus on the best available view. A wide variety of plan sizes is possible due to the arrangement of the shear walls, and the internal organisation of the apartments is kept fluid, to facilitate individual planning options. Many areas are self-shaded by the overhang of the apartment above, which provides shelter on the balconies. Sky bridges link the towers at Level 24 to create a sky lobby with unrivalled views of the fast-changing skyline.

Each residential unit is naturally ventilated and has its own low-energy variable refrigerant volume (VRV) for extra cooling. Rainwater harvesting for irrigation purposes minimises potable water consumption.

At ground level, a four-storey perimeter commercial building contains offices, shops and cafés and frames a landscaped courtyard. Entirely free from cars, the courtyard forms the heart of the development – a tranquil urban oasis. Residents enter through the courtyard via a grand entrance on Jalan Binjai, which leads to lift banks on individual floors that are shared by two apartments each. At roof level, the perimeter building provides a variety of recreational facilities for residents. Linked by shaded arcades and accessible throughout the day, they add a further level of amenity to high-rise urban living.

Structure

Bored cast in situ concrete piles up to 2 metres in diameter and 49 metres in length support the towers. Top-down construction was used for the four-storey basement, to obviate the need for costly temporary works.

In principle the main tower structural frames consist of large-span flat slab floor construction coupling shear walls that are oriented at varying angles to one another on plan. The interfaces between the flat slab floors and the walls required special built-in steel connection brackets cast inside the concrete to offer a seamless connection between the floors and the walls.

Despite their heights of up to 204 metres, the shear walls are only 600 millimetres thick throughout. To achieve the specified accuracy of construction and eliminate the need for plywood shuttering, precast concrete permanent formwork panels were utilised. These were only 65 millimetres thick, prefabricated in sizes of up to 4, 000 by 1,100 millimetres, assembled in sets of three on the ground and hoisted into position on modified strong back frames. Custom-made rubber gaskets were installed on location to seal the joints between precast panels to prevent leakage of in situ concrete backfill. The precast panels are thus an integral part of the structural system.

Three internal cores, two central and one offset, work in conjunction with the coupled shear walls to resist wind-induced sway through the diaphragm action of the floor plates. These floor plates have 'virtual' beam edge reinforcement and cantilever out 6 metres at corners. Penthouse floors are composite with a maximum span of 18 metres.

Case Study 16

Hegau Tower

Singen, Germany

HEIGHT:	67.5 M
NUMBER OF STOREYS:	18
GROSS FLOOR AREA:	17,056 M^2
PRIMARY USE:	OFFICES
START ON SITE:	1999 (DESIGN PHASE)
COMPLETION:	2008
PROJECT COST:	€24,659,000
OWNER/DEVELOPER:	GVV STÄDTISCHE WOHNUNGSBAUGESELLSCHAFT
LEAD ARCHITECT:	MURPHY/JAHN
ASSOCIATE ARCHITECT:	RIEDE ARCHITECTS AND FISCHER & PARTNER ARCHITECTS
STRUCTURAL ENGINEER:	WERNER SOBEK
MEP ENGINEER:	SCHREIBER ENGINEERS
MAIN CONTRACTOR:	ED. ZÜBLIN AG

Project Description

Hegau Tower is located in Singen, Germany, 20 kilometres west of Lake Constance and in immediate proximity to the Swiss border. The two office buildings are part of a larger development plan south of the main train station and represent the transition of this industrial town of roughly 45,000 people from production to service-based businesses.

Architecture

An extension of the regular facade beyond the building volume with screen walls generates a continuous glass sheet toward the eastern plaza and merges the two buildings into one. A facade module of 2.70 metres fosters a generous ambiance of space.

The facade not only determines the visual appearance of the building, it also forms the interface between its interior and the exterior in respect to thermal, radiative, acoustic and visual exchange. Its basic design consists of large sheets of high-performance insulated glass with a transparent appearance yet effective sun-shading values. This basic and economic principle is enhanced by flexible components that allow the envelope to react to the requirements of the inner environment and the conditions of the outer environment.

Consequently, the facade becomes an integral part of the structure and technology of the building, and a direct expression of the economical and ecological goals of its design. The building's appearance is in keeping with the urbanistic, aesthetic and technological relevance of the project for the city and region, and the time of its inception.

Raised floors provide the opportunity to revise electrical and data cabling systems as technology progresses. Lightweight partition systems allow for fast and inexpensive adaptation of space to meet changing requirements. Floor-plate depth allows for variation in office layouts from open plan to cell to combined offices. In all variations, the use of daylight is maximized through floor-to-ceiling glass. Suspended lights with a direct and indirect component provide highly energy-efficient and glare-free illumination of the workspaces.

The collaboration of architect and client has resulted in a clear, modernist expression, in full agreement with the highly functional and energy-efficient detail of the design executed.

Structure

The tower is constructed as a concrete frame with stiffening core and perimeter columns. Structure and facade are designed for flexibility in office layouts and allow open plan, cell and combined offices, as well as multiple tenants per floor.

Sustainability

Natural ventilation is provided via hopper windows operated via chain motors and held by scissor hinges. This allows the user to access fresh air for cooling, improvement of air quality and the experience of connection to the outside. The exterior sunshade on the south-west facade is a retractable curtain of stainless steel bars. Via sensors or at the user's command, the sunshade covers the entire facade, reducing the solar load to a minimum while still providing a visual connection to the outside. The shading screen, developed by the company Clauss Markisen, is highly wind resistant and has had its first large-scale application in the Hegau Tower project. It works in conjunction with automatic interior perforated louvres on the other three facades.

PVC water tubes cast into the flat concrete floor slabs cool and heat the exposed thermal mass of the structure. The mass absorbs heat during the day and is cooled both actively and passively during the night. Thirty per cent of the basic cooling is thus provided without energy- and space-consuming air convection.

The goal of the energy concept is to achieve a maximum of comfort with a minimum of resource consumption. Interactions between all heating, ventilation and cooling components and their joint response to exterior conditions form a complex and efficient system, supported by the bus-driven building control system.

Fan-assisted heating and cooling convectors receive fresh air via the facade and bring conditioned air directly in, without a detour via centralized units. Heat recovery is achieved by running the return air via a heat exchanger and transferring its energy to the water for the thermal mass system.

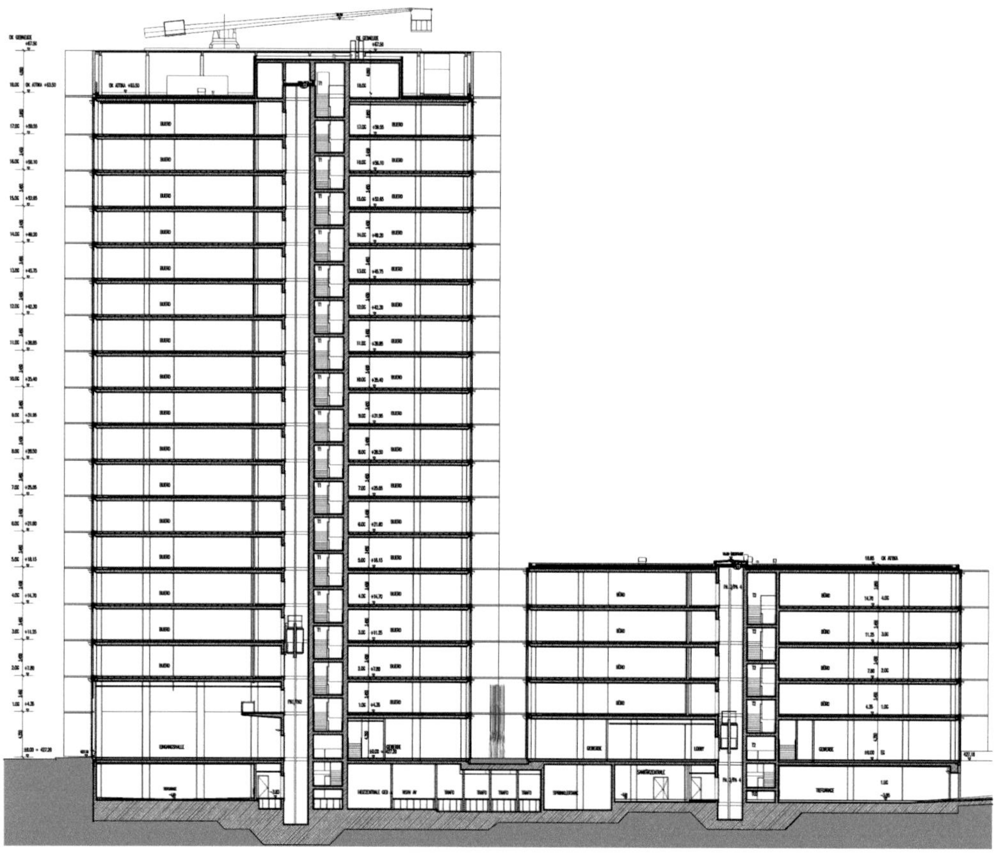

longitudinal section

0 2,5 12,5 25 m

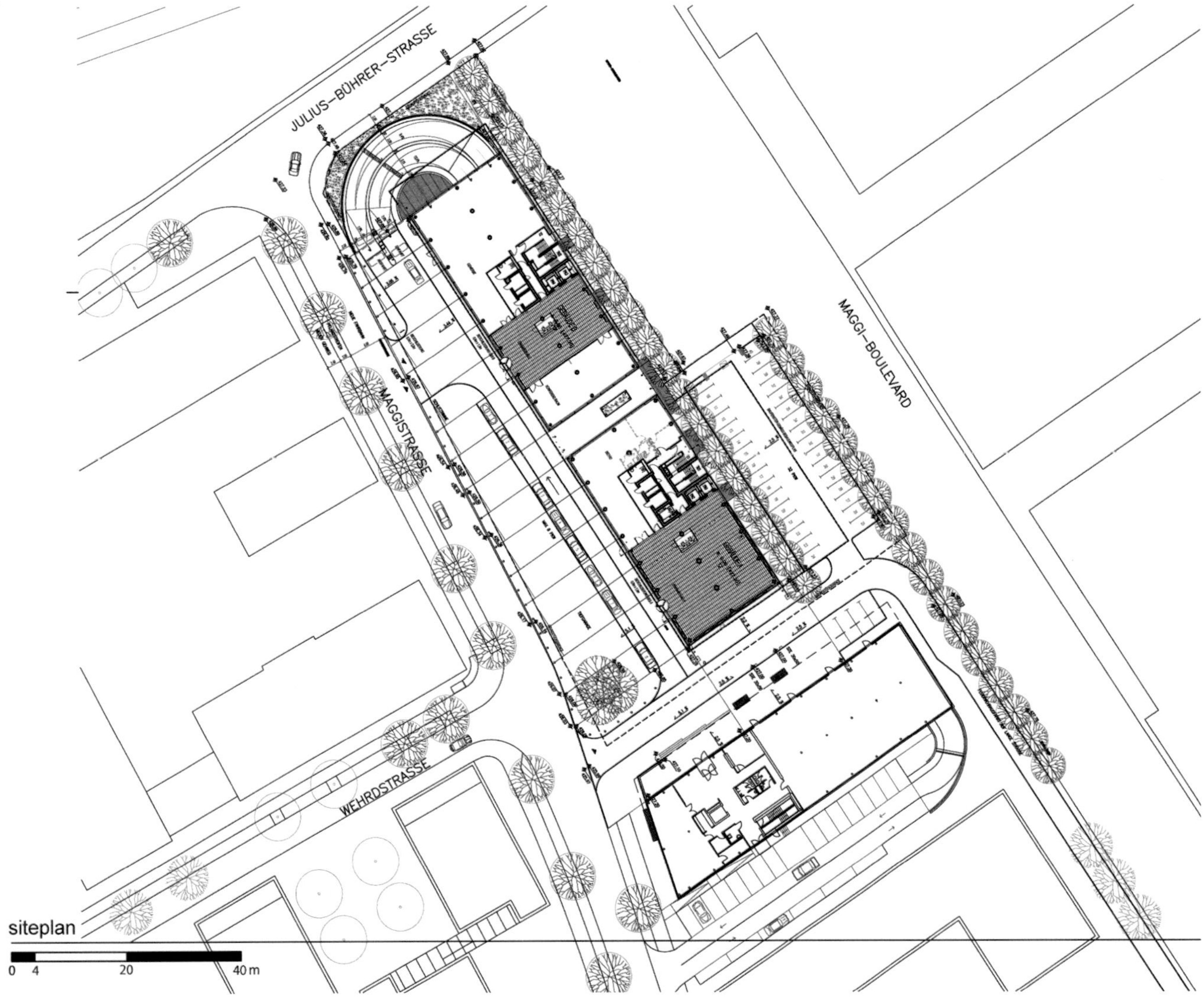
JULIUS-BÜHRER-STRASSE
MAGGISTRASSE
MAGGI-BOULEVARD
WEHRDSTRASSE
0 4 20 40 m

siteplan

Case Study 17

Heron Tower

London, UK

HEIGHT:	202 M (663 FT); 230 M (755 FT) WITH MAST
NUMBER OF STOREYS:	46
GROSS FLOOR AREA:	461,478 FT^2 / 42,873 M^2
PRIMARY USE:	OFFICES
OTHER USE:	RESTAURANT AND SKY BAR
START ON SITE:	MARCH 2008 (CONSTRUCTION); JULY 2007 (DEMOLITION AND ENABLING WORKS)
COMPLETION:	MARCH 2011
TOTAL CONSTRUCTION COST:	£242 MILLION
OWNER/DEVELOPER:	HERON INTERNATIONAL
DESIGN ARCHITECT:	KOHN PEDERSEN FOX
STRUCTURAL ENGINEER:	ARUP
MEP ENGINEER:	FOREMAN ROBERTS
MAIN CONTRACTOR:	SKANSKA
PROJECT MANAGER:	MACE
COST CONSULTANT:	DAVIS LANGDON
PLANNING:	DP9
LANDSCAPE ARCHITECT:	CHARLES FUNKE
FACADE ENGINEER:	EMMER PFENNINGER
LIGHTING:	ILLUMINATING CONCEPTS
TRANSPORTATION:	ARUP
FIRE:	ARUP

Project Description

Located in the City of London, Heron Tower occupies a prominent site at the junction of Bishopsgate and Camomile Street, defining the northern edge of the City core. Unlike the earlier generation of tall building in the City, whose monolithic forms are mute within their urban context, the Heron Tower is a transparent and articulate structure which responds to its context, tempering urban concerns, making a positive contribution to the public realm and addressing environmental requirements through its maintenance, operation and use of innovative technology. Responding to its urban context, the redevelopment of the Heron Tower site also incorporates significant improvement to circulation and access around the base of the Tower.

On a busy traffic corner, with narrow pavements, the public realm has been enhanced by opening up a pedestrian section to the north, along Houndsditch, animating the space with planting and cafés, and creating an arcade along Bishopsgate to provide a generous footpath to the busy street and address the Grade II-listed St Botolph's church opposite. An extension to the public realm is also incorporated at roof level, with dining terraces associated with the public restaurant and bar providing unrivalled views across London.

Architecture

The design of the tower provides highly flexible workspaces that support diverse tenant needs. A series of 10 office 'villages' – nine of three storeys and a double height 'village' at the top of six storeys, with a triple-height atrium at its heart – creates independent spaces which can be maintained and operated individually, providing high levels of visual connectivity and maximising daylight deep into the building, adding human scale and a sense of community. These 'villages' are structurally expressed on the northern face by the stainless steel cross-bracing and articulated to the east and west, animating the facade.

Design of the facades was informed by the orientation of the building. To the east and west the highly transparent, ventilated facade creates a bioclimatic, energy-efficient enclosure with automatic integral blinds controlling direct low angle sun. Positioned to the edge of the building to the south, an optimised core serves to protect the building from excessive heat gain. It houses 10 main double deck glazed lifts and two shuttle lifts to the roof level public restaurant and bar and incorporates a photovoltaic (PV) array – laminated units on the vertical facade of the scenic passenger lifts and plant areas. This array, covering 3,374 square metres, was the second largest PV array in the country at the time of completion and will ultimately result in a 2.2 per cent reduction in carbon emissions for the whole building.

The durability and solidity of Heron Tower is enhanced by the use of materials, combining stainless steel 'linen'-finish cladding with neutral/clear glazing. This is translated with finesse at street level, with a set-back covered three-storey arcade on Bishopsgate and full-height glazing connecting the street to the tower and the entrance lobby, defined by a dramatic 12-metre long tropical fish aquarium. The mass of the building is also stepped back at the upper levels – at restaurant and bar – cut back in three storey steps to the highest point at the south-west corner, topped by a 30-metre stainless steel mast.

Certified BREEAM 'Excellent', the environmental design of Heron Tower is a direct response to an environmental strategy based on conservation and energy efficiency. Energy use is minimised by passive design, with features such as the protective south core and interactive triple-glazed facade. In addition, the design of the building's services systems incorporates features that ensure that energy is used efficiently, with heat recovery, high-efficiency plant and low-energy cooling systems.

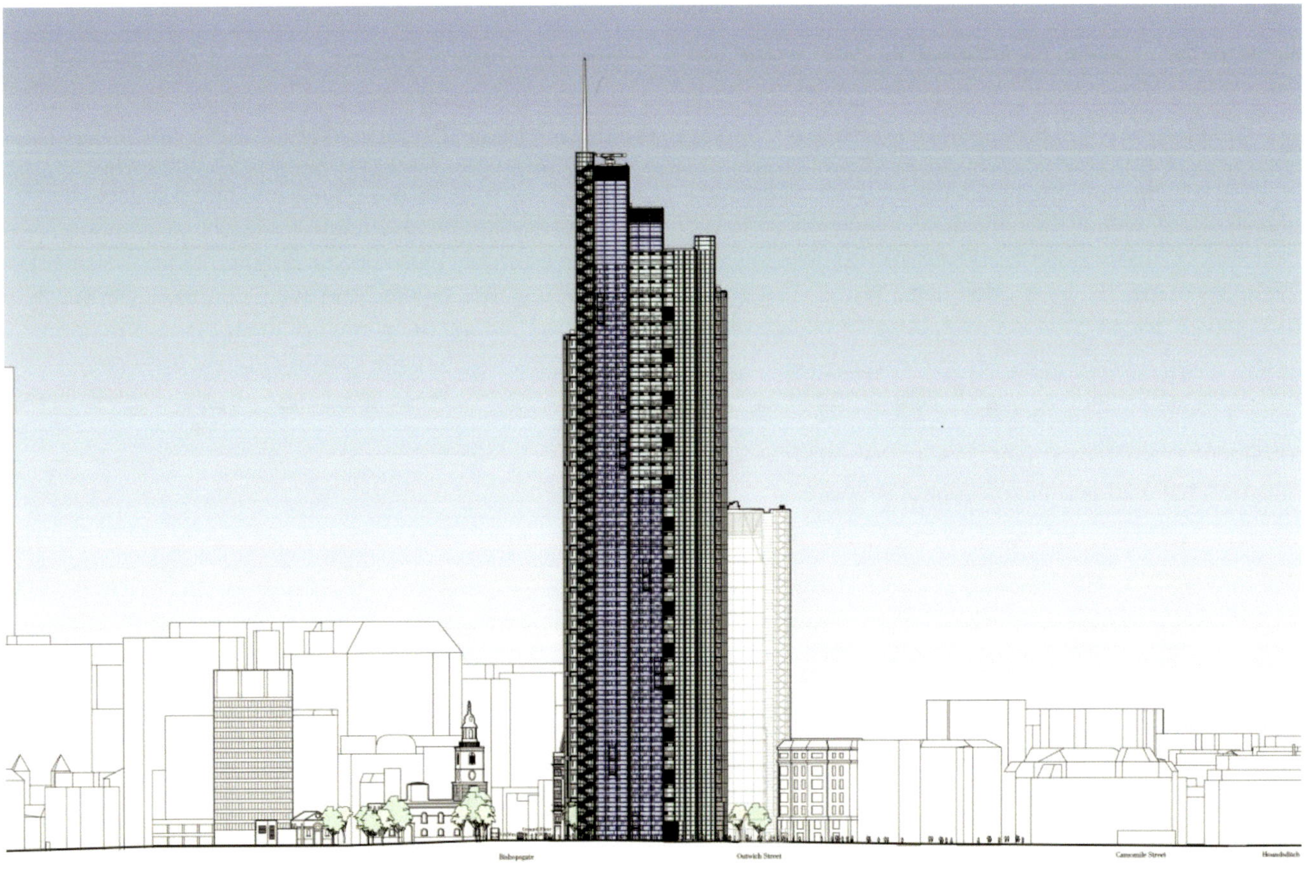

Each of the building's 12 'villages' is environmentally independent, with its own mechanical and electrical systems, life safety systems and controls so that each can be tuned to exactly the comfort patterns and values of its occupants' design comfort levels, bringing energy savings, cost benefits and allowing the refit of new technologies in the future. The building's lifts are also innovative, servicing the relatively large number of small floors efficiently. The design incorporates 'double-deck' panoramic high-speed lifts with hall-call destination control software.

Structure

Many buildings have a simple concrete core which acts as a backbone and gives the structure its stability and support. For Heron Tower, the architect's wish to achieve an open floor plate with sweeping views to the north ruled out the use of a central core. The site, too, hemmed in by roads on all sides, gave limited space for construction. The answer was a tube structure, to provide the structural stability needed by a 46-storey building, while also maximising the open floor space and therefore the lettable area.

Arup modified the tube structure by cutting out a vertical 'chunk' down one side, which is uncommon for tube-stabilised buildings. The strength of a tube structure is in the continuous outside edge, so modifications were needed to restore stability. This was achieved by a framing system that 'stitches' together the open vertical edges of the cut tube in key places. As a result, the office space is flooded with daylight and has unimpeded views of the City of London.

The tube had other advantages, such as faster construction. With no core, each level of the basement could be excavated without propping. Once the first level was done, excavation of the next level could start below it, with the tube providing the all-important stability for the frame construction above. This construction strategy meant that the basement and the structure above could proceed in tandem.

A top-down approach also meant that the load-bearing capacity of the foundation increased as construction advanced. This transformed what would normally be a problem into an advantage, by engineering the construction sequence to align the frame assembly with the construction of the basement. The foundations themselves were configured to match the load bearing demanded of them as construction advances, which made construction far less costly.

Case Study 18

Al Hamra Tower

Kuwait City, Kuwait

HEIGHT:	412.6 M
NUMBER OF STORIES:	80
GROSS FLOOR AREA:	195,000 M^2
PRIMARY USE:	OFFICES
START ON SITE:	2004
COMPLETION:	2012
PROJECT COST:	NOT AVAILABLE
OWNER/DEVELOPER:	AL HAMRA REAL ESTATE AND ENTERTAINMENT CO.
LEAD ARCHITECT:	SKIDMORE, OWINGS & MERRILL (SOM)
ASSOCIATE ARCHITECT:	AL JAZERA CONSULTANTS
STRUCTURAL ENGINEER:	SKIDMORE, OWINGS & MERRILL (SOM)
MEP ENGINEER:	SKIDMORE, OWINGS & MERRILL (SOM)
MAIN CONTRACTOR:	AHMADIAH CONTRACTING & TRADING CO.
PROJECT MANAGER:	TURNER CONSTRUCTION CO., INTERNATIONAL

Project Description

Rising 412 meters in the center of Kuwait City, Al Hamra Tower is a landmark office tower of iconic sculptural form that offers breathtaking views of the Arabian Gulf.

When construction began on the site, the owners, a joint venture between a local developer and a general contractor, had planned a 50-story tower with a four-story podium, all designed by a local architect. Then the building boom elsewhere in the region prompted the Kuwaiti authorities to change the zoning to allow for a much taller tower on the site, to compete with the super-tall, iconic structures springing up in Kuwait's neighbor countries. By the time SOM was brought in, the site was fully excavated and construction on the four-story podium was underway. As the owners wished to retain the original podium design, SOM was commissioned to design a new tower at the northern tip of the site.

Architecture

Situated in a very prominent location in the centre of the Kuwait peninsula, the Al Hamra Tower has a strong presence on the city skyline. The owners envisioned an iconic structure that would take advantage of the site and become a Kuwaiti landmark. They also required an efficient office building with a constant 12-meter deep lease span, facing towards the sea.

With its delicate glass veil reflecting the profile of the peninsula, the building resembles an enshrouded figure or a subtle, elegant modern sculpture. Its form provides transparency on the Gulf-facing north, east and west sides and near-complete opacity against the severe desert sun to the south.

During the planning process, the lease span was tested along the entire 60-meter-square perimeter of the site, and it was found that 25 percent of the floor plate needed to be taken out to meet the area requirements. Maximizing the views toward the water implied that this removal should correspond to the southern edge of the square, facing the city. In parallel, the design team ran solar and wind analyses to test the performance of different cut-out options. The solar analysis results favoured the southwest corner cut, meanwhile the wind studies showed an uneven cut would be better over a straight cut in confusing the wind.

This analytical process produced an optimal solution in a form that removed one quarter of the floor plate from the south facade, initiating at the southwest corner of the building at ground level and incrementally rotating counter clockwise along the full height of the tower.

The solid south wall helps reduce the effects of solar radiation. Openings are based on the relationship of the tower and the position of the sun. The geometry of the openings in the interior wall responds to the need to minimize solar heat gain. This wall protects the building from environmental conditions and also acts as the structural spine of the building. The point at the apex of the tower not only resolves this complex geometry of the carved, flare walls but also implies the continuation of the sculptural form infinitely upwards.

When viewed in the Kuwait City skyline, Al Hamra Tower gives the impression of a movement, twisting the space around it, rather than itself. The remarkable visual dynamism, as an epicenter of a vortex, is generated by the spiraling central void. The poetry of this tower is that it shifts the visual experience from what is there, as the concrete presence, to what is missing, as an enigmatic absence.

How a supertall building meets the ground—where there are peak structural demands, the connection of building services to the urban infrastructure, and the flow of traffic and people coming onto the site—is always a critical design challenge. Al Hamra Tower resolves all the practical challenges at the ground floor with a 20-meter tall public space that serves as the entrance lobby. The north face of the tower peels out nine meters to increase the depth of the lobby. A web-like concrete lamella structure is used to avoid buckling and to fluently bring the forces down to the foundations. The experiential qualities of the lobby—the barrel vault outline of the space and the filtering light coming through the web of concrete members—reflect the tectonics of the structure.

Structure

Concrete was chosen as the main structural material due to the building's shifting form. A four-meter deep raft foundation sits on 289 cast in situ 1.2-millimeter diameter bored piles up to 27 meters long. Shear walls at the core and the flares taper from a thickness of 1,200-millimeters to 300 millimeters. In situ floor slabs 160 millimeters thick are supported by 700-millimeter deep beams.

Formwork design was a particular challenge during construction, due to the 130-degree rotation in the tower's shape. Another was the height to which concrete had to be pumped—up to 400 meters in the case of the upper floor slabs. Nearly 200,000 cubic meters of high-quality concrete were needed to complete the structure.

Services

The tower is an all-electrical building and uses only electricity to generate heat. A central chilled water system is used for primary cooling. The refrigeration plant is located in a basement lower-level mechanical room, and consists of electric motor-driven centrifugal-type chillers and associated condenser water pumps with primary and secondary chilled water pumps to supply chilled water to all systems utilizing central chilled water. Chilled water is distributed through a chilled water piping system. One standby chiller with pumps is provided. Multi-cell induced draft fan-type package cooling towers are located on the podium roof and mounted on spring isolators and steel grillage, sized to match the refrigeration load base design. There is one standby cooling tower.

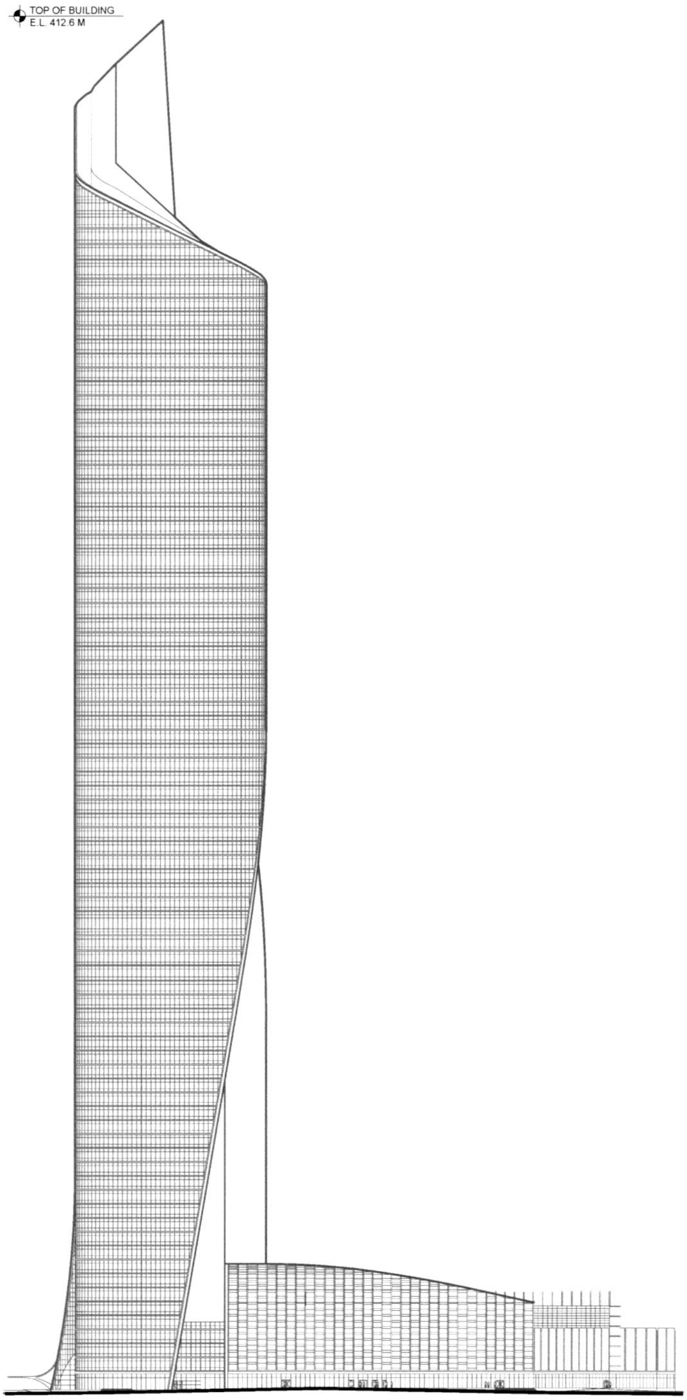

West Elevation

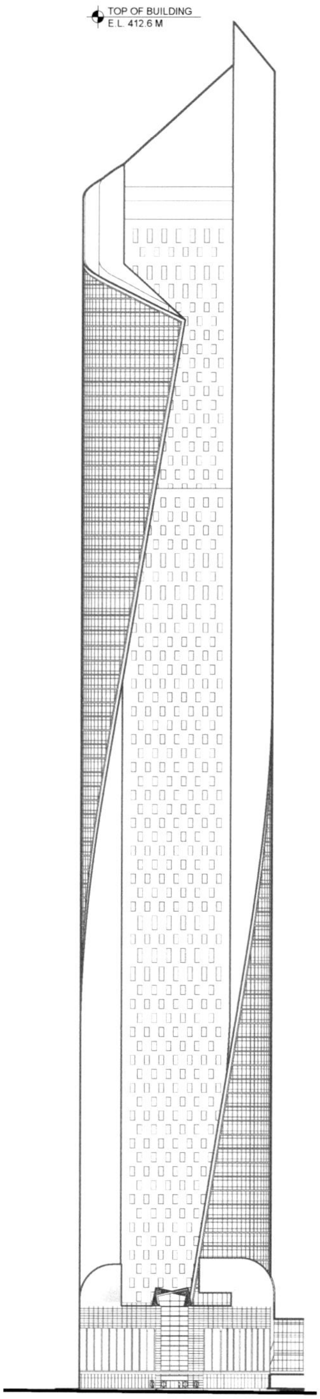

South Elevation

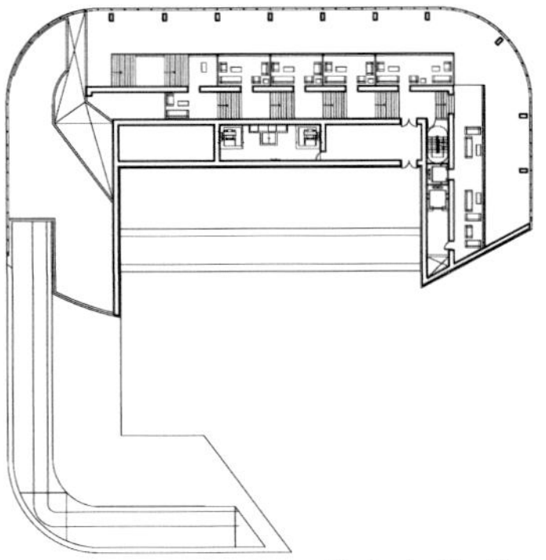

Sky Garden Floor Plan

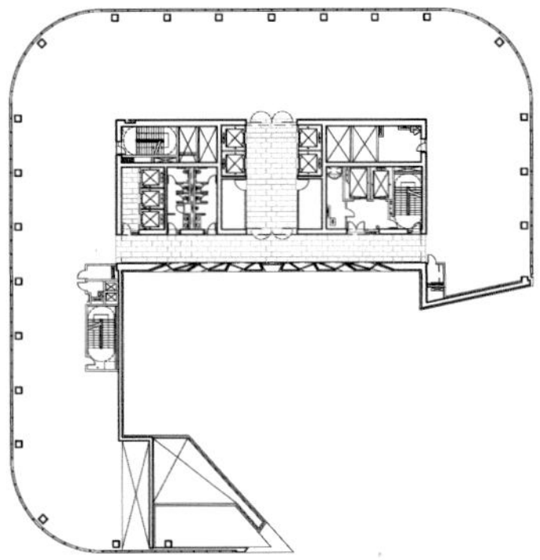

High Rise Floor Plan

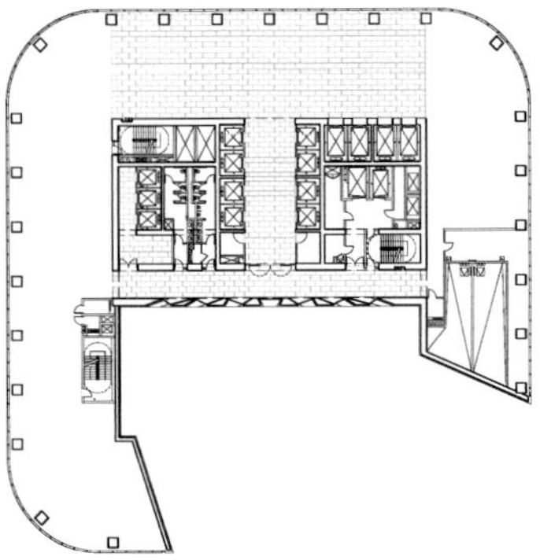

Sky Lobby 2 Floor Plan

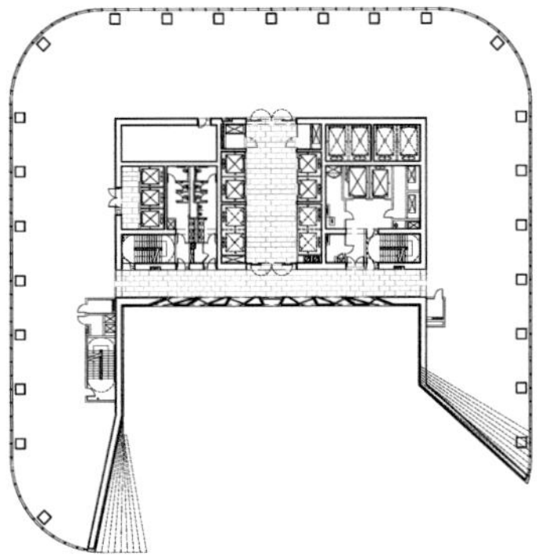

Mid Rise Floor Plan

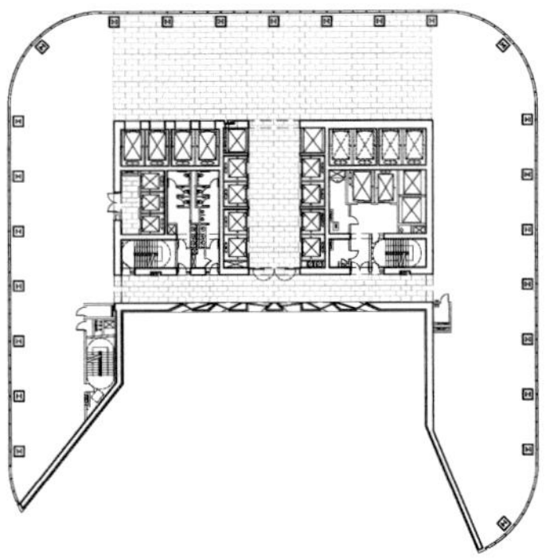

Sky Lobby 1 Floor Plan

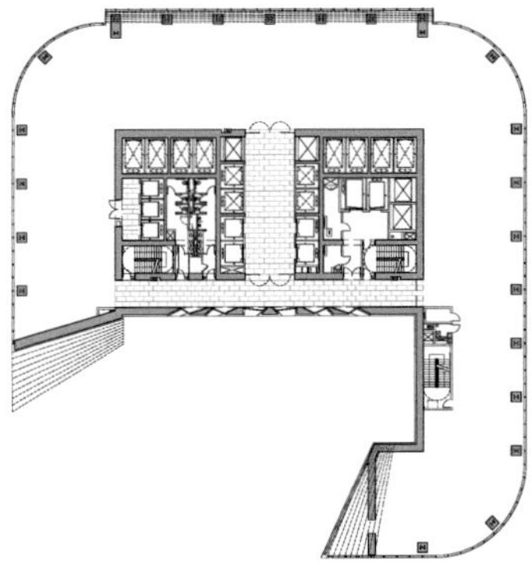

Low Rise Floor Plan

Case Study 19

Bahrain World Trade Center

Manama, Bahrain

HEIGHT:	240 M
NUMBER OF STOREYS:	42
GROSS FLOOR AREA:	120,000 M² BUILT-UP AREA
PRIMARY USE:	OFFICES, RETAIL AND HOTEL
START ON SITE:	JUNE 2004
COMPLETION:	MARCH 2008
PROJECT COST:	NOT AVAILABLE
OWNER/DEVELOPER:	CONFIDENTIAL
LEAD ARCHITECT:	ATKINS
STRUCTURAL ENGINEER:	ATKINS
MEP ENGINEER:	ATKINS
MAIN CONTRACTOR:	NASS MURRAY & ROBERTS, AND ATKINS (JV)
WIND TURBINE BRIDGE DESIGN:	RAMBOLL DANMARK A/S
WIND TURBINE SUPPLIERS:	NORWIN A/S

Project Description

The Bahrain World Trade Center is the world's first building to integrate large-scale wind turbines into its structure. In this, as well as the many other energy reduction and recovery systems it incorporates, it represents a major step forward in the development of energy-efficient and energy-sustainable structures in world terms, and particularly for the Gulf region. It has shown that commercial developments can be created with a strong environmental agenda, and in the process make sound economic sense.

The Context, Brief and Design Generators

The Bahrain World Trade Center is the focal point of a master plan to rejuvenate a 30-year-old existing hotel and shopping mall on a prestigious site overlooking the Arabian Gulf in the downtown central business district of Manama. The initial brief from the client called for the integration of the mall with the hotel, and the seamless extension of the complex towards King Faisal Highway, together with the creation of a new office complex with the necessary parking facilities.

The initial planning of the site was developed by extending the main axis of the existing shopping mall towards the sea and creating a secondary axis from the hotel, to generate the 'mall streets'. A new internal road was necessary to connect the existing basement parking with a proposed multi-storey car parking building, incorporating ramped sections to avoid interrupting internal pedestrian movement within the mall. The mall itself was repositioned in commercial terms, to offer the world's foremost brands through compact, luxurious shops in a contemporary interpretation of the intimate, traditional Arabian *souk*.

This resolution of the main approaches and axes of the development created a major axis towards the sea, reconnecting the development with Bahrain's marine heritage and island status. This in turn led to the development of twin office towers, flanking this axis and facing seawards.

At the first site inspection in October 2003, the direction and strength of the prevailing onshore wind was noticed. The idea of integrating turbines into the structure was quickly adopted, and from that point on informed the architecture to considerable degree. The form of the towers is entirely sculpted by the wind, such that they funnel and harness the daily onshore sea breezes from the Arabian Gulf to energise the three 29-metre diameter Norwin horizontal axis 225-kilowatt wind turbines. They in turn generate more than 20 per

cent of the towers' energy requirements, and save the production of some 55,000 kilograms of CO_2 per year.

Massing and Wind Dynamics

The design influences of a coastal site are inevitably marine, and thoughts of wind and sails, waves and ripples will always influence the design aesthetic. In this context there is the additional historical influence of the wind tower, a distinctive form of architecture indigenous to the Gulf, which funnels breezes from the clean air above down into the heart of the house, cooling and refreshing the interior. When massing the concept for the twin towers, it was established that over 70 per cent of Bahrain's wind energy comes directly onshore, unobstructed by buildings and perpendicular to the site—directly down the line of the new mall axis. This made the project ideal for the integration of wind turbines into the proposals.

The 42-storey, 240-metre high twin towers thus found inspiration from the concept of the traditional wind tower and its use of structure to funnel wind movement, the vast lateen sail of the traditional Arabian *dhow*, carving and curving the breeze to harness its energy and power the structure below, and the constancy and power of the wind itself. Careful computer modelling and extensive wind tunnel testing led to the towers' shape being generated and refined by the wind to create the optimum airflow around the buildings. The elliptical plan forms act as giant aerofoils; they funnel the air between them, and thus control its direction and maximise its power, enhanced by the creation of a zone of negative pressure downstream that accelerates the wind in the process. The vertical sculpting of the towers is also a function of airflow dynamics; as they taper upwards, their aerofoil sections reduce, which, when combined with the increasing velocity of the onshore breeze at increasing height, creates a near-equal regime of wind velocity on each of the three turbines. This enables them to rotate at equal speed and thus generate the same energy in constant phasing.

Understanding and utilising these phenomena have been major factors in enabling the practical integration of wind turbines into a commercial building design. Wind tunnel testing has confirmed that the shapes and spatial relationship of the towers modify the airflow, creating an 'S'-curved system in which the centre of the wind stream remains nearly perpendicular to the turbine within a 60-degree wind azimuth on either side of the central axis. The result is a much-increased potential to generate power, over longer periods and in a broader range of wind conditions, which consequently provides a considerable increase in the annual energy yield.

Sustainability

In addition to the wind turbines, the building includes a number of other active and passive systems to reduce carbon emissions. These include:

- Buffer spaces between the external environment and air-conditioned spaces; for example there is a car park deck over the south end of the mall, which reduces solar temperature and conductive solar gain.
- Significant quantities of projecting solar shading to external glass facades.
- Balconies on the end elevations, providing overhanging shading to the facade below and providing breakout spaces and sources of fresh air during the winter months.
- High-quality, technically advanced glazing systems providing outstanding thermal performance to minimise solar gains.
- Low-leakage construction, reducing the stack effect and transfer of hot and cool air, balanced by opening windows to enable mixed-mode operation in winter months.
- High levels of thermal insulation for opaque elements.
- A dense concrete core and floors within the internal environment to level loads and reduce peak demand on air and chilled water transport systems.
- Variable volume chilled water pumps that reduce the pump power required as cooling loads decrease.
- Energy recovery wheels at fresh air intakes, transferring heat from incoming air to the exhaust air, thus reducing cooling loads.
- Energy-efficient high-frequency fluorescent lighting with zoned control systems to provide flexible lighting regimes.
- Dual drainage systems that segregate foul and waste water, enabling a system of grey water recycling.
- Connection to the district cooling system, enabling the building to benefit from the economies of scale.
- Electronic taps with excess water flow restrictors.
- Reflection pools at building entrances to provide local evaporative cooling.
- Solar-powered road and amenity lighting.

Conclusion

Environmental issues have clearly been comprehensively addressed, and the commercial aspects of the project have likewise benefited. The building is a beacon for sustainability in the Gulf as well as an attractive structure in its own right, and as a consequence there is constant pressure from organisations who wish to adopt it as their commercial home, and thus be associated with what is perceived as a forward-thinking, progressive development. This not only has a positive effect upon rental rates, but increases income as well, a marked benefit to the client during the recent financial downturn. Both render the additional financial input in terms of design and construction costs more than worthwhile.

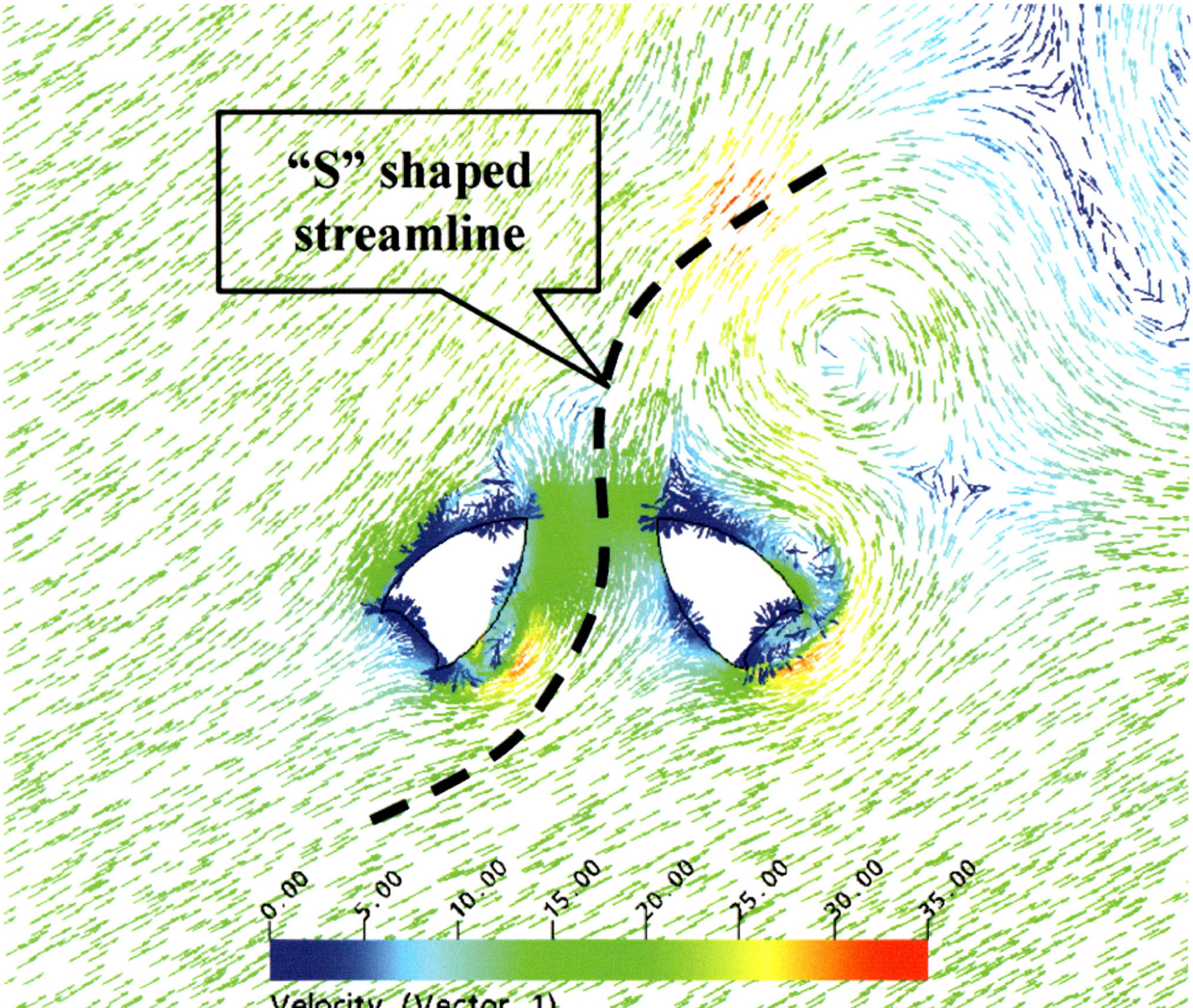
"S" shaped streamline
0.00
5.00
10.00
15.00
20.00
25.00
30.00
35.00
Velocity (Vector 1)
[m s^-1]

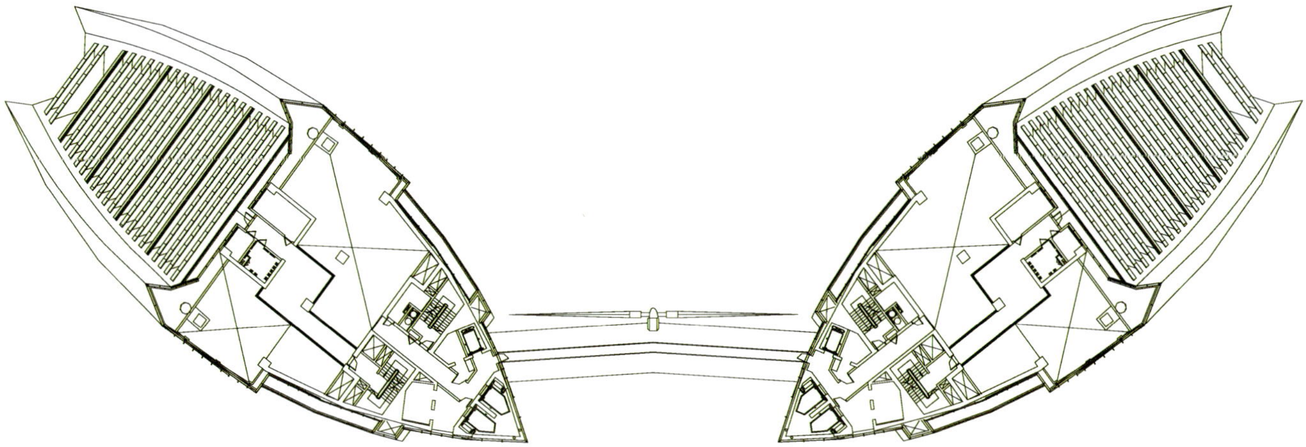

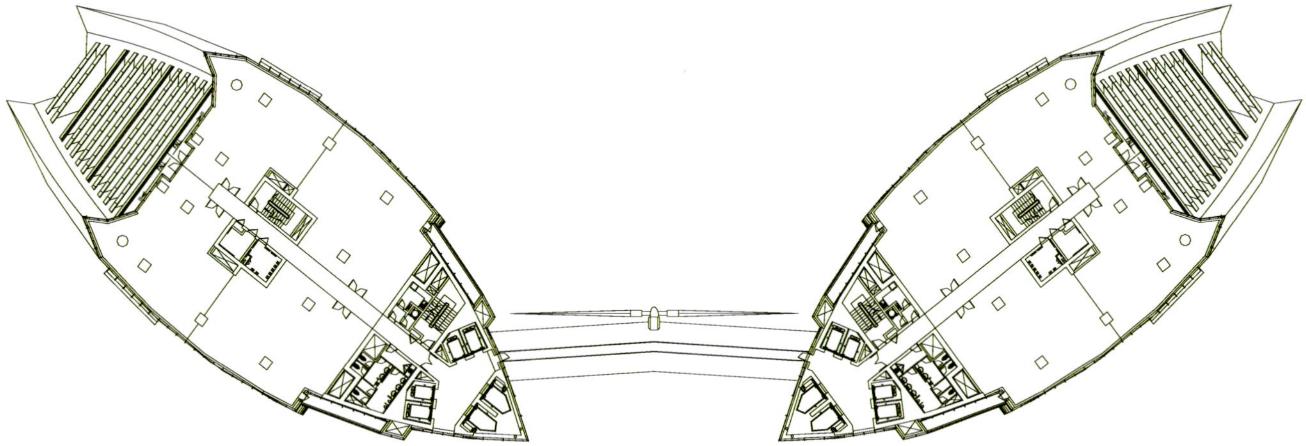

Case Study 20

Burj Khalifa

Dubai, UAE

HEIGHT:	828 M
NUMBER OF STORIES:	162 HABITABLE, 46 MAINTENANCE, 2 PARKING
GROSS FLOOR AREA:	186,000 M^2
PRIMARY USE:	MIXED USE: OFFICES, HOTEL, RESIDENTIAL
START ON SITE:	JANUARY 2004
COMPLETION:	JANUARY 2010
PROJECT COST:	NOT AVAILABLE
OWNER/DEVELOPER:	EMAAR PROPERTIES PJSC
LEAD ARCHITECT:	SKIDMORE, OWINGS & MERRILL (SOM)
STRUCTURAL ENGINEER:	SKIDMORE, OWINGS & MERRILL (SOM)
MEP ENGINEER:	SKIDMORE, OWINGS & MERRILL (SOM)
MAIN CONTRACTOR:	SAMSUNG C&T CORPORATION
CONSTRUCTION MANAGER:	TURNER CONSTRUCTION
WIND ENGINEERING:	RWDI INC.

Project Description

Dubai experienced tremendous growth and financial success during the late twentieth and early twenty-first centuries, evolving from a small trading port to the business and tourism capital of the Middle East's Gulf Region. At the height of the city's economic explosion, Dubai-based real estate developer Emaar Properties PJSC envisioned a supertall tower that would become the centerpiece of the new downtown Dubai. In 2003, as part of a competition, Emaar hired Skidmore, Owings & Merrill (SOM) to design what would become the world's tallest building—the Burj Khalifa. The construction of the Tower began in January 2004, with its official opening occurring on January 4, 2010. Today, the tower tops all three categories designated by the Council on Tall Buildings and Urban Habitat as defining the tallest building in the world, eclipsing the previous height record-holder by 319 meters.

Apart from being the tallest building in the world, the master plan for Burj Khalifa responds to the global movement toward compact, liveable urban areas. Downtown Dubai sought to create a new city center addressing issues of trade, transport, tourism, industry, and finance in response to the city's economic boom. The Burj Khalifa stands in the center of this new downtown community, spurring surrounding development and operating in similar fashion to a small vertical city.

The project consists of the supertall tower with an integrated podium structure, and an ancillary 12-story office building and four-story Pool Annex, which complement the Tower at its base and define plaza spaces on the ground. At ground level, the "Y"-shaped plan allows for each use to have its own distinct entry, with paving, landscaping, and fountains to create an individual experience. The tower's mixed-use program, consisting of hotel, residential, and office space, responds to the area's development density and provides residents, guests, and visitors with direct connections to adjacent site amenities, shopping, and mass transit systems.

Architecture

Standing tall at 828 meters, Burj Khalifa meant pushing current analyses, materials, and construction technologies—literally—to new heights. However, as such a building height had never before been attempted, it was necessary to ensure all technologies and methods used were of sound development and practice. The designers therefore employed conventional systems, materials, and construction methods, modifying and executing them in unprecedented capacities.

The building's "Y"-shaped plan was employed not only to develop an inherently stable geometry for the structure but also to yield the maximum amount of perimeter. This gave accessibility to views and daylight without allowing tenants to look into neighboring units, which would be culturally unacceptable. As the tapering tower rises, setbacks occur at the ends of each "wing" in an upward, spiraling pattern that decreases the mass of the tower as the height increases. The spiraling is reminiscent of spiraling characteristics seen on ancient Arabic obelisks and monuments. During its development, this setback structure was modeled in a wind tunnel in order to determine the safest and most effective way to minimize wind forces, and was the result of close coordination between architects and structural engineers.

Exterior cladding is comprised of aluminum and textured stainless steel spandrel panels, and was designed to withstand Dubai's extreme environment during the summer months. Vertical polished stainless steel fins were added to accentuate Burj Khalifa's height and slenderness, providing some shading on the exterior surface and catching the light at the beginning and end of the day to enhance the spectacular quality of the tower.

With over three million square feet of interior space to be designed for Burj Khalifa, planning of the building's interiors began at the earliest stages of its design. The design team worked towards attaining three main goals—recognizing and acknowledging the building's height, integrating its structural and architectural rationale, and appreciating the locale's heritage, history, and culture.

Structure

The unprecedented height of the Burj Khalifa required it to be the most innovatively engineered building possible. Design techniques, building systems, and construction practices all required rethinking, redefining, and—in many cases—reinventing applications already in place, to create a practical and efficient building. Created especially for Burj Khalifa, the structural system consists of a six-sided high-performance reinforced concrete central core with three "wings"—a system also dubbed a "buttressed core." Loads from one wing are transferred to the other wings via the six-sided core. Corridor walls extend from the central core to near the end of each wing, terminating in thickened hammer-head walls. Perimeter columns and flat plate floor construction complete the system.

Outrigger walls tie the vertical structure at the mechanical floors in order to maximize the stiffness and stability of the building. The result is an extremely efficient structure, in which the entire vertical structure of the building is used to support both gravity and lateral loads. Shaping and orientation of the tower were significantly influenced by its performance with respect to the strong winds. Numerous wind tunnel tests and design iterations were needed in order to develop solutions for optimal performance. Constructing the tower necessitated the highest single-stage pumping of concrete ever performed—an awe-inspiring 606 meters. An approximately 230-meter tall structural steel spire tops the tower.

Sustainability

The Burj Khalifa implemented new ways to increase structural and construction efficiencies while reducing material use and waste. Lessons learned from Burj Khalifa will help future supertall projects reduce the environmental impact associated with construction and raw material extraction. Sustainable elements of the Burj Khalifa include:

- *Sky-sourced ventilation* Cooler air temperatures, decreased air density, and reduced relative humidity at the top of the building allow for "sky-sourced" fresh air. When air is drawn in at the top of the building, it requires less energy for air conditioning, ventilation, and dehumidification.
- *Condensate recovery system* Burj Khalifa has one of the largest condensate recovery systems in the world. By collecting water from air-conditioning condensate, discharge is prevented from entering

the wastewater stream and the need for municipal potable water is reduced.

- *High-performance glazing* A low-emissivity glass provides Burj Khalifa with enhanced thermal insulation against the high ambient temperatures of Dubai.
- *High-voltage distribution* Conduction of electric power using higher voltages reduces energy losses and increases energy efficiency when compared to lower-voltage systems.
- *Electronic metering* Individual electric energy monitoring systems enable energy optimization of the tower's systems over its lifetime. This will result in a reduction of Burj Khalifa's energy-related environmental impact.
- *Smart lighting and mechanical controls* The building's management system results in lower operational costs, a more efficient use of building resources and services, better control of internal comfort conditions, and effective monitoring and targeting of energy consumption.
- *Stack effect controls* Thermal differences between the building's interior and exterior generate a stack effect. Burj Khalifa was designed to passively control these forces, reducing the need for mechanical means of pressurization while saving energy.

Conclusion

The design of the tower required intense efforts of collaboration between architect and engineer. Although height was central to the project's development, Burj Khalifa represents a progressive shift in design and engineering that presents new solutions to complex problems. By combining cutting-edge technologies and regional cultural inspirations, SOM created a vertical city that has become a model for the development of future urban centres, and speaks to an ever-growing global movement toward compact, liveable urban areas. Burj Khalifa represents significant achievements of a new global generation of architects and engineers, and serves as a catalyst for the future growth and development of downtown Dubai.

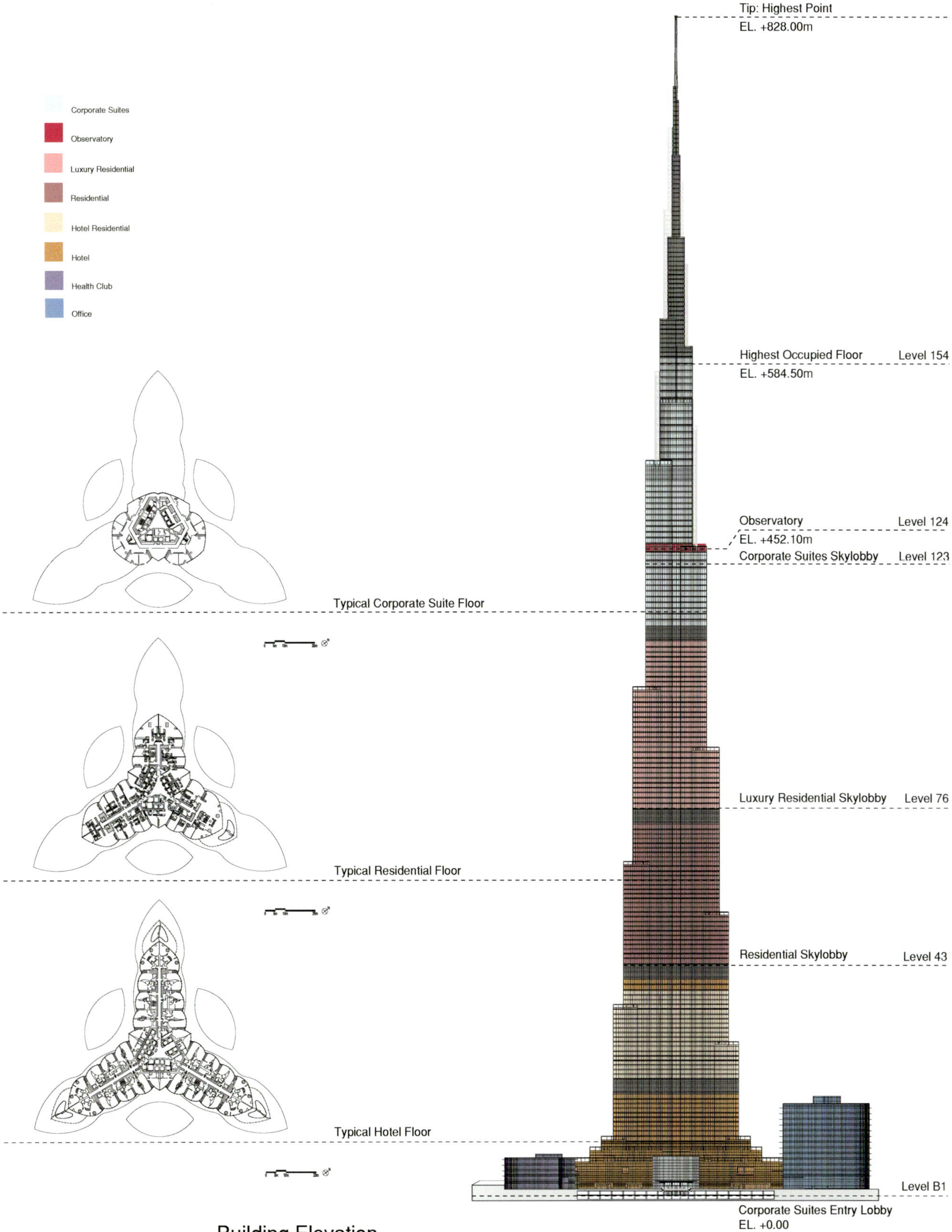

Building Elevation
1:3500

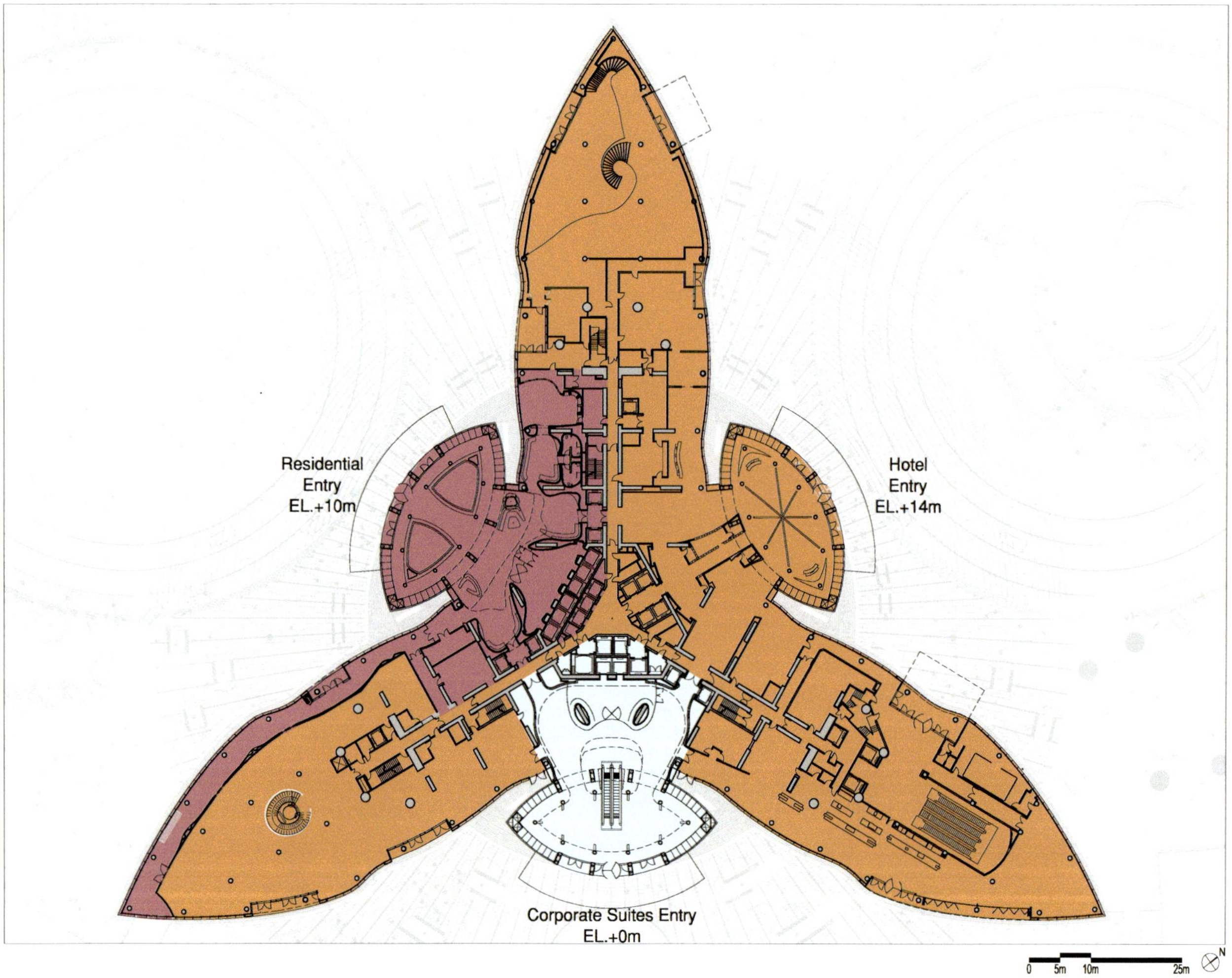
Residential
Entry
EL.+10m
Hotel
Entry
EL.+14m
Corporate Suites Entry
EL.+0m
0 5m 10m 25m
N

ROOF PLAN

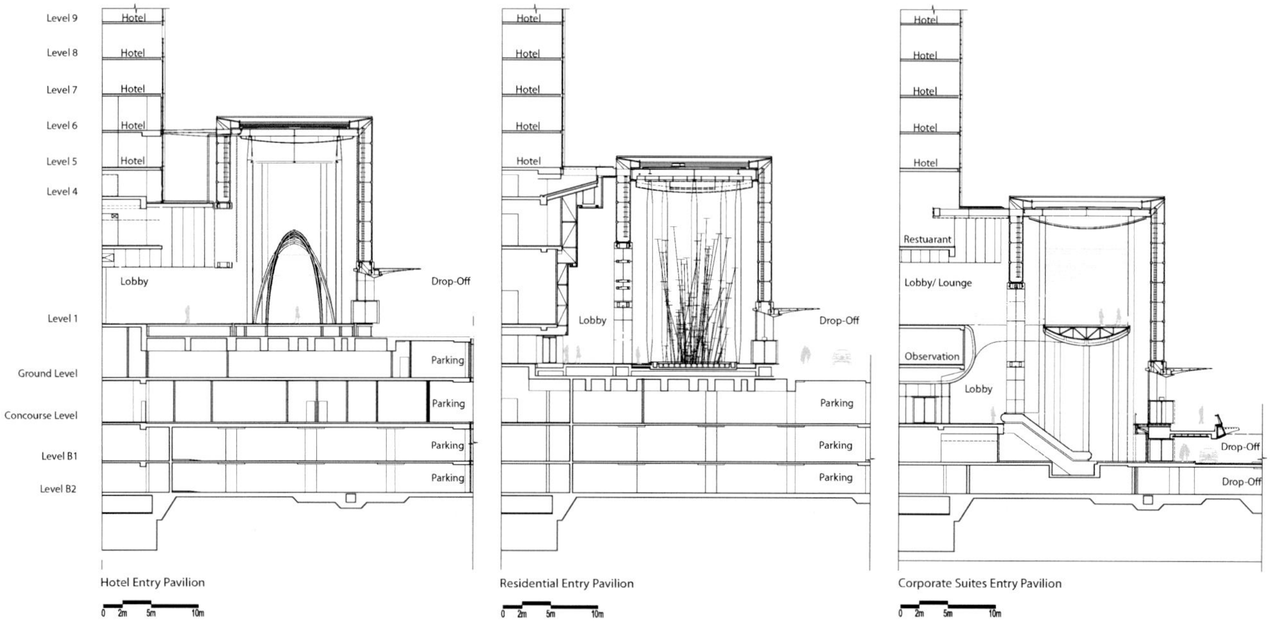

Hotel Entry Pavilion

Residential Entry Pavilion

Corporate Suites Entry Pavilion

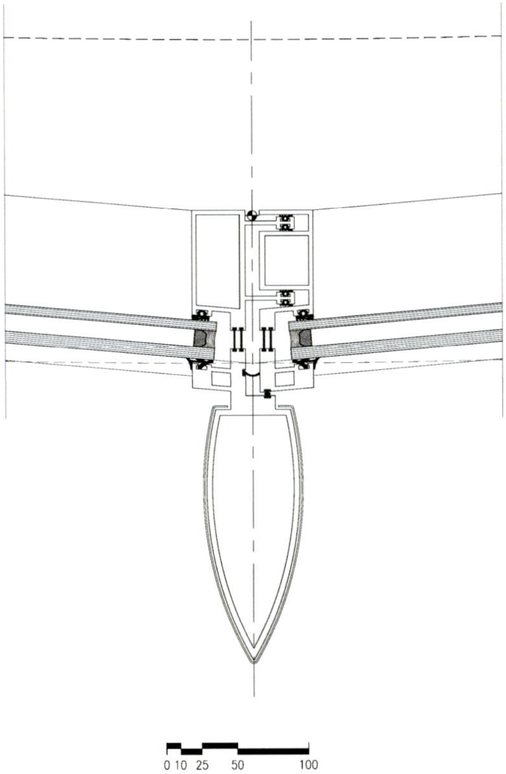
0 10 25 50 100

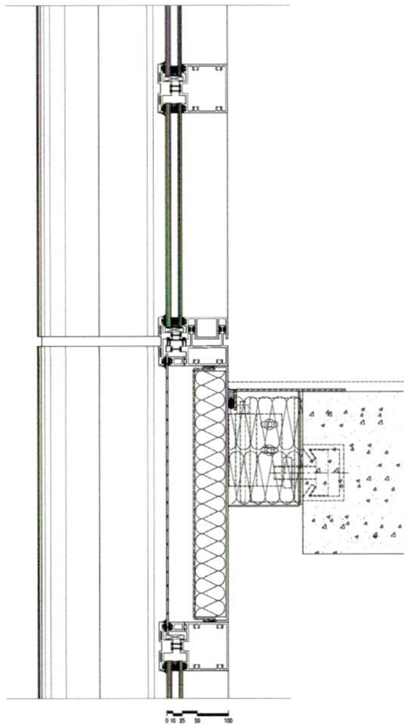

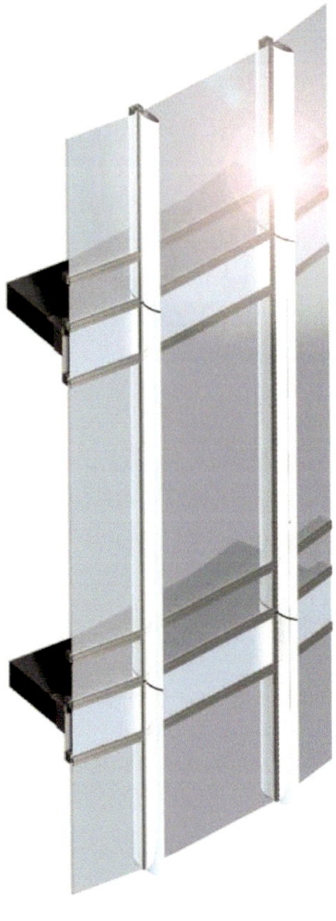

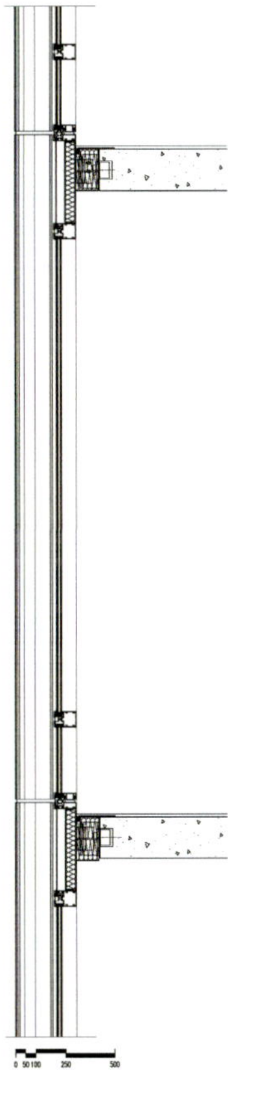

Case Study 21

Capital Gate

Abu Dhabi, UAE

HEIGHT:	160 M PLUS HELIPAD (4.5 M)
NUMBER OF STOREYS:	35 PLUS 1 BASEMENT LEVEL
GROSS FLOOR AREA:	TOTAL BUILT-UP AREA: 53,100 M^2; TOTAL OFFICE AREA: 20,900 M^2; TOTAL HOTEL AREA: 25,050 M^2
PRIMARY USE:	OFFICES, HOTEL
START ON SITE:	2007
COMPLETION:	2011
PROJECT COST:	NOT AVAILABLE
OWNER/DEVELOPER:	ADNEC
LEAD ARCHITECT:	RMJM
STRUCTURAL ENGINEER:	RMJM (DESIGN); AL HABTOOR ENGINEERING ENTERPRISES (INSTALLATION)
MEP ENGINEER:	RMJM (DESIGN); ETA (INSTALLATION)
MAIN CONTRACTOR:	AL HABTOOR ENGINEERING ENTERPRISES
STEELWORK SUBCONTRACTOR:	EVERSENDAI
PROJECT MANAGER:	MACE
LANDSCAPE ARCHITECT:	RMJM STRATA
INTERIOR DESIGNERS (HOTEL):	RPW
INTERIOR DESIGNERS (OFFICES):	RMJM

Project Description

The Abu Dhabi National Exhibition Centre is owned and operated by Abu Dhabi National Exhibitions Company (ADNEC), a modern and dynamic organisation that is also masterminding Capital Centre, a micro-city comprising 23 mixed-use towers adjacent to the exhibition venue. This stunning exhibition venue, which is also the Arabian Gulf's largest exhibition centre, illustrates ADNEC's commitment to providing Abu Dhabi, the capital of the United Arab Emirates, with world-class facilities for all manner of events in support of the ongoing development of the city as a commercial destination of repute.

RMJM has worked collaboratively with ADNEC to deliver Phases 1 to 4 of this important project, on 15 hectares of prime land adjacent to the RMJM-designed National Exhibition Centre. Developed in conjunction with ADNEC to serve and support the growing exhibition community and businesses, the design balances indoor and outdoor public spaces to seamlessly integrate a coherent mix of office, hotel, and residential space. The creation of pedestrian-friendly areas, inspired by the spaces found in New York, together with a desire to create a true visitor destination, can be best exemplified by the creation of a 35,000-square-metre outdoor exhibition space, 'Capital Plaza'.

The master plan provides views to the canal and the waterfront area. Taller buildings have been sensibly arranged to rise along the eastern edge and fall towards the south. This distinctive arrangement provides a visible and recognizable landmark from across the harbour. Capital Gate, Phase 3 of the master development, is the focal point.

The tower, which incorporates a slanting-core concept to feature an 18-degree westward lean, includes 20,000 square metres (220,000 square feet) of office space and houses Abu Dhabi's first Hyatt hotel, Hyatt Capital Gate. The tower also provides a connection to the past by integrating with the grandstand at the Abu Dhabi National Exhibition Centre, one of the world's most modern exhibition venues. Capital Gate underscores the bond between the old and the new by linking to this historic grandstand through an innovative canopy, commencing at Level 18 of the tower and sweeping across the grandstand, creating a wave-like effect.

Architecture

Capital Gate is a high-quality iconic building distinguished by a dramatic steel and glass facade with a striking organic form. With its tea lounge cantilevered out 80 metres above the ground and open-air pool deck on the 19th floor, it creates a unique presence on the skyline of Abu Dhabi and a memorable identity for the exhibition centre. A sculptural stainless steel 'splash' flows down the front and at low level forms the hotel entrance canopy, continuing over the existing grandstand and acting as a solar shading device for both the building and the grandstand seating.

A tapered internal atrium with a dynamic glass roof formed by a separate structural steel diagrid over 60 metres in height (from the 19th floor to the roof) creates a stunning internal space and brings natural light and space deep into the tower. External lighting is designed to minimise both light pollution and energy consumption, based on a combination of low-level landscape lighting with facade lighting comprising a net of compact LED clusters integrated into the design of the steel glazing system. The double-skin facade uses low-emissivity glass, a first for the UAE.

Nothing is standard about Capital Gate. Each room is different. Each pane of glass is different and every angle is different. It was designed to provide no symmetry, and to inspire those within and outside the tower. It acts as a dramatic termination to the organic free-form stainless steel mesh canopy that forms the grandstand roof and provides a visual anchor for the ADNEC Capital Centre Development as a whole within the Abu Dhabi city skyline.

Designed for mixed use, the tower features two entrances providing separate identities for the Grade-A commercial office space and the five-star hotel that is accommodated on the highest floors of the tower, benefiting from a double-storey feature sky lobby on the 19th floor. A glass and steel walkway connects the lobbies of both these facilities with the grandstand and exhibition centre.

Structure

In June 2010, Guinness World Records certified Capital Gate as the world's furthest leaning man-made tower. At 18 degrees to the west, its lean is more than four times that of the Leaning Tower of Pisa. Add the continuous twist of its form and the unique nature of every floor plate, and Capital Gate is not only a unique building, it is one of the most technically challenging structures in the world.

Some 490 thirty-metre deep piles provide the essential resistance to gravitational, wind and seismic forces. At the heart of the structure is a strong concrete core. During the tower's construction, allowance had to be made for the fact that as the floor plates were cast the core would deform, so core alignment had to be carefully calculated at each stage to ensure that its final alignment was correct.

Up to the 12th floor, the floor plates are stacked vertically one over the other. Beyond that they are staggered by up to 1,400 millimetres. Unobstructed floor plates were made possible by the adoption of an external steel diagrid as the main structural frame.

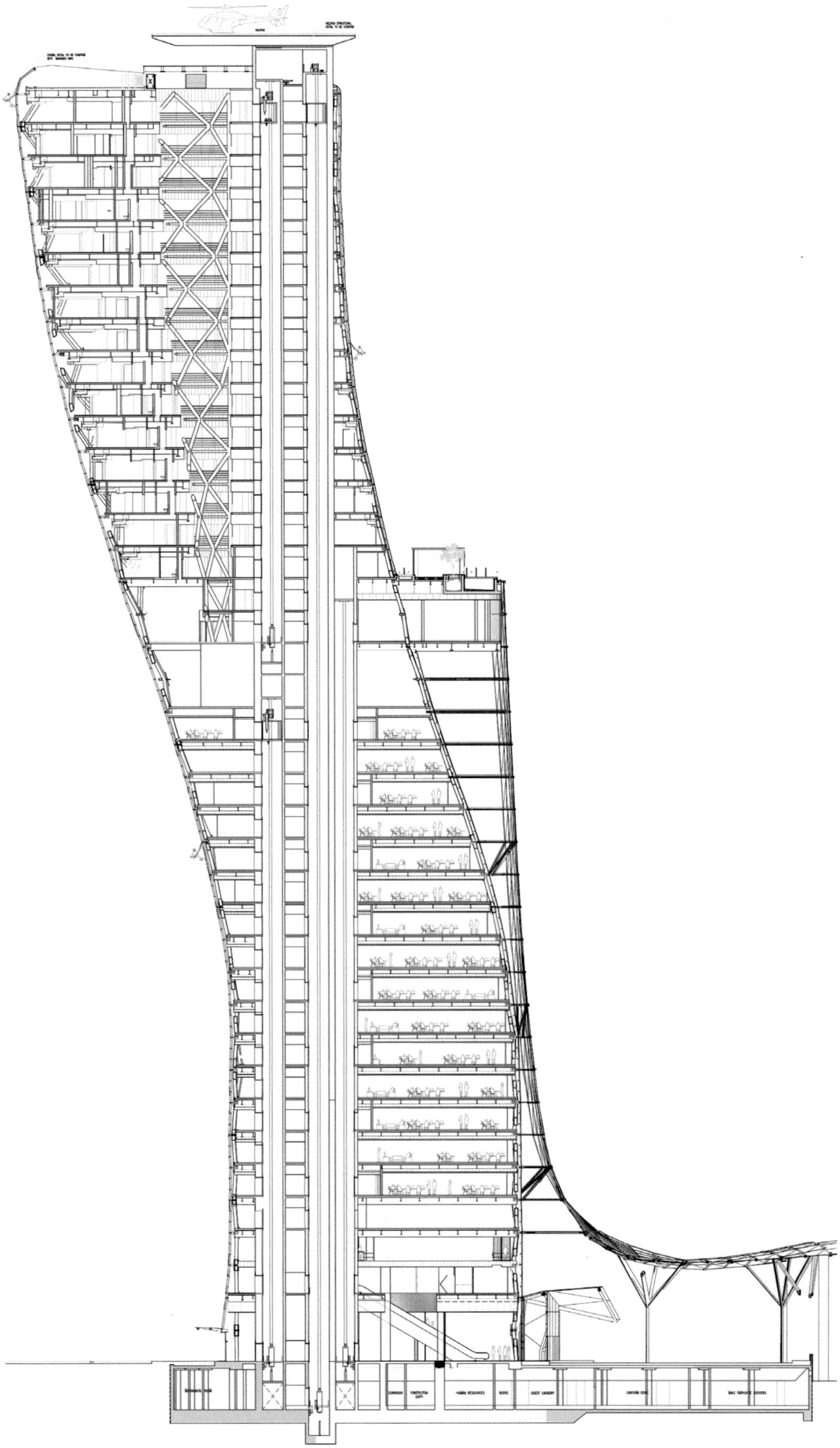

Case Study 22

O-14

Dubai, UAE

HEIGHT:	105.7 M
NUMBER OF STORIES:	23 PLUS 4 BASEMENTS FOR PARKING
GROSS FLOOR AREA:	15,979 M^2
PRIMARY USE:	OFFICES
START ON SITE:	2007
COMPLETION:	2010, OPENED MARCH 2011
PROJECT COST:	NOT AVAILABLE
OWNER/DEVELOPER:	CREEKSIDE DEVELOPMENT CORPORATION
LEAD ARCHITECT:	REISER + UMEMOTO
ASSOCIATE ARCHITECT:	ERGA PROGRESS
STRUCTURAL ENGINEER:	YSRAEL A. SEINUK, P.C.
MEP ENGINEER:	ERGA PROGRESS
MAIN CONTRACTOR:	DUBAI CONTRACTING COMPANY (DCC)
WINDOW WALL CONSULTANT:	R.A. HEINTGES & ASSOCIATES

Project Description

O-14, a 22-story tall commercial tower perched on a two-story podium, broke ground in February 2007, and comprises over 300,000 square feet (27,870 square meters) of office space for the Dubai Business Bay. O-14 is located along the extension of Dubai Creek, occupying a prominent location on the waterfront esplanade. With O-14, the office tower typology has been turned inside out—structure and skin have flipped to offer a new economy of tectonics and of space.

Architecture

Dubai's typical urban condition is one of many emerging economies; a typical tower is located directly on the street and includes a pedestrian arcade and a parking structure at the base of the building. Unfortunately, such a scenario too often creates a dead street behind the building. O-14 attempts to create a better urban condition within these typologies, as best it can within its limited site. Rather than assuming that the podium base would simply have an active front, O-14 subsumes an arcade into its shell, and produces another layer of activity higher up on the podium top. Parking is moved to four underground levels and the normally ground-level podium is elevated, thus freeing up the ground plane, and a continuous elevated pedestrian level, a "new ground," is created above the street level. The promise is that O-14 and its neighbors could produce activity on many levels, and engender new kinds of connections between the rear street and the promenade, activating the waterfront block as a kind of infrastructure for the district.

In O-14, the fenestration (or perforation) is not tied to the overall regulating geometry. In a typical office building, the subdivision of form would locate programs in a predictable way (larger windows and offices at corners, etc.). Here, the pattern rather seeks to attenuate the monotony, while still preserving the sense of the sublime and the monumental. Its deliberate lack of coordination with the floor plate engenders a randomized connection—all of this confuses legibility and scale, defeats easy reading of the building's height, and reorganizes the hierarchy of office space. Modulation of pattern works like camouflage, becoming disruptive and dematerializing the tower block. The shell's pattern changes as its relationship to the viewer changes, and in conjunction with additional patterns of light and shadow produces a sort of "virtual form." Because of the effects of this virtual form, the actual form of the building can be simplified, and be made subject to the logic of production methods, structural analysis, and economy.

Structure

O-14's concrete shell—up to 600 millimeters thick, with more than 1,300 openings making up over 40 percent of the surface area—provides an efficient structural exoskeleton that frees the core from the burden of lateral forces and creates highly efficient, column-free open spaces in the building's interior. High-strength steel mesh–reinforced self-compacting concrete was used to construct the shell, with the openings created by polystyrene void form. This exoskeleton becomes the primary vertical and lateral structure for the building, allowing the column-free office slabs to span between it and the minimal core. By moving the lateral bracing for the building to the perimeter, the core, which is traditionally enlarged to receive lateral loading in most curtain-wall office towers, can be designed for vertical loading, utilities, and transportation only. Additionally, the typical curtain-wall tower configuration results in floor plates that must be thickened to carry lateral loads to the core, yet in O-14 these can be minimized to respond only to span and vibration. Future tenants can arrange the flexible floor space according to their individual needs.

The shell acts not only as the primary structure of the building but also as a sun-screen, open to light, air, and views. Openings on the shell thus modulate according to structural requirements views, sun exposure, and luminosity. The overall pattern is not a response to a particular program (which in the tower typology is inherently variable); rather the pattern, in its modulation of solid and void, will affect the arrangement of whatever program comes to occupy the floor plates.

A one-meter gap between the main enclosure and exterior shell creates a so-called "chimney effect," a

phenomenon in which hot air has room to rise and effectively cools the surface of the glass windows behind the perforated shell. This passive solar technique is a natural component of the cooling system for O-14, reducing energy consumption and costs by more than 30 percent; just one of many innovative aspects of the building's design.

Conclusion

The project has generated extraordinary international interest in the architectural press, as it is among the very first innovative designs to be constructed among a sea of generic office towers which have become standard in Dubai's current building boom. O-14 was featured in "Impossible City," an hour-long television documentary about the recent growth in Dubai, produced by CBS News and aired in the U.S. on the Discovery Channel in October 2008. In May 2009, the tower's concrete structure was completed and the building was topped out, making O-14 one of the first towers to appear on the skyline of Business Bay, Dubai. It opened in March 2011.

Case Study 23

One World Trade Center

New York, USA

HEIGHT:	1776 FT (541 M)
NUMBER OF STORIES:	104
GROSS FLOOR AREA:	3,500,000 FT^2 (325,160 M^2)
PRIMARY USE:	OFFICES, RETAIL
START ON SITE:	APRIL 2006
COMPLETION:	2013
PROJECT COST:	NOT AVAILABLE
OWNER/DEVELOPER:	ONE WORLD TRADE CENTER LLC, A JOINT VENTURE BETWEEN THE PORT AUTHORITY OF NEW YORK AND NEW JERSEY AND THE DURST ORGANIZATION
LEAD ARCHITECT:	SKIDMORE, OWINGS & MERRILL, LLP
ASSOCIATE ARCHITECT:	N/A
STRUCTURAL ENGINEERS:	WSP CANTOR SEINUK; WEIDLINGER ASSOCIATES; SCHLAICH BERGERMANN & PARTNER (SPIRE)
MEP ENGINEER:	JAROS, BAUM & BOLLES
MAIN CONTRACTOR:	TISHMAN CONSTRUCTION
LANDSCAPE:	PETER WALKER AND PARTNERS AND MATHEWS NIELSEN LANDSCAPE ARCHITECTS

Background

The world knows what happened in Lower Manhattan on September 11, 2001. The twin towers of the World Trade Center and numerous surrounding buildings were damaged or destroyed, and more than 2,800 people were killed. The ground smoldered for months. Rescue was replaced by recovery, and there followed eight months of removing thousands of tons of debris from what became known as Ground Zero. What most people do not realize is that reconstruction of the 6.5-hectare (16-acre) site actually began during the cleanup, with initial work beginning below grade and on adjacent sites, and therefore out of sight.

The ambitious redevelopment of Ground Zero and its environs is now clearly visible from all parts of the New York City metropolitan region. One World Trade Center (1WTC), designed by Skidmore, Owings & Merrill, LLP (SOM), is rising on the northwest corner of the site. A publicly owned commercial building with 3.5 million square feet of space, the tower will be, upon completion, the tallest building in North America and an icon representing rebirth, perseverance, innovation, and urban modernism.

Master Plan

1WTC will anchor a 16-acre site, the centerpiece of which is the National 9/11 Memorial and Museum. The design, by architect Michael Arad and landscape architect Peter Walker, incorporates an eight-acre memorial plaza of which the central feature is a pair of pools—each 200 feet by 200 feet (61 by 61 meters)—marking the footprints of the original twin towers. Each reflecting pool is animated by 30-foot (9 meter) waterfalls, the largest fountains ever constructed, which empty into central voids below—a metaphor for the loss of life at the heart of the city. The Memorial Plaza, landscaped with 400 swamp white oak trees, surrounds the pools, which are lined by a bronze panel on which the names of the lost are inscribed. A steel and glass pavilion designed by Snohetta will provide entry to the Memorial Museum, located below the Memorial Plaza and reflecting pools. This will house the artifacts of 9/11 and the 1993 bombing and tell the story of those events. While the National 9/11 Memorial and Museum speak of the past and of remembrance, 1WTC represents hope for the future as it rises into the sky.

Brief

The 1WTC program is organized as follows: a 50-foot-high (15-meter) public lobby will rise from the plaza level and will be topped by a series of mechanical floors; together these form the186-foot-high (57-meter) building podium. Seventy-one office floors (floors 20–90) rise above the base to an elevation of 1,131 feet (345 meters). Mechanical floors and a three-level observation deck culminate in a metal and glass parapet, the cap of which marks 1,362 feet (415 meters) and 1,368 feet (417 meters)—the heights of the original twin towers. Several circular communications platforms rise above the parapet and surround the base of the 450-foot (137-meter), cable-stayed spire.

The tower sits atop five basement levels that are intertwined with an extensive underground system of existing and new infrastructure. These levels include lobbies for the office tower and the observation deck, mechanical areas, loading docks, parking, receiving and support areas, and connections to the transportation hub and adjacent World Financial Center—in total about 500,000 square feet (46,452 square meters). At the bedrock level, the PATH railway tracks, ventilation system, and conduits exist three dimensionally in overlapping planes at varying depths. Per the brief, the PATH train service was to remain operational, and existing structures had to be preserved throughout excavation and construction to stabilize the slurry walls holding back the Hudson River, which were left unsupported after 9/11. Threading steel members, conduits, and shafts through the maze required precision timing and coordination, not only to avoid service or construction disruptions but also to ensure that subsequent development would not be obstructed. It was often slow going; a commonly heard phrase on site was, "We were digging with spoons for just a few hours in the dark of night."

Life Safety

The building incorporates advanced life-safety systems that exceed the requirements of the City of New York Building Code and which have led the way in developing new high-rise building standards.

Safety strategies will incorporate redundant measures, including a hybrid structure: a steel moment frame surrounding the vertical concrete core. That core will encase and protect elevators, stairwells, utilities, communication and mechanical systems, and a dual-purpose emergency first responder's lift. The concrete-encased egress stairs are designed to be pressurized in the event of an emergency. There are also dual interconnected fire standpipes and extra water storage to allow for high-capacity sprinkler heads. If a standpipe is cut or broken, the interconnecting valve automatically cuts off water supply to that standpipe and redirects it to the other standpipe, ensuring that every floor has sprinkler protection. Fire stairs are 20 percent wider than code and are fully pressurized, equipped with emergency lighting with both generator and battery back-up, in addition to photo-luminescent markings and exit signs mounted close to the floor.

Sustainability

Sustainability and energy-saving measures will be far-reaching. 1WTC will use 30 percent less water than the New York City Building Code allows for this type of building. A system for collecting the rainwater that falls within the site boundaries is being put in place; this reclaimed water will be used for landscape irrigation and make-up water for the cooling system.

Tenants will be educated on building infrastructure, from simple, metal walk-off grilles at major entrances to MERV 16 high-efficiency particulate filters and gas-phase filtration to serve the outside air intake system and air-handling units on each floor. Individual electrical supply meters will encourage tenants to reduce energy consumption and meet the building's goal of a 25 percent reduction in energy consumption.

The tower will be partially powered by 12 hydrogen fuel cells, expected to generate 3.1 megawatts of power for 1WTC. The building is set to achieve Gold certification for core and shell as part of the United States Green Building Council (USGBC) Leadership in Energy and Environmental Design (LEED) program. The building is designed to promote, support, and encourage tenants to achieve USGBC LEED certification for their interior fit-out, and to that end will incorporate a sustainable exemplar space demonstrating the potential of 1WTC.

Entrances, identified by glass canopies and highly transparent glass cable-net wall systems, penetrate the base on all four elevations. The main lobby wraps around the central core and is filled with daylight from the entrances on the east and west sides and clerestory windows in the north and south walls. As the tower rises from the base it tapers, and its square edges are progressively chamfered, thereby transforming the square into eight tall isosceles triangles in elevation (four that point up and four that point down, alternately).

Design Concept

The tower's footprint measures 61 by 61 meters (200 by 200 feet) and is set back from the site's northwest corner. The podium wall base is 186 feet (57 meters) tall and consists of vertical laminated glass fins and horizontal stainless steel slats. The entire wall is internally illuminated by LEDs. More than 4,000 glass fins, each measuring 13 feet 4 inches by 2 feet (3.96 by 0.6 meters), are positioned at varying angles in a regular pattern over the height of the podium. This pattern both accommodates ventilation for the mechanical levels behind the podium wall, and, in combination with a reflective coating, refracts and transmits light to create a dynamic, shimmering glass surface.

Structural System

The tower's structure is designed around a massive, redundant steel moment frame consisting of beams and columns connected by a combination of welding and bolting. Paired with a massive concrete core shear wall, the moment frame lends substantial rigidity and redundancy to the overall building structure while providing column-free interior spans for maximum flexibility. Due to union rules, a peculiarity of the New York construction market, the building of the concrete core followed the erection of the steel floor framing, utilizing an innovative self-jacking lift system for the concrete formwork. The result is brute strength veiled in prismatic elegance.

1WTC is the first New York City project in which concrete with a characteristic compressive strength of 15,000 psi has been used. Before this project, the standard for high-rise buildings in New York was maxed out at 8,000 psi. Since then, 12,000 psi has been used in numerous projects in the city, including Seven World Trade Center (which sits just across Vesey Street to the north of the site of 1WTC, and which was the first building reconstructed on the site after 9/11).

At the halfway point, the 104-story tower forms a perfect octagon; it culminates in a square glass parapet rotated 45 degrees from the base. The resulting crystalline form will capture an ever-evolving display of refracted light. As the sun moves through the sky or onlookers move around the tower, the surfaces will appear like a kaleidoscope, changing throughout the day as light and weather conditions vary.

The curtain wall begins at the 20th floor and continues to the observation deck. Based on a 1.52-by-4.06-meter (5-by-13.3-foot) module, the panels are floor-to-floor high-performance insulated units weighing as much as 1,500 pounds (680 kilograms) each. A thermally broken unitized wall system was developed that incorporates insulated glazing units (IGUs) of a new, monumental scale capable of withstanding the wind pressure experienced by a supertall building while also meeting stringent security requirements.

The Manitowoc cranes that positioned the structural steel sections were also called into service to hoist the large corner curtain-wall panels into place. These units, starting at the 20th floor, just above the podium, are the heaviest, due to blast-resistance requirements for more robust connections, reinforcing, and laminated inner lites.

Conclusion

The tower will reach a height of 417 meters (1,368 feet) above ground. A communications platform ring rising above the parapet and a 450-foot (137-meter), cable-stayed spire, designed in collaboration with artist Kenneth Snelson and structural engineer Dr. Hans Schober of Schlaich Bergermann and Partner, will crown the project. The spire reaches a symbolic 541 meters (1,776 feet).

Near the top of the spire, LED lamps will create a sweeping beacon, reminiscent of the light ships that were used to identify a port to ships coming over the horizon. More than a gesture, the beacon recalls New York's maritime past and beams a safe-harbor welcome that will be seen as far away as 26 miles. In its subtlety, it signals a resurrection.

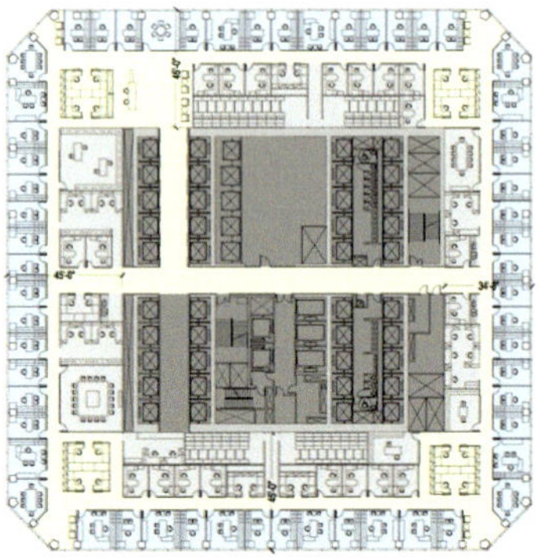

Typical Low-Rise Floor Plan

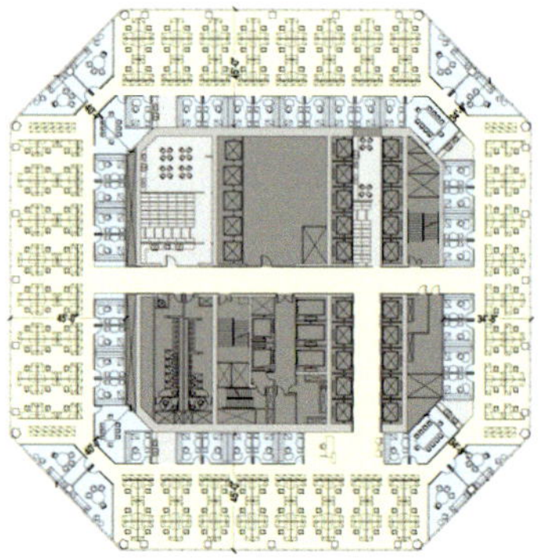

Typical Mid-Rise Floor Plan

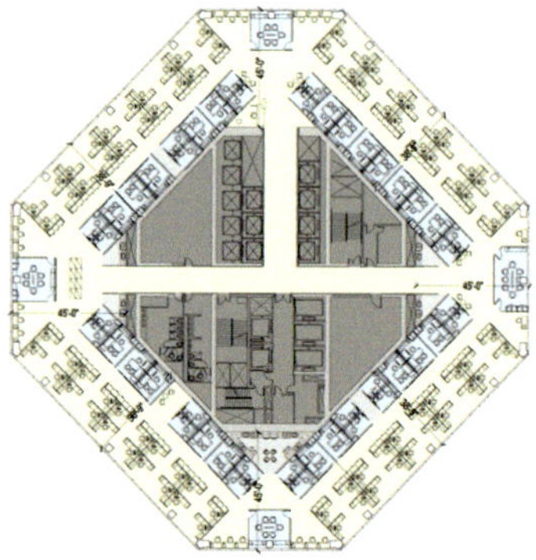

Typical High-Rise Floor Plan

Case Study Credits

Case Study 1: Aqua Tower
Studio Gang Architects
Images on pages 364–7: Steve Hall © Hedrich Blessing
All others: © Studio Gang Architects

Case Study 2: Bank of America Tower
COOKFOX Architects
All images: © COOKFOX Architects

Case Study 3: Hearst Tower
Foster + Partners
All images: © Nigel Young/Foster + Partners

Case Study 4: Manitoba Hydro Place
Kuwabara Payne McKenna Blumberg Architects (KPMB)
Photographers: Eduard Hueber, Gerry Kopelow, Tom Arban
Diagrams: © Bryan Christie

Case Study 5: One Madison Park
CetraRuddy (Cetra/CRI Architecture PLLC)
All images: © David Sundberg/Esto

Case Study 6: Veer Towers
Murphy/Jahn Architects
All images: © Rainer Viertlbock

Case Study 7: Guangzhou International Finance Center (GZIFC)
Wilkinson Eyre Architects
Image on page 398: © Christian Richters
Image on page 400: © Jonathan Leijonhufvud
All drawings: © Wilkinson Eyre Architects

Case Study 8: International Commerce Centre (ICC)
Kohn Pedersen Fox Associates
All images: © Kohn Pedersen Fox Associates

Case Study 9: Regent Place: Lumière Residences and Fraser Suites
Foster + Partners
Gerard Evenden
All images: © Nigel Young/Foster + Partners

Case Study 10: Marina Bay Sands
Safdie Architects
Moshe Safdie
Images on pages 414 and 416: © Timothy Hurley
Image on page 415 and all drawings © Safdie Architects

Case Study 11: Missing Matrix Building: Boutique Monaco
Mass Studies
Zongxoo U
Images on pages 420 and 422: © Yong-Kwan Kim
Images on page 421 and 423: © Mass Studies

Case Study 12: Mode Gakuen Spiral Towers
Nikken Sekkei Ltd
Images on page 424: © Osamu Murai
Images on pages 425 and 428: © Kenichi Suzuki

Case Study 13: The Met
WOHA
Image on page 430: © Tim Griffith
Image on page 432 (top): © Patrick Bingham-Hall
Image on page 432 (bottom): © Kirsten Bucher

Case Study 14: Shanghai Tower
Gensler
All images: © Gensler

Case Study 15: The Troika
Foster + Partners
All images: © Nigel Young/Foster + Partners

Case Study 16: Hegau Tower
Murphy/Jahn Architects
All images: © Rainer Viertlbock
All drawings: © Murphy/Jahn Architects

Case Study 17: Heron Tower
Kohn Pedersen Fox Associates
All images: © Kohn Pedersen Fox Associates

Case Study 18: Al Hamra Tower
Skidmore, Owings & Merrill (SOM)
Image on page 452: © Pawel Sulima
Image on page 453: © Tim Griffith
All others: © SOM

Case Study 19: Bahrain World Trade Center
Atkins
All images: © Atkins

Case Study 20: Burj Khalifa
Skidmore, Owings & Merrill (SOM)
Image on page 464: SOM | Nick Merrick © Hedrich Blessing
All others: © SOM

Case Study 21: Capital Gate
RMJM
All images: © RMJM/ADNEC

Case Study 22: O-14
Reiser + Umemoto
Image on page 476: © John Hauser
Image on page 477 (top): © Sebastian Optiz
All others: © Reiser + Umemoto

Case Study 23: One World Trade Center
Skidmore, Owings & Merrill (SOM)
Image on page 480: © 2012 Tower 1 Joint Venture LLC.
All others: © SOM

Index

References in bold indicate tables and in italics indicate figures.

For Product Safety Concerns and Information please contact our EU representative GPSR@taylorandfrancis.com Taylor & Francis Verlag GmbH, Kaufingerstraße 24, 80331 München, Germany

Batch number: 10733483

Printed by Printforce, the Netherlands